The Mitochondria
of
Microorganisms

The Mitochondria of Microorganisms

DAVID LLOYD

Reader in Microbiology,
University College, Cardiff, Wales

1974

ACADEMIC PRESS

London   New York   San Francisco

A Subsidiary of Harcourt Brace Jovanovich, Publishers

ACADEMIC PRESS INC. (LONDON) LTD.
24/28 Oval Road,
London NW1

United States Edition published by
ACADEMIC PRESS INC.
111 Fifth Avenue
New York, New York 10003

Library of Congress Catalog Card Number: 74 5680
ISBN: 0 12 453650 6

PRINTED IN GREAT BRITAIN BY
PAGE BROS (NORWICH) LTD, NORWICH

Preface

It has become widely recognized over the past ten years that many of the unsolved problems of the "supramolecular" organization, structure, function, and biogenesis of organelles of eukaryotic cells may be most easily approached by using microorganisms as systems for study. This book represents an attempt to summarize the spectacular successes of those engaged in such studies in the field of mitochondriology.

A complete description of the molecular mechanisms involved in the functioning and development of mitochondria requires answers to the following questions (Lloyd, 1969; Jayaraman, 1969; Linnane *et al.*, 1972b):

1. How are the individual components arranged in mitochondria to give functional "supramolecular" complexes; what composes the hierarchy of organization which constitutes a functional organelle?
2. Where are the individual components synthesized?
3. How are the components transported to the site of integration?
4. How are the components specifically integrated to give functional assemblies?
5. What is the nature of the control processes involved in the regulation of rates and extents of syntheses, transport and assembly?
6. How are these processes organized temporally; is there a strict sequence of biochemical events involved in synthesis, is there a strict order of assembly and integration processes? Are new components inserted continuously or discontinuously in synchronous waves with the same time constants or different time constants?
7. How are the fully-assembled organelles modulated by intracellular controls which define the expression of functional integrity; how are these control processes temporally ordered?
8. Where does the information which specifies the structure and regulation of mitochondrial components reside, and what is the mechanism for its ordered temporal expression?
9. How do existing mitochondria grow, and to what extent can existing membrane be modified without net increase in mass? Does turnover of existing membrane components contribute to this process and what control mechanisms relate processes of degradation to biosynthetic events? How is this balance ordered temporally?
10. What determines the mitochondrial growth rate and the numbers of mitochondria per cell? Do mitochondria divide and if so, how is the replication of mt DNA temporally related to mitochondrial division? How

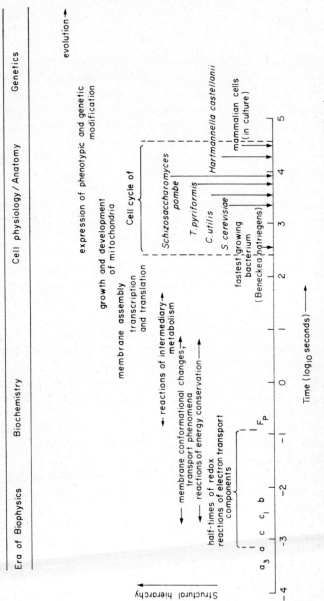

Fig. 1. The time scale of events involved in mitochondrial function, organization and development.

closely is the division of mitochondria related to cell division? Is the replication of mitochondrial DNA confined to a short interval relative to the length of the cell-cycle?

11.   Are any of the events occurring during mitochondrial biogenesis synchronized in all individuals of the mitochondrial population of a single cell.

It is evident that all these questions are closely related, and that the investigation of assembly processes may yield unexpected elucidation of mitochondrial function at the molecular level. Indeed the whole area of research should be regarded as a continuum of disciplines which focus at different levels of organization and on differing time scales (Fig. 1).

Abbreviations

The symbols, abbreviations and conventions used in this book are those recommended by The Biochemical Journals' instructions to authors. Other abbreviations used are as follows:

ANS 1-anilino-8-naphthalene sulphonate
CCCP carbonyl-cyanide m-chlorophenylhydrazone
CD circular dichroism
CLAM m-chlorobenzhydroxamic acid
DCCD dicyclohexylcarbodiimide
DCPIP 2,6-dichlorophenol indophenol
DNP 2,4-dinitrophenol
EGTA ethylene glycol-bis (β-amino-ethyl ether) N,N^1-tetra-acetic acid
EPR electron paramagnetic resonance
F_1F_2 synchrony indices in the first and second divisions of a synchronous culture, where $F = (N/N_0) - 2^{t/g}$ (N = number of organisms at time t, N_0 = number of organisms at zero time, g = mean generation time.
FCCP carbonyl cyanide p-trifluoromethoxyphenylhydrazone
HEPES N-2-hydroxyethylpiperazine-N^1-2-ethanesulphonic acid
MOPS morpholinopropane sulphonate
ORD optical rotatory dispersion
PMS phenazine methosulphate
SHAM salicylhydroxamic acid
SMPs submitochondrial particles
T_m mid point of melting curve
TES N-tris (hydroxymethyl)methyl-2-aminoethanesulphonic acid
TMPD NNN^1N^1-tetramethyl-p-phenylenediamine
TTC 2,3,5-triphenyl tetrazolium chloride
TTFA thenoyl trifluoroacetone (4,4,4-trifluoro-1 (2-thienyl)-1,3-butanedione)
TTFB tetrachlorotrifluoromethyl benzimidazole

Contents

Part I

Structure and Function

ix

Part II

Biogenesis

Part III

Evolution

Acknowledgements

The author wishes to thank all those who have provided unpublished material, reprints and permission to reproduce figures and tables. I would like to thank Professor D. E. Hughes for his continuing interest and encouragement. Thanks are also due to Drs. A. J. Griffiths, G. Turner, P. R. Avner, and R. T. Rowlands for helpful suggestions and for reading sections of the manuscript, and to Margaret, Lorraine, Lynda and Helen for various types of secretarial assistance. I am also indebted to the staff of Academic Press for their cooperation, and for the tolerance and patience shown by my family during the preparation and writing of the book.

David Lloyd.

Cardiff, August, 1974.

PART I

STRUCTURE AND FUNCTION

1

The Ultrastructure of Microbial Mitochondria

I. General Features of Mitochondrial Structure

The general features of mitochondrial organization revealed in thin sections of osmium-fixed tissue, consist of an outer membrane which limits the organelle and an inner membrane which is periodically invaginated to form the characteristic cristae mitochondriales, (Palade, 1953). The intracristal spaces communicate with the intermembrane space between inner and outer membranes, and these spaces together constitute the outer compartment. The inner membrane and cristae form the boundary layer of the inner compartment or matrix (Fig. 2).

This basic plan of mitochondrial organization is seen in mitochondria of all eukaryotic cell types, although considerable variation in shape and size of the organelle can occur. The extent of development of its ultra-structure, particularly that of the inner membrane system, is the most variable feature.

This classical model of mitochondrial structure has been confirmed in freeze-etched and negatively-stained preparations (Moor and Mühlethaler, 1963; Horne and Whittaker, 1962); these techniques provide alternative methods of obtaining information and do not necessitate the extensive chemical treatment and consequent possibility of artefacts inherent in processing prior to the preparation of thin sections. Thus electron-dense negative stains such as sodium phosphotungstate or ammonium molybdate freely penetrate the outer membrane of isolated mitochondria which have been fixed in osmium tetroxide. The outer compartment then appears dark in electron micrographs, with the cristae clearly revealed against the un-stained matrix (Fig. 3). Freeze etching also clearly demonstrates the presence of two distinct membrane systems and their enclosed compart-ments (Fig. 4).

Modifications of conventional techniques of electron microscopy have provided data on ultrastructure almost to the practical limits of resolution of present-day equipment. Thus Sjöstrand and Barajas (1968) found that mitochondrial membranes appear to consist of particulate components; after rapid and gentle fixation with glutaraldehyde, particles of from 4 to

3

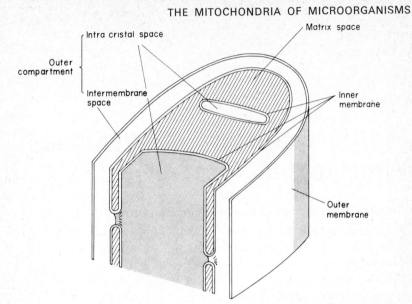

Fɪɢ. 2. Diagram illustrating the main features of mitochondrial structure.

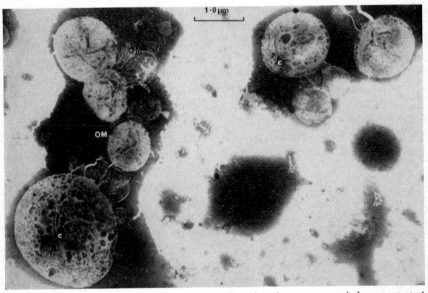

Fɪɢ. 3. Negatively-stained yeast mitochondria. The phosphotungstate stain has penetrated the outer membranes (OM) and fills the intermembrane spaces to reveal the cristae (c). (Reproduced with permission of W. Bandlow.)

10 nm diameter were revealed; these are in the size range expected for individual protein molecules or their subunits or lipoprotein complexes.

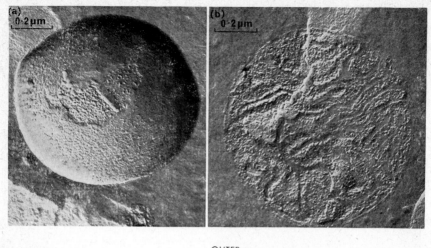

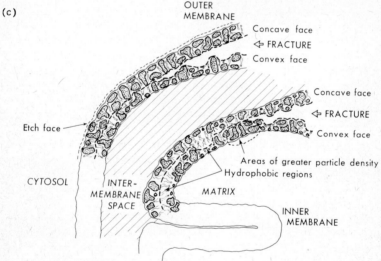

FIG. 4. (a) Freeze-etched guinea-pig mitochondria showing the outer surface (etch face) of the outer membrane and its fracture face. (b) Fracture faces of the cristae. (c) Diagram of surfaces seen in freeze-etched mitochondria. (Reproduced with permission from Packer, 1972.)

Whereas the average thicknesses of mitochondrial membranes in micrographs of conventionally-treated tissues lie between 5 and 7 nm, the newer technique gave a value of 15 nm. A staggered three-dimensional arrangement of particles rather than a strictly two-dimensional array was revealed by direct observation of profile views of the membranes. These workers suggested that the continuous lipid-bilayer backbone proposed in the unit-membrane concept (Robertson, 1960), or in the Davson-Danielli

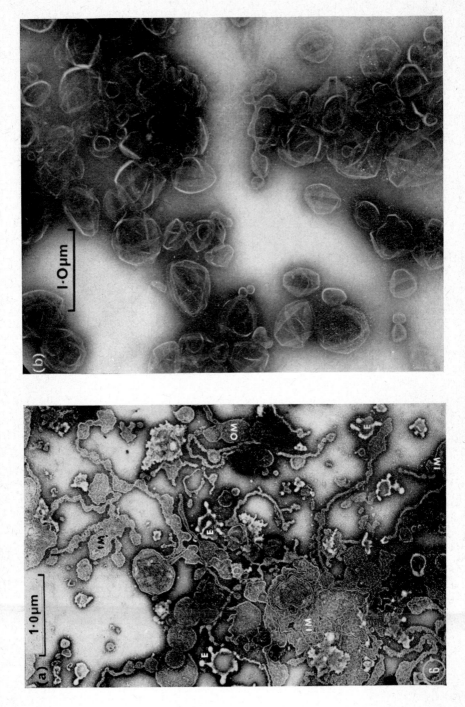

(1943) model, may result from artifactual modification of membrane structure due to extensive denaturation of membrane proteins (Sjöstrand and Barajas, 1968). Freeze etching also reveals detailed ultrastructure not recognized by the usual procedures of electron microscopy (Fig. 4) including a fibrous network in the matrix, a smooth etch face (surface) of the inner mitochondrial membrane, and a granular fracture face (interior) of both inner and outer membranes with particles in the size range 5–15 nm (Hackenbrock, 1968a, b; Wrigglesworth *et al.*, 1970). Critical-point drying has also demonstrated a "mitochondrial skeleton" in the matrix space; this consists of fibrous strands 30 nm in diameter surrounded by a matrix gel (Pihl and Bahr, 1970). Negative staining of unfixed partly-fragmented mitochondria indicates striking morphological differences between inner and outer membranes (Fig. 5). The outer membranes are smooth while the inner (matrix facing) surface of the inner membrane is covered with regular spherical knobs about 9 nm in diameter connected to the membrane by stalks 4–5 nm long, (Fernandez-Moran, 1962).

There is no convincing evidence for the presence of these inner membrane subunits as stalked projections in membranes subjected to processes other than negative staining, indeed the smooth etched-faces of freeze-etched mitochondrial inner membranes suggests that these may be extruded from the interior of the membrane by the stain (Wrigglesworth *et al.*, 1970).

Other structures observed in thin sections of mitochondria include intracristal rods (Hall and Crane, 1971), and intramitochondrial inclusions in the matrix space (for instance granules of calcium phosphate, Greenawalt *et al.*, 1964). Mitochondrial ribosomes can be demonstrated in electron micrographs (André and Marinozzi, 1965), as can mitochondrial DNA (Nass and Nass, 1963a, b). RNA synthesis has been demonstrated in mitochondria by electron microscopic radioautography of cells exposed to [^{3}H]-uridine (see Nagata, 1972, for a review).

II. Ultrastructure of Microbial Mitochondria

A. Ultrastructure of the Mitochondria of Protozoa

Early studies of protozoa with the light microscope revealed the presence of mitochondria as particles of variable size and shape, generally randomly

---

FIG. 5. (a) Non-fixed, negatively-stained fragmented rat liver mitochondrial fraction to show the three main types of membrane present. Some irregular pieces of membrane show numerous projecting 9 nm subunits at the edges (inner membrane or cristae, IM). Some round pieces of membrane show no projecting subunits (outer membrane, OM). The white pieces with rounded edges are contaminating endoplasmic reticulum (E). (b) Mitochondrial outer membranes purified from guinea pig liver mitochondria, fixed with osmium tetroxide before negative staining. (Reproduced with permission fron Parsons (1967) and Parsons and Williams (1967).)

distributed through the cytoplasm, but sometimes aggregated in the periphery of the organisms near the plasma membranes (Fauré-Frémiet, 1910a; Alexeieff, 1916). There may be as many as 3×10^5 mitochondria in a single amoeba (*Pelomyxa carolinensis*, Andresen, 1956), and these may reach 8 μm in length (Torch, 1955). In many species of protozoa the cristae consist of finger-like projections (microvilli) which terminate at blind-ends without making a second contact with the limiting portion of the inner membrane; these tubules are often so densely packed ("labyrinthine") that the matrix is barely distinguishable (Sedar and Rudzinska, 1956).

Examples of this unusual mitochondrial organization (Pitelka, 1963), are provided by species of amoeba such as *Amoeba proteus* (Flickinger, 1968), *Dictyostelium discoideum* (Gezelius, 1959), *Acanthamoeba castellanii* (Bowers and Korn, 1968), by ciliates like *Paramecium multimicronucleatum* (Sedar and Porter, 1955), *Spirostomum ambiguum* (Burton, 1970), and *Tetrahymena pyriformis* (Turner *et al.*, 1971; Schwab-Stey *et al.*, 1971; Schwab and Schwab-Stey, 1972; Krebs *et al.*, 1972), and also by species of heliozoa (Anderson and Beams, 1960). The arrangement of the tubular cristae is seen best in negatively-stained preparations (Fig. 6); in *Tetrahymena* mitochondria which have been partially disrupted by the stain a single row of inner membrane subunits line the matrix-facing surface of the tubules (Schwab-Stey *et al.*, 1971). These particles are not seen in sections or freeze-etched preparations and may thus have "popped out" of the membrane during treatment with phosphotungstate (Sjöstrand and Barajas, 1970). Physical continuity between smooth endoplasmic reticulum and outer mitochondrial membranes has been noted in both sections and negatively-stained preparations of mitochondria from *Tetrahymena pyriformis* (Franke and Kartenbeck, 1971). In preparations of *Spirostomum ambiguum* two rows of inner membrane subunits occur along the tubules (Burton, 1970). Paracrystalline arrays have been found within the matrix of mitochondria of the peritrichious ciliate *Carchesium polypinum* (Zagon, 1970). In some ciliates e.g. *Chilomonas paramecium* (Anderson, 1962) the mitochondria show considerable variation in shape and some are highly branched. In obligately anaerobic parasitic flagellates (Anderson and Beams, 1959; Grimstone, 1959) and ciliates (Nath and Dutta, 1962) mitochondria are either so poorly developed that they cannot be recognized in micrographs, or are entirely absent. Many zooflagellates (e.g. trypanosomes and leishmanias) have a single long slender mitochondrion which is differentiated near to the base of the flagellum to form the DNA-containing kinetoplast (Fig. 7). Both the kinetoplast and mitochondrion show transversely orientated cristae (Clark and Wallace 1960). DNA has also been located in mitochondria of *Physarum polycephalum* (Stockem, 1968) and *Didynium nigripes* (Schuster, 1963, 1965).

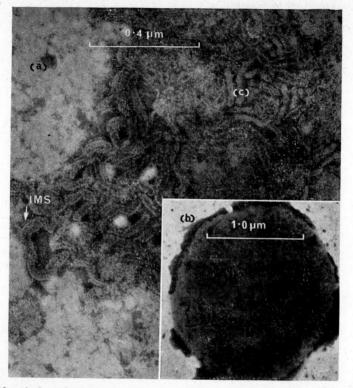

Fig. 6. Negatively-stained mitochondria isolated from *Tetrahymena pyriformis*. (a) Tubular cristae (c) lined with inner-membrane subunits (IMS). (b) Tubular cristae *in situ* (Reproduced with permission from Turner *et al.*, 1971.)

Some examples of the variability of mitochondrial ultrastructure found in protozoa are shown in Fig. 8 (Hofer *et al.*, 1972). A method for staining the mitochondrial cristae of protozoa has been developed by Childs (1973) who showed that diaminobenzidine gives an electron dense product after osmium-uranyl acetate fixation in *Hartmannella culbertsoni* (Fig. 9).

The ultrastructure of mitochondria in hybrid amoebae produced by nuclear transplantation has been studied (Flickinger, 1973).

Special preparative techniques enable the processes of mitochondrial transcription and translation to be observed in Tetrahymena mitochondria (Charret and Charlier, 1973).

B. Ultrastructure of the Mitochondria of Algae

Some small marine phytoflagellates (*Micromonas pusilla*. and species of the genus Heteromastix, Manton 1959; Manton *et al.*, 1965) have a single sausage-shaped mitochondrion near the base of the flagellum. As is the

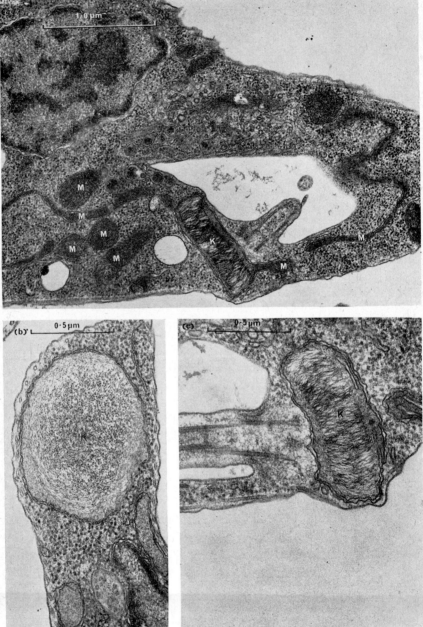

FIG. 7. (a, b, c). Sections of *Crithidia fasciculata* showing the structure of the single mitochondrion (M) and its kinetoplast (K). Fixation was with glutaraldehyde and osmium, staining was with lead citrate and uranyl acetate. (Reproduced with permission of B. Brooker and the British Museum (Natural History).)

case with mitochondria in *Chlamydomonas* (Sager and Palade, 1957), this mitochondrion has a few widely-separated flattened cristae which scarcely interrupt the central lumen. Similar profiles are seen in sections of *Chlorella* and *Euglena* (Leedale *et al.*, 1965), and the colourless chlorophytes such as *Prototheca* (Lloyd, 1966a), *Polytoma uvella* (Lang, 1963a, b), and *Polytomella* species (Lloyd *et al.*, 1970b; Moore *et al.*, 1970). The normal mitochondrial structure seen in all members of the Chrysophyceae, Phaeophyceae and Xanthophyceae shows an array of closely crowded tubular villi extending well into the lumen.

The simple appearance of sectioned mitochondria often belies a great three-dimensional complexity of overall shape. Thus Lang (1963b) suggested that the mitochondria of *Chlamydomonas*, *Eudorina*, *Pandorina* and *Volvox*, consist of "long sinuous elements, forming a network". Scale models constructed on the basis of electron microscopy of serial sections shows that only a small proportion of the mitochondria of *Chlamydomonas reinhardii* have a simple spherical or elliptical shape (Arnold *et al.*, 1972). Most of the mitochondria are elongated and show branches and constrictions (Fig. 10; Schötz *et al.*, 1972). A single mitochondrion can reach lengths far greater than the diameter of the cell, and the profiles seen in section even at opposite poles of the cell are often parts of the same mitochondrion. Consequently, serial section reconstruction is the only reliable method of determining the numbers of mitochondria present in this organism. From observations on two cells completely analysed in consecutive sections Arnold *et al.* (1972) concluded that these organisms contained 9 and 14 mitochondria respectively.

A similar investigation of *E. gracilis* has shown that giant mitochondria with multiple branchings are formed at one stage of the cell cycle (Osafune, 1973). Crystalloid inclusions, composed of hexagonally-packed fibrils or tubules occur in mitochondria of *Porterioochromonas* (Tsekos, 1972) and in those of *Ochromonas* (Aaronson, 1973). A twisted structure in the mitochondrial matrix of *Polysiphonia*, a red alga, is sensitive to DNAase (Tripodi *et al.*, 1972). Quantitative electron microscopic radioautography has demonstrated the synthesis of RNA from [^{3}H]-orotic acid within the mitochondria of *Ochromonas* (Gibbs, 1970). During zoosporogenesis of the green alga *Oedogonium*, bristle-like projections occur on the cristae (Pickett-Heaps, 1971).

C. Ultrastructure of the Mitochondria of Fungi

Whilst in general the structure of fungal mitochondria (which are ubiquitous throughout more than fifty genera) fits the classical model (Moore and McAlear, 1962), some exceptional features have been noted, for instance the presence of tubular cristae in the slime mold *Badhamia utriculus*, in the basidiomycetes *Coprinus disseminatus* and *Lycoperdon*

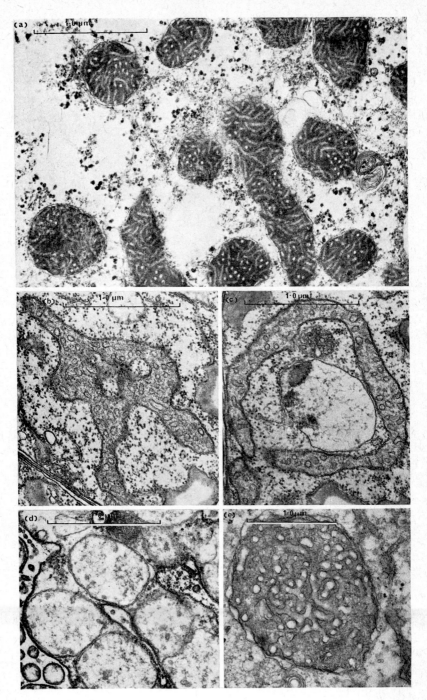

gemmatum (Nicklowitz, 1957; Heitz, 1959) and in *Thraustotheca roseum* (Goldstein *et al.*, 1964). The inner membranes of negatively-stained *Neurospora* mitochondria bear rows of inner membrane subunits (Stoeckenius, 1963; Malhotra and Eakin, 1967), as also do those of *Pythium ultimum* (Marchant and Smith, 1968a). The outer surfaces of mitochondria in hyphae of *Pythium ultimum* bear parallel columns of two rows each of particles 5–9 nm wide, and 12–18 nm long (Bracker and Grove, 1971a). These columns, seen both in sectioned and freeze-etched material, are orientated parallel to the long axes of the mitochondria with alternately arranged subunits, and a surface view of the membrane presents a paracrystalline appearance. The long axes of the particles are nearly perpendicular to the plane of the membrane, and apparently attached to it, and have a tubular ultrastructure (Fig. 11). It is suggested that these surface structures may play a contractile role in the dynamic pleomorphism of mitochondria (i.e. motility independent of cytoplasmic streaming) observed in phase-contrast microscopy (Bracker and Grove, 1971a). The continual changes of shape and distribution of fungal mitochondria has been eloquently summarized in the often-quoted passage from Ritchie and Hazeltine (1953): "under optimum conditions they resemble an aggregation of vigorous earthworms. In addition to twisting, turning and progressive movement . . . a mitochondrion in *Allomyces* can shorten and thicken or lengthen and become slender or a lump may pass along its length like a rat being swallowed by a snake. Mitochondria can be seen to swell, coil, branch, fragment, coalesce, put out pseudopods, even to get tied in knots. They can change their form until they resemble bubbles, strings of beads, dumb-bells, lemons or snow shoes". The large number of mitochondria in *Allomyces* contrasts with the single mitochondrion found in the zoospore of *Blastocladiella emersonii* (Cantino, 1965).

Continuity between cytoplasmic membranes and outer mitochondrial membranes has recently been observed in *Pythium aphanidermum* and *Aspergillus niger* (Bracker and Grove, 1971b). Points of close physical association, contact, thread-like continuity and direct luminal continuity were distinguished in sections (Fig. 11). It is suggested that these points of continuity between smooth endoplasmic reticulum and the outer mitochondrial membranes may represent sites of membrane interaction and facilitate exchange between adjacent membrane compartments.

FIG. 8. Sections of protozoa showing the varying extents of mitochondrial organization. (a) Mitochondria of *Tetrahymena pyriformis* showing numerous tubuli within an electron-dense matrix. (b, c) Branched (b) and curved (c) mitochondria of *Labyrinthula coenocystis* with short mitochondrial tubuli and less electron density of the matrix space. (d) Round and ovoid mitochondria of *Allogromia laticollaris*. Only a few short mitochondrial tubules are visible in the light matrix space. (e) Mitochondrion of *Amoeba dubia*. ((a–d) reproduced with permission of Hofer *et al.*, 1972; (e) by courtesy of C. J. Flickinger.)

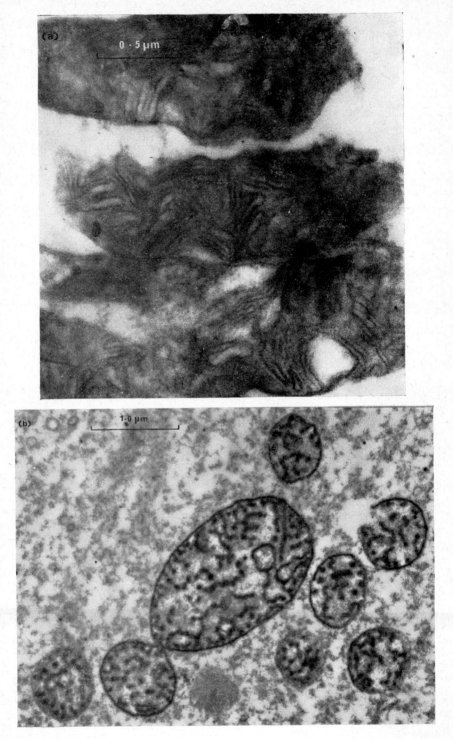

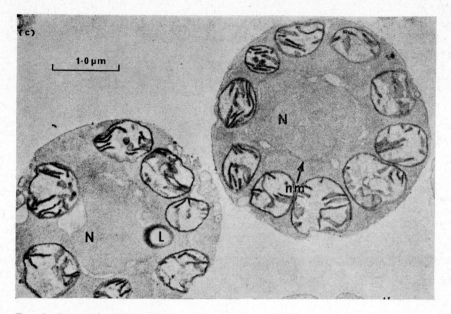

FIG. 9. Diaminobenzidine-osmium stained mitochondrial membranes in sections of (a) *Polytomella caeca*, (b) *Hartmannella culbertsonii*, and (c) *Candida utilis*. N, nucleus; nm, nuclear membrane; L, lipid granule. (Reproduced with permission of R. Cooper, G. E. Childs and from Keyhani, 1972 respectively.)

Similar features have been described for mammalian and plant mitochondria (Franke and Kartenbeck, 1971; Morré *et al.*, 1971). The ultrastructure of mitochondria of *Neurospora crassa* has also been investigated by freeze etching (Tewari *et al.*, 1972).

D. Ultrastructure of the Mitochondria of Yeasts

Early claims for the cytological demonstration of mitochondria in yeasts, using tetrazolium salts or Janus green, were finally confirmed by the electron microscopic observations of Agar and Douglas (1957), Vitols *et al.* (1961) and Hirano and Lindegren (1961, 1963). Numerous observations since that date (Marchant and Smith, 1968b) have confirmed the presence in aerobically-grown yeasts of organelles with conventional ultrastructure; best preservation of membrane ultrastructure, in for instance *Rhodotorula mucilaginosa*, is achieved using permanganate fixation (Fig. 12), although many other details (e.g. cytoplasmic ribosomes) are lost (Osumi and Kitsutani, 1971).

Estimates of mitochondrial numbers per cell in yeast vary from 15–29 in a diploid strain and 7–17 in a haploid strain (determined by serial sections and stereology) (Mahler, 1974) to one, (Fig. 13; Hoffmann and Avers,

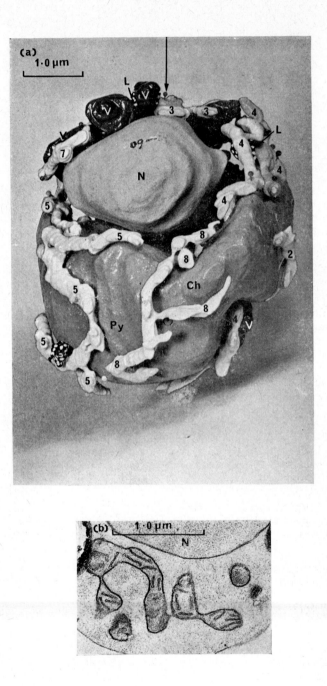

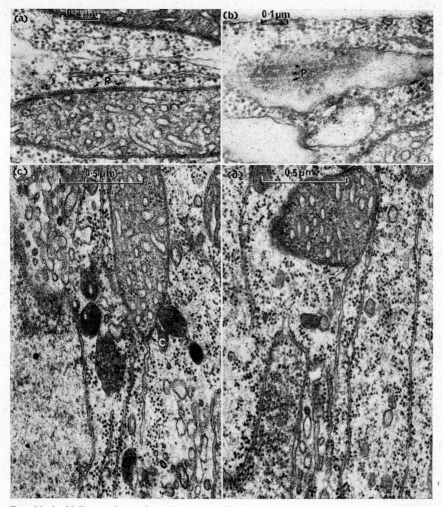

Fig. 11. (a, b) Rows of peg-shaped particles (P) on the outer surface of outer membranes of mitochondria of *Pythium ultimum*. (a) Longitudinal section, (b) mitochondrial surface intercepted tangentially. (c, d) Continuities (C) between mitochondrial outer membranes and endoplasmic reticulum-like double membranes in *Pythium aphanidermatum*. (Reproduced with permission from Bracker and Grove, 1971 a, b.)

Fig. 10. (a) Scale model of a cell (+ gamete) of *Chlamydomonas reinhardii* constructed on the basis of serial section electron microscopy, showing the location of the mitochondria very close to the surface of the chloroplast. N = nucleus; Ch = chloroplast; Py = pyrenoid; V = vacuoles; L = lipid bodies; 2, 3, 4, 5, 7, 8 mitochondria; points to the insertion of the flagella. (b) Part of a mitochondrion of *Chlamydonomas reinhardii* showing the sinuous shape, and the constrictions and branchings characteristic of the mitochondria of this flagellate. N = nucleus. All the mitochondrial profiles in this electron micrograph belong to the same organelle. (Reproduced with permission from Schotz *et al.*, 1972 and Arnold *et al.*, 1972.)

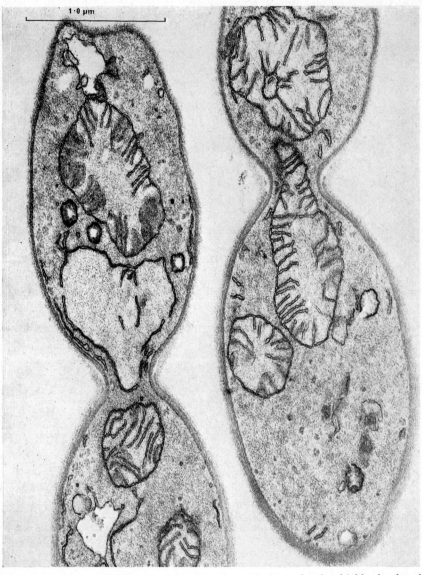

FIG. 12. Section of KMnO₄-fixed *Rhodotorula mucilaginosa* showing highly developed mitochondria. (Reproduced with permission of M. Osumi.)

1973). About 12% of the total volume of the cell is occupied by mitochondria in both diploid and haploid strains (Mahler, 1974). In freeze-etched cells of *Saccharomyces cerevisiae* mitochondria appear to be 1–2 µm long and 0·5–1·0 µm thick, shaped like sausages or dumb-bells, and lie near the cytoplasmic membrane (Moore and Mühlethaler, 1963; Moor, 1964). The face of the outer membrane is penetrated by slits up to 100 nm long and 19 nm wide. High resistance of the matrix to ice-crystal formation suggests high solute content. Thread-like mitochondria can be seen protruding into the buds of growing organisms (Matile *et al.*, 1969) where

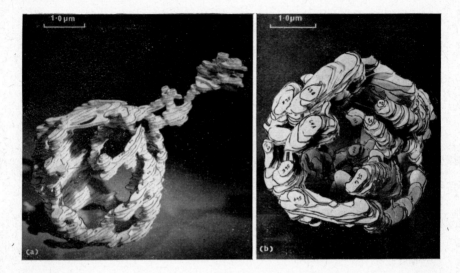

Fig. 13. Models of the single mitochondrion of *Saccharomyces cerevisiae*. A cell harvested from a synchronous culture growing aerobically in the presence of 3% glucose. Model assembled from tracings of 305 separate mitochondrial profiles in 58 consecutive sections (average thickness of section 75 nm). The smaller protuberant portion extends into a bud, but continuity with the mother-cell mitochondrial structure is evident. B. Cell grown in the presence of 2% ethanol. Model assembled from tracings of 443 separate mito-chondrial profiles in 58 thin sections. (Reproduced with permission from Hoffmann and Avers, 1973).

they divide (Thyagarajan *et al.*, 1961). In the fission-yeast, *Shizosaccharo-myces pombe* mitochondria can reach lengths almost equal to those of the organisms (10 μm) and sometimes contain mesosome-like membranes (McCully and Robinow, 1971) or "nucleoids" (Osumi and Sando, 1969). Taeter (1972) measured the average number of mitochondria per cell in *S. pombe* at 54; each occupied 0·022 μm³. The presence of inner mem-brane subunits in negatively-stained yeast mitochondria was first reported by Shinagawa *et al.* (1966).

Mitochondrial ribosomes have been demonstrated in uranyl acetate- and lead citrate-stained sections of mitochondria from *Candida utilis* (Vignais *et al.*, 1970). Rows of mitochondrial ribosomes (diameter 18 nm) lay attached to the matrix side of the inner membranes and cristae (see p. 372).

MtDNA fibrils occur at electron-transparent areas of *S. cerevisae* mito-chondria (Osumi, 1969).

Cytochrome *c* oxidase may be specifically located in *C. utilis* mito-chondria by electron microscopy by reaction with diaminobenzidine followed by osmium fixation, Fig. 9 (Keyhani, 1972).

III. Changes in Mitochondrial Ultrastrucutre

A. Alterations in Mitochondrial Ultrastructure During the Cell Cycle

The static pictures of mitochondrial morphology revealed by all the commonly available techniques of electron microscopy cannot provide a great deal of information on the dynamics of the constantly-shifting and ever-changing ultrastructure of mitochondria. Attempts at reconstruction of postulated sequential events involved in cytological changes from these static representations are always open to criticism. The interpretation of thin sections in three-dimensional terms is also a hazardous proposition which can only be reliably achieved by the time-consuming method of serial sectioning. Thus many of the estimates of mitochondrial numbers per cell are certainly not reliable where not reinforced by independent data. Perhaps the most valuable approach to problems of this sort is still the use of the phase contrast or interference microscope, although the resolution of these methods, even when used with vital staining com-pounds (e.g. Janus green, tetrazolium or thallium salts) is only capable of elucidating gross morphological events such as changes in size, shape, numbers and distributions of mitochondria within cells (Lindegren, 1971, 1972; Lindegren and Lindegren, 1971, 1973; Watrud and Ellingboe, 1973a). It is possible that electron microscopy at 1–3 mV may provide a useful extension to this resolution of structures in fixed (Fig. 14) or even in living cells.

It is now evident that apart from slow large-scale variations, mito-chondria also undergo dramatic alterations of gross ultrastructure on a time scale of the order of seconds. Thus, during energization, mito-chondria pass from the condensed conformation (corresponding to a low energy state in which the outer compartment has a large volume and the cristae appear densely osmiophilic) into the orthodox conformation (high energy state) (Fig. 15; Burgos *et al.*, 1964; Hackenbrock, 1966; Green *et al.*, 1968a, b; Packer, 1970). Differences in conformational ultra-structure have been noted in several types of microbial mitochondria *in situ*, but whether these reflect an alteration in the functioning of the organelles in energy production is not yet ascertained; indeed mito-chondrial polymorphism was first observed (for instance in amoebae, Pappas and Brandt, 1959; Pappas, 1959) long before the present upsurge of interest in the phenomenon. Thus some mitochondria of *Pelomyxa*

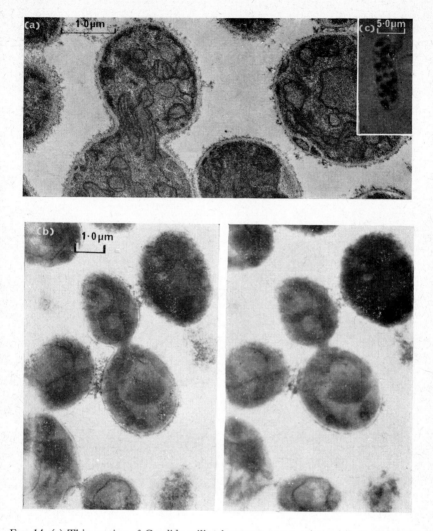

FIG. 14. (a) Thin section of *Candida utilis* (chemostat-grown, glycerol limited organisms grown at dilution rate of 0·2 h⁻¹) permanganate-fixed and stained with lead citrate. (b) High voltage electron micrographs of similar cells. The 0·5 μm section was photographed at 1·0 mV using specimen tilt ($\pm 8°$) to obtain the pair of micrographs which should be viewed with a stereo viewer. Continuity and branching suggests that many of the profiles seen in the thin section are of the same organelle. (Reproduced with permission of M. Davison and P. B. Garland.) (c) Intact cell of *Shizosaccharomyces pombe* grown in the presence of 1% glucose and incubated 3 h with 0·05% 3(4,5 dimethylthiazolyl 1-2)2,5 diphenyl tetrazolium bromide in growth medium and photographed using light microscopy.

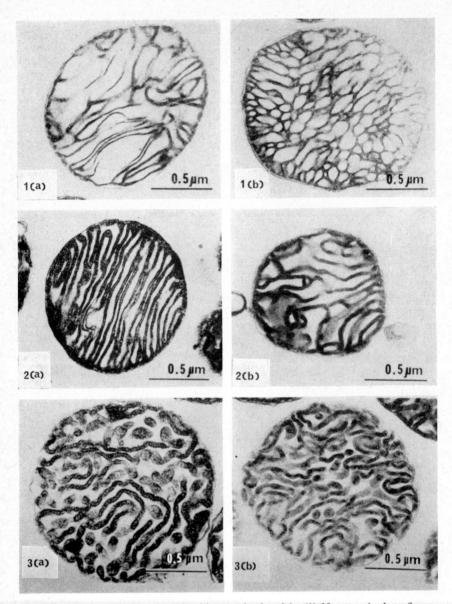

FIG. 15. Conformational states of beef-heart mitochondria. (1) Nonenergized configuration obtaining in the presence of respiratory inhibitors (antimycin A plus rotenone) and rutamycin: (a), ATP and P_i also present: (b), ATP present without P_i. (2) Energized configuration obtaining in the presence of oxidizable substrate: (a), ascorbate plus TMPD; (b), pyruvate plus malate. (3) Energized-twisted configuration obtaining in the presence of substrate and P_i: (a), ascorbate plus TMPD; (b), pyruvate plus malate. (Reproduced with permission from Green et al., 1968b.)

carolinensis possess the simple tubular cristae characteristic of many protozoa, while other mitochondria in the same cell display more complex organization. After glutaraldehyde-formaldehyde fixation and osmium staining of *Amoeba proteus*, mitochondria with darkly- and lightly-staining matrices coexist alongside one another (Fig. 16, Flickinger, 1968). This heterogeneity may arise either (a) as an osmotically-induced artefact during processing, (b) as a consequence of different metabolic states of individual mitochondria at the time of fixation, (c) as a consequence of mitochondria being at different phases in a growth and division

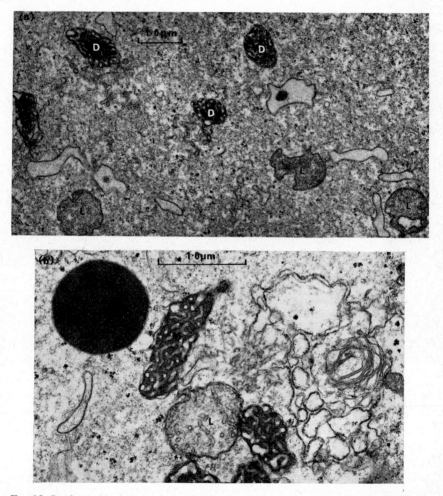

Fig. 16. Conformational states of mitochondria *in vivo* in *Amoeba proteus* after fixation with glutaraldehyde-formaldehyde. (a, b) The matrix of some mitochondria (D) appears much more dense than that of others (L). (Reproduced with permission from Flickinger, 1968.)

cycle, or (d) as a coexistence of separate and independent populations of mitochondria within the same cell. The diversity of "conformational states" of mitochondria of *Tetrahymena pyriformis* observed *in vivo* by Sato (1960), and by Elliott and Bak (1964) has also been shown in isolated mitochondria (Schwab-Stey *et al.*, 1971).

A variety of conformational states has also been observed in mitochondria after density gradient separation from homogenates of aerobically

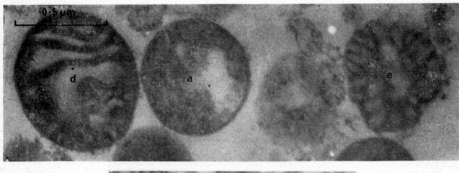

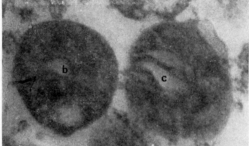

FIG. 17. Conformational states of mitochondria isolated from aerobically-grown *Saccharomyces carlsbergensis* during the phase of glucose derepression. (a)–(e) Orthodox to condensed conformations. (Reproduced with permission from Cartledge and Lloyd, 1972a.)

grown, glucose derepressed *Saccharomyces carlsbergensis* (Fig. 17; Cartledge and Lloyd, 1972a). No attempt has yet been made to relate observed alterations of mitochondrial ultrastructure to changes in the metabolic activity of mitochondria isolated from microorganisms; this approach has been confined to mitochondria from mammalian sources (Hackenbrock, 1968a, b; Pollack and Munn, 1970). Histochemical staining for cytochrome oxidase activity suggests the presence of mixed populations of mitochondria in a single yeast cell (Avers *et al.*, 1965).

Changes in structural diversity of mitochondria at different stages of the cell cycle have been widely observed. The frequency of occurrence in

Pelomyxa illinoisensis of mitochondria with complex cristae patterns declines from prophase through metaphase and becomes very low in interphase (Daniels and Breyer, 1965). These patterns typically consist of a parallel zig-zag configuration of cristae tubules which occupy about 30% of the total area of mitochondrial profile.

Chlorella pyrenoidosa contains only a single mitochondrial reticulum which is inherited by random partitioning between daughter cells (Atkinson, 1972). Mitochondrial volume increases smoothly in proportion to the cell, and is maintained within the limits of 2·3 and 3·1% of total cell volume. (John *et al.*, 1973).

Changes in ultrastructure of *Chlamydomonas reinhardii* have been followed through the cell cycle in synchronous cultures (Osafune *et al.*, 1972a, b; Osafune, 1973). The fusion of small mitochondria to give giant organelles is followed by fragmentation to give smaller forms and these changes were accompanied by alterations in respiratory activity. Temporary associations of mitochondria with chloroplasts were also observed.

In a classical analysis of mitochondrial weight distribution based on electron microscopic evidence, Bahr and Zeitler (1962) showed that differences of shape and size of rat liver mitochondria seen in sections represent differences between organelles at different stages in a growth division cycle. Division of mitochondria was first observed in *Spirostomum ambiguum* (Fauré-Frémiet, 1910b). Changes in numbers and shapes of mitochondria in the fission yeast *Shizosaccharomyces pombe* suggest that growth and division of mitochondria is closely linked to cell growth and division (Osumi and Sando, 1969) although no statistical analyses were performed. A more extensive investigation of growing hyphae of *Neurospora crassa*, employing methods of quantitative electron microscopic analysis of sections, suggested that dumb-bell and cup-shaped forms (Fig. 18) predominate at certain regions of the elongating mycelium. These regions are zones of synchronous mitochondrial division (Hawley and Wagner, 1967). The division of the mitochondria of *Physarum polycephalum* is preceded by the division of the "nucleoid" (Guttes *et al.*, 1969). This central rod-shaped body consists of an axial DNA-containing region and a peripheral region composed of RNA (Kuroiwa, 1973a, b). Autoradiographic analysis shows that one half of the RNA synthesized in the central body is transmitted to the mitochondrial matrix in 120 min.

B. Alterations in Mitochondrial Ultrastructure During the Life Cycle

Many eukaryotic microorganisms exhibit extremely marked changes of form during their complex life cycles; these examples of cytodifferentiation involve massive changes in membrane structure and function, and many

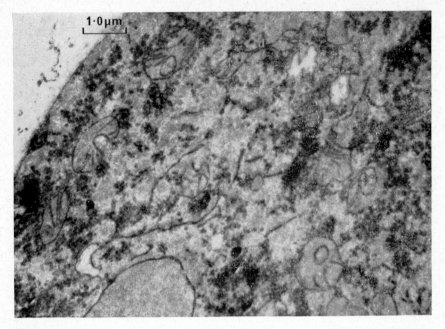

FIG. 18. Dumbell-shaped mitochondria (d) in *Neurospora crassa*. This section is of the zone where synchronous mitochondrial division occurs. (Reproduced with permission from Hawley and Wagner, 1967.)

examples may be cited although few have been studied biochemically (Baldwin and Rusch, 1965).

Acanthamoeba castellanii as a vegetative amoeba has numerous well developed mitochondria with a honeycomb of tubular cristae, but on encystment these mitochondria decrease in size, the cristae become less organized, and intramitochondrial inclusions (300–500 nm in diameter) of unknown composition become very pronounced (Vickerman, 1960; Fig. 19). In the microcysts of the cellular slime mould, *Polysphondylum pallidum*, numerous mitochondria are seen to have condensed, osmiophilic matrices (Hohl *et al.*, 1970). Encystment has been studied in a number of different species of amoebae (Griffiths, 1970).

Possible morphological changes of mitochondria during amoeba–flagellate transformation in *Tetramitus* or *Naegleria* have not been studied critically.

In the cellular slime mould, *Dictyostelium discoideum*, aggregation of myxamoebae to form multicellular structures is accompanied by degeneration of mitochondria to give empty vesicles which are similar in size, shape and distribution to the original mitochondrial population (Ashworth *et al.*, 1968). However, at the culmination of the process of fruit-body

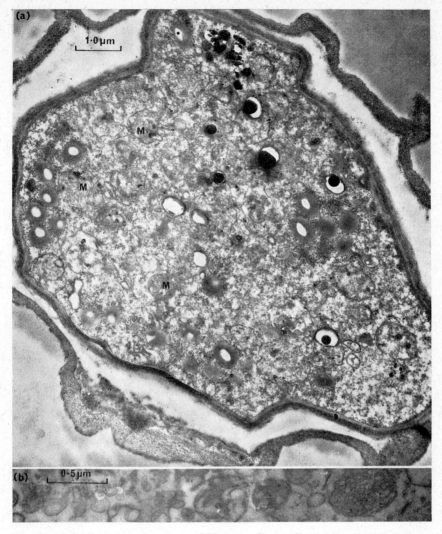

Fig. 19. (a) Section of a mature cyst of *Hartmannella castellanii*. Mitochondria (M) are still present, but apparently lack the complex internal organization of those organelles observed in the organisms. (b) Mitochondria isolated from mature cysts. (Reproduced with permission of M. Stratford and A. J. Griffiths.)

formation, mitochondrial ultrastructure again resembles that of the vegetative amoebae.

The Sporozoa are parasitic protozoa without obvious means of loco-motion and with complex life cycles. In a malaria parasite, *Plasmodium berghei*, typical mitochondria are absent, but a concentric double-membrane structure situated in the periphery of the organism has been suggested as a functional substitute (Rudzinska and Trager, 1959). In *Plasmodium cathermerium* mitochondria are present in oocysts and sporozoites, but are not evident in merozoites (Duncan *et al.*, 1960; Myer and Olivera-Musacchio, 1960). In *Gregarina rigida* mitochondria are hardly visible in the trophic form, but appear as tubular organelles in the cephalont of the epimerite embedded in the host cell (Anderson, 1967).

Trypanosomes undergo marked morphological changes during the course of their life cycles which take them from insect vectors to animal hosts (Vickerman, 1965), and the single mitochondrion of the organism shows extreme variation in ultrastructural complexity. In the bloodstream form of the sleeping sickness trypanosomes (*T. gambiense* or *T. rhodesiense*) the cristae are poorly developed until a stumpy form of the organism is produced which can infect the tetse-fly. Culture forms (which may correspond to the midgut stages of the insect vector) have well developed inner mitochondrial membranes (Fig. 20).

Repeated mechanical transmission (syringe passage) of *Trypanosoma evansi* or *Trypanosoma equinum* between successive mammalian hosts (thus eliminating the role of the insect vectors) gives rise to dyskinetoplastic forms which have lost fibrillar material from the kinetoplast and undergone a marked degeneration of mitochondrial ultrastructure (Vickerman, 1965).

The life cycle of the phytoflagellate, *Polytomella agilis*, can be divided into several distinct stages; well developed mitochondria can be shown in sections of the motile vegetative and pre-cyst forms, but although still detectable in mature cysts, often appear degenerate (Moore *et al.*, 1970) despite some claims (due to inadequate fixation) that they disappear altogether during the differentiation process (Gittleson *et al.*, 1969).

Zoospores of the aquatic phycomycete *Blastocladiella emersonii* have a feature unique amongst the fungi of possessing only a single mitochondrion in close association with lipid granules (Cantino, 1965). Germination by one of two alternative pathways can lead to the formation of either of two forms of sporangia which have many mitochondria with conventional ultrastructure. The processes of spore formation and germination throughout the phycomycetes, ascomycetes and basidiomycetes involve complex processes of differentiation, (Sussman, 1965), but claims for the absence of mitochondria from spores may be the result of poor fixation, (Hawker and Abbott, 1963). Young zygotes before and after nuclear fusion of haploid yeast conjugants show mitochondrial dedifferentiation (Fig. 21; Smith *et al.*, 1972). A detailed description of the ultrastructural changes seen during ascospore formation in *Saccharomyces cerevisiae* has been made by Osumi *et al.* (1966) and Illingworth *et al.* (1973); a notable feature was the appearance and disappearance in

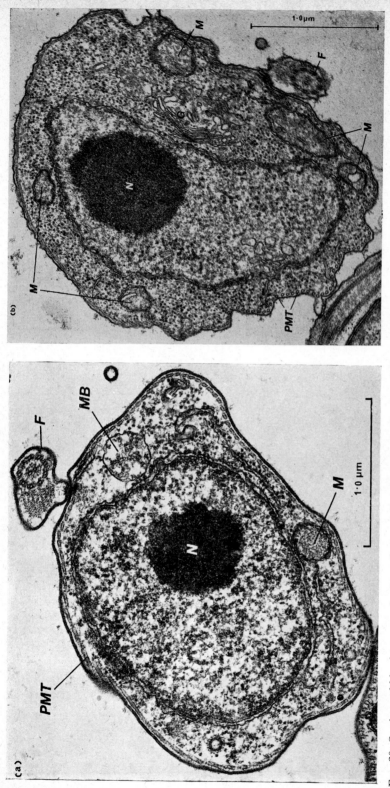

Fig. 20. Sections of bloodstream and culture forms of *Trypanosoma brucei*. (a) Slender bloodstream form transverse section through region of nucleus. Note single mitochondrial profile (*F*—flagellum; *M*—mitochondrion; *MB*—microbody-like organelle which probably houses glycerophosphate oxidase; *N*—nucleus; *PMT*—pellicular-microtubules. (b) Transverse section of trypanosome taken from culture 36 h after inoculation with bloodstream forms. The mitochondrion has become a reticulum represented in this section by several peri-nuclear profiles with plate-like cristae (*M*); (*F*—flagellum; *N*—nucleus; *PMT*—pellicular microtubules). (Reproduced with permission of K. Vickerman.)

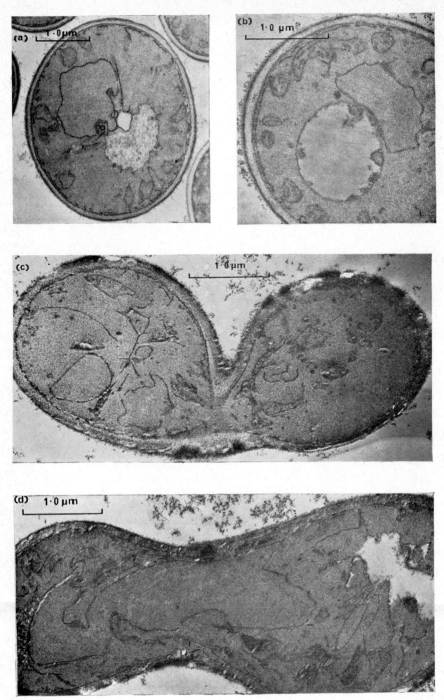

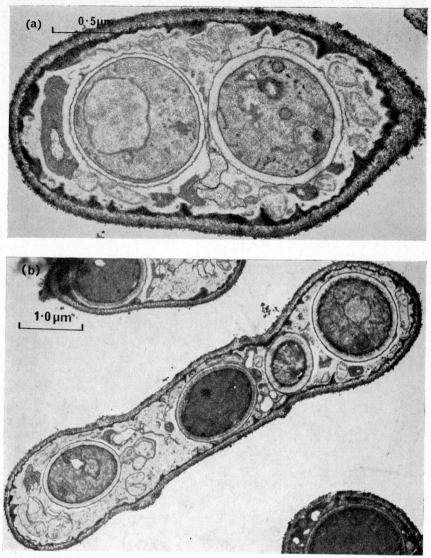

FIG. 22. Mitochondria in ascospores of *Shizosaccharomyces pombe*. (a) Two ascospores are in process of maturation. (b) Electron micrograph of an ascus containing 4 ascospores, each of which appears to show a different degree of maturation. Note persistent mitochondria (M) in the epiplasm (E), and in some spores. (Reproduced with permission from Yoo *et al.*, 1973.)

FIG. 21. Changes in mitochondrial ultrastructure during conjugation in *Saccharomyces cerevisiae*. (a) *S. cerevisiae* strain 41 showing normal cell structures including compact peripheral mitochondria. (b) Cell of strain 4a with structures similar to strain 41. (c) Cross (41 × 4a): cells of the two parental strains fused in early zygote formation. The two nuclei are seen in close proximity prior to fusion. There is proliferation of endoplasmic reticulum and the mitochondria are elongated and variously disorganized. (d) Zygote showing nuclear fusion and disorganized mitochondria. (Reproduced with permission from Smith *et al.*, 1972.)

the mitochondria of a fibrous DNA-containing structure during sporulation. The sporangiospores of *Rhizopus* (Hawker and Abbott, 1963), the conidia of *Botrytis cinerea* (Hawker and Hendy, 1963) of *Neurospora crassa* (Weiss, 1965) and of the yeast *Nadsonia fulvescens* (Kawakami, 1961) and the ascospores of *Schizosaccharomyces pombe* (Fig. 22; Yoo *et al.*, 1973) all contain mitochondria which are not unlike those of the vegetative cells or hyphae produced on germination; during outgrowth these organelles increase greatly in numbers (Kawakami, 1961). Similar features are also typical of basidiospore germination in *Schizophyllum commune* (Aitken and Niederpruem, 1970).

C. Changes in Mitochondrial Ultrastructure as a Result of Genetic Modification

The original description of the effect of mutagenic acridine dyes on the kinetoplasts of Trypanosomatidae (Werzbitzski, 1910) has been followed by a great deal of work on several different species. Spontaneous and irreversible loss of the organelle has been reported in strains of *Trypanosoma evansi* (Fig. 23; Hoare, 1954) and in this state these strains can be maintained for some time. "Akinetoplastic strains" can also be obtained by trypaflavin treatment (Mühlpfordt, 1963a, b), and in this case the organelle was reformed when drug-treatment was stopped. However it seems that modification rather than complete eradication of the organelle occurs, and use of the term "dys-kinetoplastic" rather than "akinetoplastic", has been suggested (Marmur *et al.*, 1963). The formation of an atypical kinetoplast (Guttman and Eisenman, 1965) is accompanied by the reduction in the organization of the mitochondrial cristae (Kusel *et al.*, 1967) both in *Crithidia fasciculata* and in *Leishmania tarentolae* (Trager and Rudzinska, 1964).

The ultrastructure of acriflavine-induced *petite* mutants of yeast was first investigated by Yotsuyanagi (1962b). After growth with 4% glucose it was claimed that, whereas segregational (*p*) mutants had mitochondria similar in structure to those of respiratory-competent yeasts, both cytoplasmic (ρ^-) *petites* and double mutants (with nuclear and cytoplasmic mutations) had structurally-modified mitochondria. However, this clear difference has not been confirmed in more recent studies. Acriflavin-induced *petite* mutants, analyzed histochemically after growth with 1% glucose for succinate dehydogenase (SDH) and cytochrome oxidase (CO) activities divided into four phenotypes: (a) reduced CO, wild-type SDH; (b) reduced CO, high SDH; (c) no detectable CO, high SDH; and (d) no detectable CO, wild-type SDH (Avers *et al.*, 1965). Differences of mitochondrial ultrastructure were noted in all four phenotypes and electron microscopic localization of cytochrome oxidase showed that reaction product was found only on mitochondrial membranes of respi-

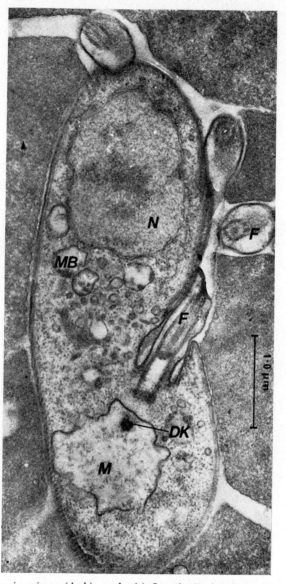

Fig. 23. *T. Evansi* equinum (dyskinetoplastic). Longitudinal section of posterior portion of trypanosome. The kinetoplast is represented by a small remnant (*DK*) only, lying within the swollen but acristate mitochondrion (*M*). (*F*—flagellum; *MB*—microbody-like organelle; *N*—nucleus). (Reproduced with permission of K. Vickerman.)

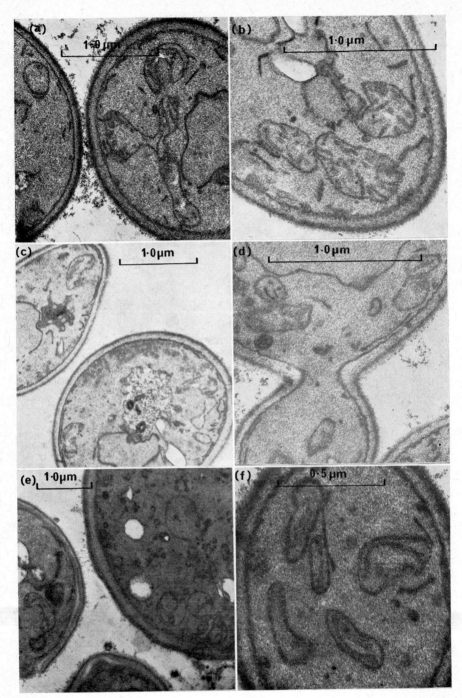

ratory-competent cells. Both reactive and unreactive mitochondria were found to coexist within the same cell in mutants with only partial respiratory ability, and this situation was stable through many serial transfers over a period of a year or more. Unfortunately some of these studies were with diploid organisms and thus respiratory-deficient strains could not be unequivocally established as cytoplasmic *petites*. Crosses between wild-type yeasts (with cristate mitochondria) and *petites* containing non-cristate mitochondria (characterized either by concentric loops of inner membrane orientated parallel to the long axis of the mitochondria or simply non-invaginated inner membrane) produced zygotes having both types of mitochondria (Federman and Avers, 1967). The frequency of abnormal mitochondrial profiles decreased in successive generations. *Petites* of *Saccharomyces cerevisiae* produced by *p*-nitrophenol treatment also showed a meagre development of cristae (Osumi and Katoh, 1967). The importance of the glucose concentration in the growth medium in determining the ultrastructure of a mutant of *Saccharomyces* unable to synthesize aconitase has been demonstrated by Bowers *et al.* (1967). Growth in the presence of 2% glucose gave organisms with non-cristate mitochondria, whereas growth in a chemostat with glucose as the limiting nutrient gave cells with mitochondria which were indistinguishable from the wild-type organelles. *Petites* induced in a respiratory competent strain (1D) of *Saccharomyces cerevisiae* by acriflavine, ultraviolet-radiation or phenyl ethanol showed fewer mitochondrial profiles having cristae than either the wild-type or a spontaneous mutant (Fig. 24; Smith *et al.*, 1969) when grown on melibiose, a fermentable but non-repressing substrate. This suggested that perhaps the spontaneous mutant arose from a minimal change, while mutagens caused multiple lesions. Quantitative analysis of the increased frequency of non-cristate or grossly aberrant mitochondria in sections of standard thickness of the various mutants was achieved, despite difficulties due to variation of mitochondrial ultrastructure in the wild-type. In another strain (11) no grossly aberrant mitochondria were found in the wild-type or in a mutant strain, but the decreased amount of inner membrane in the *petite* was evident. *Petites* produced from a 1D × 11 cross by treatment with phenyl ethanol had similar reduction of cristate profiles to that seen with the parent strain *petites;* a 2:2 segregation for aberrant mitochondria suggested nuclear control over mito-

FIG. 24. Mitochondria in yeast mutants. (a) Section of parent strain ID. (b) Section of parent strain 11. (c) Ultraviolet-induced *petite* mutant of strain ID. The mitochondrial profiles are poorly defined, but some are distinctly acristate. (d) Phenyl-ethanol induced *petite* mutant of strain 11, showing reduction in the number of cristae in the mitochondrial profiles. (e, f) Phenyl-ethanol-induced *petite* mutant of strain 2B (resulting from 11 × 1D cross). Some aberrant mitochondria are evident in (e) and some normal cristate profiles (f) are present. (Reproduced with permission of Smith *et al.*, 1969.)

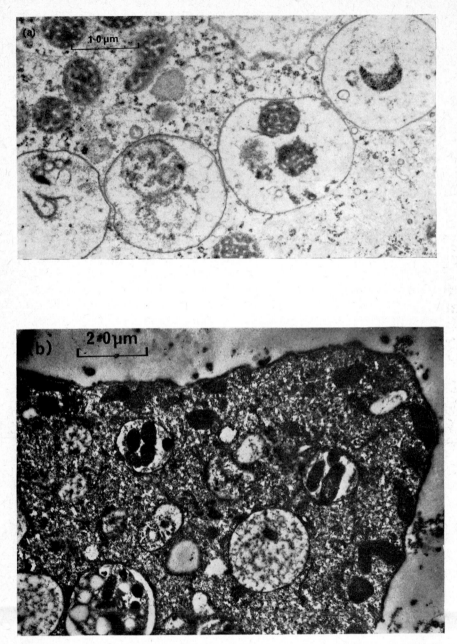

Fig. 25. Autophagic vacuoles containing mitochondria. (a) Mitochondria in autophagic vacuoles of *Hartmannella culbertsonii* at different stages of digestion. (b) Mitochondria in autophagic vacuoles in stationary-phase culture of *Tetrahymena pyriformis*. (Reproduced with permission of (a), G. E. Childs, and (b) from Turner and Lloyd, 1971.)

chondrial structure. No significant differences in ultrastructure of mito-chondria of various *petite* strains produced by different mutagenic treatments could be recognized. Contrary to the reports of Avers *et al.* (1965), Smith *et al.* (1969) were unable to find mitochondria so markedly altered in gross ultrastructure that their characteristics could be used as distinctive morphological markers for the ρ^- condition. Freeze-fracture studies on mitochondria from respiratory-deficient yeast mutants which have no detectable mtDNA, revealed that the distribution, clustering and size of membrane particles was not different from that of the wild-type (Packer *et al.*, 1973).

Shizosaccharomyces pombe gives viable respiratory-deficient mutants on treatment with acriflavine or ethidium bromide only infrequently, but segregational respiratory-deficient mutants can be obtained (Heslot *et al.*, 1970b). Statistical analysis of the frequency of occurrence of six different categories of ultrastructurally-abnormal mitochondria in the wild-type respiratory-competent strain 972 h⁻ showed that in an average of six mitochondria per cell section, about half appear as normal mitochondria, while the other half show reduced organization of cristae. A benzi-midazole-resistant mutant had fewer mitochondria, but the proportions of cristate and non-cristate mitochondria was not drastically altered. Normal mitochondria comprised only 3% of the total number observed in a cobalt-resistant respiratory deficient mutant, and another similar mutant showed no normal mitochondrial profiles at all. A stable acri-flavine-induced *petite* of *S. pombe* (Bulder, 1964a) also contained fewer mitochondria than the wild-type; 70% of the mitochondria lacked cristae (Heslot *et al.*, 1970b) and there were a high proportion (25%) of mito-chondria with transverse cristae. Unfortunately the organisms used in these studies were taken from colonies growing on solid media rather than from liquid cultures. Some uncertainty is thus introduced as to the condition of individual cells with respect to age, growth rate, and effective nutrient status. Analysis of gross ultrastructural changes which accompany production of *petites* has also been performed with *Saccharomyces fragilis* (McClary and Bowers, 1968). Structures resembling mesosomes have been described in mitochondria of yeast *petites* (Yotsuyanagi, 1966) and in those of *poky* mutants of *Neurospora* (Malhotra, 1968). Inner membrane subunits do not present an unusual appearence in oligomycin-resistant mutants of yeast (Watson and Linnane, 1972).

D. Changes in Mitochondrial Ultrastructure as a Result of Modification of the Environment

1. Changes produced on starvation

The population of mitochondria in *Tetrahymena pyriformis* is kept at a constant minimum number during exponential growth; cell divisions and mitochondrial divisions are closely coordinated (Elliott and Bak, 1964). As the growth rate of the ciliates declines when the culture enters the stationary phase of growth, mitochondrial numbers per cell increase and the organelles change in shape from elongated ellipsoids to spheres. Intramitochondrial inclusions appear, and some mitochondria become incorporated into autophagic vacuoles (Fig. 25). The progress of this process of autophagy was observed during the ageing of the stationary-phase cells. Mitochondrial degeneration has also been investigated during starvation of *Tetrahymena pyriformis* (Levy and Elliott, 1968). In this case exponentially-growing organisms were harvested and resuspended in a phosphate-buffered inorganic medium. Mitochondria become sur-rounded by a double membrane of unknown origin, and as digestion by acid hydrolases proceeds, this structure eventually becomes a single membraned autophagic vacuole. Mitochondria not sequestered in this way also showed degenerative changes including the formation of myelin-whorls at one pole. Alteration of mitochondrial ultrastructure also occurs during the starvation of the giant amoeba *Pelomyxa carolinensis* (Daniels and Breyer, 1968). Whereas in the non-starved amoeba the cristae tubules lie at random in the matrix, many mitochondria in starved organisms have enlarged tubules aligned in a zig-zag pattern. The frequency of occurrence of these altered mitochondrial forms increases over a period of starvation of up to two weeks until 60% of the total population shows the zig-zag form. Microfilaments are also seen in the mitochondrial matrices of starved amoebae. No mention was made in this report of the occurrence of autophagosomes. Autophagic digestion has never been clearly dis-tinguished in cytological studies of yeast undergoing starvation (Matile *et al.*, 1969), although complex membrane structures and myelin-like bodies are produced (Hinkelmann and Kraepelin, 1970). During carbon deprivation, mitochondria of a streptomycin-bleached strain SM–LI of *Euglena gracilis* var. bacillaris undergo marked morphological changes, and their distribution within the cells is altered (Brandes *et al.*, 1964). Con-striction and elongation lead to the production of bizarre configurations, and the mitochondria become lined up in the periphery of the cytoplasm. They often become surrounded by Golgi-like vesicles which fuse to isolate the mitochondrion in a membrane-bound acid phosphatase-containing vacuole; these events are followed by a regression in which characteristic mitochondrial features (outer membranes and cristae) are progressively disarrayed and become unrecognizable. In the terminal stages of digestion the cytolysosomes contained multivesicular bodies and amorphous bodies. Fragmentation of a labile reticulum composed of a fused network of thread-like mitochondria (0·2–0·5 μm thick) in this strain of *Euglena* has been observed by Leedale and Buetow (1970). The formation of a single giant sheath-like mitochondrion just within the periphery of the cell occurs after changing the carbon source from L-glutamate+ DL-malate to DL-lactate (Calvayrac, 1970).

2. Effects of growth under conditions of specific nutrient limitation

The morphology of mitochondria in a choline-deficient strain of *Neurospora* (*Chol-1*) is influenced by the level of choline in the culture medium (Luck, 1965b). Mitochondria in organisms grown at 1 μg/ml choline are significantly larger than those of cells grown at 10 μg/ml; a shift in choline concentration produced a gradual change in appearance of mitochondria which affected all the mitochondria synchronously. The number of cristae per unit mitochondrial area is unaffected. Hypertrophy of mitochondria has also been described in *p*-aminobenzoic acid-deficient *Rhodotorula aurantiaca* (Volkova and Meissel, 1967). Copper at limiting concentrations or at higher concentrations than normal, leads to abnormal mitochondrial morphology in *C. utilis* (Keyhani, 1973). Zinc deficiency of *Rhodotorula gracilis* leads to a decrease in numbers and organization of mitochondria (Cocucci and Rossi, 1972). Morphological changes have also been observed in biotin-deficient yeasts (Dixon and Rose, 1964), but difficulties with fixation procedures made assessment of possible morphological mitochondrial lesions impossible. Thiamine-deficiency leads to production of swollen mitochondria in rat heart sarcosomes (Arcos *et al.*, 1964) and riboflavin deficiency in mice produces giant hepatic mitochondria which divide frequently during the recovery process (Tandler *et al.*, 1969). Elongated mitochondria with reorientated cristae are found in Mn-deficient mouse liver (Hurley *et al.*, 1970). The ultrastructural effects produced on nutrient limitation have not yet been extensively explored in eukaryotic microorganisms, although these provide many systems which could be used for highly controlled experimentation.

3. Effects of growth at different oxygen tensions

Although the appearance of mitochondria in facultatively anaerobic species of yeast under different conditions of nutrient status and oxygen tension has been extensively documented, much of the earlier literature is confused and misleading principally for two reasons. (1) It is important to appreciate that the chemistry of fixation and staining processes used in electron microscopy is not fully understood, and that the marked changes in membrane composition known to result from altered growth conditions often lead to the formation of membrane structures which cannot be fixed and stained by conventional methods. (2) The phenotypic expression of mitochondrial ultrastructure is dependent upon many parameters, and a full appreciation of the interaction between the many variables involved is probably still lacking. Factors known to influence the structure and physiology of yeast mitochondria include (a) oxygen tension, (b) presence of fermentable or non-fermentable carbon sources, (c) concentration of fermentable carbon-source, (d) presence of lipid supplements (sterols and unsaturated fatty acids) and, (e) growth rate.

(a) Anaerobic growth. The long debate over the presence or absence of mitochondrial structures in anaerobically-grown yeast has recently been clarified (Table I). Earlier work reporting the complete absence of mito-chondria from anaerobically-grown *Candida utilis* (Linnane *et al.*, 1962) and *Saccharomyces cerevisiae* (Wallace and Linnane, 1964; Polakis *et al.*, 1964; Osumi, 1965; Osumi and Katoh, 1967) is now known to be based on inadequate procedures of electron microscopy. Again it is only recently that adequate precautions to maintain cultures under strictly anaerobic conditions have been ensured, and for this reason some of the earlier claims to have found typical mitochondria in anaerobically-grown yeast must be treated with caution (Morpugo *et al.*, 1964; Polakis *et al.*, 1965; Swift *et al.*, 1967); the minimal criteria for strictly anaerobic growth require that both succinate-cytochrome *c* oxidoreductase and cytochrome oxidase be undetectable in a cell-free extract of the organisms thus produced (Cartledge and Lloyd, 1972b).

The technique of freeze-etching, a method independent of those which rely on chemical reactions between fixative, stains and membrane com-ponents, has revealed that strictly anaerobically-grown yeast does contain mitochondria–like organelles (Fig. 26; Matile *et al.*, 1969). After harvesting in the presence of protein synthesis inhibitors, yeast (with barely detect-able levels of activity of cytochrome oxidase (Criddle and Schatz, 1969)), does possess organelles of diameters between 0.2 and 0.8 μm and with inner and outer membranes (Plattner and Schatz, 1969). The inner membranes are invaginated to form cristae, and the outer membranes exhibit 10×100 nm slits. Even when grown in the absence of lipid supplements these features are still seen, and a similar picture of mito-chondrial organization emerges from the examination of sphaeroplasts after processing by the method of freeze-substitution (Plattner *et al.*, 1970). The presence of mitochondria-like structures in *Saccharomyces carlsbergensis* after anaerobic lipid-limited growth with 2% maltose has been demonstrated by more conventional methods. Lightly-fixed cells (glutaraldehyde or paraformaldehyde) were disrupted by shaking with glass beads and were subsequently post-fixed with aldehydes and osmium tetroxide. Staining of sections with uranyl acetate and lead citrate revealed mitochondrial structures between 0.4 and 0.8 μm diameter, with densely-staining matrices and few cristae (Damsky *et al.*, 1969). These authors found that permanganate fixation does not preserve these mitochondrial membranes, possibly because of their low sterol content. Glutaraldehyde

Table I. Mitochondria-like organelles in anaerobically-grown yeasts.

	Techniques of electron microscopy	Size	Ultrastructure	References
A. In *S. cerevisiae* grown with:				
2% Glucose + lipids	Chemical fixation of intact cells.	—	cristae	Morpugo et al. (1964)
5% Glucose (no lipids)	Freeze etching of intact cells.	0·3–2·0 μm	no cristae	Matile et al. (1969)
0·8% Glucose + lipids	Freeze etching of intact cells.	0·2–0·8 μm	cristae	Plattner and Schatz (1969)
0·8% Glucose (no lipids)			cristae	
2% Maltose (no lipids)	Chemical fixation following light prefixation and gentle disruption.			
0·3% Glucose + lipids	Negative staining of intact cells.	0·4–0·8 μm	few cristae	Damsky et al. (1969)
10% Glucose (no lipids)			cristae	Plattner et al. (1970)
5% Galactose + lipids	Chemical fixation of mitochondria isolated from prefixed protoplasts.	0·5 μm	cristae	Watson et al. (1970, 1971)
5% Glucose + lipids			well developed cristae	
5% Glucose (no lipids)			poorly developed cristae	
			no obvious cristae	
B. In *S. carlsbergensis* grown with:				
10% Glucose + lipids	Chemical fixation of subcellular fractions.	0·2–1·0 μm	cristae	Cartledge and Lloyd (1972b)

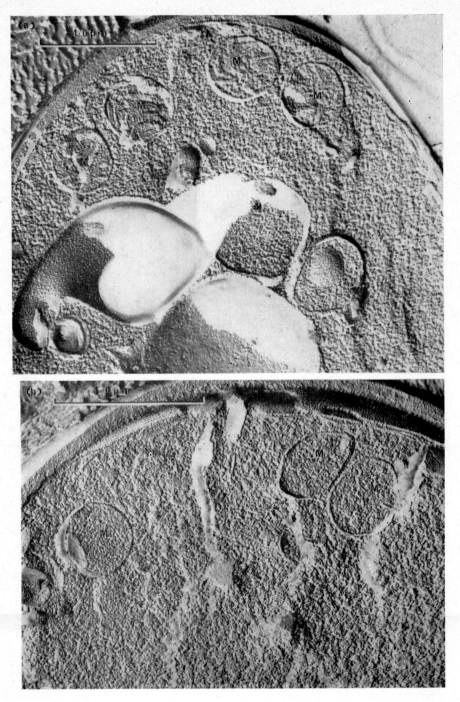

prefixed sphaeroplasts of *S. cerevisiae* broken in a French Press also showed mitochondria-like structures (Watson *et al.*, 1970, 1971). These investigators observed changes in morphology of organelles after lipid supplementation (ergosterol and Tween 80) or after changing the glucose levels in the medium. Thus lipid-supplemented organisms showed well-developed cristae similar to, but not identical in appearance with the respiratory-competent mitochondria of aerobically-grown cells. Catabolite repression produced little change in these structures, and they were quite stable to the isolation procedures normally used for the extraction of mitochondria. However the organelles of lipid-depleted cells were so fragile that their extraction as intact entities required stabilization by prior fixation of the sphaeroplasts; they showed dense granular matrices enclosed by inner and outer membranes. The inner membranes were not invaginated to form cristae. Both lipid-supplemented and lipid-depleted anaerobically grown cells have mitochondrial ribosomes (Watson, 1972). Considerable cristate development (Fig. 27) is seen in mitochondria-like structures isolated from lipid-supplemented, glucose-repressed, anaerobically grown *Saccharomyces carlsbergensis* (Cartledge and Lloyd, 1972b).

Freeze-etched preparations reveal that the attainment of the stationary phase due to ergosterol limitation during anaerobic growth of *Saccharomyces cerevisiae* is accompanied by degeneration of mitochondria into spherical or rod-shaped bodies without cristae and with an altered morphology. However these are still clearly recognizable as mitochondria and not vacuoles (Matile *et al.*, 1969). These structural changes are accompanied by the formation of true vacuoles, and this had led to some confusion in electron micrographs prepared by conventional sectioning techniques.

A summary of the morphological characteristics of mitochondria-like structures in anaerobically-grown yeasts is presented in Table II. Cells of *S. cerevisiae* grown anaerobically for 8 or 9 generations in a completely chemically defined (lipid-supplemented) medium have been subjected to detailed morphometric examination (Table II, A. Goffeau, E. Mrena and A. Claude, personal communication) after glutaraldehyde fixation. It is to be noted that whereas anaerobically-grown cells contain mitochondria-like structures which are 10 times larger (on a volume basis) than the mitochondria of aerobically-grown cells, the latter contain 20 times more mitochondria. This suggests that the total mitochondrial volume in the anaerobic cells was only about one-half that in the aerobic organisms.

FIG. 26. Freeze-etched preparations of *Saccharomyces cerevisiae* (a) grown aerobically, and (b) grown anaerobically. Both show mitochondrial structures (M) but cristae are not evident in the organelles of the anaerobically-grown yeast. (Reproduced with permission of Dr. H. Moor.)

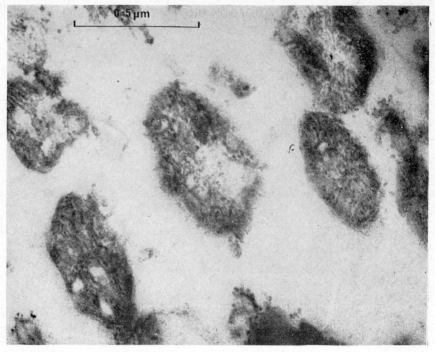

FIG. 27. Mitochondria-like organelles isolated from *Saccharomyces carlsbergensis* grown anaerobically in the presence of 10 % glucose. (Reproduced with permission from Cartledge and Lloyd, 1972b.)

Table II. The numbers and dimensions of mitochondria in sections of aerobically-grown cells of *S. cerevisiae* compared with those of the mitochondria-like organelles observed in cells grown anaerobically with lipid supplementation (A. Goffeau, E. Mrena and A. Claude, personal communication.)

	In aerobically-grown cells	In anaerobically-grown cells
1. Average surface of a cell section	7·83 μm²	6·77 (± 0·36) μm²
2. Yeast cell as a sphere: diameter	3·16 μm	2·94 μm
volume	16·5 μm³	13·27 μm³
3. Average surface of a mitochondrion	0·097 μm²	0·437 (+ 0·01) μm²
4. Mitochondrion as a sphere: diameter	0·35 μm	0·75 μm
volume	0·023 μm³	0·22 μm³
5. Average number of mitochondria per section	10·52	1·22 (± 0·08)
6. Average total mitochondria a surface per section	1·02 μm²	0·54 μm²
7. Cell surface/total mitochondrial surface	7·68	12·66
8. Total volume of mitochondria	2·15 μm³	1·05 μm³
9. Number of mitochondria/cell	94·57	4·80

MtDNA fibres are more apparent in the mitochondria-like structures of the anaerobic cells than in fully-functional mitochondria. If the total cellular mtDNA content is not drastically altered during anaerobic repression (see Chapter 5), then the amount of DNA per mitochondrion works out to be 20 times more in those of the anaerobically-grown cells than in those of the aerobic yeasts.

It is remarkable that no critical examination for mitochondrial equivalents in anaerobically-grown protozoa has been described. Only one strictly anaerobic fungus *Aqualinderella* has been discovered (Emerson and Held, 1969), although many fungi are facultative anaerobes (e.g. some *Mucor* species and animal pathogens, Bartnicki-Garcia and Nickerson 1961; Clark-Walker, 1972). In a facultatively-anaerobic mutant of *Neurospora crassa* (Howell *et al.*, 1971) grown anaerobically, mitochondrial numbers were reduced (as seen in glutaraldehyde-fixed mycelia); they appeared swollen and possessed few cristae. Reduction of mitochondrial numbers and organization progressed with increasing time of anaerobic growth. Aerobically-grown mutant and wild-type had indistinguishable mitochondrial ultrastructure.

(b) Growth at limiting oxygen tensions. Growth of the obligately aerobic yeast *Candida parapsilosis* at limiting oxygen tension in a chemostat produced altered mitochondrial morphology as revealed in permanganate-fixed whole cells; the state of organization of the cristae was decreased (Kellerman *et al.*, 1969). As this change was to some extent less marked when ergosterol and unsaturated fatty acids were included in the medium, it is again questionable whether the apparently altered mitochondrial ultrastructure was a consequence of poor fixation rather than a true reflection of mitochondrial disorganization. No other reports of changes in mitochondrial structure at controlled oxygen tensions in obligately aerobic eukaryotic microorganisms have appeared; human lymphocytes cultured *in vitro* show little alteration in mitochondrial numbers per cell under 5, 20 or 50% O_2 (Andersen *et al.*, 1970).

(c) Respiratory adaptation. The theory of *de novo* genesis of mitochondria in cells grown anaerobically and then exposed to oxygen (Linnane *et al.*, 1962; Wallace and Linnane, 1964) was based on the finding, now known to be incorrect, of the complete absence of mitochondria-like structures from the anaerobic cells. The currently favoured hypothesis, originally proposed by Marquardt (1962b), is that mitochondria pre-existing in the anaerobically-grown yeast, differentiate during respiratory adaptation and acquire many of the enzyme activities characteristic of the fully-functional mitochondria. The structures seen in anaerobic cells ("promitochondria" or "mitochondrial precursors") would thus form the basic skeleton of

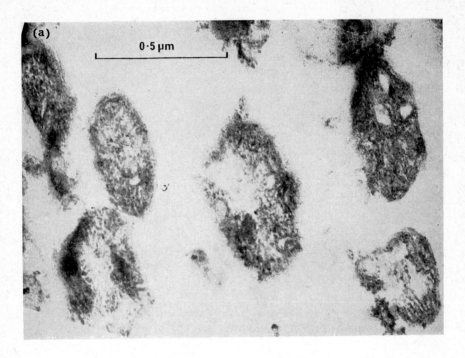

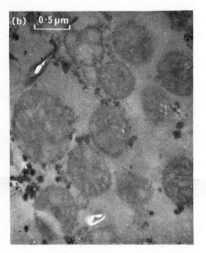

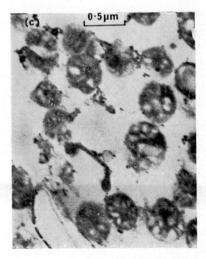

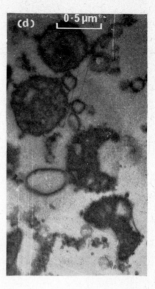

FIG. 28. (a) Mitochondria-like structures isolated from glucose-derepressed anaerobically grown *Saccharomyces carlsbergensis* and mitochondria isolated from cells adapted to aerobiosis for (b) 10 min, (c) 30 min, and (d) 3 h. (Reproduced with permission from Cartledge and Lloyd, 1973.)

membrane systems into which the newly-synthesized enzymes must fit. Cells taken at various stages from cultures of glucose-repressed, lipid-depleted cells undergoing respiratory adaptation, contain forms of mitochondria apparently at different stages of progress toward the ultrastructure of the fully aerobic phenotype (Watson *et al.*, 1970). In anaerobically grown glucose derepressed *S. carlsbergensis* extensive development of inner membranes was apparent after 10 min adaptation (Fig. 28). However this type of analysis does not furnish unequivocal evidence for the straightforward conversion of a promitochondrion into a mitochondrion. Some evidence for the continuity of radioactively-labelled promitochondria and cytochrome *c*-oxidase-containing membranes of adapted cells has been provided by a combination of autoradiography and histochemical staining of thin sections for the electron microscope, although again the evidence is not unequivocal (Plattner *et al.*, 1970). Other workers (Osumi and Katoh, 1967) have described the transient formation of an extensive membranous endoplasmic reticulum system as a prelude to the formation of fully developed mitochondria (in the case of wild-type organisms) or aberrant mitochondrial structures (in a *petite* strain).

(*d*) *Effect of carbon substrates: glucose repression.* Differences in mitochondrial organization also occur between yeasts grown in the presence of different carbon sources. Aerobic growth with non-fermentable substrates like glycerol or acetate yield cells with highly-developed mitochondria with abundant cristae (Marquardt, 1962a, b; 1963; Polakis *et al.*,

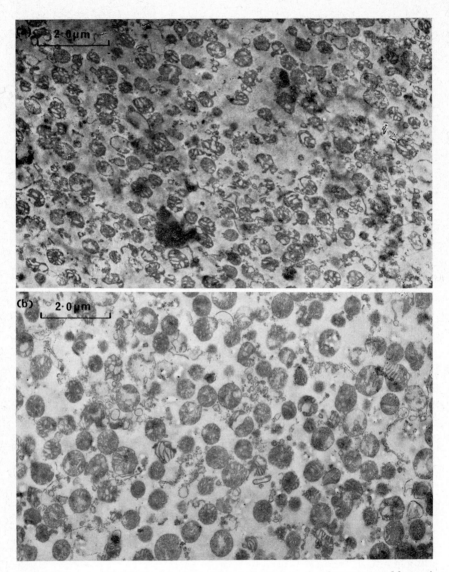

FIG. 29. The two populations of mitochondria isolated from *Saccharomyces carlsbergensis* during glucose derepression separated by equilibrium density centrifugation on buffered sucrose density gradients containing 2mm—MgCl₂. The buoyant densities of the fractions were, in (a) 1·21 g/cm³, and in (b) 1·235 g/cm³. (Reproduced with permission from Cartledge and Lloyd, 1972a.)

1964; Yotsuyanagi, 1962a); *petite* mutants do not grow rapidly under these conditions. Fermentable substrates such as glucose, when present at high concentrations in the medium, are not oxidized completely but rather fermented (aerobic fermentation or glucose-repressed respiration) and in this case mitochondria are few in number, large and show poor cristate development (Jayaraman *et al.*, 1966; Clark-Walker and Linnane, 1967; Matile *et al.*, 1969; Neal *et al.*, 1970, 1971). The utilization of glucose from the culture leads to derepression, a phase of aerobic respiration and to the formation of many small highly developed mitochondria (Jayaraman *et al.*, 1966), possibly after the breakdown of some of the original mitochondrial population. Morphological differences between mitochondria isolated from repressed and derepressed cells (or sphaeroplasts), are also observed (Neal *et al.*, 1970, 1971). At an intermediate stage in derepression, mitochondria isolated from *Saccharomyces carlsbergensis* fall into two distinct populations after density-gradient centrifugation (Fig. 29), but this observation does not necessarily reflect the existence of two physiologically different classes of mitochondria *in vivo* (Cartledge and Lloyd, 1972b).

Growth with galactose, a low concentration of glucose (Polakis *et al.*, 1965), or melibiose (Marchant and Smith, 1968b), does not lead to a diminution in mitochondrial organization in aerobic cultures.

4. Effects of growth with antibiotics and inhibitors

Of the many anti-bacterial antibiotics known to specifically inhibit mitochondrial protein synthesis, only the effects of chloramphenicol have been widely investigated at the ultrastructural level. A marked reduction in the organization of mitochondrial cristae of yeast occurs during glucose-repressed aerobic growth in the presence of high concentrations (4 mg/ml) of this antibiotic (Huang *et al.*, 1966; Clark-Walker and Linnane, 1967); some strains of yeast are sensitive to as little as 0·05 µg/ml (Wilkie *et al.*, 1967). Mitochondria of almost normal appearance are seen in chloramphenicol-inhibited glucose-derepressed chemostat cultures (Kellerman *et al.*, 1971). Obligately aerobic yeasts also show marked mitochondrial ultrastructural changes when grown with chloramphenicol. In *Rhodotorula glutinis*, 500 µg/ml of the drug led to the formation of "looped" cristae and after seventy generations reticular cristae are formed, (Fig. 30; Smith and Marchant 1968). Similar effects have been reported for *Rhodotorula mucilaginosa* (Osumi and Kitsutani, 1971). Large abnormal mitochondria have been found in cells of *Candida parapsilosis* after growth with 4 mg/ml chloramphenicol (Kellerman *et al.*, 1969). Gross morphological changes in mitochondria were not found when the obligately aerobic fungus *Pythium ultimum* was grown with 100 µg/ml chloramphenicol, although respiratory-deficient hyphae were produced (Marchant and Smith, 1968a).

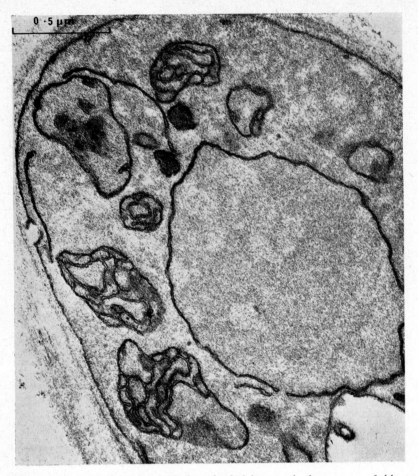

Fig. 30. Section of KMnO₄-fixed *Rhodotorula glutinis* grown in the presence of chloramphenicol (500 μg/ml). (Reproduced with permission from D. G. Smith and R. Marchant.)

The only lesion observable by electron microscopy was the apparent absence of inner-membrane subunits in negatively-stained preparations. In the colourless alga *Polytomella caeca* no ultrastructural defect was produced, and inner membrane subunits were still observed after growth with 1 mg/ml chloramphenicol, despite the physiological impairment of the mitochondrial respiratory system (Lloyd *et al.*, 1970b). Increased mitochondrial numbers accompanied the reduction in size of these organelles when growth ceased two cell generations after the addition of 500 μg/ml chloramphenicol to exponentially-growing cultures of the ciliate protozoon *Tetrahymena pyriformis* (Turner and Lloyd, 1971). The normally highly organized tubular cristae of the mitochondria of this species were greatly reduced in numbers and became degenerate in appearance (Fig. 31). Similar changes (including increased mitochondrial numbers) can be induced by growth in the presence of ethidium bromide (20 μg/ml) (Charret, 1972).

Respiratory adaptation of *Saccharomyces cerevisiae* in the presence of cycloheximide or Blasticidin S produces dedifferentiated mitochondria and an extensive endoplasmic reticulum (Osumi and Ubukata, 1970).

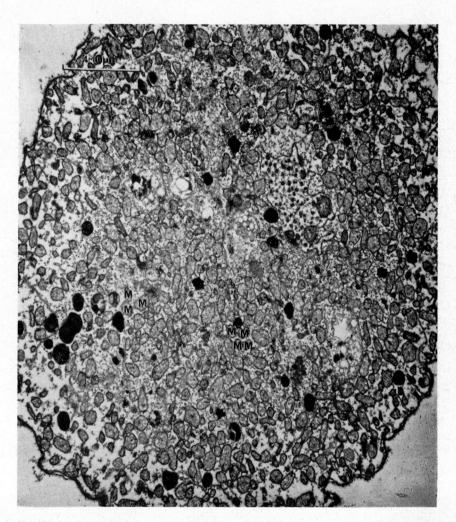

FIG. 31. Section of *Tetrahymena pyriformis* from a culture grown for 38 h in the presence of 500 μg/ml chloramphenicol. The organisms show an approximately three-fold increase in mitochondria, but these are small and show little of the complex tubular cristae seen in normal organisms. (Reproduced with permission from Turner and Lloyd, 1971.)

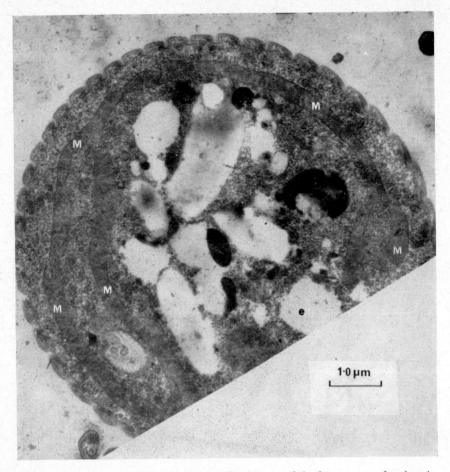

FIG. 32. Transverse sections of *Euglena gracilis* after growth in the presence of antimycin
A showing the giant ring-shaped mitochondrion(M). (Reproduced with permission from
Calvayrac *et al.*, 1971.)

Growth of *Candida parapsilosis* with euflavine leads to degeneration of the
mitochondrial cristae (Kellerman *et al.*, 1969). Ethidium bromide present
in the growth medium leads to the production of ultrastructurally-modified
mitochondria in *Acetabularia* (Heilporn and Limbosch, 1971). Growth of
Neurospora crassa in the presence of acridine or ethidium bromide leads
to the accumulation of paracrystalline inclusions similar to those seen in
the mitochondrial mutant *abn*-l (Wood and Luck, 1971).

Induction of dyskinetoplastic trypanosomes with greatly reduced mitochondria has been noted in *Crithidia fasciculata* (Kusel *et al.*, 1967;
Steinert and van Assel, 1967a; Hill and Anderson, 1969), and similar

effects have been reported after treatments of a number of trypanosomal species with trypanocidal agents e.g. prothidium (Ray and Malhotra, 1960), ethidium, bromide, berenil (Brack *et al.*, 1972a,b), and stilbamidines (for short survey see Delain *et al.*, 1971). Griseofulvin-treated hyphae of *Pythium ultimum* and *P. debarynum* possess smaller mitochondria than untreated controls (Bent and Moore, 1966). Growth of a bleached strain of *Euglena gracilis* in the presence of antimycin A (0·5 µg/ml) leads to fusion of the mitochondria to form a sheath-like structure in the periphery of the cells (Fig. 32; Calvayrac and Butow 1971; Calvayrac *et al.*, 1971). Agents known to lead to production of giant mitochondria in mammalian systems include triphenyl boron (Harris and Leone, 1966), azo dyes (Tandler *et al.*, 1969), triamcinolone acetonide (Bullock *et al.*, 1971), and cuprizone, a chelator of Cu^{2+} (Suzuki, 1969).

2

Subcellular Fractionation of Eukaryotic Microorganisms: the Isolation of Mitochondria and their Subfractionation

I. Introduction

For many purposes the only valid object of experimental investigation is the intact living organism; mitochondrial functions are so inextricably interwoven with those of the other parts of the cell that a complete description of mitochondrial physiology will eventually necessitate analyses at many different levels of organization including that of the whole cell. Nevertheless from many viewpoints, most of our present knowledge of mitochondrial physiology has been made possible by studies of isolated organelles: work initiated in the pioneering experiments of Bensley and Hoerr (1934) and Claude (1946), and followed up by Hogeboom *et al.* (1948). The two philosophies of subcellular fractionation have recently been compared and contrasted by deDuve (1971). On the one hand there is the preparative approach in which the goal is to prepare a fraction containing a population of organelles in a reasonable state of "integrity" and with a minimum of contaminating membranes. On the other hand there is the analytical methodology arising from the approach of Schneider and Hogeboom (1951) who stressed "the need of establishing balance sheets in which the summation of the activities of the tissue fractions is compared with that of the whole tissue". The two approaches are both necessary and complementary, as the assessment of "purity" of a particulate preparation can only be achieved by analytical fractionation. Even so the presence of lysosomes and peroxisomes (deDuve and Baudhuin, 1966; de Duve and Wattiaux, 1966) in the crude mitochondrial fractions used by workers in the fields of respiration and phosphorylation is nearly always overlooked as a matter of convenience.

The basic principles involved in the subcellular fractionation of eukaryotic microorganisms are no different from those for the fractionation of mammalian tissues; and although many investigators have attempted purification of organelles from microbial extracts, there have been hardly any critical studies of an analytical nature. Special difficulties encountered when dealing with microbial systems include the refractory nature of the cell walls of algae, fungi and yeasts, and the highly differentiated cyto-plasmic organization of many microbial species. Even when a satisfactory breakage method has been devised, the extract produced often contains a wide diversity of organelles, some of unknown or poorly defined functions; for many of these marker compounds or enzymes have not yet been described. This situation provides a great challenge to the most refined techniques now at our command.

II.Methods of Disintegration of Microorganisms

The conditions employed for the disruption of a microorganism are influenced by the necessity to minimize damage to intracellular organelles. Thus many of the methods used for the quantitative release of enzymes from microorganisms (Hughes *et al.*, 1971), especially those which rely on the development of high liquid shearing forces, are not suitable for the release of "intact" mitochondria. The more fragile protozoa may be dis-rupted by gentle hand homogenization in a Kontes or Dounce homogenizer, under conditions of more carefully controlled hydrodynamic shear in the Chaikoff Press (Emanuel and Chaikoff, 1957), or by explosive decom-pression (Hunter and Commerford, 1961). Shaking with acid-washed glass beads (0·15–0·30 mm diameter) for periods of less than 1 min in a high-speed shaker has been used to disrupt the more resistant protozoa. Organisms with walls of high tensile strength may be broken by methods based on the generation of solid shear (hand grinding or treatment in a mechanical mill). A survey of methods employed for various species of eukaryotic microorganisms which have been shown to release a proportion of relatively undamaged mitochondria, is presented in Table III; practical and theoretical details are covered at length in the review of Hughes *et al.* (1971).

Chemical methods of breakage, including treatment with enzymes, have many advantages over mechanical methods, in that the sphaeroplasts or weakened structures produced may be more gently disrupted than the cells from which they are derived. For instance osmotic lysis of sphaero-plasts of yeasts prepared by using the digestive enzymes of *Helix pomatia*, provided the first really satisfactory preparations of yeast mitochondria (Duell *et al.*, 1964; Ohnishi *et al.*, 1966a), and very gentle mechanical breakage of sphaeroplasts is still the method of choice prior to analytical fractionation (Cartledge and Lloyd 1972a, b). The main disadvantage of the use of enzymic methods is their unpredictability arising from differ-

Table III. Isolation of phosphorylating mitochondria from microorganisms. (BSA, bovine serum albumin; PVP, polyvinylpyrolidone.)

A. Mitochondria from algae

Organism	Growth medium	Cell yield	Method of disruption	Isolation buffer	Yield of mitochondria	Comments
Prototheca zopfii (Lloyd, 1965, 1966a; Lloyd and Venables, 1967)	Acetate (propion-ate) mineral salts-thiamine pH 7.2 30°C	26 g/mol acetate	5 min hand grinding at 0°C as a paste with 0·15 mm glass beads	1 M-sucrose, 20 mM-tris, 1mM-EDTA, pH 7·4		R.C. ratios up to 5·5 for α-ketoglutarate
Euglena gracilis (streptomycin-bleached strain SM-LI; Buetow and Buchanan 1964, 1965)	Ethanol (acetate) defined (in dark) pH 6:8 23–28°C	7×10^6 cells/ml	20–60 s hand grinding at 2°C as a paste with 0·28 mm glass beads	0·25 M-sucrose, 24 mM tris, 0·1 mM-EDTA, pH 7·4	23 mg protein per 10^9 cells	Carry out oxidative phosphorylation but show no respiratory control
Euglena gracilis (bleached strain Z, Sharpless and Butow 1970a)	Glutamate + malate (succinate, ethanol) semi-defined (in dark) pH 4·1 27°C	5×10^5 cells/ml $\equiv 4$ g wet wt/litre	20 s shaking with 0·5 mm glass beads in an MSK homo-genizer (Bronwill Scientific Co).	0·3 M-sorbitol, 25 mM-HEPES, 1 mM-EDTA, pH 7·0	1·5–2·5 mg protein per g wet wt of cells	Carry out oxidative phosphorylation but do not consistently show respiratory control
Astasia longa (strain J, Buetow and Buchanan 1969)	Acetate-defined medium (in dark) pH6:8. 23–28°C		45 s hand-grinding at 2°C	0·25 M-sucrose (pH 7·4)	18–28 mg protein/10^9 cells	Carry out oxidative phosphorylation but show no respiratory control
Polytomella caeca Pringsheim (Lloyd and Chance 1968; Lloyd *et al.*, 1968, 1970b)	Acetate (propionate)-defined medium pH 6·0 25°C	0·15 mg dry wt per ml	Gentle hand-homogenization in all glass homogenizer (Kontes Glass Co., Vineland, N.J., U.S.A.) to 50 % breakage	0·32 M-sucrose, 29 mM tris, 2 mM-EGTA, pH 7·4		Carry out oxidative phosphorylation show respiratory control ratios up to 3·5 for α-keto-ketoglutarate

B. Mitochondria from protozoa

Tetrahymena pyriformis (strain W; Kobayashi, 1965)	Protease-peptone-yeast extract, pH 7.0–7.2, 25°C	2×10^5 cells/ml	Gentle homogenization in Teflon homogenizer	0.35 M-mannitol, 0.05 % BSA, 1 mM-tris HCl, 0.1 mM-EDTA, pH 7.0–7.2		Respiratory control ratios of up to 3 with succinate or glutamate
Tetrahymena pyriformis (strain ST; Turner *et al.*, 1971)	Protease-peptone-liver extract pH 6.0 29°C	5–7×10^4 cells/ml	Gentle homogenization in all glass homogenizer (Kontes, type B plunger) to 90 % cell breakage	0.35 M-mannitol 0.1 % BSA, 0.1 mM-EGTA, 10 mM-tris-HCl, pH 7.2	10 mg protein/litre culture	Respiratory control ratios of 2 for succinate or α-ketoglutarate
Hartmannella castellanii Neff (Lloyd and Griffiths, 1968)	Mycological peptone pH 6.0 30°C		Gentle homogenization in all glass homogenizer (Kontes, type B plunger) to 90 % cell breakage	0.3M-manitol 10 mM-KCl, 10 mM-tris HCl, 5 mM-K phosphate, 0.2 mM-EDTA, pH6.9	12 mg protein/litre culture	Respiratory control ratios up to 2.5 for malate
Dictyostelium discoideum (Erickson and Ashworth, 1969)	glucose-peptone-yeast extract 23°C	5×10^6 cells/ml	Dounce homogenizer to 95 % breakage	0.25 M-mannitol 5 mM-tris HCl, pH 7.4		Oxidative phosphorylation not tested, no respiratory control
Crithidia fasciculata (Toner and Weber, 1967; 1972)	glucose-proteose-peptone-liver extract, complex medium	1.5×10^8 cells/ml	Grinding with SiC to 80 % disruption	0.35 M-sucrose, 2.5 mM-Mg acetate, 0.1 mM-EGTA 0.15 % BSA, 20 mM-MOPS, pH 7.7		Respiratory control ratios of 2 with succinate

Organism	Growth medium	Cell yield	Method of disruption	Isolation buffer	Yield of mitochondria	Comments
Crithidia fasciculata (Kusel and Storey, 1972)	Glycerol-complex medium	2×10^8 cells/ml	Digitonin treatment + 30 s homogenization with Polytron blender (Bronwill Sci. Co.)	0·25 M-mannitol, 10 mM-MOPS, 0·2% PVP, 3% BSA, 4·5 mM-ascorbate, 250 μM-MgCl₂ 250 μM-EDTA, pH 8·0		Respiratory control ratios of 2–3 with succinate or α-glycerophosphate
Physarum polycephalum (plasmodia) (Barnes *et al.*, 1973)	Medium containing 5 mg protohaem/litre		MSE Atomix blender	0·5 M-sucrose, 10 mM-TES buffer pH 7·5. Resuspended in this buffer + 5 mM-K₂HPO₄, 2 mM-MgCl₂		No respiratory control

C. Mitochondria from fungi

Organism	Growth medium	Cell yield	Method of disruption	Isolation buffer	Yield of mitochondria	Comments
Alomyces macrogunus (Bonner and Machlis, 1957)	0·5% glucose defined medium		Grinding with ground Pyrex glass 6–8 min in cold mortar	0·5 M-sucrose, 0·1 M-K phosphate, 1 mM-EDTA, pH 7·4		Low P/O ratios (0·47) with α-ketoglutarate
Neurospora crassa, strain Chiltona (Weiss, 1965) (Conidia and hyphae)	Sucrose-yeast-extract-salts medium		Sphaeroplasts (prepared using snail enzymes after glutathione treatment) disrupted gently in a glass homogenizer	0·44 M-sucrose 50 mM-tris HCl, pH 7·4	2–5 mg protein/litre culture	Respiratory control not tested

Organism	Medium		Method	Buffer	Protein	Comments
Neurospora crassa (wild-type 74A, Weiss *et al.*, 1970)	Sucrose-defined medium		Specially constructed grinding mill	0·44 M-sucrose, 10 mM-tris acetate, 2 mM-EDTA, 0·2% BSA, pH 7·3	2·5 mg protein per g wet wt	Respiratory control ratios up to 3 with pyruvate + malate
Neurospora crassa (wild-type SY 7A, Hall and Greenawalt, 1964) (Conidia and hyphae)			Gifford-Wood Eppenbach Micromill, top speed, 1 min (Model MV 6-3) 0·75 mm gap	0·25 M-sucrose, 5 mM-EDTA, 0·15% BSA		Respiratory control not tested
Neurospora crassa (Wild-type SY 7A, Cholineless mutant 34486, Greenawalt, *et al.*, 1967; Hall and Greenawalt, 1967) (Conidia and hyphae)	Citrate-defined medium		Sphaeroplasts (prepared by method of Duell *et al.*, 1964) broken in hand homogenizer	0·25 M-sucrose, 5 mMEDTA, 0·15% BSA		Respiratory control ratios up to 2:3 with pyruvate + malate
Neurospora crassa or *N. sitophila* and "*poky*", *abn-1* and *abn-2* mutants (Luck, 1967)	Citrate-defined medium	10 g wet wt/litre	Hand grinding in mortar with sand	0·44 M-sucrose, 1 mM-EDTA, pH 7·6		Phosphorylate and show respiratory control, but not further investigated
Neurospora crassa wild-type and *poky* mutant (Lambowitz *et al.*, 1972a)	Sucrose-defined medium		Snail enzyme treatment followed by gentle homogenization	0·3 M-sucrose, 8 mM-NaH$_2$PO$_4$, 8 mM-tris, 5 mM-MgCl$_2$, 0·7 mM-EDTA, pH 7·2		Respiratory control ratios 3–6 with NAD-linked substrates in wild type. No RC in *poky*
Aspergillus oryzae Iwasa (1960a, b)	1·5% sucrose -salts medium 30°C		Homogenization 15–25 s	0·5 M-sucrose, 10 mM-EDTA, 6 mM-K citrate		Phosphorylate only with succinate P/O = 1·5

Organism	Growth medium	Cell yield	Method of disruption	Isolation buffer	Yield of mitochondria	Comment
Aspergillus niger Van Tieghem IMI 59374 (Watson and Smith 1967a, b)	Glucose-defined medium pH 4·5 27°C	20–30 g wet wt per litre	Motor-driven all glass homogenizer	0·5 M-mannitol 4 mM-EDTA, pH 7·0	1 mg protein/ g. wet wt of cells	Respiratory control ratios up to 5 with α-ketoglutarate
Aspergillus niger strain resistant to nitroaryl compounds (Higgins and Friend, 1968)	5% glucose-mineral salts		Virtis homogenizer 45,000 RPM 30 s, 0°C phosphate, pH 7·8	0·25 M-sucrose, 10 mM-EDTA, 50 mM-K phosphate, pH 7·8		Respiratory control ratios > 3 with succinate
Aspergillus oryzae Baden RCST 754 (Watson *et al.*, 1969)	(As for *A. niger*, Watson and Smith, 1967a)				12–16 mg protein per 10 g mycelium	Respiratory control ratios up to 5 with citrate
Aspergillus oryzae (Kawakita, 1970a, b)			Mechanical breakage in a specially-constructed roller-mill	0·5 M-mannitol, 2·5 mM-EDTA, 0·1% BSA, 10 mM mercaptoethyla-mine HCl	5 mg protein per 10 g mycelium	Respiratory control ratios up to 2·77 with α-ketoglutarate
D. Mitochondria from yeasts						
Saccharomyces cerevisiae Commercial baker's yeast (Utter *et al.*, 1958)			Shaking with glass beads 5–10 s 0°C. 6000 Hz	1% NaCl, 1 mM-EDTA, pH 6·8		P/O ratio of 0·62 with succinate

Organism / Reference	Growth medium	Yield	Disruption method	Medium	Protein yield	Results
Saccharomyces cerevisiae (Vitols and Linnane, 1961)		10 g wet wt per litre	Shaking with glass beads in a Waring Blender, 15,000 RPM, 3–4 min 2°C	0·5 M-sucrose, 10 mM-tris, 5mM-EDTA, pH 7·4	300 mg protein from 100 ml packed cells	P/O ratios up to 1·6 for succinate or pyruvate + malate
Saccharomyces cerevisiae (Harden and Young strain; Duell *et al.*, 1964)	Glucose-complex medium		Sphaeroplasts prepared using snail enzymes after pretreatment with mercaptoethyl-amine and EDTA. Disrupted in a Vortex mixer	0·25 M-sucrose 50 mM-K-phosphate, 1 mM-EDTA, pH 6·8	20–25 mg/g cells	Respiratory control ratios up to 3·8 with α-ketoglutarate
Saccharomyces cerevisiae (wild-type diploid strain Fleishmann; Jayaraman *et al.*, 1966)	1% glucose–0·2% yeast-extract-mineral salts or 0·8% glucose–1% yeast extract mineral salts 30°C		Method of Duell *et al.*, 1964	20% sucrose, 50 mM-K phosphate, 1 mM-EDTA, pH 6·8		Respiratory control ratios up to 2·8 with α-ketoglutarate
Saccharomyces cerevisiae (strain DTXII; Kováč *et al.*, 1968)	0·25% glucose, 0·375% yeast extract-mineral salts		Sphaeroplasts prepared using snail enzymes after preincubation with 0·5 M-mercaptoethanol (pH 9·3 with tris). Disruption by 8 s homogenization in blender	0·44 M-mannitol (or sucrose), 1 mM-EDTA–0·1% BSA, pH 7·6 (0·8 M-sorbitol, better)		Respiratory control ratios up to 3 with α-ketoglutarate
Saccharomyces cerevisiae (Strains D-261, D311-3A, G-31; Mattoon and Balcavage, 1967; Balcavage and Mattoon, 1968; Guarnieri *et al.*, 1970)	0·1% glucose + 3% ethanol–1% yeast-extract–2% peptone pH 4·5	8–12 g dry wt cells/litre	Gifford-Wood Minimill (mV-6-3) 6500 RPM gap setting "40" 6 min	0·6 M-mannitol, 0·1 mM-EDTA, 0·4% BSA controlled at pH 6-8 by adding HEPES or histidine during grinding	3 mg protein per 10 g cells	Respiratory control ratios up to 6 with α-ketoglutarate

Organism	Growth medium	Cell yield	Method of disruption	Isolation buffer	Yield of mitochondria	Comments
Saccharomyces cerevisiae (Commercial baker's yeasts; Spenser et al., 1971)	—	—	Shaking in Braun ball mill (Shandon, Scientific Co. Ltd., London) 15 s with CO_2 cooling	0·25 M-sucrose, 20 mM-K phosphate, 1 mM-EDTA, 10 mM-$MgCl_2$, pH 6·8.		Respiratory control of up to 6 with α-ketoglutarate
Saccharomyces carlsbergensis and *Candida utilis* (Ohnishi et al., 1966a; 1966b)	K-lactate (2%)-yeast-extract (0·5%)-mineral salts pH 4·4	—	Sphaeroplasts broken by osmotic shock	0·65 M-mannitol, 0·1 mM-EDTA, pH 6·5	5–7 mg protein/g wet wt cells	Respiratory control ratios of 5–6 with α-ketoglutarate
Saccharomyces carlsbergensis (NCYC 74; Schuurmans-Stekhoven, 1966a, b)	0·06% glucose-lactate-0·5% peptone mineral salts	—		1 M-sorbitol, 10 mM-tris phosphate, 5 mM-EDTA, 0·05% BSA, pH 7·0		Respiratory control ratios up to 4 with malate + pyruvate
Candida utilis (Linnane et al., 1962)	Complex medium		As for *S. cerevisiae* (Vitols and Linnane, 1961)			No details of P/O ratios given
Candida utilis (NCYC 193; Light and Garland, 1971)	Glycerol ('133 mM) glucose (55·6 mM)-defined medium	7·2 mg dry wt of cells/ml	Sphaeroplasts prepared by method of Duell et al., 1964 Lysed osmotically in 50 mM-K phosphate 50 mM-sucrose, 1 mM-EDTA, pH 5·8	Sucrose conc. conc. of disruption buffer raised to 0·25 M by addition of sucrose	2–4 mg protein/g wet wt cells	High P/O ratios, figure for R.C. ratios not presented

Organism	Growth medium	Concentration	Method	Comments
Endomyces magnusii (Kotelnikova and Zvjagilskaja, 1966)			Mechanical homogenization at high speed for 1 min 40 s	P/O ratios of 0·8–1·4 for succinate 2·0 for NADH
Shizosaccharomyces pombe (strain 972h⁻, Heslot et al., 1970a)	Glycerol or glucose 160 mM–2% yeast-extract-2% peptone pH 4·5, 30°C	1–5 mg dry wt/ml	Method of Balcavage and Mattoon (1968)	Respiratory ratios control of up to 2 with succinate and NADH
Candida lipolytica (Volland and Chaix, 1970)	Hexadecane-minimal medium		Sphaeroplasts prepared using snail enzymes	Respiratory control ratios of up to 6·5 with α-ketoglutarate

ences in cell wall composition in different species or strains, or in organisms grown under different conditions; often the efficiency of cell wall degradation is too low for practical purposes. Several modifications of the original methods have been introduced which in some cases overcome these problems (Kováč et al., 1968; Lebeault et al., 1969). The use of a commercially available colloid mill (Mini-mill, Gifford-Wood, Hudson, New York) for the extraction of reasonably intact mitochondria from yeast under carefully controlled conditions has been described in detail by Guarnieri et al. (1970). A number of home-made devices for the breakage of fungal hyphae also produce high quality mitochondrial preparations (see Table III); it appears that the disruption of mycelial forms is somewhat easier than that of single cells, as only a few breaks or tears in the hyphal walls are necessary for the release of much of their contents.

Nevertheless the application of many commercially available digestive enzymes of microbial origin provides fresh opportunities for the development of suitable methods of sphaeroplast preparation. Enrichment cultures of soil organisms (provided with cell walls as the sole carbon source) have provided extracellular hydrolases which can be used to digest walls of yeasts (Jones et al., 1969) and fungi (Villanueva and Garcia Acha, 1971), not susceptible to snail digestive enzymes. Two other methods of potential value have recently been developed. The first of these is the production of "wall-less" mutants of fungi (Gaertner and Leef, 1970) and algae (Hyams and Davies, 1972) which may be more easily disrupted either mechanically or enzymically than the wild-type organisms. It is also now feasible to produce cells with modified walls by growth in the presence of an analogue of a cell wall constituent, e.g. Schizosaccharomyces pombe grown with 2-deoxyglucose (70 μg/ml) apparently lacks the α (1→3) glucan which usually makes the organism resistant to the action of snail enzymes (Birnboim, 1971; Poole and Lloyd, 1972).

A similar approach has produced modified cell surfaces in Tetrahymena pyriformis; in this case the inclusion of ergosterol in the growth medium produces organisms which may be rapidly lysed by the titration of the incorporated sterol with digitonin (Conner et al., 1971).

No entirely satisfactory method of extraction of algal mitochondria has yet been evolved; e.g. in Prototheca zopfii the extremely rigid cell walls are also very resistant to enzyme treatment (Lloyd and Turner, 1968).

III. Fractionation Methods

A. Characterization of Organelles by Analytical Fractionation

The success of the method chosen for breakage can only be fully judged by analytical fractionation of the extract produced. The analytical method "par excellence" consists of density-gradient centrifugation in a continuous gradient; by this means advantage may be taken, either of the differing densities of distinct species of particles (isopycnic centrifugation, equilibrium-gradient centrifugation), or of differences in their size (kinetic- or rate-gradient centrifugation). A combination of equilibrium and rate-centrifugation may also sometimes be used to advantage. Centrifugation may be carried out in tubes in swinging-bucket rotors, or in one of the several types of zonal rotor now available (Anderson, 1966; Beaufay, 1966). These methods lend themselves to assessment of frequency-distribution curves of certain physical properties, such as density or sedimentation coefficient, from which characteristics of the particle population including its size distribution can be calculated (deDuve, 1967; Poole et al., 1971b). The classical method of differential centrifugation is a much less useful approach to fractionation in analytical terms, as the resolution available is very poor by comparison; for instance so-called mitochondrial fractions prepared by this method invariably also contain lysosomes and peroxisomes. The use of marker enzymes for the characterization of the various classes of organelle present in a fraction depends on the postulate that certain enzymes have a single location in a particular subcellular particle. The specific activity of the enzyme must also be the same in all the individual organelles of the population, although this is only approximately the case. It is very unlikely that every mitochondrion has exactly the same composition, and it becomes increasingly possible, as techniques improve, to detect heterogeneity within a given population (see p. 72). Having determined the distribution of enzymes through the gradient, it is useful to be able to correlate these with the occurrence of various particles as revealed by the examination of fractions in the electron microscope; a complete evaluation of number, size, and shape of the individual particles is possible using special techniques which are quantitatively and statistically valid (Baudhuin et al., 1967).

It cannot be stressed too firmly that quantitative analytical fractionation requires that enzyme recoveries are calculated and expressed; it is not sufficient to present results only in terms of specific activities, as this can be very misleading (deDuve, 1967).

Although the rationales of these procedures and the processing of data so produced have been so elegantly worked out and expressed, and so carefully applied to fractionation of mammalian tissues, almost without exception the rules have been broken by workers involved in fractionation of microorganisms. It has been amply demonstrated that analytical fractionation properly conducted provides a vast amount of otherwise unobtainable information, and that the pursuit of particle purity in the absence of this information is an extremely hazardous occupation. The history of cell fractionation abounds with the erroneous attribution of

Table IV. Some examples of analytical subcellular fractionations of eukaryotic microorganisms (see also Lloyd and Cartledge, 1974).

Algae	References
Ochromonas malhamensis	Lui *et al.* (1968)
Polytomella caeca	Cooper and Lloyd (1972)
	Cartledge *et al.* (1971)
Euglena gracilis	Graves *et al.* (1972)
Astasia longa	Bégin-Heick (1973)

Protozoa	
Crithidia fasciculata	Edwards and Lloyd, unpublished results
Tetrahymenea pyriformis	Müller *et al.* (1966)
	Müller *et al.* (1968)
	Lloyd *et al.* (1971a)
	Müller (1972)
	Poole *et al.* (1971a,b)
Hartmannella (Acanthamoeba) castellanii	Müller (1969)
	Müller and Møller (1969)
	Morgan and Griffiths (1972)
	Morgan *et al.* (1973)
Dictyostelium discoideum	Weiner and Ashworth (1970)
Tritrichomonas foetus	Müller (1973)

Fungi	
Neurospora crassa	Matile (1971)
Coprinus lagopus	O'Sullivan and Casselton (1973)

Yeasts	
Saccharomyces cerevisiae (*petite* mutant)	Perlman and Mahler (1970a)
	Perlman and Mahler (1970b)
Saccharomyces carlsbergensis (aerobically-grown, 0·8 % glucose) (anaerobically grown, 10 % glucose + lipids	Cartledge *et al.* (1970a) Cartledge and Lloyd (1972a, c) Cartledge *et al.* (1970b) Cartledge and Lloyd (1972b)
(anaerobically grown, glucose derepressed and adapting to aerobiosis)	Lloyd *et al.* (1970a) Cartledge *et al.* (1972) Cartledge and Lloyd (1973)
Shizosaccharomyces pombe	Poole and Lloyd, unpublished results

enzyme activities to particles which had been incompletely purified, and the characterization of the contaminating species by strictly-conducted analytical procedures often led to the description of hitherto unrecognized organelles. In the case, for instance, of a highly-differentiated protozoon, where electron microscopy provides evidence of a great degree of sub-cellular complexity of membrane organization (many of these organelles not having been characterized biochemically) there is a pressing need for the application of these principles. The situation is made difficult because the whole library of marker enzymes for the membrane systems of mammalian cells is not necessarily applicable. However it is already apparent that the most commonly used marker enzymes can often be used as indicators of mitochondria, lysosomes and peroxisomes, plasmalemma, Golgi membranes and other types of endoplasmic reticulum, even though more specialized structures present difficulties. Examples of the application of analytical fractionation techniques to eukaryotic microorganisms are presented in Table IV, and examples of enzymes distributions in fraction-ated homogenates of *Saccharomyces carlsbergensis* are shown in Figs. 60 and 74. Methods other than those based on centrifugation, such as free-flow electrophoresis (Stahn *et al.*, 1970), phase distribution (Albertsson, 1960), or membrane filtration (Glaumann and Dallner, 1970) have not yet been applied to eukaryotic microorganisms.

B. Preparation of Subcellular Fractions Containing Mitochondria

Methods used for the release of "intact" mitochondria from microorgan-isms have the following features in common: (1) Cell breakage by the gentlest method consistent with the efficient release of the organelles (some sacrifice of yield for quality is necessary). (2) Rapid removal of the mitochondria from the extract at $0°$–$4°C$ (the low temperature reduces metabolic activity to a minimum and minimizes action of hydrolases which may be present). (3) The use of isolation media buffered to 0.25–0.6 osmolarity with sucrose, mannitol or sorbitol and kept in the pH range 6.8–7.6 with low concentrations of buffer (conditions which lead to aggluti-nation of subcellular particles must be avoided). (4) The use of minimal centrifugation speeds for the sedimentation of the mitochondria. A summary of methods used for the preparation of fractions containing phosphorylating mitochondria from various microorganisms is presented in Table III. Only in a few cases have density gradient methods been used to further purify these preparations. From what we know of the properties of likely contaminating particles, rate-zonal separation affords a rapid method for achieving further purification. It has the added advantage over isopycnic density gradient fractionation in that high osmotic pressures and consequent dehydration, distortion and damage to the membranes may be avoided when shallow stabilizing gradients are used. Gradient media other than sucrose or mannitol may be preferable; Ficoll (Pharmacia, Uppsala, Sweden) is the most popular high molecular weight gradient medium. Disadvantages of further purification on gradients include increased handling times, increased dilution of particles, and non-representative recovered sample (the largest mitochondria are usually the most free from contaminating membranes). It has recently been shown that plant mitochondria extensively purified from crude mitochondrial fractions by equilibrium-density centrifugation through sucrose gradients after slow swelling in 5 mM–KCl–5 mM–$MgCl_2$–10 mM–potassium phosphate at pH 7·2 can be slowly readjusted to isosmolar conditions without impairment to their excellent phosphorylation characteristics (Douce and Bonner, 1972). During this procedure the latency of matrix marker enzymes is also preserved; this method has not been tested with microbial mitochondria.

Yeast cells may be stored at 4°C for up to 9 days without impairing seriously the quality of mitochondria subsequently isolated (Balcavage et al., 1970). Isolated mitochondria may be stored at 77°K for 30 days or more without loss of Q_{O_2}, or impairmen to P/O or RC ratios, morphology, or capacity for adenine nucleotide translocation. They must be frozen rapidly (shell freezing) and thawed rapidly (in hot water) until just a little frozen material remains.

IV. Criteria for Assessment of Isolated Mitochondria

A. Functional Integrity

The success of the isolation procedures employed can be assessed by a number of well-defined biochemical tests of mitochondrial function. Satisfactory preparations should satisfy certain minimal criteria as follows:

(a) The mitochondrial fraction should show negligible respiration in the absence of added respiratory substrates.

(b) It should be possible to show high P/O ratios, although phosphorylation with succinate as substrate may be efficient even when extensive damage to membranes has been incurred.

(c) The respiration rates should be low in the absence of added ADP (State 4) and should show an immediate increase on adding ADP (State 3). Exhaustion of ADP should be accompanied by an abrupt return to the State 4 rate, and repeated cycles may then be demonstrated. A gradual increase of rate on adding the phosphate acceptor is not acceptable, as this does not constitute true respiratory control as defined by Chance and Williams (1956).

(d) The ATPase activity should be low and stimulated by uncoupling agents. High ATPase may indicate the presence of contaminating membranes (e.g. flagella, cilia, plasmamembranes), and these ATPases are often Mg^{2+}-stimulated. The inclusion of EDTA (or more specifically EGTA which chelates Ca^{2+} but does not impair mitochondrial membranes by extraction of Mg^{2+}) minimizes these activities and may increase the respiratory control ratios.

(e) It should be possible to demonstrate the reduction of intramito-chondrial nicotinamide nucleotides on addition of succinate or ascorbate + TMPD (i.e. reversal of electron transport). This response is abolished on addition of uncouplers.

(f) Intact mitochondria retain their endogenous nicotinamide nucleo-tides, cytochrome c and all the enzymes occurring in the intermembrane and matrix compartments. Thus acceleration of nicotinamide-nucleotide linked oxidations by the addition of NAD^+ or $NADP^+$ is indicative that endogenous cofactors have leaked out of mitochondria during their preparation. Cytochrome c addition does not stimulate the respiratory rates of intact mitochondria.

(g) Latency of mitochondrial enzymes can be demonstrated in intact mitochondria. The apparently low activities of nicotinamide nucleotide-linked dehydrogenases, aconitase, fumarase etc. in intact mitochondria, which are increased on mechanical damage or detergent treatment, reflect the low permeability of the intact inner mitochondrial membrane to substrates and cofactors (Chappell and Greville, 1963).

(h) Stimulation of oxidation rates of citrate or isocitrate by the addition of L-malate indicates the integrity of the permeability control mechanisms of the inner mitochondrial membrane (Chappell, 1964).

These criteria have been evolved for the evaluation of integrity of mitochondria isolated from higher plants and animals, and it becomes increasingly evident as methods of obtaining mitochondria from micro-organisms improve, that most if not all of these tests are still applicable, irrespective of the source of the organelles. One notable exception to this general rule is the dissimilarity of many microbial and plant mitochondria as compared with mammalian mitochondria with respect to their ability to oxidize exogenous NADH. Whereas carefully isolated mammalian mitochondria do not oxidize this substrate (Lehninger, 1951) mito-chondria from some fungi, yeasts and algae (but not those from some protozoa) have specialized mechanisms for doing so (see Chapter 3). It seems likely that other differences between various types of mitochondria will be revealed as research continues, and some of the accepted criteria for integrity may not apply universally. The ability to show respiratory control is perhaps the most useful and most easily demonstrated property of good quality mitochondria; some practical hints have been summarized by Chappell and Hansford (1969). The highest values so far reported for yeast mitochondria (of around 4 for pyruvate + malate and 6 for α-keto-glutarate oxidation, see Table III) are not nearly as high as those deter-mined for some insect, mammalian or plant mitochondria. It seems likely that this discrepancy does not necessarily indicate a lack of functional integrity, but rather an inherently less well-coupled yeast system. Another possibility is that non-phosphorylating pathways of electron transport

D

alternative to the main respiratory chains may play a significant role in the overall electron flux. This is especially true for instance in *Tetrahymena* mitochondria which have a complex branched-chain electron transport system (Lloyd and Chance, 1972). Another problem in the isolation of mitochondria with good respiratory control is the presence of endogenous uncouplers in some extracts. For instance unsaturated fatty acids produced by the action of phospholipases (Eichel, 1960), may accumulate in organisms when growth ceases or during encystment (Tibbs and Marshall, 1969). The addition of bovine serum albumin, which has been used to reverse the action of uncouplers (Weinbach and Garbus, 1966), is not always effective in these circumstances.

Analytical fractionation may be used as a method for assessment of integrity of mitochondria. The sedimentability of cytochrome *c* oxidase is the most frequently used criterion, although it is evident that extensive comminution of the inner membrane must occur before any non-sedimentable portion of this enzyme is produced. Other "soluble" matrix enzymes (e.g. citrate synthase or glutamate dehydrogenase) provide a better indication of slight damage to the permeability barrier normally retaining these enzymes (although the possible occurrence of cytosolic enzymes of similar activities to those of the matrix must be considered). Non-sedimentable cytochrome *c* provides another indication of leaky mitochondria. Departures from the Gaussian distribution of mitochondrial markers through a linear density gradient, or displacement with respect to density of inner and outer membrane markers across the mitochondrial peak may also reflect damage. Sometimes this type of disruption may be incurred during extensive centrifugation in equilibrium-density experiments (Wattiaux *et al.*, 1971).

B. Extent of Contamination

Crude mitochondrial fractions invariably contain many organelles other than mitochondria; indeed the usual methods of isolation produce a fraction that contains all the subcellular particles with sedimentation coefficients in 0·25 M–sucrose of between 5×10^3 and 10^5 S. Thus, when working with crude fractions, there is always a possibility that properties ascribed to the mitochondria are rather properties of the contaminants, or arise by interaction of mitochondria with these contaminating organelles. It is essential therefore to check that mitochondria are the main components of the fraction and to assess the extent of contamination. Examination in the electron microscope, as well as giving an indication of the degree of preservation of mitochondrial integrity, also affords a method of detecting the presence of contaminating subcellular particles. But the identification of the various fragments of membranes is extremely difficult especially in the case of the smaller contaminants. Many of the particles usually designated as damaged mitochondria may have originated from completely unrelated organelles. It is essential that the fields examined are truly representative of the mitochondrial pellet, and the usual methods of sample treatment for thin sectioning prior to electron microscopic examination do not ensure random distribution; stratification of pellets after fixing and during dehydration and embedding should be avoided by thorough resuspension between each process, or the procedure of Millipore filtration of Baudhuin *et al.* (1967) for random sampling should be adopted.

A more powerful and complementary approach is that of analytical fractionation. By analyzing the relative content of the fractions for various marker enzymes, it is possible to obtain an indication of the proportion of the enzymes characteristic of various organelles present in the mitochondrial fraction as a fraction of the total units of enzyme in the original homogenate. Possible contaminants include whole cells, fragments of

Table V. Various subcellular structures which may be contaminants of crude mitochondrial fractions.

Organelle or membrane	Constituents or enzymes often used as markers
Nuclei	DNA, NAD pyrophosphorylase
Nucleoli	RNA polymerase(s)
Chloroplasts	Chlorophyll
Lysosomes	Acid hydrolases
Peroxisomes	Catalase, urate oxidase, D-amino acid oxidase
"Microsomal" membranes (endoplasmic reticulum and Golgi-derived membranes) microtubules	RNA (rough e.r.), ADPase, UDPase, cytochromes b_5 and P-450, thiamine pyrophosphatase and glycoprotein glycosyl transferase (Golgi)
Plasmamembranes	5'-nucleotidase, alkaline phosphatase, ADPase, ATPase (K^+-, Na^+-and Mg^{2+}-stimulated, oligomycin- and Dio-9-insensitive, ouabain-sensitive), cholesterol
Reserve materials	Polysaccharides (amylopectins, amylose, glycogen, starch), Lipids, Polyphosphate
Eye spots	Carotenoids
Cilia and flagella	ATPases, contractile proteins
Cell Walls	Laminarinase, invertase, acid p-nitrophenol phosphatase (in yeasts)
Many other specialized structures (Woronin bodies, lomasomes, spongiosomes etc.)	Poorly characterized biochemically

cell walls, chloroplasts, nuclei, storage granules, lysosomes, lomasomes, Woronin bodies, peroxisomes, plasmalemma, polysomes, endoplasmic reticulum, Golgi vesicles, eye spots, cilia, flagella, and the more specialized organelles which abound in highly differentiated microorganisms. Possible marker enzymes and characteristic components of some of these subcellular particles are listed in Table V. Where one of these components is detected as a significant contaminant of a mitochondrial fraction, a whole family of enzymes characteristic of that organelle will also be present. Perhaps the most disturbing feature of the failure to recognize the hazards of equating the terms "mitochondrial fractions" and "mitochondria" in the literature is the obvious presence of a second organelle in crude mitochondrial fractions, the peroxisome, which has diverse oxidative (but no energy-conserving) functions.

C. Heterogeneity of Mitochondria

Some of the possibilities for the origins of the heterogeneity of mitochondria frequently observed in equilibrium-density or rate separations are listed in Table VI. Many claims to have demonstrated heterogeneity of isolated rat liver mitochondria separated by differential centrifugation (Laird *et al.*, 1952; Novikoff *et al.*, 1953; Gear, 1965a, b; Lusena and

---

Table VI. Possible origins of the heterogeneity of mitochondria from microorganisms.

A. Heterogeneity resulting from damage during preparative procedures
   1. During cell disruption
   a. From shear forces
   b. From osmotic forces
   c. By enzymic damage to membranes (autodigestion or by enzymes used for cell wall digestion)
   2. During centrifugation
   a. From liquid shear forces (e.g. during loading of zonal rotors)
   b. From osmotic forces (during gradient sedimentation)
   c. From hydrostatic pressures (especially in zonal rotors with great path lengths)
   3. Resulting from ageing of mitochondria in the extracts prior to fractionation

B. Heterogeneity arising from heterogeneity of organisms
   a. From diverse cell types (growth conditions not steady-state, systems undergoing cellular differentiation)
   b. From the normal scatter of cell "ages" found in an exponentially-growing culture. (i.e. temporal position with respect to the cell-cycle)

C. Real physiological and biochemical heterogeneity at a subcellular level
   a. From coexisting biochemically distinct mitochondrial populations?
   b. From coexisting functionally distinct mitochondrial populations (e.g. state of energization)
   c. From mitochondria at different stages of the growth and division cycle of mitochondria?

Depocas, 1966), isopycnic density-gradient centrifugation (Kuff and Schneider, 1954; Pette, 1966; Pollak and Munn, 1970), and by rate-zonal centrifugation (Swick et al., 1967: Schuel et al., 1969) have been criticized. Reported heterogeneities may always have resulted from preparative damage by shear, osmotic forces or by hydrostatic pressure. Beaufay and Berthet (1963) and Beaufay et al. (1964) concluded that the bimodal density distribution of rat liver mitochondria could be attributed to density changes caused by high concentrations of sucrose. Wattiaux and Wattiaux-DeConinck (1970), Wattiaux et al. (1971), and Wattiaux(1974) have shown that one cannot centrifuge subcellular particles in a density gradient with impunity at any speed in any rotor, as they are seriously affected when the hydrostatic pressure becomes too high.

The hydrostatic pressure (P) in the centrifuge tube depends on the angular velocity (ω), the density of the medium (ρ), the minimum and maximum radial distances $(x_0$ and $x)$:

$$p = \omega^2 \rho \left(\frac{x^2 - x_0^2}{2}\right)$$

Perhaps the most convincing demonstration of what appears to be physiological heterogeneity in rat liver mitochondria is that by Wilson and Cascarno (1972) who showed that separation on a rate basis in isosmotic Ficoll-sucrose gradients gave different distribution patterns with respect to cytochrome contents, succinate dehydrogenase, NADH dehydrogenase and α-glycerophosphate dehydrogenase. Care was taken to check that the differences found in different fractions were not the results of damage or fragmentation of the mitochondria and did not arise from microsomal contamination.

One of the problems confronting the workers with animal tissues, that of tissue heterogeneity (with regard to cell types), does not complicate the problem when microbial cells are fractionated except where a differentiating system (for instance fungal spores attached to hyphae) is under investigation. However an extension of this line of reasoning reveals that heterogeneity of cell types also arises by way of the differentiation which occurs during normal cell growth. This differentiation process is the stepwise sequence of changes in enzyme constitution which takes place during the normal cell cycle. So even in an exponentially-growing culture of an organism which undergoes division by simple fission, where all nutrients are in excess, or in a steady-state chemostat culture where a single nutrient limits growth, there exists a complete spectrum of cell sizes representing the different stages of the cell cycle. It has been demonstrated unequivocally that the composition of the inner mitochondrial membrane is variable with respect to cytochrome c oxidase and succinate dehydrogenase activities (Poole and Lloyd, 1973) and also in cytochrome content (Poole *et al.*, 1974) through the cell cycle of the fission yeast, *Schizosaccharomyces pombe*. Thus cells at different stages in the division cycle are biochemically heterogeneous with respect to mitochondrial membrane components. Elimination of this source of possible mitochondria heterogeneity has required the use of synchronous cultures, and has led to the confirmation at a biochemical level of the existence of a growth cycle of mitochondria. Heterogeneity of mitochondria from glucose-repressed *S. pombe* after centrifugation through an isosmotic gradient is presented in Fig. 33.

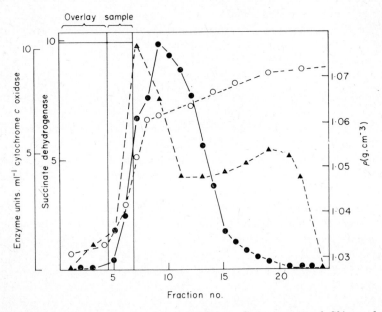

Fig. 33. Heterogeneity of mitochondria isolated from glucose-repressed *Shizosaccharomyces pombe*. Sphaeroplasts were gently disrupted in a Chaikoff press, and after removing the unbroken organisms, the homogenate containing 9·75 mg protein was subjected to rate separation on an isosmotic gradient (Ficoll–0·25 M–sucrose–10 mM–tris–0·4 mM– EDTA at pH 7·4). Centrifugation was at 10,000 RPM for 6 min (1·56 × 10^5 g min at the sample zone: $_0\int^t \omega^2 dt = 10^9$ rad^2 s^{-1}). The relative activities of cytochrome *c* oxidase (●) and succinate dehydrogenase (▲) are quite different in the large mitochondria which have sedimented furthest, from those in the small mitochondria which have only just left the sample zone.

The problem is more complex in organisms showing more irregular patterns of vegetative growth, e.g. in the budding yeasts where multiple budding and overlap of generations sometimes occur.

Enzymic heterogeneity of mitochondrial populations separated on a density basis from *S. cerevisiae* (Neal *et al.*, 1970, 1971) and from *S. carlsbergensis* cultures during glucose derepression (Cartledge and Lloyd, 1972a) may arise (a) as a result of heterogeneity of cell "ages" and/or (b) from the heterogeneity of individual cells with respect to facility for glucose derepression through the cell cycle. Biochemical heterogeneity of mitochondria isolated from *S. carlsbergensis* has also been clearly demonstrated in experiments where mitochondria were separated on a size basis by rate-zonal centrifugation (M. Statham, P. Wynn and D. Lloyd, unpublished results).

V. Sub-fractionation of Mitochondria

A. Separation of Inner and Outer Membranes

The separation of the two mitochondrial membranes has been achieved by Parsons *et al.* (1966), Sottocasa *et al.* (1967), and by Schnaitman and Greenawalt (1968), with mammalian liver mitochondria. Some of the enzyme locations determined in these investigations are presented in Fig. 34.

1. The inner and outer membranes of yeast mitochondria

Separation of the two membranes of mitochondria from *S. cerevisiae* by the method of Sottocasa *et al.* (1967) has been reported by Accoceberry and Stahl (1972). The mitochondria were lysed in a hypotonic medium containing ATP; this leads to contraction of the inner membrane. After brief treatment with an Ultra-Turrax homogenizer, centrifugation through a density gradient gave three major bands. The lightest band showed a high specific activity of kynurenine hydroxylase and consisted chiefly of outer membranes, although some inner membrane enzymes (cytochrome *c* oxidase, NADH-cytochrome *c* oxidoreductase and succinate dehydrogenase) were present in this fraction. The second band contained mitochondria which had not yielded to the separation process. The most dense band contained all three oxidoreductases assayed, at high specific activities, and no detectable kynurenine hydroxylase, and was thus a fairly pure preparation of inner membrane. Cardiolipin is specifically located in the inner membranes of yeast mitochondria (Jakovcic *et al.*, 1971). Monoamine oxidase (a useful outer membrane marker in mammalian systems) is not detectable in mitochondria from *S. cerevisiae*. NADPH-cytochrome *c* reductase which has been used as a marker enzyme for the estimation of the extent of microsomal contamination of mitochondrial membranes in mammalian systems (Parsons *et al.*, 1966) cannot be used for this purpose in many microorganisms: e.g. this enzyme is multilocational in *S. carlsbergensis* (Cartledge and Lloyd, 1972a) and in *Tetrahymena pyriformis* (Lloyd *et al.*, 1971a), and is found associated with both mitochondrial and

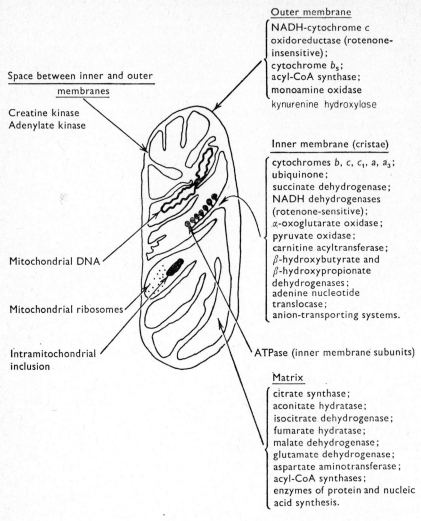

Outer membrane
[NADH-cytochrome c
  oxidoreductase (rotenone-
  insensitive);
  cytochrome b_s;
  acyl-CoA synthase;
  monoamine oxidase
  kynurenine hydroxylase

Space between inner and outer
          membranes

Creatine kinase
Adenylate kinase

Inner membrane (cristae)
[cytochromes b, c, c_1, a, a_3;
  ubiquinone;
  succinate dehydrogenase;
  NADH dehydrogenases
  (rotenone-sensitive);
  α-oxoglutarate oxidase;
  pyruvate oxidase;
  carnitine acyltransferase;
  β-hydroxybutyrate and
  β-hydroxypropionate
  dehydrogenases;
  adenine nucleotide
  translocase;
  anion-transporting systems.

Mitochondrial DNA

Mitochondrial ribosomes

Intramitochondrial
inclusion

ATPase (inner membrane subunits)

Matrix
[citrate synthase;
  aconitate hydratase;
  isocitrate dehydrogenase;
  fumarate hydratase;
  malate dehydrogenase;
  glutamate dehydrogenase;
  aspartate aminotransferase;
  acyl-CoA synthases;
  enzymes of protein and nucleic
  acid synthesis.

Fig. 34. Diagram of submitochondrial localization of enzymes. (Reproduced with permission from Hughes *et al.*, 1970.)

"microsomal" membranes. It is present at especially high activity in the microsomal fraction of tetradecane-grown *Candida tropicalis* together with cytochrome *P450* and a specific hydroxylase (Gallo *et al.*, 1973). Cytochrome *c* peroxidase, alcohol dehydrogenase, and aldehyde dehydrogenase were also found in this fraction. Cytosolic enzyme markers in yeast, not present (at least under some conditions of growth) in mitochondria include NAD-linked glutamate dehydrogenase (Perlman and Mahler, 1970a; Hollenberg *et al.*, 1970a; Mahler *et al.*, 1971a), NADP-

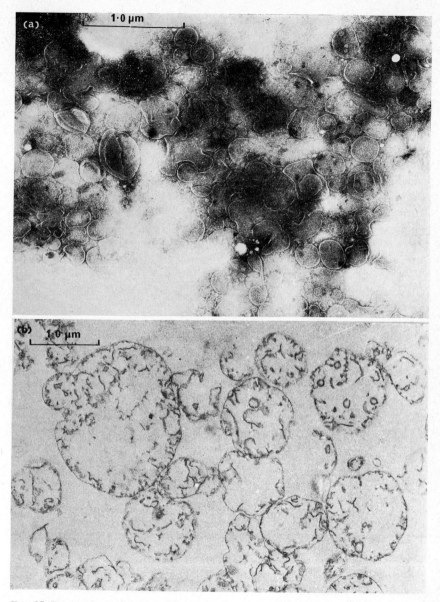

FIG. 35. Preparations of outer and inner membranes of mitochondria from *Saccharomyces cerevisiae*. (a) Negatively-stained outer membranes; (b) section of inner membranes. (Reproduced with permission from Bandlow, 1972.)

linked glutamate dehydrogenase, and aspartate aminotransferase (Hollenberg *et al.*, 1970a). NADP-linked isocitrate dehydrogenase and citrate
synthase have been used as mitochondrial matrix markers (Mahler *et al.*,
1971a).

Bandlow (1972) used valinomycin $+K^+$ for the swelling step, and then
shrank the inner membranes by addition of ATP. Differential centrifugation after gentle hand homogenization gave three pellets (sedimenting
at 6×10^4 *g* min, $2\cdot25 \times 10^5$ *g* min and $6\cdot3 \times 10^6$ *g* min respectively).
The first pellet consisted mainly of unbroken mitochondria, the other two
pellets were purified by equilibrium centrifugation on sucrose gradients.
A band equilibrating at $1\cdot19$ g/cm³ contained inner membranes, as it was
rich in succinate-cytochrome *c* oxidoreductase, cytochrome *c* oxidase and
malate dehydrogenase; antimycin A-insensitive NADH-cytochrome *c*
oxidoreductase, kynurenine hydroxylase and adenylate kinase were
absent from this fraction. Intact mitochondria showed a density of
$1\cdot173$ g/cm³ in these experiments; ruptured mitochondria which has lost
matrix markers and adenylate kinase were found at $\rho = 1\cdot144$ g/cm³.
Outer membranes ($\rho = 1\cdot084$ g/cm³) contained kynurenine hydroxylase
and antimycin A-insensitive NADH-cytochrome *c* oxidoreductase. Pulse
labelling with radioactive Leu *in vivo* in the presence of cycloheximide
was followed by mitochondrial isolation and separation of the membranes.
The outer membrane contained none of the radioactivity associated with
incorporation due to the activity of the mitochondrial protein synthesis
system, whereas about one-third of the inner membrane protein is produced on mitochondrial ribosomes. The outer membranes accounted for
about 6·8% ,and the inner membrane about 30%, of the total protein of
yeast mitochondria. The morphological characteristics of these inner and
outer membranes is shown in Fig. 35.

2. Separation of inner and outer membranes of mitochondria from Neuros- pora crassa

The combined swelling, shrinking and sonication procedure may also be
used for the separation of the mitochondrial membranes of *N. crassa*
(Cassady and Wagner, 1968, 1971; Neupert and Ludwig, 1971). Gradient
purified mitochondria were first swollen in 10 mM–tris phosphate buffer
(pH 8·0), and after 30 min, the mitochondrial suspension was adjusted to
0·6 M–sucrose–0·6 mM–ATP–0·6 mM MgSO₄. Mild sonication was
followed by centrifugation through a discontinuous sucrose gradient
(Cassady and Wagner, 1971). The inner membranes accumulated at the
1·0–1·9 M–sucrose interface, whereas the outer membranes did not
sediment beyond the 0·1–1·0 M–sucrose interface. Marker enzymes
employed were succinate–cytochrome *c* oxidoreductase and kynurenine

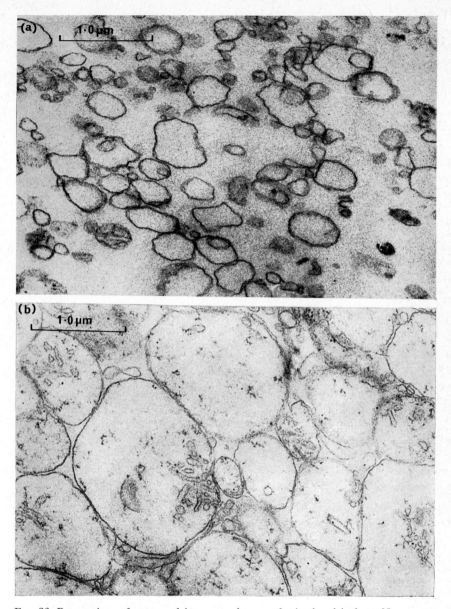

FIG. 36. Preparations of outer and inner membranes of mitochondria from *Neurospora crassa*. (a) Section of outer membranes; (b) section of inner membranes. (Reproduced with permission from Neupert and Ludwig, 1971.)

hydroxylase respectively. The distinctive distribution of aceto-hydroxy-acid reductoisomerase (an enzyme of the isoleucine-valine pathway) suggests that it is localized in the matrix space. A similar location for all four enzymes of this pathway was confirmed in experiments using the digitonin method of mitochondrial membrane separation (Cassady *et al.*, 1972). The characteristic morphology of the separated membranes in electron micrographs was used to further confirm their identities.

In the experiments of Neupert and Ludwig (1971) the distribution of a red carotenoid pigment (neurosporaxanthin) was similar to that of kyn-urenine hydroxylase; the inner membrane-containing fractions showed succinate-cytochrome c oxidoreductase activity and contained cyto-chromes $a + a_3$ and b. The outer membranes (Fig. 36 gave only one band when analysed by polyacrylamide gel electrophoresis, whereas similar treatment to the inner membranes revealed at least 20 different bands. *In vivo* or *in vitro* incorporation of radioactive amino acids by the mito-chondrial system of protein synthesis occurred only into the inner membranes.

3. Separation of inner and outer membranes of mitochondria of Tetrahymena pyriformis

Swelling, shrinking and mild sonication also resulted in the partial separation of the membranes of mitochondria of *T. pyriformis*, which were then isolated by centrifugation through a discontinuous gradient in a BXIV zonal rotor (Turner, 1969; G. Turner and D. Lloyd, unpublished observations). The inner membranes ($\rho = 1.23$ g/cm³) contained all the cytochromes, and succinate-cytochrome c oxidoreductase was a con-venient marker. The organism does not possess monoamine oxidase, and the most easily assayed outer membrane enzyme was antimycin A-insensitive NADH-cytochrome c oxidoreductase. A b-type cytochrome was also detected in the fractions enriched in outer membranes.

B. Preparation of Phosphorylating Sub-mitochondrial Particles

Procedures for the isolation of phosphorylating SMPs from commercially grown baker's yeast have been described (Schatz and Racker, 1966; Schatz, 1967). The organisms were disrupted by shaking with 0·45–0·50 mm diam. glass beads for 20 s in a buffer consisting of 0·25 M–mannitol–20 mM–tris SO₄–1 mM–EDTA (pH 7·4) at 4000 cycles/min in a Bronwell MSK Mechanical Cell Homogenizer (Bronwill Sci. Div., Rochester, New York). The mitochondrial suspension subsequently isolated was sonicated in 10 mM–tris SO₄ (pH 7·4) at a protein concentra-tion of 20 mg/ml for 40 s at 0–4°C in a 20 kHz ultrasonic disintegrator.

Removal of the intact mitochondria was followed by high speed centrifugation to sediment the phosphorylating SMPs. Approximately 15–20% of the mitochondrial protein is recovered in this fraction. Particles prepared in this way are more stable on storage at $-55°C$ than are intact mitochondria. After negative staining, they appear in the electron microscope as vesicles with diameters between 0·05 and 0·2 μm. They have high specific activities of succinate and α-glycerophosphate oxidases, oxidize ethanol in the presence of NAD$^+$ and purified alcohol dehydrogenase, and do not oxidize pyruvate + malate. P/2e ratios for succinate, α-glycerophosphate and ethanol were 0·55, 0·45 and 0·38 respectively. Energy conservation site-specific assays indicated that both sites II and III were functional. Another method which is based on the use of a Micro Mill (Gifford Wood Co., Hudson, New York) has been detailed by Mattoon and Balcavage (1967). Similar methods do not yield stable phosphorylating SMPs from *Candida utilis* (Ragan and Garland, 1971).

3

Respiration and the Respiratory Components in Mitochondria of Microorganisms

I. Introduction

The organization of respiratory chain components and the energy-coupling reactions of oxidative phosphorylation in the inner mitochondrial membrane have recently been reviewed by Chance (1972). The sequence of electron carriers which mediate the flow of electrons from respiratory substrates to oxygen are shown in Fig. 37(a). The three sites of energy conservation (Sites I, II and III) are also the sites of action of inhibitors: thus amytal, rotenone and piericidin A inhibit electron transport in the vicinity of Site I, antimycin A and hydroxyquinoline-N-oxide act at Site II, and cyanide and carbon monoxide inhibit at Site III. Chemical fractionation procedures have been developed which split the chain into four complexes (Fig. 37(b)). Early work on the nature and sequence of the components involved in electron transport has been summarized by Keilin (1966), and the methodology employed to study the reactions of the respiratory chain in intact mitochondria or in whole cell suspensions has been detailed by Chance (1964) and by Chance and Williams (1956) who also introduced the terminology applied to the various metabolic states of respiring mitochondria (Table VII). Significant developments in the techniques for the identification of different flavoproteins and iron-sulphur proteins, and the *in situ* measurement of redox potentials of these and of the cytochrome components, has enabled the representation of the electron carriers on a potential diagram (Fig. 38; Chance, 1972). This diagram shows three groups of electron carriers of fixed mid-potential (Groups I, II and III) interspersed with three energy-transducing carriers of variable potential. The identity of the component responsible for energy conservation at Site I is not yet clear.

Whereas most of the information on the components involved in mito-chondrial respiration has been accumulated through the use of prepara-tions from mammalian sources, the experimental advantages of using microbial systems have long been recognized (Keilin, 1925). Studies on the organization of the cytochrome system of yeast cells originated by Keilin and extended by Chance (1959a, b) and Chance and Spencer (1959) clearly indicated that suspensions of intact microorganisms could provide a great deal of insight into the kinetics of interaction of respiratory chain components *in vivo* during respiration.

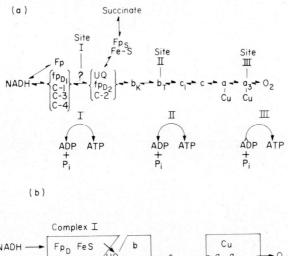

FIG. 37. (a) The electron transfer components of the respiratory chain arranged as a continuous sequence from NADH (low potential) to oxygen (high potential). The components on the substrate side of Site I are Fp_L, the highly fluorescent lipoate dehydrogenase flavoprotein; Fp_{D1}, the NADH dehydrogenase flavoprotein, and iron-sulphur proteins here given as C-1, C-3, and C-4. On the oxygen side of Site I are Fp_S the succinate dehydrogenase flavoprotein, with associated iron-sulphur proteins (Fe–S); UQ, ubiquinone; Fp_{D2}, the fluorescent flavoprotein; C-2 iron-sulphur protein; cytochromes b_K and b_T. On the oxygen side of Site II are the four cytochromes, c_1, c, a, and a_3 with associated copper. (Reproduced with permission from Chance, 1972.) (b) The four complexes and their sequential arrangement in the electron transfer system as deduced from chemical fractionation and reconstitution experiments in Green's laboratory.

The development of methods for the release of "intact" mitochondria from a number of different eukaryotic microorganisms, has led over the past decade to several detailed investigations of the energy-yielding processes in these organelles. Most of the work has been carried out on yeasts, and at present the work on other fungi, algae and protozoa is very incomplete. Studies at the whole cell level have however provided a good deal of information on a wide variety of species, particularly with

Table VII. Metabolic states of mitochondria and the associated oxidation-reduction levels of the respiratory enzymes. (Reproduced with permission from Chance and Williams, 1956.)

		Characteristics				Steady-state percentage reduction of components				
State	[O$_2$]	ADP level	Substrate level	Respiration rate	Rate-limiting substance	a	c	b	Flavo-protein*	DNPH
1	>0	low	low	slow	ADP	0	7	17	21	~90
2	>0	high	~0	slow	substrate	0	0	0	0	0
3	>0	high	high	fast	respiratory chain	<4	6	16	20	53
4	>0	low	high	slow	ADP	0	14	35	40	>99
5	0	high	high	0	oxygen	100	100	100	150	100

* These values are based upon the amount of flavoprotein that is reduced upon addition of antimycin A to the mitochondria in State 2 which is $\frac{2}{3}$ the State 5 value.

regard to their haemoprotein content, and these data provide useful starting points for the subcellular fractionation studies which are now urgently required. In many cases controversy and confusion have arisen as a result of inadequate regard for the necessity for the rigorous control of growth conditions in order to obtain reproducible results; investigators have in the past almost invariably tended to underestimate the adaptability of microorganisms and of their mitochondria. It is this adaptability which now provides the major attraction of microorganisms as systems for the study of the flexibility of mitochondrial composition

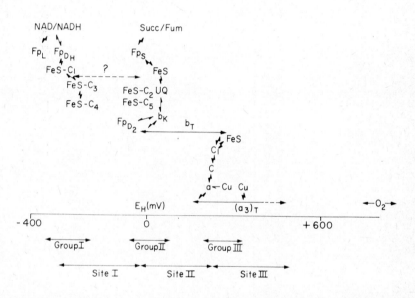

Fig. 38. The electron carriers of the respiratory chain arranged as groups of fixed potential (Groups, I, II and III) and individual components of variable mid-potential: cytochromes b_T and $(a_3)_T$ for Sites II and III and (?) for Site I. The other components are designated in Fig. 37(a). (Reproduced with permission from Chance, 1972.)

and function, for the elucidation of mechanisms involved in electron transport and energy conservation, and for investigation of the steps in the synthesis and assembly of the membrane components involved. This chapter provides a comparative account of respiratory processes in microbial mitochondria isolated after growth of organisms under the usual conditions of laboratory culture employed, with only passing reference to considerations of the very considerable modifications of mitochondrial energy metabolism which can be brought about by genetic or environmental alterations.

Table VIII. Oxygen uptake rates (State 3, ng atoms/mg protein/min), respiratory control ratios (R.C.), and ADP/O ratios for mitochondria isolated from algae and protozoa.

Substrate	PROTOTHECA ZOPFII (Lloyd, 1965; Lloyd and Venables, 1967) Qo_2	R.C.	ADP/O	POLYTOMELLA CAECA (Lloyd et al., 1968) Qo_2	R.C.	ADP/O	EUGLENA GRACILIS (SM-LI) Buetow and Buchanan, 1965 Qo_2	P/O	(z) Sharpless and Butow, 1970a Qo_2	P/O
α-ketoglutarate	60	5·5	3·3	22	3·5	3·1	70	1·91		
citrate	18	1·8	2·1				44	1·54		
glutamate (+malate)	40	4·0	3·0				38	1·66		
pyruvate (+malate)	32	1·2	2·5				37	0·38		
NADH	170	1·5	—						177	2·34
ascorbate (+TMPD or PMS)									74	0·22
succinate	120	1·6	1·6	110	1·2	1·7	72	0·91	81	0·74
malate	60	3·8	2·0				59	1·71		
propionyl-CoA	40	1·4	2·2	9	—	—				
β-OH propionate	43	2·2	2·9	35	2·9	2·9				
malonic semialdehyde	48	1·4	2·7	13	2·7	2·0				
DL lactate							91	0·96	*140	2·0

Substrate	TETRAHYMENA PYRIFORMIS						HARTMANNELLA CASTELLANII			CRITHIDIA FASCICULATA		
	(W) Kobayashi, 1965			(ST) Turner et al., 1971			Lloyd and Griffiths, 1968			Kusel and Storey, 1972; Toner and Weber, 1972		
	Q_{O_2}	R.C.	ADP/O	Q_{O_2}	R.C.	ADP/O	Q_{O_2}	R.C.	ADP/O	Q_{O_2}	R.C.	ADP/O
α-ketoglutarate	120	2·7	1·9	86	2·2	1·3–2·3	22	1·5	1·3			1·2
citrate	100	2·0	1·4	16	—	—	22	—	—			
glutamate (+malate)	100	2·9	2·2	36	—	—	18	1·3	1·7			
pyruvate (+malate)	50	1·4	1·2				36	1·7	1·4			1·0
NADH	20	—	—	4	—	—	44	1·5	1·0			
ascorbate (+TMPD or PMS)	140	1·3	0·8	186	1·6	0·8–1·5	66	1·7	0·7	63	1·2	0·3
succinate	250	2·3	1·2	25			56	1·7	1·7	140	2·6	1·0
malate	110	2·3	1·6				50	2·5	2·0			1·0
β-hydroxybutyrate	90	1·6	1·6	38			28	1·6	2·0			
DL-lactate	60	1·9	1·4	13								
ethanol				18			12	—	—			
α-glycerophosphate	20	—	—				12	—	—	150	2·4	1·1

* Results similar for D- or L-lactate.

Table IX. Oxygen uptake rates (State 3, ng atoms/mg protein/min), respiratory control ratios (R.C.), P/O, and ADP/O ratios for mitochondria isolated from fungi.

Substrate	ASPERGILLUS NIGER (Watson and Smith, 1967a)			ASPERGILLUS ORYZAE (Kawakita, 1970a, b)			NEUROSPORA CRASSA (Hall and Greenawalt, 1967)			NEUROSPORA CRASSA (Weiss et al., 1970)			NEUROSPORA CRASSA (Lambowitz et al., 1972b, c)			
	Qo_2	R.C.	ADP/O	Qo_2	R.C.	P/O	Qo_2	R.C.	P/O	Qo_2	R.C.	ADP/O	Qo_2	R.C.	ADP/O	P/O
α-ketoglutarate	100	5	2·4–2·8	100	2·7	2·0–2·7	23	2·0	3·0				280 (+malate)	5·8	2·1	1·9
citrate	80	2	1·7–2·3	90	2·2	1·7–2·4	43	2·0	3·0				(+malate)	2·7	1·59	1·75
isocitrate							25	1·7	1·6							
glutamate	60	3	2·2–2·6													
pyruvate+malate	60	2·5	2·0–2·4				6	2·3	1·9	220	3·0	1·9	550	3·2	1·52	1·59
NADH	100	3	1·4–1·8	160	2·1	0·9–1·0	115	1·15	0·8	622	2·4	1·2	900	2·4	1·18	1·44
NADPH	100	1·5	0·8–1·0	200	1·1	0·9–1·0	140	1·36	0·6	126	1·3	0·9				
ascorbate+TMPD										2076	1·3	—	990			0·77
succinate	120	4	1·5–1·8	150	1·8	1·6–1·8	80	1·7	1·5	286	1·8	1·2	520	1·4	1·30	1·31

Table X. Oxygen uptake rates (State 3, ng atoms/mg protein/min), respiratory control ratios (R.C.), and ADP/O ratios for mitochondria isolated from yeasts.

| | SACCHAROMYCES CEREVISIAE | | | | | | | | SACCHAROMYCES | | |
| | Duell et al., 1964 | | Kováč et al., 1968 | | Spencer et al., 1971 | | | Ohnishi et al., 1966a | | |
Substrate	Q_{O_2}	R.C.	Q_{O_2}	P/O	Q_{O_2}	R.C.	ADP/O	Q_{O_2}	R.C.	ADP/O
α-ketoglutarate	220	16·7	340	1·5	280	5·9	2·2	260–410	5·5–6	2·3–2·6
citrate			490	1·5	390	3·5	1·2	200–450	1·5–1·7	1·7–1·8
isocitrate (+malate)	170	5·0			100	2·0		140–200	1·2–1·5	1·6–1·8
pyruvate (+malate)	230	2·4	420	1·5	330	2·2	1·65	150–350	1·6–2·0	1·6–1·8
NADH	300	2·1			1050	1·5	1·8	60–110	3–3·6	1·5–1·8
NADPH										
ascorbate (+TMPD)								500–700	1·2–1·5	0·8–1·0
succinate	300	2·9	280	1·4	390	1·65	1·6	250–380	1·4–1·7	1·6–1·8
α-glycerophosphate	146	1·8			570	1·4		80–300		
D(−) lactate	160	1·1	490	1·0				500–750	1·1–1·5	0·9–1·1
L(+) lactate	82	5·8	180	1·0				150–200	1·1–1·3	0·9–1·1
ethanol	212	1·7	460	1·4	60	1·0		110–300	1·4–1·5	1·6–1·8

(continued)

Table X. (continued)

Substrate	CARLSBERGENSIS			CANDIDA UTILIS					SHIZOSACCHAROMYCES POMBE	
	Schuurmans-Stekhoven, 1966a, b			*Light and Garland, 1971		†Katz et al., 1971			Heslot et al., 1970a	
	Qo_2	R.C.	ADP/O	Qo_2	ADP/O	Qo_2	R.C.	ADP/O	Qo_2^*	R.C.
α-ketoglutarate	340–700	1·6–2·3	2·0–2·9	120–140	3·8					
citrate				65–115	2·4					
isocitrate (+malate)										
pyruvate (+malate)	400–700	1·4–3·9	2·0–2·7	145–205	3·0					
NADH	920–1920	2·3–2·7	1·5–2·0	370–420	1·65	214	2·7	2·9	170	2·1
NADPH	260–660	1·4–2·0	1·8–2·1			230	2·0	3·3		
ascorbate (+TMPD)										
succinate	320–600	1·7	1·6–1·8	40– 90	—				83	1·0
α-glycerophosphate	340–600	1·6–2·2	1·4–2·1	390–480	1·8				39	1·8
D(−) lactate									39	1·0
L(+) lactate										
ethanol	440–1120	1·2–1·5	1·2–1·4	135–160	2·6	230	3·1	3·6		

* mitochondria from glycerol-limited chemostat cultures.   † mitochondria from ethanol-limited chemostat cultures.

II. Respiration of Mitochondria Isolated from Microorganisms

A preliminary study of the respiratory metabolism of isolated mitochondria usually involves manometric or polarographic determinations of
oxygen uptake rates in the presence of different respiratory substrates,
together with assays of the efficiency of oxidative phosphorylation as
determined by P/O ratios. A summary of results obtained with various
species of microorganisms is presented in Tables VIII–X. Tightly-
coupled phosphorylation is reflected by the dependence of oxygen uptake
rates on the presence of a phosphate acceptor (respiratory control, Chance
and Williams, 1956); several of the examples given are of phosphorylating
preparations which do not show this property. Although failure to satisfy
this criterion of functional integrity suggests damage during isolation, in
some cases there may be other explanations for the lack of respiratory
control, e.g. presence of contaminating ATPases of extra-mitochondrial
origin, or rapid electron flux through alternative nonphosphorylating
pathways.

A. Mitochondria from Algae (see Table VIII)

1. Euglena gracilis

Oxidative phosphorylation was first demonstrated in algal mitochondria
by Buetow and Buchanan (1964, 1965) who showed that, after disruption
of a streptomycin-bleached strain of *Euglena gracilis* by grinding with
glass beads, it was possible to isolate mitochondria which would oxidize
NADH, NADPH, succinate, L-malate, L-glutamate, α-ketoglutarate,
lactate and L-malate + pyruvate. The P/O ratios were about 1·0 for
lactate and succinate, and approached 2·0 for L-malate, L-glutamate and
α-ketoglutarate. Pyruvate alone was not oxidized, but when 0·2 mM–
malate was added to the pyruvate, oxidation and phosphorylation with a
P/O ratio of 1·24 was observed. This low concentration of malate did not
itself give measurable oxygen uptake rates. β-Hydroxybutyrate was not
oxidized; citrate and isocitrate were oxidized only very slowly. These
mitochondria also showed oxidative phosphorylation (P/O = 0·4) with
externally added NADH: no phosphorylation accompanied the oxidation
of NADPH. DNP was inhibitory to both phosphorylation and oxidation
over a wide range of concentrations; thus the classical uncoupler failed to
produce the expected stimulation of respiration. Complete inhibiton of
phosphorylation was attained at 50 μM–DNP, whereas oxidation was
inhibited 37%; further increases of concentration did not produce further
inhibition of respiration. Phosphorylation was completely inhibited by
0·07 mM–KCN, whereas respiration was only 76% inhibited, increasing
the inhibitor concentration to 0·4 mM gave 92% inhibition of respiration.

Amytal (1 mM) reduced malate oxidation by 74% and phosphorylation by 87% and at a concentration of 2 mM completely inhibited respiration. Complete inhibition of malate oxidation was observed in the presence of 0·5 μM–rotenone. About 10% of the succinoxidase activity was completely antimycin A-insensitive, whereas phosphorylation was eliminated when this inhibitor was present at much lower concentrations (0·9 μM).

Sharpless and Butow (1970a), working with a permanently bleached strain (Z), have confirmed the presence of all three sites of phosphorylation. Mitochondria from cells grown with glutamate and malate as carbon sources were examined by assay procedures specific for each of the three possible phosphorylation sites. Phosphorylation corresponding to Site I was assayed by the method of Schatz and Racker (1966) with UQ_1 as acceptor of electrons from NADH. The low P/2e ratio (0·16) measured by this method was about one-third that for NADH oxidase. This confirmed the implication of the presence of Site I phosphorylation implicit in the result of Buetow and Buchanan (1965) who had previously shown that P/O ratios for succinate were about 1 unit lower than with NAD-linked substrates. Site II phosphorylation was demonstrated by the method of Lee et al. (1967) by measuring phosphorylation which accompanied the reduction of ferricyanide by succinate in the presence of rotenone and cyanide, and in the presence and absence of antimycin A. Site III phosphorylation accompanied the oxidation of ascorbate + PMS in the presence of antimycin A; the P/2e ratio was 0·35. In the presence of KCN this value fell to 0·09. Further confirmation of the presence of two coupling sites in the region between succinate and O_2 was obtained by allowing ferricyanide and O_2 to compete for reducing equivalents, and measuring P/2e ratios as a function of electron flux to the artificial acceptor. The P/2e ratios so obtained decreased linearly with a decrease in the fraction of reducing equivalents spanning the two phosphorylation sites on their pathway to oxygen. When extrapolated to the point at which all electrons are used to reduce ferricyanide, the value of the P/2e ratio was somewhat less than half that for the oxidase pathway. Phosphorylation at both sites was completely inhibited by 3·3 μM–CCCP. As similar results were also obtained with Euglena gracilis bacillaris (Sharpless and Butow, 1970a), it is evident that the low P/O ratios obtained for mitochondria from this strain by Buetow and Buchanan (1965) do not reflect the absence of any of the three sites, but suggest a low overall efficiency of phosphorylation due possibly to mitochondrial damage.

2. Astasia longa

Low P/O ratios for succinate (0·43–0·67) have also been found for mitochondria from Astasia longa, the colourless counterpart of Euglena

(Buetow and Buchanan, 1969). Oxidation of NADH, L-malate and α-ketoglutarate occur in particulate preparations from this organism. It is evident that the isolation of mitochondria from these Euglenophyta is no easy task, and methods of extraction require further development.

3. Polytomella caeca

A much more fragile phytoflagellate, *Polytomella caeca*, after growth with acetate yields mitochondria which show a degree of respiratory control and P/O ratios of 3·1 and 1·7 for α-ketoglutarate and succinate oxidations respectively (Lloyd *et al.*, 1968).

4. Prototheca zopfii

The first demonstration of respiratory control in algal mitochondria was with organelles from acetate-grown *Prototheca zopfii* (Lloyd, 1965). In this case the addition of small known amounts of ADP produced almost immediate acceleration to the State 3 rates, and these rates in successive phosphorylation cycles were very similar, indicating a high degree of mitochondrial integrity. The only exception to this rule was found with NADH, In this case, although addition of ADP produced 50% stimulation of the State 4 rate, there was no return to State 3. P/O ratios determined manometrically (Lloyd, 1966a) and polarographically (Lloyd, 1965) indicated values of 1·5–1·6 for succinate, about 2·0 for malate, and 2·6–3·3 for α-ketoglutarate; site specific assays were not attempted. Uncoupling of oxidative phosphorylation was observed in the presence of 0·1 mM–DNP (Lloyd, 1966b). Over 90% inhibition of NADH oxidase by mitochondrial fragments was produced by 0·1 mM–KCN or 20 μM–antimycin A. Rotenone (0·08 μM) produced 50% inhibition, as did 1 mM–Na amytal. Na-azide (1 mM) gave 65% inhibition. A number of rotenone derivatives were also found to be very inhibitory to NADH oxidase activity (Lloyd, 1966b). Rotenone, antimycin A and KCN also inhibit the respiration of acetate by whole cell suspensions of *P. zopfii*.

Thus the indications are, that the main phosphorylating electron transport pathways of algae (in the few cases where mitochondria have been examined) are quite conventional with respect to phosphorylation sites and inhibitor sensitivities.

A particularly attractive feature of algae as systems for experimentation is the adaptability of their mitochondria under various conditions of growth (see p. 218).

B. Mitochondria from Protozoa (see Table VIII)

1. Tetrahymena pyriformis

There have been several reports of the preparation of particulate fractions from *Tetrahymena pyriformis* which oxidize succinate, NADH and

β-hydroxybutyrate (Eichel and Rem, 1963; Nishi and Scherbaum, 1962). These preparations seemed to be considerably damaged, as they showed very low P/O ratios. Much more satisfactory isolation was achieved by Kobayashi (1965), who showed that mitochondria isolated from Strain W carried out tightly-coupled oxidative phosphorylation with a wide range of substrates. Respiratory control values of up to 2·9 were obtained with glutamate. The P/O ratios were however rather low. A P/O ratio of 0·8 for the oxidation of ascorbate + TMPD suggests the presence of Site III between a c-type cytochrome and O_2. Succinate oxidation gave a high respiratory control ratio, but the P/O ratio was only just greater than 1, and NAD-linked substrates usually showed P/O ratios of between 1 and 2, although values of up to 2·5 were sometimes observed for glutamate oxidation. It was of note that DL-lactate gave a P/O ratio of 1·4, but the two stereoisomers were not tested separately. NADH and α-glycero-phosphate were not oxidized. The oxidative and phosphorylation proper-ties of the mitochondria were stable for more than 4 h at 0°C, but the mitochondria could not be kept overnight without considerable loss of activities; 0·3–0·5 M–mannitol gave the best P/O and R.C. ratios, and the addition of bovine serum albumin conferred some protection from un-couplers present in the extract (Eichel, 1960). More than 80% inhibition of O_2-uptake with succinate, glutamate or ascorbate + TMPD was produced in the presence of 0·2 mM–KCN; 20 mM–azide gave 81, 50 and 100% inhibition with each of these three substrates respectively.

Very high concentrations of antimycin A were necessary for inhibition of electron transport; a concentration of 0·2 mM gave 92% and 58% inhibition of succinate and glutamate oxidations respectively. Amytal (5 mM) gave 64% inhibition of glutamate oxidation. Optimal concentra-tions of uncouplers for the stimulation of respiration with glutamate or succinate as substrates were as follows: DNP 0·05–0·1 mM, pentachloro-phenol 1 μM, chlorpromazine 0·15 mM; no uncoupling was observed with dicumarol. High titres of oligomycin (120 μg/mg mitochondrial protein) were necessary to produce 50% inhibition of the State 3 rate of succinate oxidation; tributyl tin (3 μM) accelerated both State 3 and State 4 respiration rates, and thus acts more like an uncoupler than an energy transfer inhibitor in this system.

It is impossible to achieve satisfactory extraction of mitochondria from many of the strains of *Tetrahymena* popular in the laboratories of cell biologists. Massive accumulation of polysaccharide reserve materials (for instance in Strains GL, T, W) gives rise to the production of extremely viscous cell-free homogenates. These organisms also possess a wide range of highly active lysosomal acid hydrolases, and damage to lysosomal mem-branes during mitochondrial preparation must be minimized (Lloyd *et al.*, 1971a). Mitochondria which show respiratory control with α-ketoglutarate and succinate have also been extracted from Strain ST (Turner *et al.*, 1971). Although stimulation of isocitrate oxidation occurred on adding a low concentration of sodium malate, respiratory control was not demonstrated with this or other NAD-linked substrates. As in Kobayashi's (1965) experiments, DL-lactate was oxidized; ethanol, sarcosine, and α-glycerophosphate were oxidized slowly; whilst NADH was not oxidized by the best mitochondrial preparations. This last property could be used as an index of the mitochondrial integrity. Again a high titre of antimycin A (0·1 μmole/mg mitochondrial protein) was necessary to completely inhibit succinoxidase; thus *Tetrahymena* mitochondria are about 1000-fold less sensitive to this inhibitor than are rat liver mitochondria. Rotenone was also required at very high concentrations (20 nmoles/mg protein) to give between 80 and 90% inhibition of α-ketoglutarate oxidation; piericidin A was more effective, 0·1 nmole/mg protein completely eliminated oxygen uptake with this substrate. Although the respiration of whole cells is relatively insensitive to KCN, this inhibitor gives 99% inhibition of both succinate and α-ketoglutarate (State 3) oxidation rates at a concentration of 1 mM. There is considerable variation in the cyanide sensitivity of different strains of *Tetrahymena* under different growth conditions (see Danforth, 1967 for review); figures of from 0% to 80% in the presence of 1 mM–KCN have been reported.

The lipids of mitochondria from *T. pyriformis* have been fractionated (Jonah and Erwin, 1971). Neutral lipids plus free fatty acids constituted 30% of the total lipid extracted, glycerol phospholipids 50·9%, and the remaining 19·1% was accounted for by a poorly characterized lipid fraction ("acetone eluate") containing 5 or 6 components tentatively identified as ceramide-like lipids. The principal neutral lipid was identified as the pentacyclic triterpenoid tetrahymenol. The glycerol phospholipids of mitochondria detected included phosphatidyl ethanolamine, phosphatidyl choline, cardiolipin and glyceryl-2-aminoethyl phosphonolipid. A higher concentration of polyunsaturated fatty acids was found in the mitochondrial lipids than in those of whole cells.

2. Hartmannella castellanii

Mitochondria from the amoeba, *Hartmannella castellanii* Neff, do not oxidize pyruvate or aspartate; α-glycerophosphate, glutamate and ethanol were oxidized very slowly (Lloyd and Griffiths, 1968). Malate, malate + pyruvate, succinate, α-ketoglutarate, NADH, β-hydroxybutyrate and ascorbate + TMPD were all oxidized, and State 4 to State 3 transitions could be repeated until oxygen concentration became limiting. P/O ratios greater than 2 were not obtained with any substrate. The preparations were stable for over 5 h at 4°C; some respiratory activity was lost on storing overnight, and stored preparations showed no respiratory control. The oxidation of malate was not inhibited by rotenone or piericidin A; that of NADH was not inhibited by rotenone or antimycin A. Sodium amytal (3 mM) did not show greater inhibitory activity with NAD-linked substrates than with succinate, but seemed rather to act as a general inhibitor of electron transport. It was suggested that Site I phosphorylation is lacking in this amoeba during exponential growth on mycological peptone. However, higher P/O ratios and rotenone- and piericidin A-sensitivity of oxygen uptake with NAD-linked substrates have been shown in mitochondria isolated from organisms harvested in the stationary phase of growth (Evans, 1973). Further studies on inhibition of mitochondrial respiration have been reported by Fouquet (1973a, b).

3. Physarum polycephalum

The preparation of mitochondria from plasmodia of this slime mould has been described (Barnes et al., 1973); the mitochondria were not coupled. They oxidized NADH via a rotenone-insensitive pathway, but the oxidation of malate + glutamate was rotenone-sensitive. Sensitivites of substrate oxidations to antimycin A, 2-heptyl-4-hydroxyquinoline-N-oxide, malonate and KCN were measured, and these showed no unusual features.

4. Dictyostelium discoideum

Mitochondria isolated from the myxamoebae of this cellular slime mould oxidized succinate, NADH, and ascorbate + 2,6-dichlorophenolindophenol (Erickson and Ashworth, 1969). The preparations shows very low malate oxidase activity, and this was stimulated by the addition of NAD^+. This observation and the failure to detect respiratory control suggests that the organelles had suffered some damage during isolation. Both the succinate and NADH oxidase systems were completely inhibited by 0·1 mM–KCN and by antimycin A (0·3 μmole/mg protein). High concentrations of rotenone (15 μmoles/mg protein) and amytal (8 μmoles/mg protein) inhibited the NADH oxidase by only 55% and 22% respectively whereas the succinoxidase activity was amytal-insensitive.

5. Crithidia fasciculata

Mitochondrial preparations from mechanically disrupted Crithidia fasciculata carried out oxidative phosphorylation with NADH and succinate (P/O ratios 0·5–1·6 and 0·6–0·7 respectively), but showed no tightly coupled phosphorylation (Toner and Weber, 1967). No stimulation of respiration was produced on addition of the uncouplers gramicidin or CCCP. Substrate-dependent respiration was cyanide- and antimycin A-sensitive. Improvements in the isolation procedure have recently yielded mitochondria which do show respiratory control (Toner and Weber, 1972). The optimal conditions for isolation necessitate the inclusion of 3 mM–Mg^{2+} and the omission of Cl^-, SO_4^{2-} and F^- from the buffer; acetate buffer was least inhibitory to the degree of ADP stimulation of succinoxidase activity. The temperature of the polarographic assay was also critical; much higher R.C. ratios were obtained at 15°C than at 30°C, an effect most likely to be due to the presence of an active ATPase. A high degree of integrity of the preparations was indicated by the presence of a permeability barrier to oxaloacetate, which inhibited succinoxidase activity only after sonication. Whether or not NADH oxidation reflects membrane damage has not yet been critically assessed. Rutamycin inhibited State 3 succinoxidase rate and this inhibition was released by the addition of 2 μM–CCCP. Low efficiencies of phosphorylation were also noted with DL-α-glycerophosphate, α-ketoglutarate and malate as substrates. Another method of isolation using digitonin treatment followed by a 30 s homogenization also yields mitochondria which show respiratory control (ratios up to 3·4) with endogenous substrate, succinate and L-α-glycerophosphate (Kusel and Storey, 1972). Uncoupling of phosphorylation with exogenous substrates was achieved by the addition of 2·5 μM–FCCP. Site specific assays indicated that the low efficiency of phosphorylation (P/O ratios for succinate 0·9–1·1) does not arise from the lack of a phosphorylation site, as both Sites II and III were shown to be operative. The authors suggest that the low P/O ratios may arise as a consequence of the simultaneous operation of an alternative non-phosphorylating pathway.

C. Mitochondria from Fungi (see Table IX)

The mechanical disruption of fungal hyphae results in the extrusion of cytoplasmic contents through the torn hyphal walls; surprisingly high quality mitochondria are released, considering the extent of liquid shear generated by these procedures. The mitochondria produced often show P/O ratios approaching the theoretical values of 3 for NAD-linked substrates, 2 for succinate, and 1 for ascorbate + TMPD. Values of between 1·4 and 1·8 which accompany the oxidation of externally added NADH by mitochondria from *Aspergillus niger* (Watson and Smith, 1967a) are somewhat higher than those reported for other fungal mitochondrial fractions. Values of less than 1 for NADH oxidation in mitochondria isolated from *A. oryzae* (Kawakita, 1970b) and also for mitochondria from

Neurospora crassa (Hall and Greenwalt, 1967) may suggest a different pathway of phosphorylating electron transport associated with the oxidation of externally added NADH in these species or, alternatively may simply reflect mitochondrial damage.

The former alternative seems likely, as there are clear differences between the P/O ratios obtained for oxidation of exogenously-added and endogenously-generated NADH in these two species. This distinction is abolished when the permeability barrier for nicotinamide nucleotides (the inner mitochondrial membrane) has not retained its integrity. Marked differences in the sensitivities of α-ketoglutarate oxidase and NADH oxidase to two inhibitors acting in the vicinity of Site I (rotenone and amytal) in mitochondria from *A. niger* (Watson and Smith, 1967b) also indicate that the pathways of oxidation of internal and external NADH are not the same. Antimycin A (0.2 μM) or KCN (0.4 mM) produced almost 100% inhibition of respiration with succinate, α-ketoglutarate or NADH as substrates. Oligomycin (2 μg/mg protein) inhibited the State 3 respiration rates with NADH or α-ketoglutarate as substrates, and this inhibition was released on the addition of 0.13 mM–DNP (Watson and Smith, 1967a). Tri-*n*-butyl tin chloride (10 μM) gives an oligomycin-like effect with mitochondria from *A. oryzae* (Kawakita, 1970a, b); with these mitochondria pentachlorophenol (10 μM) acts as an inhibitor rather than as an uncoupler. Mitochondria from *N. crassa* show no oxidative phosphorylation when incubated in the presence of 0.1 mM–atractylate or 0.1 mM–oleate (Hall and Greenawalt, 1967). Mitochondria from *A. oryzae* appear to have a higher antimycin A- and cyanide-resistant respiration with succinate, citrate, α-ketoglutarate or NADH as substrates (Kawakita, 1970a, b) than those from *A. niger*. NADPH is oxidized by way of a phosphorylating electron transport chain in mitochondria from *N. crassa* (Weiss *et al.*, 1970) via the antimycin A- and cyanide-sensitive sites. Rotenone does not inhibit NADPH oxidation and phosphorylation at Site I appears to be by-passed.

The results of Lambowitz *et al.* (1972b) in contrast to those of Weiss *et al.* (1970) show that rotenone or amytal have little inhibitory effect on the respiration of NAD-linked substrates by *Neurospora* mitochondria; a difference in growth conditions may explain this discrepancy. Pyrolnitrin, an inhibitor of electron transport in the flavin region of the respiratory chain (Wong *et al.*, 1971) inhibits the NAD oxidase of *Neurospora* mitochondria (Lambowitz *et al.*, 1972b) and uncouples oxidative phosphorylation (Lambowitz and Slayman, 1972). Results with a variety of substrates and inhibitors led these authors to the conclusion that most of the oxidative phosphorylation takes place at Sites II and III. Site I is quite inefficient, as judged by the small difference in phosphorylation efficiency between NAD-linked substrates and succinate, and also by the low level of reduction of endogenous NAD$^+$ during respiration with succinate + ATP (27% in *Neurospora* mitochondria as compared with as much as 75% in animal mitochondria).

Rapid deterioration of succinoxidase activity of isolated mitochondria is accompanied by altered kinetic properties (West and Woodward, 1973). The systemic fungicide oxathiin inhibits electron transport between succinate and ubiquinone in mitochondria from *Ustilago maydis* (Mathre, 1971; Ulrich and Mathre, 1972).

D. Mitochondria from Yeasts

1. Mitochondria from Saccharomyces spp.

A major improvement in the quality of mitochondria isolated from yeast was brought about by the introduction of methods for the preparation of yeast sphaeroplasts. A great enhancement both of oxidative rates and respiratory control ratios was reported for mitochondria prepared by controlled osmotic lysis of sphaeroplasts of *Saccharomyces cerevisiae* over those prepared by use of the Nossal shaker (Duell *et al.*, 1964); this and concurrent studies with lactate-grown *S. carlsbergensis* (Ohnishi *et al.*, 1966a) set a new standard for yeast mitochondrial preparations (see Table X). These mitochondria were able to oxidize various intermediates of the TCA cycle such as citrate, isocitrate, α-ketoglutarate, succinate and pyruvate (in the presence of a catalytic amount of malate). Zinc dimethyl-dithiocarbamate is a preferential inhibitor of succinate oxidation (Briquet and Gofféau, 1973). Externally added NADH, D- and L-lactate, ethanol and TMPD + ascorbate were also actively oxidized and respiratory control was obtained with all these substrates; especially notable were the high R.C. ratios obtained with NADH and α-ketoglutarate (Fig. 39). The

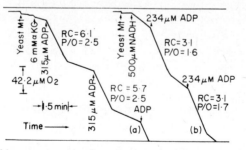

Fig. 39. Polarographic measurement of respiration of mitochondria from *Saccharomyces carlsbergensis* with α-ketoglutarate (αKG) or NADH as substrates. The reaction medium contained 0·5 M-mannitol, 10 mM-potassium phosphate buffer (pH 6·5), 20 mM-tris-maleate buffer (pH 6·5), 10 mM KCl and 0·1 mM-EDTA. (a) Final protein concentration, 0·28 mg per ml of reaction mixture; (b) 0·13 mg per ml. Other additions were indicated by *arrows* in the figure. The reaction was carried out at 25°C. Yeast Mt. = yeast mitochondria. (Reproduced with permission from Ohnishi *et al.*, 1966a.)

P/O ratios obtained by the polarographic method corresponded well with those measured manometrically in the presence of a hexokinase trap. Low P/O ratios for all NAD-linked substrate oxidations compared with those for succinate oxidation, suggested that phosphorylation at Site I was not present. These results were in accord with the earlier observations of Vitols and Linnane (1961) that Site I phosphorylation was also absent from the mitochondria of *S. cerevisiae*, but contrast with the early spectrophotometric analyses of Chance (1959a, b) which indicated the operation of all three sites in intact cells. Lack of energy conservation at Site I was also evident in experiments with SMPs (Mackler *et al.*, 1962; Schatz and Racker, 1966). The high P/O value (2·3–2·6) for α-ketoglutarate suggests the involvement of substrate level phosphorylation. The results also suggested that the oxidation of both D- and L-lactate, like that of ascorbate + TMPD involves Site III only. The oxidation of all substrates tested was completely sensitive to KCN (2·1 mM). Antimycin A (2·9 nmoles/mg protein) blocked respiration with α-ketoglutarate, NADH, or succinate, but gave only partial inhibition of respiration with D- and L-lactate (possibly affecting the oxidation of pyruvate produced from these substrates), and had no affect on the oxidation of ascorbate + TMPD. DNP or pentachlorophenol were both effective as uncouplers at 0·1 mM concentrations. High concentrations of oligomycin (20 μg/mg protein) depressed the State 3 oxidation rates to the State 4 level; this inhibition was relieved by the addition of uncouplers. Tri-*n*-butyl tin chloride (0·015 μM) mimicked the effect of oligomycin, but at higher concentrations acted as an uncoupler. Synthalin (decamethylene diguanide) stimulates succinoxidase activity in the presence of phosphate, this effect is reversed by K^+ or Mg^{2+} (Gómez–Puyou *et al.*, 1973). Carefully prepared yeast mitochondria still showed traces of respiratory control after storage at 4°C for 3 days, and only a slight decrease in P/O ratios was observed over this period. Substrate-level phosphorylation appears to be the rate-limiting step in α-ketoglutarate oxidaton, and respiratory control was more stable in mitochondria stored at 4°C with this substrate than with other substrates tested. A study of the functional relationship of the exogenous and endogenous NAD systems in mitochondria from *S. carlsbergensis* shows that added NAD and NADH do not penetrate the inner membrane into the matrix space, although external NADH is rapidly oxidized (von Jagow and Klingenberg, 1970). Transhydrogenation between external and internal NADH and NAD across the inner membrane is not observed. By making use of the impermeability of the inner membrane to added ferricyanide it was shown that two different NADH dehydrogenases are present on the outer- and inner-facing sides of the inner membrane; these are separately responsible for the oxidation of exogenous and endogenous NADH respectively. Both dehydrogenases are connected to the cytochrome chain via the ubiquinone pool. The localization of some other dehydrogenases was also determined by the ferricyanide method and a scheme for the "sidedness" of the inner mitochondrial membrane was proposed on the basis of these results (Fig. 40). The polyene antibiotic Filipin II alters membrane permeability of yeast mitochondria and this leads to loss of endogenous confactors and inactivation of primary dehydrogenases (Balcavage *et al.*, 1968).

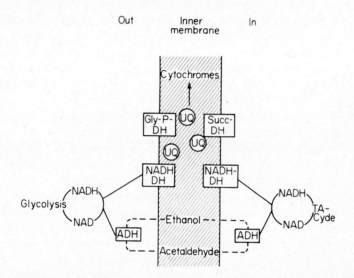

Fig. 40. Sidedness of localization of various membrane bound dehydrogenases at the inner mitochondrial membrane of *Saccharomyces carlsbergensis*. The scheme includes a postulated ethanol-acetaldehyde shuttle connecting the intra- and extra-mitochondrial NAD-system. Gly-P-DH = glycerophosphate dehydrogenase, Succ-DH = succinate dehydrogenase NADH-DH = NADH dehydrogenase, ADH = alcohol dehydrogenase, TA-cycle = tricarboxylic acid cycle. (Reproduced with permission from von Jagow and Klingenberg, 1970.)

Concurrent studies on mitochondria from *S. carlsbergensis* (Schuurmans-Steckhoven, 1966a) grown with glucose + lactate as carbon sources conflict with those of Ohnishi *et al.* (1966a) with respect to the P/O ratios determined for NAD-linked oxidations. In these experiments values of between 2 and 3 were obtained with ethanol, L-malate + pyruvate, and citrate as substrates, thus implicating all three phosphorylation sites in the oxidation of these substrates. Like NADH, externally added NADPH was oxidized via two phosphorylation-coupled steps only. Antimycin A inhibited almost completely the oxidation of L-malate + pyruvate, succinate, α-glycerophosphate, NADH and NADPH, but the oxidation of L-lactate was only inhibited partially by this agent. The oxidation of NAD-linked substrates other than ethanol was not however inhibited by

E

rotenone (in agreement with the results of Ohnishi *et al.*, 1966b) and this was surprising in view of the close association of the rotenone-sensitive site with Site I phosphorylation in mitochondria from two other yeast species, *Candida utilis* and *Endomyces magnusii* (Ohnishi *et al.*, 1966b; Kotelnikova and Zvjagilskaja, 1966; Light and Garland, 1971) and in mammalian mitochondria. The endogenous respiration of whole cell suspensions of *S. carlsbergensis* and *S. cerevisiae* is also completely rotenone insensitive, whereas that of *C. utilis* and *E. magnusii* is 80–90% inhibited by 30 μM–rotenone (Ohnishi *et al.*, 1966b). However these results refer only to a particular set of growth conditions, as Ghosh and Bhattacharrya (1971) and Ohnishi (1970) have demonstrated that mitochondria isolated from *S. carlsbergensis* after aeration of the cells as a non-proliferating suspension are still rotenone-insensitive but do carry out energy con-servation at Site I. It should be stressed that the reaction of rotenone is not absolutely specific for Site I, as it also inhibits alcohol dehydrogenase (Balcavage and Mattoon, 1967).

Mitochondria obtained from *S. cerevisiae* harvested in the stationary phase after growth in a semisynthetic medium containing peptone, yeast autolysate and glucose (Kováč *et al.*, 1968) resembled those obtained by Ohnishi *et al.* (1966b) from exponentially grown *S. carlsbergensis* in their oxidative activities, phosphorylation efficiencies, osmotic stability and reactions to the presence of DNP and oligomycin. Three sites of phos-phorylation occur in mitochondria obtained from stationary-phase cultures of *S. cerevisiae* and *S. carlsbergensis* under conditions of growth described by Mackler and Haynes (1973). Aeration of *S. carlsbergensis* in phosphate buffer also produces cells which possess all three sites (Ohnishi, 1970).

Kováč *et al.* (1970a) have shown that the respiration of aerobically grown yeast can be inhibited up to 67% by oligomycin and this inhibition was relieved by DNP. It is often useful to be able to study mitochondrial energy conservation *in vivo* in this way.

More recently mechanical methods of disruption have been employed which yield mitochondria with at least as great a degree of functional integrity as those extracted from enzymically-prepared sphaeroplasts (Guarnieri *et al.*, 1970; Spenser *et al.*, 1971); these methods are especially useful in cases where a yeast is not susceptible to the action of the snail enzymes.

Some evidence suggests that a malate-stimulated citrate and isocitrate transport system similar to that reported for animal mitochondria (Chappell and Haarhoff, 1967) is involved in substrate uptake by yeast mitochondria (Spenser *et al.*, 1971). Thus swelling experiments indicated that malate stimulates both citrate penetration and respiration of mito-chondria from *S. cerevisae*. A difference in the specificity of the yeast and mammalian tricarboxylate carrier systems was noted with respect to the uptake of propane 1, 2, 3 tricarboxylate; whereas this anion does not penetrate the mammalian mitochondrion (Chappell and Robinson, 1968) it enters the yeast organelle by way of a malate-stimulated route (Spencer et al., 1971). Prevention of mitochondrial swelling in the presence of ammonium salts of various respiratory substrates by fluorocitrates may indicate that the site of action of these inhibitors is at the level of substrate uptake (Brunt et al., 1971).

The phosphate transfer system of mitochondria from S. carlsbergensis has been investigated by Ohnishi et al. (1967). Exogenous ADP is phosphorylated faster than endogenous ADP even at low temperatures (a result contrary to those obtained with mammalian mitochondria). The specific atractyloside-sensitive translocation of adenine nucleotides shows similar properties to that in mammalian mitochondria.

A survey of mitochondria isolated from a wide variety of organisms has revealed two types of organelles based on their response to added Ca^{2+} (Carafoli and Lehninger, 1971). The first group includes mitochondria obtained from many vertebrate tissues which possess a specific carrier system for Ca^{2+}, and which are characterized by an extremely high affinity for Ca^{2+} and a rapid respiratory response which is accompanied by rapid cation uptake and proton ejection. Organelles of the second type exhibit low affinity for Ca^{2+}, and respiratory responses to Ca^{2+} and ion transport are slow or essentially absent.

In a study on the effect of divalent cations on the respiratory activity of yeast mitochondria, Šubík and Kolarov (1970) found that, in contrast with animal mitochondria, Ca^{2+} is not accumulated through an energy-dependent process. Succinate or citrate (but not NADH) oxidations are inhibited competitively by Ca^{2+} and Mg^{2+}. The effects of Co^{2+}, Zn^{2+} and Cd^{2+} were also studied.

Carafoli et al. (1970) showed that the growth and respiration of yeast is only slightly stimulated by the presence of Ca^{2+} in the culture medium. Mitochondria isolated from S. cerevisiae or C. utilis showed neither energy-linked Ca^{2+} transport nor high affinity Ca^{2+} binding. Both types of mitochondria showed metabolically-independent low affinity Ca^{2+} binding (Km = 10–20 μM, concentration of sites 40–50 nmoles/mg mitochondrial protein (see also Ghosh and Bhattacharyya, 1971). Mitochondria from S. cerevisiae were found to contain about 10 nmoles of endogenous Ca^{2+}/mg protein in a tightly bound or sequestered form. Thus it appears that yeast mitochondria lack a specific Ca^{2+} carrier.

However, Balcavage et al. (1973) have shown that mitochondria prepared from S. cerevisiae and C. utilis can carry out respiration-linked ion-uptake when Ca^{2+} is present at concentrations in the range 1–10 mM; these very high concentrations overcome the apparent lack of a natural

Ca^{2+} carrier. Under these conditions proton ejection, two-fold respiratory stimulation and an increased steady-state oxidation level of cytochrome b was observed (Fig. 41). The Ca^{2+} uptake and proton ejection are strongly inhibited by uncouplers and by antimycin A. An eight-fold stimulation of Ca^{2+} uptake by Pi was noted; no appreciable ATP-driven Ca^{2+} uptake could be detected. Respiration-driven cation transport by yeast mitochondria shows a narrow range of ion specificities, as is the case with liver mitochondria; Ca^{2+}, Sr^{2+} and Mn^{2+} are all active, whereas Mg^{2+} and Na^+ are not. The physiological significance of Ca^{2+} transport at high Ca^{2+} concentrations is uncertain.

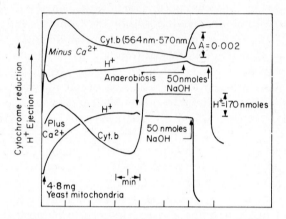

FIG. 41. Effect of Ca^{2+} on redox levels of cytochrome b in *Saccharomyces cerevisiae* mitochondria. Mitochondria (4·8 mg protein) were added to 5·0 ml of reaction medium containing 0·6 M–mannitol, 4 mM–HEPES buffer, 3·3 mM–sodium succinate and, when present, 3 mM–Ca^{2+}. The spectrophotometer cuvette contained a Clarke oxygen electrode and a pH electrode. The vessel was continuously stirred with a small magnetic stirring bar. Temperature was 25°C. (Reproduced with permission from Balcavage *et al.*, 1973.)

The impermeability to protons of yeast mitochondria incubated anaerobically in a K^+-containing medium was shown by Kováč *et al.* (1972). Uncoupler in the presence of valinomycin induced proton permeability. The swelling of yeast mitochondria in K acetate was similar to that of mammalian mitochondria except that the volume changes produced were significantly smaller. Energy-dependent K^+ transport induced by valinomycin also showed similar characteristics to that process in mammalian mitochondria.

2. Mitochondria from Candida utilis

Mitochondria prepared from *C. utilis* grown in a chemostat under conditions of ammonium, magnesium or glycerol-limited growth were shown to possess rotenone and piercidin A sensitivity, and P/O ratios indicated the presence of Site I energy-conservation (Light and Garland, 1971); phosphate-limited cells yielded uncoupled mitochondrial suspensions. Another report indicated that mitochondria from *C. utilis* harvested during exponential growth on synthetic medium with ethanol as carbon and energy source are rotenone-insensitive, whereas those prepared from ethanol-depleted stationary-phase cultures have acquired the rotenone-sensitive component (Katz, 1971; Katz *et al.*, 1971); studies on the P/O ratios obtained with pyruvate + malate as respiratory substrates indicate that, as in ethanol-limited chemostat cultures, ethanol-depleted batch culture cells have three phosphorylation sites associated with the oxidation of NAD-linked substrates. Ohnishi (1972) has suggested that the results obtained by Katz (1971) and by Katz *et al.* (1971) may reflect a by-pass mechanism at the mitochondrial level between a phosphorylating and non-phosphorylating electron transfer pathway at Site I for the oxidation of NADH generated in the matrix space of the mitochondria. Thus the non-phosphorylating pathway may be used when the cells are growing rapidly under optimal growth conditions, whereas the pathway involving Site I is utilized only when cells are growing under sub-optimal conditions or after growth has ceased in batch cultures. The acquisition of piericidin sensitivity and Site I phosphorylation is accompanied by increases in the e.p.r. signals of iron-sulphur centres, 1, 2 and 3 (at 13°K) and in NADH-dehydrogenase activity (Cobley *et al.*, 1973). These changes occur in the presence of high concentrations of iron in the media, and are clearly different from those studied in iron-limited chemostat cultures (Light *et al.*, 1968; Light and Garland, 1971).

Adenine nucleotide translocation in *C. utilis* mitochondria has been investigated under optimal conditions (mitochondria suspended in 0·63 M–mannitol, 2 mM–EDTA or EGTA, 10 mM–MOPS buffer at pH 6·8 and at 0°C, (Lauquin and Vignais, 1973). Translocation is an exchange-diffusion process; the whole pool of internal adenine nucleotides is exchangeable. ADP is the most readily exchangeable nucleotide. The rate of mitochondrial ADP exchange, but not its K_m value (2 μM), depends on the conditions of cellular growth. At 0°C this exchange occurs at about 3–4 nmoles ADP/min per mg protein for mitochondria from cells grown in the presence of 1·5% glucose. This value rises to 11·5 nmoles when the carbon source is 3% ethanol. The Q_{10} of the process is about 2 between 0°C and 20°C. Other exchangeable adenine nucleotides include ATP, dATP and the methylene and hypophosphate analogues of ADP; unlike mammalian mitochondria, *C. utilis* mitochondria are able to transport UDP by a carboxyatractyloside-sensitive process. When phosphate and substrate are present in the aerated mitochondrial medium (i.e. under conditions of oxidative phosphorylation) added ADP is ex-

changed with internal ATP. The ATP/ADP ratio in the extra-mitochondrial space was higher than that in the intra-mitochondrial space; the
difference in the calculated phosphate potentials in the two spaces was
0·9–1·7 kcal/mole. Atractyloside, carboxyatractyloside, bongkrekic acid
and palmityl-CoA all inhibited mitochondrial adenine nucleotide translocation as they do in mammalian mitochondria, but two to four times
less effectively. Inhibition by atractyloside or palmityl–CoA is competitive
with respect to ADP, but that due to bongkrekic acid or carboxyatractyloside is non-competitive. Carboxyatractyloside and atractyloside inhibitions
are additive. The apparent K_d for the binding of (^{35}S)-carboxyatractylo-
side and (^{14}C)-bongkrekic acid is 10–15 nM and the concentration of sites
0·4–0·6 nmole/mg protein in both cases. Binding of (^{35}S)-carboxy-
atractyloside is competitively displaced by atractyloside and *vice versa*.
The amount of bound (^{14}C) ADP, which is atractyloside removable from
mitochondria depleted of their endogenous adenine nucleotides, is 0·08–
0·16 nmole/mg protein.

Downie and Garland (1973b) have shown that, as predicted by Mitchell's
chemiosmotic hypothesis, mitochondria translocate two protons at each
energy conservation site for each pair of reducing equivalents transferred.
C. utilis was cultured under conditions which produced cells with three,
two or one phosphorylation sites between NADH and oxygen (organisms
from glycerol-limited, sulphate-limited, and from Cu-limited cultures
respectively). Measurements of the $\rightarrow H^+/O$ ratios in mitochondrial
fractions from these three cell-types gave values of 6·0, 3·0–4·0 and 1·9–2·1
for NAD-linked substrates. Other substrates gave values of zero in the
case of the organisms from Cu-limited cultures.

3. Mitochondria from Shizosaccharomyces pombe

A mitochondrial preparation from *Shiz. pombe* grown with glucose, yeast
extract and peptone, oxidizes NADH, succinate, α-glycerophosphate and
ascorbate + TMPD (Heslot *et al.*, 1970a). NADH and succinate oxidations are sensitive to antimycin A and NaCN but not rotenone. The
respiration of glucose by whole cells grown on glycerol was completely
inhibited by antimycin A (1·66 μM) or NaCN (1 mM) but unaffected by
1·2 μM rotenone. Glucose oxidation by organisms grown aerobically with
glucose was also rotenone insensitive, and in this case the Sites II and III
specific inhibitors gave incomplete inhibition (23% and 13% of the
control respiration rates respectively). Synthalin (20 mg/l, an inhibitor of
energy transfer perhaps specifically at Site III) gave 75% inhibition of
growth on glycerol but did not affect growth on glucose. A similar effect
was produced by DNP (25 mg/l). Growth on glycerol was also much more
sensitive to azide or antimycin A than that on glucose.

III. Electron Transport Components of Mitochondria Isolated From Microorganisms

A. Cytochromes

1. Respiratory cytochromes of algae

Studies on mitochondrial cytochromes of algae have as yet been restricted to those colourless strains and mutants which show little or no photosynthetic activity (Table XI); a wider survey still awaits the solution of the formidable technical difficulties involved in obtaining mitochondria free from chloroplast fragments. No critical attempts at analytical fractionation of photosynthetic species have been reported. Studies of algal respiration at the level of the whole cell suggest that some of the processes of terminal oxidation are somewhat different from those occurring, for instance in mammalian cells (Lloyd, 1974), and that more detailed investigations would prove rewarding.

(a) *Prototheca zopfii.* The most recent study of the obligately-heterotrophic alga, *P. zopfii* (Epel and Butler, 1970a) shows that glycerol-grown cell suspensions contain seven cytochromes (two c-type, three b-type and cytochromes $a + a_3$) all of which are clearly resolved at 77°K in dithionite-reduced versus oxidized spectra (Fig. 78, p. 234). The α-band of cytochrome oxidase showed a maximum at 599 nm and this composite band was resolved by reducing cells in the presence of methanol and CN^-. Methanol shifted the absorption maximum of cytochrome a from 598 to 603 nm and permitted dithionite to reduce the cytochrome a_3-CN^- complex to give a 595 nm absorption band. In the absence of CN^- the Soret region showed a double-peak due to a-type cytochromes (439 and 447 nm); treatment with CN^- (which does not complex with cytochrome a) led to a shift to 445 nm of the band due to cytochrome a_3. The b-type cytochromes were isolated spectroscopically through the use of antimycin A. Difference spectra (obtained between cells respiring ethanol in the presence of this Site II inhibitor and aerated cells) showed reduction of cytochromes b_{555}, b_{559} and b_{564} (fused Soret band at 428 nm), while the c- and a-type cytochromes remained oxidized. The two c-type cytochromes (fused Soret band at 410 nm) were distinguished by subfractionation of mitochondria. Soluble cytochrome c_{549} is easily released from isolated mitochondria leaving cytochrome c_{551} along with the three b-type cytochromes and cytochrome oxidase in a membrane fraction.

The stimulation of endogenous respiration of this organism by CN^- and azide was shown to result from an increased auto-oxidation rate of a b-type cytochrome (possibly cytochrome b_{564}) after its reaction with these inhibitors. Similar stimulation by CN^- of the reactivity of low spin peroxidases to O_2 has been shown in plant tissues (Yamazaki *et al.*, 1968).

Table XI. Respiratory cytochromes of algae. RT indicates spectral data obtained at room temperature, LT at liquid N_2 temp; IM in mitochondria-rich fractions.

Organism	Cytochrome λ_{max} (nm)			Comments	References
	c-type	b-type	a-type		
Prototheca zopfii	549 551	559 565	(595) 604	RT	Lloyd (1966a)
	551 555	559 564	599 (447) (439) 447	LT	Webster and Hackett (1965) Epel and Butler (1970a)
Chlamydomonas reinhardii (pale green mutant)	551	563		RT low content of a-type cytochrome detected as pyridine haemochromogen	Chance and Sager (1957) Hiyama et al. (1959)
Euglena gracilis (bleached strain Z) (bacillaris)	551 555	558 561 568	593 (444) 607 (453)	LT IM	Raison and Smillie (1969) Sharpless and Butow (1970a)
	556		605	RT IM	Perini et al. (1964)
Astasia longa (Pringsheim)	547 552	554 557 563	604 (448) (440)	LT IM	Webster and Hackett (1965)
Polytoma uvella	550	560	605 (444)	RT	Webster and Hackett (1965)
Polytomella agilis (Arago)	550	556 565	605 (444)	LT	Webster and Hackett (1965)
Polytomella caeca (Pringsheim)	549	562	605 (440)	LT IM	Lloyd and Chance (1968)

(b) *Astasia longa.* Liquid N_2-difference spectra of whole cell suspensions of *A. longa* (the colourless counterpart of *Euglena gracilis*) also suggest the presence of 2 c-type and 3 b-type cytochromes (Webster and Hackett, 1965). The position (λ_{max} 605 nm) and shape of the α-band of cytochrome oxidase differ from that in higher animals and plants, yeast, or in other colourless algae. In whole cells and in isolated mitochondria b-type cytochromes remain oxidized in the endogenously-reduced state. CO-difference spectra of isolated mitochondria revealed the presence of cytochrome a_3 (λ_{min} 444 nm, λ_{max} 431 nm) together with another CO-binding pigment (see Table XVIII, p. 154). Mitochondrial suspensions were not capable of oxidizing reduced mammalian cytochrome c, again suggesting the presence of an unusual cytochrome oxidase system.

(c) *Euglena gracilis.* Of the four cytochromes detected in light-grown wild type *E. gracilis* strain bacillaris (λ_{max} 552, 561, 605 and 556 nm) only cytochrome a 605 and c 556 were detected in the dark-grown wild-type, or in the Albino (W_3 and W_8) mutants (which lack chloroplasts completely), or in the Yellow (Y_3) mutant which has only a rudimentary plastid (Perini *et al.*, 1964). The cytochrome with the absorption at 552 nm was enriched in a plastid fraction from normal green cells, whereas a six-fold enrichment of cytochrome 556 was obtained in a "small-particle" fraction from the Albino mutant together with the cytochrome a 605, diaphorase and succinate dehydrogenase. Purified reduced cytochrome 556 showed a double α-band (at 554·5 and 556 nm) and was reoxidized by oxygen in the presence of yeast mitochondria at one tenth of the rate for reduced mammalian cytochrome c, but three times more rapidly than reduced cytochrome c 552. The respiration of sonicated cell suspensions of normal cells of *E. gracilis* W_3 with succinate was 95% inhibited by 1·0 mM–CN^- but only 15% inhibited by a 19 : 1 ratio of $CO:O_2$. Cytochromes a 605 and c 556 were reducible in whole cells by succinate or acetate, and the α-band of cytochrome a 605 disappeared more rapidly on aeration than that of the other cytochrome. The addition of CN^- (0·5 mM) led to the disappearance of the absorption band at 605 nm but did not affect that at 556 nm. Perini *et al.* (1964) suggested on the basis of these results that although cytochromes a 605 and c 556 are mitochondrially located, that the function of the a-type cytochrome was not clear in view of the CO-insensitive respiration; the autooxidizable cytochrome c 556 may provide an alternative terminal oxidation pathway.

Difference spectra of *E. gracilis* (Z strain, streptomycin-bleached), and of particulate preparations rich in mitochondria, indicated the presence of cytochromes b 561 and a 609 reducible by succinate, oxidized by O_2 and reacting with a soluble cytochrome c (Raison and Smillie, 1969). Measurement of steady-state reduction levels suggested that these three pigments were part of an electron transport chain in a sequence $b \rightarrow c \rightarrow a$. Although respiration was only 60% inhibited by 1 mM–CN$^-$, oxidation of the c- and a-type cytochromes was completely abolished at this inhibitor concentration. As the alternative electron transport pathway was also insensitive to antimycin A inhibition, it was suggested that cyanide-insensitive respiration is mediated through an autoxidizable cytochrome b. Further resolution of the mitochondrial cytochromes of this strain of *E. gracilis* has been achieved by Sharpless and Butow (1970a, b) who found seven substrate-reducible cytochromes including components analogous to cytochromes $a + a_3$ two c-type cytochromes and three b-type cytochromes. The α-band of cytochrome oxidase was anomalous as it showed a λ_{max} at 607 nm and a shoulder at 593 at 77°K; the Soret band was also split (λ_{max} 444 and 453 nm). Reaction of reduced mitochondrial suspensions with CO gives a trough at 445 nm characteristic of cytochrome a_3; an absorption maximum at 593 nm is probably also due to this pigment, but at least one other CO-reaction component is also present (see Table XVIII). Spectra in the presence of antimycin A enabled absorption maxima due to b-type cytochromes to be distinguished from those of c-type cytochromes. One of the b-type cytochromes was not fully reduced in the presence of antimycin A; another (closely related to coupling Site II) showed greater absorption in the presence of this inhibitor than in anaerobic syspensions. Either cytochrome b 561 or cytochrome b 558 may react with CN$^-$ and slowly with CO. Of the two c-type cytochromes (λ_{max} 555 and 551 nm; Soret bands at 422 and 419 nm respectively), the former is the soluble cytochrome designated by Perini *et al.* (1964) as "*Euglena* cytochrome 556". This cytochrome has been purified from *E. gracilis* (Meyer and Cusanovitch, 1972). It has a covalently bound haem, and the atypical nature of the pyridine ferrohaemochrome of this cytochrome is consistent with the proposal that the haem is bound to the peptide chain *via* a single thioether linkage. The only quantitative difference found between spectra of mitochondrial suspensions oxidizing different substrates was that antimycin A produced 3–4 times more reduction of the three b-type cytochromes with succinate than with D-lactate.

(*d*) *Polytomella caeca.* Mitochondria isolated from *P. caeca* (Lloyd and Chance, 1968) contain cytochromes b, c_1, c, a and a_3. Electron transport is sensitive to the classical inhibitors, succinoxidase is inhibited by 94% in the presence of 10 mM–KCN and to a similar extent by 4 nmoles of antimycin A per mg protein. In intact mitochondria the oxidation of endogenously-generated NADH (substrate citrate) was more sensitive to inhibition by rotenone or Piericidin A than that of externally added NADH: this difference was abolished in disrupted mitochondrial preparations. The site of antimycin A inhibition was located between cytochromes b and c.

(e) *Polytoma uvella and Polytomella agilis.* The colourless flagellates *P. uvella* and *P. agilis* both have cytochromes b, c and $a + a_3$, Webster and Hackett (1965). The NADH-oxidase of *P. uvella* mitochondria is sensitive to KCN, antimycin A, rotenone, diphenylamine and amytal. Difference spectra of acetone-extracted whole cells shows the presence of a dithionite-reducible CO reacting haemoprotein with λ_{max} 570, 540 and 414 nm in addition to cytochrome oxidase (trough at 445 nm, shoulder at 430 nm, see Table XVIII, p. 154) the 414 nm Soret band was also found in CO-spectra of isolated mitochondria. This CO-binding pigment was not a peroxidase, but may have been a denatured cytochrome b.

(f) *Chalamydomonas reinhardii.* A report that the pale green mutant of *C. reinhardii* has no detectable cytochrome a (Chance and Sager, 1957) has been checked, and it is evident that the very low content of this cytochrome (detected as the pyridine haemochromogen) suggests the possibility of a novel pathway of mitochondrial electron transport (Hiyama *et al.*, 1959).

(g) *Chlorella spp.* Emerson (1927) observed that *Chlorella* respiration is resistant to CO and stimulated by CN^-; as much as 60% of the oxygen uptake of this organism proceeds uninhibited at high concentrations of CN^- (Syrett, 1951). Addition of cupric ions followed by fluoride, does however lead to cessation of oxygen consumption (Hassal, 1967, 1969).

2. Cytochromes of protozoa (see Table XII)

(a) *Tetrahymena pyriformis.* The presence of unusual a-type cytochromes in the ciliate protozoon, *T. pyriformis*, was first noted by Baker and Baumberger (1941) who identified absorption maxima at 587–590 nm and 616–618 nm with cytochromes a_1 and a_2 respectively. In subsequent investigations (Ryley, 1952; Keilin and Ryley, 1953, Eichel, 1956; Kobayashi, 1965; Van de Vijver, 1966) it was reported that cell-free homogenates or mitochondria were unable to oxidize reduced mammalian cytochrome c despite the sensitivity of substrate oxidations to azide, CO and CN^-, the classical inhibitors of cytochrome c oxidase. The absence of a cytochrome oxidase band at 604 nm from absorption spectra of whole cells (Ryley, 1952) was apparently confirmed by examination of isolated mitochondria, (Kobayshi, 1965), and subsequent investigations (Turner *et al.*, 1969; Perlish and Eichel, 1968, 1969) reported detailed spectrophotometric analyses of cytochrome components. The work of Turner *et al.* (1971) with phosphorylating mitochondria clearly indicated that

Table XII. Cytochromes of protozoa. RT indicates spectral data obtained at room temperature, LT at liquid N_2 temperature: IM in mitochondria-rich fractions.

Organism	Cytochrome			Comments	Reference
	λ_{max} (nm)				
	c-type	b-type	a-type		
Tetrahymena pyriformis Strain ST	549 553	556 560	604 617 (447)	LT IM	Turner *et al.* (1971) Lloyd and Chance (1972)
Paramecium aurelia Stock SI	551	556	607	LT IM	Kung (1970)
Chilomonas paramecium	(cytochromes *c* and *a* + a_3 present)				Hutchens (1940)
Hartmannella castellanii Strain Neff	551	560–563	606 (444)	RT IM	Lloyd and Griffiths (1968)
Dictyostelium discoideum	(cytochromes *b*, *c*, *a* + a_3 present)			IM	Erickson and Ashworth (1969)
Physarum polycephalum	547	553 560	601	LT IM	Barnes *et al.* (1973)
Entamoeba histolytica	No detectable cytochromes				Hilker and White (1959)
Trichomonas foetus	No detectable cytochromes (cyt *b*?)				Ryley (1955) Suzuoki and Suzuoki (1951a, b)
Toxoplasma gondii	(cytochromes *b*, *c* and *a* present)				Fulton and Spooner (1960)

Organism								Conditions	References
Crithidia fasciculata			555	561			605	RT IM	Lwofff (1934); Kusel and Weber (1968); Hill and White (1968a); Hill and Anderson (1970); Kusel and Storey (1973a)
							(445)		
	551		555	557	(438)		601	LT IM	Edwards and Lloyd (1973)
							444		
Crithidia oncopelti	550			558			605	RT	Ryley (1955); Fulton and Spooner (1959); Srivastava (1971)
Trypanosoma cruzi (Culture forms)				565			608		Ryley (1956); Fulton and Spooner (1959)
				560			600		
			556				604		Baernstein (1953)
Trypanosoma rhodesiense (Culture forms)		553		570	605				Ryley (1962); Fulton and Spooner (1959); Grant et al. (1961)
(Bloodstream forms)			No cyts detected						
Trypanosoma vivax / *congolense* / *gambiense*			No cyts detected						
				560	600				Ryley (1951); Ryley (1956); Fulton and Spooner (1959)
Trypanosoma lewisi (Bloodstream form)	550	553		559	605				Ryley (1951); Fulton and Spooner (1959)
	550				600				

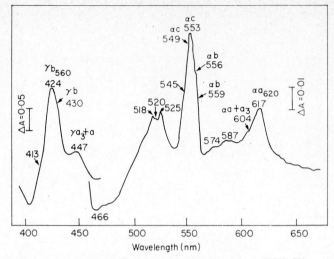

FIG. 42. Difference spectrum of *Tetrahymena* mitochondria at 77°K. The mitochondrial suspension contained 9·0 mg of protein/ml. Reduction was achieved by adding excess of $Na_2S_2O_4$ to the sample cuvette contents; oxidation of the reference suspension was by aeration; both cuvettes were then immersed in liquid N_2. Cuvettes had a light-path of 2 mm and the spectral band width was 2 nm. (Reproduced with permission from Lloyd and Chance, 1972.)

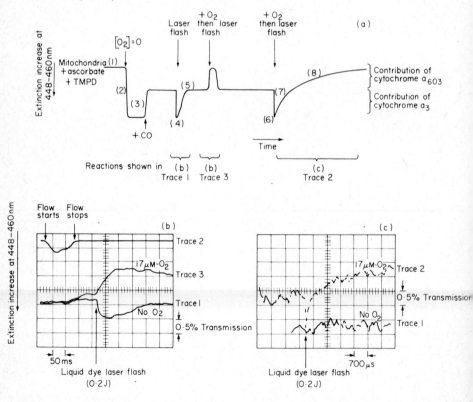

haem a is present, that the absorption maximum at 448 nm represents the Soret band of a cytochrome oxidase, and that the haemoprotein with a broad absorption band at around 620 nm undergoes reduction during the transition to anaerobiosis.

The dithionite reduced-oxidized spectrum of *Tetrahymena* mitochondria (purified by centrifugation through a sucrose gradient to equilibrium density) at 77°K is shown in Fig. 42. The interpretation of this spectrum (Lloyd and Chance, 1972) is aided by considering the spectra of cytochromes purified from *T. pyriformis*. Cytochrome b_{560} shows absorption maxima at 560, 529 and 424 nm, while cytochrome c_{553} has maxima at 553, 523 and 414 nm (Yamanaka et al., 1968).

Treatment of mitochondrial suspensions with ultrasound released cytochromes c_{553} and b_{560} into the soluble fraction. Further fractionation of the residual membranes revealed the presence of cytochrome c_{549} λ_{max} 549, 520 and 418 nm) and cytochrome b_{558} (λ_{max} 558, 524, and 432 nm). The spectrum shown in Fig. 42 also indicates the presence of a-type cytochromes; the major absorption maximum in the α-region was at 617 nm with a shoulder at 604 nm, while the Soret maximum was at 447 nm. Other components absorbing at 545, 574 and 587 nm were also revealed. Mammalian cytochrome a_3 reacts in the reduced or oxidized state with CN^-; the reduced CN^- complex is easily autooxidized, while the oxidized CN^- complex cannot be easily reduced

FIG. 43. Kinetics of photolytic decomposition of CO-liganded cytochrome oxidase and the subsequent reactions with O_2 in mitochondria from *Tetrahymena pyriformis*. (a) Diagram of idealized trace indicating sequence of reactions producing changes of extinction in experiments shown in (b) and (c). Addition of ascorbate + TMPD to a mitochondrial suspension gives a steady-state reduction of cytochromes $a + a_3$ corresponding to the trace at position (1). Attainment of anaerobiosis gives an extinction increase (2) caused by cytochrome reduction. Mixing with CO leads to a 50% decrease in extinction (3) resulting from formation of reduced cytochrome a_3-CO complex. Flash photolysis dissociates this complex and restores the extinction to the value (4) obtaining before CO addition; in the absence of O_2 recombination of cytochrome a_3 and CO occurs (5). Mixing with O_2 100 ms before photolysis allows both cytochromes a_3 and a_{603} to become oxidized after the flash, but resolution of the very fast initial reaction is not attained. Better time resolution indicates an extinction increase (6) resulting from the dissociation of the reduced cytochrome a_3-CO complex, followed by oxidation of released reduced cytochrome a_3, during which extinction decreases rapidly to a value (7) corresponding to that before photolysis. This reaction is followed by a further slower extinction decrease (8), which corresponds with that expected for the reoxidation of reduced cytochrome a_{603}. (b) Trace 1, extinction changes produced during anaerobic photolysis; trace 2, flow-velocity trace (downward displacement indicates period of mixing with oxygenated buffer); trace 3, extinction changes produced on mixing with oxygenated buffer followed by photolysis. Only overall reactions are shown; this scale is not adequate for satisfactory resolution of kinetics. (c) Trace 1, position of trace after anaerobic photolysis; trace 2 extinction changes produced on mixing with oxygenated buffer followed by photolysis. All cuvettes (light-path 6 mm) contained mitochondrial protein (3 mg/ml), respiratory substrate 10 mM–ascorbate + 1 mM–TMPD, 300 μm–CO and 17 μm–O_2. The temperature of incubation was 24°C. (Reproduced with permission from Lloyd and Chance, 1972.)

(Keilin and Hartree, 1939). Cytochrome a does not react with CN^-. Spectrophotometric separation of cytochrome a and a_3 have been achieved by exploitation of these properties (Yonetani, 1960) although other interpretations of these data are also possible (Wainio, 1970). Similar studies with *Tetrahymena* mitochondria reveal an absorption maximum due to cytochrome a_{603} (Lloyd and Chance, 1972). Cytochrome a_{620}, a pigment peculiar to this species, is rapidly oxidized by O_2 in anaerobic mitochondrial suspensions, and shows a high affinity for O_2 (Turner *et al.*, 1971). It resembles cytochrome a_3 in that it reacts in the reduced form with CO and in both oxidized and reduced states with CN^-. The broad absorption band in the α-region due to this component largely obscures the presence of an absorption maximum at 600–610 nm characteristic of cytochromes $a + a_3$, but these species can be demonstrated in the Soret region as two kinetically distinct species contributing equally to the total absorption at around 448 nm (Fig. 43). The contribution of cytochrome a_{620} in this region has not been precisely ascertained, but conditions which lead to complete loss of this unstable species do not diminish the Soret absorption band; this can be clearly demonstrated in CO difference spectra.

Steady-state levels of reduction of electron transport components in mitochondria isolated from *T. pyriformis* are presented in Table XIII (Turner *et al.*, 1971). State 5 reduction for b-type cytochromes by α-ketoglutarate or by pyruvate was only 60% of that observed with succinate, whereas cytochromes c and a were equally reduced with all three substrates. This suggests the presence of two pools of cytochrome b, one of which is reducible only by succinate. Extra reduction of cytochromes b was produced on dithionite addition. Substrates other than α-ketoglutarate or pyruvate (including succinate, DL-lactate, citrate and malate) gave no steady-state reduction of intramitochondrial NAD, indicating that these substrates were either unable to gain access to the mitochondrial matrix, or were not NAD-linked. Tightly-coupled mitochondria respiring in the presence of succinate, α-ketoglutarate or pyruvate showed changes in the steady-state levels of reduction of cytochrome b and NAD on addition of ADP; both components became more oxidized. Possible pathways of terminal oxidation in mitochondria from *T. pyriformis* are presented on p. 151.

(*b*) *Paramecium sp.* The cytochrome spectra of another ciliate, *Paramecium*, have not been examined in such detail, but are quite different from those of *T. pyriformis*. Study of this organism is difficult, as it is often grown in the presence of bacteria, and some strains also contain bacterial-like endosymbionts. However spectra of anaerobic whole-cell suspensions (Sato and Tamiya, 1937) show bands at 523, 551, 555 and 608 nm, and mitochondria isolated from *P. aurelia* also show similar

Table XIII. Steady-state levels of reduction of electron transport components in mitochondria isolated from *Tetrahymena pyriformis*. (Reproduced with permission from Turner *et al.*, 1971.)

One hundred per cent reduction of a component taken as anaerobic reduction level minus level in the absence of exogenous substrate (State 5 – State 2). ADP concentration was 0·1 mM. Absence of a figure indicates that if any reduction occurred it was below the level of detection. No reduction of the 620 nm absorbing species (620–650 nm) was detectable in States 3 or 4 with any of the substrates tested, and no extra reduction of this pigment was seen on adding dithionite to anaerobic suspensions. Temperature of incubation 25°C.

Substrate Electron transport component	10 mM-Succinate		10 mM-2-Oxoglutarate			10 mM-Pyruvate		
	cyt b	cyt c	NAD	cyt b	cyt c	NAD	cyt b	cyt c
Wavelength pair (nm)	560– 575	553– 540	340– 374	560– 575	533– 540	340– 374	560– 575	553– 540
Reduction State 4 (%)	35·6	—	84·5	29·7	—	85	25	—
Reduction State 3 (%)	22·4	—	72·7	21·8	—	76	10	—
Not reduced in State 5 but reduced by excess dithionite %	35·0	—	—	55·0	10	—	50	10

characteristics (Kung, 1970) together with an additional uncharacterized absorption maximum at 588 nm. The endogenous respiration rates of *P. caudatum*, *P. aurelia*, and *P. calkinski* are only about 50% inhibited by very high concentrations (1–10 mM) of CN^- or azide (see review by Danforth, 1967). This finding suggests the operation of a pathway of terminal oxidation alternative to cytochrome *c* oxidase. The fragmentary nature of data on these organisms provides a starting point for further investigations.

(c) *Hartmannella castellanii*. Difference spectra of mitochondria from this amoeba, scanned during the oxidation of succinate, show extensive reduction of *b*-type cytochromes during State 4 respiration; only when anaerobiosis (State 5) was attained did *a*- and *c*-type cytochromes become reduced (Lloyd and Griffiths, 1968). Antimycin A addition enabled isolation of the substrate end of the respiratory chain; in the presence of this inhibitor *b*-type cytochromes were fully reduced whereas *c*- and *a*-type cytochromes were fully oxidized. The presence of a diversity of *b*-type cytochromes was evident: changes in the α-absorption maxima at around 560 nm accompany the use of different substrates as reductants. Thus NADH gave a maximum at 556 nm, succinate produced a peak at 560–563 nm, and malate gave an α-band with a maximum at 565 nm. This organism presents an attractive system for further more detailed studies.

(d) *Dictyostelium discoideum*. Mitochondria isolated from myxamoeba of this cellular slime mould contain cytochromes *b* and *c* and $a + a_3$. More of the *b*- and *c*-type cytochromes were detectable on reduction with dithionite than with succinate; 60% and 28% extra reduction of cytochromes *b* and *c* respectively was produced with the chemical reductant (Erickson and Ashworth, 1969).

(e) *Physarum polycephalum*. Cytochrome *c* from this slime mould has an oxidation-reduction mid-point potential of $+ 257$mV at pH 7·0 and a mol. wt of $12,500 \pm 1500$ (Colleran and Jones, 1973). Spectra of isolated mitochondria revealed the presence of two *b*-type cytochromes [λ_{max} (77°K) at 560 nm and 553 nm], and one *c*-type cytochrome at 547 nm (Barnes *et al.*, 1973).

(f) *Haemoflagellates*. The culture forms of several of the genera of trypanosomes restricted to invertebrates e.g. *Crithidia* ($=$ *Strigomonas*) *oncopelti* and *C. fasciculata* have received a good deal of attention. Mitochondria isolated from *C. fasciculata* contain substrate reducible cytochromes $a + a_3$, *b* and *c* (Fig. 44; Kusel and Weber, 1968; Hill and White, 1968; Edwards and Lloyd, 1973; Kusel and Storey, 1973a, b) and a CO-binding

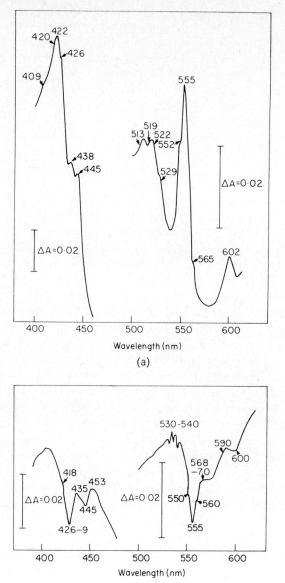

FIG. 44. (a) Low temperature difference spectrum of *Crithidia fasciculata* (3×10^9 organisms/ml). Reduction was achieved in the presence of 9 mM–glycerol; the oxidized sample contained 100 mM–H_2O_2. Path length 2 mm, spectrum band width 1 nm, temperature 77°K. (b) Low temperature CO-difference spectrum. Both cuvette contents were allowed to become anaerobic in the presence of 9 mM–glycerol, then the sample cuvette was sparged with CO for 2 min. Other experimental conditions as in (a). (Reproduced with permission from Edwards and Lloyd, 1973.)

pigment designated cytochrome o (Hill and Anderson, 1970, but see p. 153). Intact cells and isolated mitochondria contain protohaem, haem c and haem a in the same molar proportions (Hill and White, 1968). Cytochrome c_{555} has been purified and characterized as a basic protein (isoelectric point pH 9·9) with a redox potential mid point of $+ 280$ mV at pH 7·0 and mol wt of 13, 200 (Kusel $et\ al.$, 1969); the reduced form has absorption maxima at 420, 525 and 555·5 nm, and the value of ϵ for the α-band is 29·7mM^{-1} cm^{-1}. The pyridine haemochromogen showed an α-band at 553 nm. The cytochrome contains 1 to 2 residues of ϵ-N-trimethyl lysine per molecule and reacts with beef-heart cytochrome c oxidase at only one sixth of the rate of purified beef-heart cytochrome c_{550} (Hill $et\ al.$, 1971a).

C. oncopelti examined in a hand spectroscope showed a strong band at 558 nm, weaker β-bands at 528-530 nm and an extremely weak band at 605 nm (Ryley, 1955). The spectroscopic examination of this organism is made easy, as unlike C. fasciculata, it has no haemin requirement. It has recently been suggested that cytochrome o acts as a terminal oxidase during exponential growth and cytochrome a_3 is synthesized only in the stationary phase (Srivastava, 1971). However these claims have not been backed by studies of action spectra (see p. 153). Cytochrome c purified from C. oncopelti has a pyridine haemochrome peak at 553 nm (lying between that of protohaem-containing protein and cytochromes c). The haem could not be removed by acid-acetone treatment, but was removed by mercuric chloride in acid. Like Euglena cytochrome 558 (Meyer and Cusanovitch, 1972), C. fasciculata cytochrome 555 (Hill and White, 1968; Kusel $et\ al.$, 1969) appears to have a 2 (4) vinyl-4(2)-ethyl substituted haem (Pettigrew and Meyer, 1971). Its atypical spectrum may arise from an unusual attachment of the haem (through a single cysteine, leaving one vinyl side-chain free), as the amino acid sequence in the vicinity of the prosthetic group is altered by an alanine insertion (Pettigrew, 1972). As further work proceeds on the sequencing of purified protozoal cytochromes, it will be interesting to see whether these features making C. oncopelti cytochrome 555 so different from other eukaryotic cytochromes c are restricted to a few genera or are more widespread. A preliminary study of cytochromes c from three other insect trypanosomatids (Trypanosoma cruzi, T. rhodesiense and Leptomonas sp. suggest that they are similar in spectral and other properties to that from C. fasciculata (Hill $et\ al.$, 1971b).

Although studies on mitochondria isolated from mammalian trypanosomes have not yet been attempted, there is a wealth of information from experiments with intact cells which suggests that this interesting and important group of protozoa provides several unusual systems for further study. Ryley (1951) showed that the bloodstream form of T. lewisi has a complete cytochrome system and that its respiration is blocked by 0·46 mM—

CN^- and by CO, this latter inhibition being light-reversible. *T. cruzi* (culture form) showed an intense α-band at 553–565 nm and a weak band at 608 nm due to *a*-type cytochrome (Ryley, 1956). No trace of a conventional cytochrome *c* band was found even at liquid nitrogen temperature. Spectroscopic examination of bloodstream forms of *T. rhodesiense, T. equiperdum* or *T. congolense* failed to reveal any cytochrome bands (Ryley, 1956); the respiration of the first named species were cyanide-insensitive, whereas that of *T. congolense* was only slightly sensitive to this inhibitor. A further study of *T. rhodesiense* (Ryley, 1962) revealed that although the blood-stream forms contain no detectable cytochromes and have cyanide-insensitive respiration, the culture forms show cytochrome bands at 553–570 nm and 605 nm, and their respiration is blocked by 0·46 mM–KCN. A survey of the respiration of seventeen different species of trypanosomes also showed major differences between various groups (Fulton and Spooner, 1959) with regard to cytochrome complement and sensitivity to respiratory inhibitors (see also review of Baernstein, 1963). The respiratory physiology of the culture forms is markedly influenced by culture conditions (Evans and Brown, 1972).

3. Cytochromes of Fungi

The visible absorption spectra of forty-five different species of fungi were observed both at room temperature and at 77°K in the presence of dithionite using a hand spectroscope (Boulter and Derbyshire, 1957). Typical cytochrome spectra, similar to that of a reduced suspension of baker's yeast, were found in all these species, although there was no clear evidence for the presence of cytochrome c_1. Other early papers on cytochromes in fungi include those on the slime mould, *Physarum polycephalum* (Ohta, 1954), *Ustilago zeae* (Grimm and Allen, 1954) and *Fusarium lini* (Kikuchi and Barron, 1959). More recent detailed investigations (see Table XIV), bring to light interesting variations between different species of fungi especially with regard to the number of spectrally-distinct *b*- and *c*-type cytochromes they contain. For instance, within all the species of three orders of the Oomycetes, Gleason and Unestam (1968a, b) detected two types of cytochrome *b* and only one cytochrome *c*, whereas the Chytridiomycetes contain three *b*-type and two *c*-type cytochromes. One of these *c*-type cytochromes may be analogous to cytochrome c_1. Thus the Oomycetes resemble higher plants in that they lack cytochrome c_1, a component which *is* found in Ascomycetes and Chytridiomycetes. The other remarkable feature of the cytochrome spectra of the Phycomycetes is the variable position of the absorption maximum of the α-band of cytochrome oxidase in different species (Gleason and Unestam, 1968a, b).

Neurospora crassa contains two distinct species of cytochrome *c*; an

Table XIV. Cytochromes of fungi. RT indicates spectral data obtained at room temperature, LT at liquid N₂ temperature; IM in mitochondria-rich fractions.

Organism	Cytochromes λ_{max} (nm)			Comments	Reference
	c-type	b-type	a-type		
Phycomycetes					
Allomyces	548 551	556 559 561	603	LT	Gleason (1968)
	553	560 564	606	RT	
Blastocladiella	548 552	556 559 563	603	LT	Gleason and Unestam (1968a)
	553	560 564	606	RT	
Monoblepharella	548 552	555 559 563	608	LT	
	553	560 564	609	RT	
Aphanomyces					
Pythium					
Saprolegnia	549	555 562	602	LT	Gleason and Unestam (1968b)
Apodachlya	551	557 564	605 (446)	RT	Unestam and Gleason (1968)
Leptomitus					
Sapromyces					
Minderella					
Ascomycetes					
Neurospora crassa	548	556 563	606 (442)	LT	Weiss et al. (1970)
	545 550	554 561	601 (440) 446)	LT IM	Lambowitz et al. (1972a)
	(412) (417)	(428)			
	546 552	556 562		LT IM	von Jagow et al. (1973)
Basidiomycetes					
Shizophyllum commune	550–552	560–562	603–605 (443–445)	RT	Neiderpruem and Hackett (1961)
Deuteromycetes					
Aspergillus oryzae	550	564	604	RT IM	Tamiya (1928) / Kawakita (1970a, b)
	551	560	605	RT IM	Watson and Smith (1968)
Aspergillus niger	549	560	604 (443)	RT	
Myrothecium verrucaria	546	560? 563	599–602	LT IM	Kidder and Goddard (1965)

unbound precursor, c_{II} containing a lysine-72 residue which is converted into cytochrome c_I, which contains an ϵ-trimethyl lysine-72 (DeLange et al., 1969; Scott and Mitchell, 1969), and is the biologically functional species. The cytochromes of mitochondria isolated from N. crassa have been investigated by Lambowitz et al. (1972a, b, c) and by von Jagow et al. (1973). When mitochondria are aerated in the presence of succinate + antimycin A, two b-type cytochromes are reduced, while the c- and a-type cytochromes remain oxidized. Under these conditions the pigment with the α-absorption maximum at 561 nm shows a greater absorption relative to that at 554 nm, when compared with the relative peak heights seen in spectra of reduced-oxidized mitochondrial suspensions. This observation confirms their separate identities as two distinct b-type cytochromes. After extraction of the soluble cytochrome c_{545} from the mitochondria, reduction of the residue by means of ascorbate + TMPD reveals the presence of cytochrome c_{550} which is analogous to cytochrome c_1. A membrane-bound cytochrome with λ_{max} 545 nm may represent either unextracted cytochrome c_{545}, or a third c-type cytochrome. Further aspects of the cytochrome chain of this organism are discussed in relation to the changes resulting from the poky mutation (see p. 145). Cytochrome c (Heller and Smith, 1966) and cytochrome c oxidase (Weiss et al., 1971; Birkmayer, 1971a, b) have been purified from wild-type N. crassa (see also p. 420). Cytochromes c have also been purified from Ustilago sphaerogena (Neilands, 1952), and from Physarum polycephalum (Yamanaka et al., 1962).

4. Cytochromes of aerobically-grown yeast (see Table XV)

The discovery of cytochromes in baker's yeast by Keilin in 1925 was followed by studies of the differences in cytochrome content between top and bottom yeasts (Euler and Fink, 1927; Euler et al., 1927). Although the haematin content of both types of yeast was similar, only the top yeasts showed the characteristic bands in the reduced state due to cytochromes a, b, and c at 603, 564 and 550 nm respectively; bottom yeast contained the cytochromes "a_1" (589 nm) and "b_1" (557 nm), now known to be characteristic of anaerobically-grown organisms (see p. 160). A spectroscopic survey of many different strains and species of yeasts was carried out by Fink (1932). The presence of cytochrome c_1 was first reported by Keilin and Hartree in 1949 who, using their newly introduced technique of low-temperature spectroscopy, observed an extra band at 552 nm in Candida utilis. This cytochrome was also observed in baker's yeast by Lindenmayer and Estabrook (1958), and as a functional electron carrier in the respiratory chain by Chance and Spencer (1959).

Difference spectra of mitochondria isolated from S. carlsbergensis during respiration (with citrate as substrate) show partial reduction of cyto-

Table XV. Cytochromes of aerobically-grown yeasts. RT indicates spectral data obtained at room temperature, LT at liquid N_2 temperature; IM in mitochondria-rich fractions.

Organism	Cytochrome λ max (nm)			Comments	Reference
	c-type	b-type	a-type		
Saccharomyces cerevisiae	549	564·5	603·5	RT	Keilin (1925)
	551	563·5	604	RT	Boulter and Derbyshire (1957)
	549 552	561	601	LT	Lindenmayer and Estabrook (1958)
	547 554	559·5		LT	Kawai and Mizushima (1973)
Saccharomyces oviformis M2	546·8(α₁) 535(α₃) noα₂			LT (purified haemoprotein)	Kawai and Mizushima (1973)
Saccharomyces carlsbergensis	550	563	605 (444)	RT IM	Ohnishi et al. (1966a)
	549 555	560 (563)	599	LT (shoulder at 563)	Cartledge et al. (1972)
Shizosaccharomyces pombe	541(α₂) 548(α₁) 553	559	607	LT, split α band c	Heslot et al. (1970a)
	541	554 560 563 (b_T)	600–605	LT	Claisse (1969)
	548			LT	Poole et al. (1974)
Candida utilis	545	558 (b_K) 562 (b_T)	599 (445)	LT	Keyhani and Chance (1971)
	549	561·5 (b_K), 563, [558, 565] (b_T)		RT IM	Sato et al. (1972)
Candida lipolytica	547 551	557 564	603	LT IM glucose- or hexadecane-grown	Volland and Chaix (1970)
Candida krusei	546·1(α₁), 535·4(α₃), noα₂			LT (purified haemoprotein)	Kawai and Mizushima (1973)
Rhodotorula glutinis	550	560	603 (444)	RT	Matsunaka et al. (1966a)
Debaryomyces kloekeri	547 (α₁), 536(α₃) noα₂			LT (purified haemoprotein)	Kawai and Mizushima (1973)

chromes b and $c + c_1$, and of flavoprotein during State 4 respiration; extensive reduction of all the respiratory carriers occurred when anaerobiosis (State 5) was attained (Ohnishi *et al.*, 1966). When antimycin A (2 nmoles/mg protein) was added and the suspension aerated, c- and a-type cytochromes became oxidized whereas cytochrome b and flavoprotein remained reduced. Respiration of D- or L-lactate does not involve the antimycin A-sensitive site; L-lactate oxidation involves cytochrome b_2 (λ_{max} 557, 528 and 524 nm) as an electron carrier (Appleby and Morton, 1959; Rippa, 1961; Baudras and Spyridakis, 1971). Further studies on the steady-state redox levels of cytochromes of mitochondria from lactate-grown *S. carlsbergensis* confirmed the absence of a control point due to Site I phosphorylation. Energy dependent reversed electron transfer from lactate via cytochrome c to cytochrome b and ubiquinone has been demonstrated (Ohnishi *et al.*, 1967, Fig. 45).

At 77°K the α-band of cytochrome c observed in intact *Shizosaccharomyces pombe* is split to give an α_2-satellite at 541 nm (Claisse, 1969; Heslot *et al.*, 1970a), a phenomenon first observed with plant and animal cytochromes c by Keilin and Hartree (1949) but not usual in yeast cytochromes

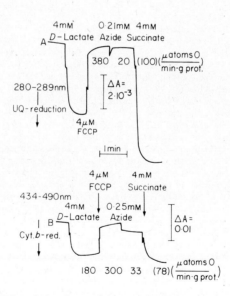

FIG. 45. Response of UQ and cytochrome b in reverse electron transport in mitochondria from *Saccharmyces carlsbergensis* with D-lactate as the substrate. The respiratory rates in the different states are given for control. The value in brackets is the respiration rate with succinate in the uncoupled state, temperature of incubation, 20°C, 0·67 mg mitochondrial protein/ml. (a) Recording of UQ absorption; optical pathlength = 2 mm. (b) Recording of cytochrome b absorption; optical pathlength = 5 mm. (Reproduced with permission from Ohnishi *et al.*, 1967.)

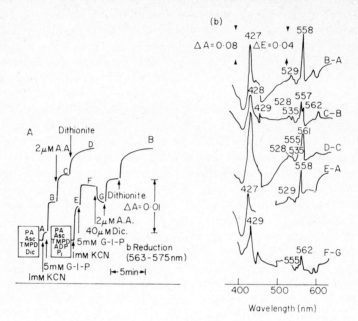

Fig. 46. (a) The effect of substrate on the degree of reduction of the *b* cytochromes in coupled and uncoupled mitochondria isolated from *Candida utilis*. The mitochondria were suspended in the assay medium at $2\cdot1$ mg protein per ml and supplemented with 3 μM-piericidin A, 2 mM-sodium ascorbate and 60 μM-TMPD. To the suspension (I) 40 μM dicumarol was then added, and to the suspension (II) $1\cdot0$ mM-ADP and 3 mM-phosphate were added to permit ATP formation. The reagents were added at the indicated times and final concentrations PA, piericidin A; Asc, ascorbate; Dic., dicumarol; G-1-P, glycerol-1-phosphate; AA, antimycin A. (b) The absorption spectra of *b* cytochromes in *Candida utilis* mitochondria at 77°K. The mitochondria were suspended at $7\cdot1$ mg protein per ml and treated as given in legend to Fig. 46(a). Spectrum B-A: the reference sample was withdrawn at the designated point A (in (a)), where the sample was treated with piericidin A, ascorbate, TMPD, dicumarol and KCN, and injected into the spectrophotometer cuvette which had been pre-cooled to liquid nitrogen temperature. The measuring sample was similarly treated, but was withdrawn 1 min after the addition of 5 mM-glycerol-1-phosphate (point B in (a)). Therefore this shows the spectrum of *b* cytochrome reduced when glycerol-1-phosphate was added (cytochrome b_K). Spectrum C-B (mainly cytochrome b_T): the reference material was the same as the measuring sample in Spectrum B-A (withdrawn at the point B in (a)). The measuring sample was withdrawn 1 min after the addition of antimycin A in the presence of glycerol-1-phosphate (point C in (a)) and similarly treated as described in the legend of Spectrum B-A. Spectrum D-C: the reference material was the same as the measuring sample in Spectrum C-B. The measuring sample was withdrawn 4 min after the addition of dithionite in the presence of glycerol-1-phosphate and antimycin A (point D in (a)). Spectrum E-A (cytochrome b_K) sample was withdrawn at the point A (in (a)), and similarly treated as the reference sample in Spectrum B-A. The measure sample was withdrawn at the point E (in (a)) where the suspension was treated with 3 μm piericidin A, 2 mM-ascorbate, 60 μm-TMPD, $1\cdot0$ mM-ADP, 2 mM-phosphate and 30 seconds later 1 mM-KCN. Spectrum F-G (cytochrome b_T) the reference sample was withdrawn at the point G (in (a)) where 40 μM-dicumarol was added in the presence of glycerol-1-phosphate. The measuring sample was withdrawn before the addition of dicumarol (point F in (a)). (Reproduced with permission from Sato *et al.*, 1972.)

(Claisse, 1969). Two b-type cytochromes were observed in *C. lipolytica* grown either on hexadecane or glucose (Volland and Chaix, 1970).

Three types of b cytochromes have been demonstrated in mitochondria from *C. utilis* (Sato *et al.*, 1972; Fig. 46). Cytochrome b_K (α-band at 561·5 nm at room temperature) is readily reduced either by anaerobiosis or by cyanide treatment in the presence of glycerol-1-phosphate or succinate both in coupled or uncoupled mitochondria. Cytochrome b_T has a double α-band (at 565 nm and 558 nm) and is readily reduced either by anaerobiosis or by cyanide treatment in the presence of glycerol-1-phosphate or succinate in coupled mitochondria; in uncoupled mitochondria it is slowly reduced after anaerobiosis and this rate of reduction is enhanced by addition of antimycin A. Thus the oxidation state of this cytochrome is energy-dependent. The third b cytochrome (b_{563}) is reduced slowly after anaerobiosis in uncoupled mitochondria, but faster than the cytochrome b_T. The mid-point potentials of cytochromes b_T, b_{563} and b_K are approximately -50mV, $+5$mV, and $+65$mV respectively. This demonstration of the presence of functionally distinct b-type cytochromes confirms and extends the previous observations of Chance (1952) and Kováč *et al.* (1970b) with intact yeast cells. A functional distinction between cytochromes b_K and b_T has also been observed in *S. carlsbergensis* (Cartledge *et al.*, 1972), in *Candida utilis* (Keyhani and Chance, 1971), and in *Shizosaccharomyces pombe* (Poole *et al.*, 1974).

Isolation of cytochrome c from yeast was first reported by Keilin (1930) and crystallization was achieved by Hagihara *et al.* (1956a, b). Investigations on the purified cytochrome include those of Estabrook (1956), Armstrong *et al.* (1961), Okunuki (1961), Paléus and Tuppy (1961), and Margoliash (1962). It has been suggested that Glutamine-16 may be an invariant residue in mitochondrial cytochromes c (Lederer, 1972; Lederer *et al.*, 1972). The presence of two isoenzymes of cytochrome c in a haploid yeast was first shown by Clavilier *et al.* (1964). These two cytochromes have different amino acid compositions (Slonimski *et al.*, 1965; Stewart *et al.*, 1966). A third (non-trimethylated) cytochrome c is also present (Foucher *et al.*, 1972). Cytochrome c oxidase has been purified from *C. krusei* (Yumsui *et al.*, 1966), *S. oviformis* (Sekuzu *et al.*, 1967) and from *S. cerevisiae* (Sekuzu *et al.*, 1964; Duncan and Mackler, 1966; Tzagoloff, 1969a; Shakespeare and Mahler, 1971; Mason and Schatz, 1973a; Rubin and Tzagoloff, 1973a).

B. Flavoproteins

Spectral changes observed in mitochondrial preparations in the region 450–520 nm cannot be specifically assigned to flavoproteins as was formerly thought, as it has become clear that the spectral characteristics of non-haem iron proteins overlap with those of flavoproteins. The mitochondrial concentration of non-haem iron is high (e.g. 5 nmol/mg protein, Clegg and Garland, 1971) relative to flavoprotein (e.g. 0·25 nmol/mg protein) and their extinction coefficients are not dissimilar over the wavelength range of the "iron flavoprotein" trough in reduced-oxidized spectra.

1. Flavoproteins of yeast mitochondria

By growing *C. utilis* under various conditions of iron or sulphate limitation the mitochondrial concentration of non-haem iron or acid-labile sulphide may be drastically reduced (Light *et al.*, 1968; Clegg and Garland, 1971; Haddock and Garland, 1971). Spectroscopic changes in such iron-deficient SMPs can then in some cases be interpreted uniquivocally in terms of either flavoproteins or non-haem iron proteins. Seven components contributing to the "iron-flavoprotein" trough were identified other than cytochrome *b*, ubiquinone and succinate dehydrogenase (Ragan and Garland, 1971; Fig. 47). Correlation of analyses of FMN, FAD, non-haem

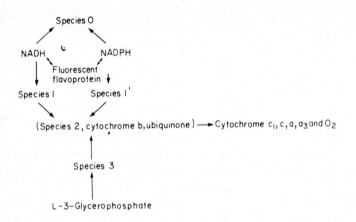

Fig. 47. Reaction of flavoprotein species 0, 1, 1′, 2 and 3 in ETP preparations from glucose-limited *Candida utilis*. It is not established that species 2 is the first oxidant of species 1, 1′ and 3, and cytochrome *b* and ubiquinone are included in parentheses to indicate the uncertainties of sequences in this region. (Reproduced with permission from Ragan and Garland, 1971.)

iron and acid labile sulphide of SMPs from phenotypically modified cells with the magnitude of the spectroscopic changes observed, enabled the identification of five components: species 1a, the flavin of NADH dehydrogenase ferroflavoprotein; species 1b, the iron-sulphur component of NADH dehydrogenase ferroflavoprotein; species 1^1, the flavin of an NADPH dehydrogenase; species 2, an iron-sulphur or ferroflavoprotein component; species 3, the flavin of glycerophosphate dehydrogenase. Two additional components were a fluorescent flavoprotein, probably lipoamide dehydrogenase, and a b-type cytochrome reducible by NADH or NADPH but not reoxidizable by the respiratory chain. Electron transport particles from iron- or sulphate-limited cells did not contain detectable amounts of species 1b or 2; both species were detectable after subjecting non-growing cell suspensions to iron or sulphate "recovery". The *in vivo* recovery of species 2 but not species 1b was inhibited by cycloheximide. The recovery of species 1b correlates with the recovery of Site I energy conservation, whereas the recovery of species 1b with species 2 correlates with the recovery of piericidin A sensitivity. The oxidation of NADH and NADPH by the respiratory chain is sensitive to piericidin A, and a iron-sulphur protein common to both pathways (species 2) is suggested as the inhibitor-sensitive component. Redox potential measurements indicate that Site I energy conservation occurs between the levels of species 1 (a and b) and species 2.

Studies on the NADH dehydrogenase include those on mitochondria and membrane preparations from *C. utilis* and *S. cerevisiae* (Biggs *et al.*, 1970); the enzyme complex has been purified from *C. utilis* by Tottmar and Ragan (1971).

Demaille *et al.* (1970) have described the action of fuscine, an inhibitor which acts in the flavoprotein-iron sulphur region of the chain at a point distinct from that at which rotenone or piericidin inhibit. This inhibitor shows similar titration curves with mammalian and yeast mitochondria. It inhibits both pyruvate and succinate oxidation by mitochondria from *S. cerevisiae*, *C. utilis* and Fe-limited *C. utilis*. EPR measurements suggest that it acts on the succinate side of the non-haem iron of succinate dehydro-genase; it is suggested that inhibition of NAD-linked substrate oxidation results from close association of the non-haem iron proteins of Fp_s and Fp_d.

2. Flavoproteins of mitochondria from other organisms

Various types of flavoprotein have been characterized by simultaneous spectrophotometric and fluorimetric measurements (Chance, 1966; Chance and Schoener, 1966). These methods have been applied to intact mito-chondria from *Polytomella caeca* (Lloyd and Chance, 1968, Fig. 48). The possibility of interference from iron-sulphur proteins in these experiments cannot be discounted (Ragan and Garland, 1969) but estimations of the flavin content of intact mitochondria correlated well with data obtained for acid and enzymically-extractable flavins. It was evident that the flavoprotein involved in the oxidation of internally generated NADH is distinct from that oxidizing externally added NADH.

C. Iron-sulphur Proteins

Studies on the iron-sulphur proteins of microbial mitochondria by EPR

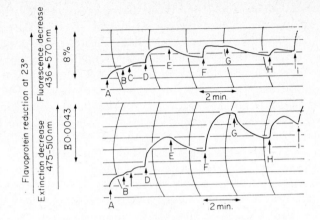

Fig. 48. Reduction and oxidation of flavoproteins in mitochondria from *Polytomella caeca*. Simultaneous traces of fluorescence (excitation at 436 nm, measurement at 570 nm) and extinction changes (475–510 nm) of the suspension of mitochondria (5 mg of protein/ml) in 100 mM–KCl–5 mM–MgCl₂–50 mM–tris HCl buffer, pH 7·4. The mitochondria were pretreated with 3 µM–pentachlorophenol followed by 10 µg of antimycin A. Other additions were made as follows: A, ethanol (3 mM): B, β-hydroxybutyrate (3 mM); C, glutamate (3 mM); D, malate (3 mM); E, rotenone (15 µg); F, succinate (3 mM): G, malonate (10 mM); H, NADH (1 mM); I, dithionite (excess). The total volume was 3·3 ml. (Reproduced with permission from Lloyd and Chance, 1968.)

spectroscopy have been confined almost entirely to those with various species of yeast.

1. The iron-sulphur proteins of Candida utilis mitochondria

Resolution of the EPR signals due to iron-sulphur proteins in the cytochrome *b–c* region and NADH dehydrogenase region of the respiratory chain has been achieved by recording spectra after incubation of SMPs with either succinate + glycerol-1-phosphate or NADH until anaerobiosis was attained (Fig. 49; Ohnishi *et al.*, 1970, 1971). The additional EPR signals detected below 30°K which had peaks at g = 2·09, 2·05, 1·93, 1·89 and 1·86 could be specifically assigned to species of iron-sulphur proteins of the NADH dehydrogenase region of the respiratory chain and were also observed when SMPs were reduced with NADH in the presence of piericidin A (Fig. 50). Similar signals were also found below 30°K in SMPs isolated from bovine heart muscle (Ohnishi *et al.*, 1971) and in purified NADH-UQ reductase from beef-heart (Orme-Johnson *et al.*, 1971); four separate iron-sulphur centres have been identified. Adaptation of the methods of redox-titration to EPR absorbance measurement (Wilson *et al.*, 1970) has enabled the resolution of these four separate iron-sulphur centres in *C. utilis* (Ohnishi *et al.*, 1972a). Addition of various oxidation-reduction mediators to an anaerobic suspension of SMPs revealed the

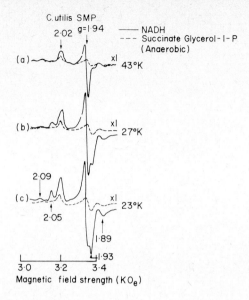

FIG. 49. EPR spectra of iron-sulphur proteins in *Candida utilis* SMPs reduced with different substrates under anaerobic conditions, measured at 43°, 27°, and 23°K. SMPs (56·1 mg of protein per ml) were incubated with 1·7 mM–NADH (solid line), or with 17 mM–succinate plus 9 mM–glycerol-1-phosphate (dotted line). After transferring into EPR tubes, reaction mixtures were incubated for 10 min at 100m temperature. (Reproduced with permission from Ohnishi *et al.*, 1971.)

presence of different components of differing redox potentials and their half-reduction potentials could be determined (Fig. 51). A large gap (190 mV) occurred between the half-reduction potentials of iron-sulphur centres 1 and 2 and a large difference in the steady-state reduction levels of these two components was observed during ethanol respiration by intact mitochondria. This suggests that energy conservation for Site I phosphorylation occurs between centres 1 and 2.

2. The iron-sulphur proteins of S. cerevisiae *mitochondria*

Schatz *et al.* (1966) and Sharp *et al.* (1967) reported that no EPR signal of iron-sulphur proteins associated with the NADH dehydrogenase region of the respiratory chain was detectable at 77°K in SMPs prepared from *S. cerevisiae*. The spectra obtained on reduction with either succinate + glycerol-1-phosphate or NADH were identical at 77°K and at temperatures below 30°K confirming the absence of any iron-sulphur proteins associated with the NADH dehydrogenase portion of the chain (Ohnishi *et al;* 1971). A signal of unknown origin was seen at around g = 2·02.

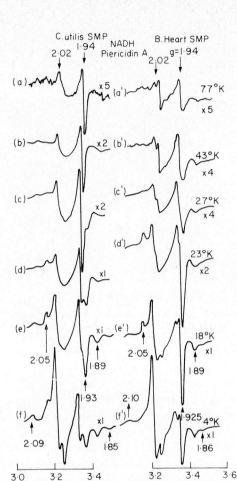

Fig. 50. EPR spectra of iron-sulphur proteins in the NADH dehydrogenase region of the respiratory chain in *Candida* and bovine heart SMPs at temperatures below 77°K. Iron-sulphur proteins in these particles were reduced with NADH in the presence of piericidin A. SMPs (56·1 mg of protein per ml) isolated from *C. utilis* cells which were grown in a synthetic medium containing 25 μM–FeCl₃ were previously incubated with piericidin A (0·5 nmole per mg of protein) for 10 min in an ice bath. Bovine heart SMPs (38·4 mg of protein per ml) were previously incubated with piericidin A (1·2 nmoles per mg of protein) in a similar fashion. In both cases, NADH oxidase activity was inhibited up to 98 % and thus about 1 min was allowed prior to anaerobiosis. Both suspensions were frozen in liquid nitrogen in less than 40 s after the addition of 1·7 mM–NADH. EPR operating conditions were: field modulation, 100 kHz; modulation amplitude 12 Oersted (Oe); microwave power, 0·092 mwatt; microwave frequency, 9·02 GHz; temperature, 77°, 43°, 27°, 23°, 18°, or 4°K, as indicated. Relative instrument gain is shown in the figure. (Reproduced with permission from Ohnishi *et al.*, 1971.)

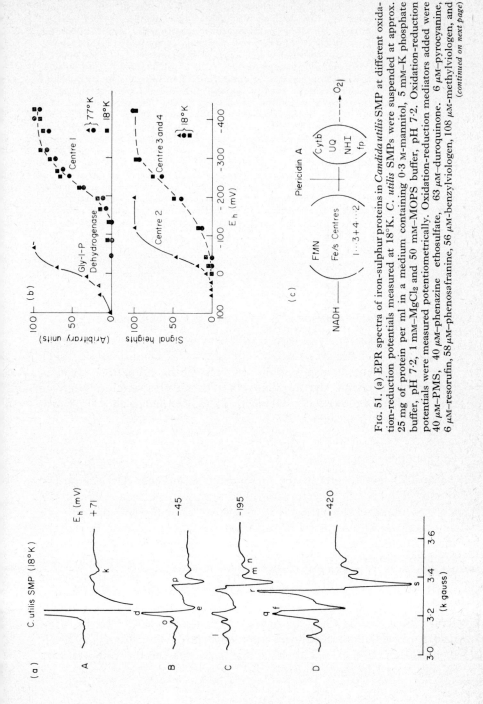

FIG. 51. (a) EPR spectra of iron-sulphur proteins in *Candida utilis* SMP at different oxidation-reduction potentials measured at 18°K. *C. utilis* SMPs were suspended at approx. 25 mg of protein per ml in a medium containing 0·3 M-mannitol, 5 mM-K phosphate buffer, pH 7·2, 1 mM-MgCl₂ and 50 mM-MOPS buffer, pH 7·2. Oxidation-reduction potentials were measured potentiometrically. Oxidation-reduction mediators added were 40 μM-PMS, 40 μM-phenazine ethosulfate, 63 μM-duroquinone. 6 μM-pyrocyanine, 6 μM-resorufin, 58 μM-phenosafranine, 56 μM-benzylviologen, 108 μM-methylviologen, and
(continued on next page)

F

25 μM–2-OH-naphthoquinone. The oxidation-reduction potential of the suspension system was lowered by stepwise additions of small aliquots of 0·1 M–NADH solution. Oxidation-reduction potentials shown in the figure are relative to the standard hydrogen electrode. The EPR operating conditions were: modulation amplitude, 12 gauss; microwave power, 0·205 mwatt; microwave frequency, 9·02 GHz; time constant, 0·001 s; scanning rate, 1 kgauss per min; temperature 18°K. The ordinate is the first derivative of the microwave absorption in an arbitrary unit. Small Roman letters are placed along the spectra vertically above or below the field position of the resonances typical for various components according to Orme-Johnson *et al.* (1971). (b) The oxidation-reduction potential dependence of the EPR signals of the iron-sulphur proteins of *Candida utilis* SMPs. Experimental conditions as in (a). At 77°K, the $g = 1.94$ signal was measured for both iron-sulphur proteins associated with glycerol-1-phosphate (and succinate) dehydrogenase and centre 1 in the NADH dehydrogenase. The principal absorption band (r–sp) was used at 18°K for the titration of both centre 1 and centre 2. Centres 3 and 4 were titrated using the signal heights of troughs m and n. The signal heights of different iron-sulphur proteins were normalized because of different temperatures and instrumental gains used. (c) Schematic representation of the respiratory chain in *C. utilis* SMPs in the region between NADH and cytochrome *b*. The half-reduction potentials of centres 1, 3 + 4, and 2 are –240, –210 and –50 mV respectively. (Reproduced with permission from Ohnishi *et al.*, 1972a.)

3. The iron-sulphur proteins of S. carlsbergensis mitochondria

The EPR signals obtained with SMPs from *S. carlsbergensis* are similar to those obtained for *S. cerevisiae* (Ohnishi, 1970; Ohnishi *et al.*, 1971). The development of iron–sulphur proteins of the respiratory chain of this organism during the respiratory adaptation of anaerobically-grown cells has been followed (T. Ohnishi, T. G. Cartledge and D. Lloyd, unpublished results, see p. 177).

D. Quinones

Table XVI lists reports of the presence of ubiquinone (UQ) in various species of eukaryotic microorganisms. Changes in the redox state of UQ which accompany transitions from State 4 to State 3 respiration of yeast mitochondria with a variety of substrates have been observed directly by spectrophotometric means and these redox changes were confirmed by extraction and analysis (Ohnishi *et al.*, 1967). These investigations also provided evidence for energy-linked reverse electron transfer from lactate to UQ. Redox changes in mitochondrial UQ on addition of succinate or NADH have also been studied in *Neurospora crassa* (Drabikowska and Kruszewska, 1972). The total amount of redox active UQ of the mammalian mitochondrial electron transport chain is functionally and kinetically homogeneous and is not divided into substrate-specific compartments (Kröger and Klingenberg, 1973).

Table XVI. Ubiquinone in eukaryotic microorganisms.

	Organism	Type of UQ	Amount µmoles/g dry wt unless stated otherwise	References
Algae	*Prototheca zopfii*	UQ_7	0·018	Lloyd (1966a)
	Cladophora sp	UQ_9		Lester and Crane (1959)
	Ochromonas malhamensis	UQ_{10}	0·6 mg/g dry wt	Gale et al. (1964)
	Euglena gracilis	UQ_9	grown autotrophically (light) or heterotrophically (dark)	Vakirtzi-Lemonias et al. (1963)
		UQ_9		Threlfall and Goodwin (1964)
		UQ_8	192 µg/g dry wt; grown heterotrophically (light); 168 µg/g dry wt	Fuller et al. (1961)
	Astasia klebsii	UQ_9	0·14	Vakirtzi-Lemonias et al. (1963)
Protozoa	*Tetrahymena pyriformis*	UQ_9	15·2% total lipid	Taketomi (1961)
	Crithidia fasciculata	UQ_9	0·23	Vakirtzi-Lemonias et al. (1963)
				Kusel and Weber (1965)
	Crithidia oncopelti	UQ_9	0·40	Vakirtzi-Lemonias et al. (1963)
	Trypanosoma rhodesiense	UQ_9	culture form 1·27 µg/g lipid; bloodstream form 299 µg/g lipid	Threlfall et al. (1965)

Table XVI. (continued)

Organism	Type of UQ	Amount μmoles/g dry wt unless stated otherwise	References
Fungi			
Mucor corymbifer	UQ$_9$	0·20	Lester and Crane (1959)
Neurospora crassa	UQ$_{10}$, THUQ$_{10}$		Lester et al. (1958, 1959), Gloor et al. (1958)
Aspergillus fumigatus	UQ$_{10}$	0·6	Lester and Crane (1959), Levate et al. (1962)
Aspergillus flavus			Packter and Glover (1960)
Aspergillus flavus-oryzae	THUQ$_{10}$		Levate et al. (1962)
Aspergillus terreus			
Penicillium stipitatum			
Ustilago maydis	UQ$_{10}$ + UQ$_9$	0·20, 0·02	Erickson et al. (1960)
Claviceps purpurea	UQ$_{10}$		Anderson et al. (1964)
Yeasts			
Saccharomyces cerevisiae (Grown aerobically)	UQ$_6$	0·35	Lester and Crane (1959), Lester et al. (1958, 1959), Gloor et al. (1958), Mahler et al. (1964c), Sugimura et al. (1964).
(Grown anaerobically)	Not detectable		Lester and Crane (1959), Lester et al. (1958, 1959), Sugimura and Rudney (1960)
(Respiratory-deficient mutants grown aerobically)	UQ$_6$	0·08, 0·11	Mahler et al. (1964c), Sugimura et al. (1964)
Saccharomyces cavalieri	UQ$_6$		
Saccharomyces fragilis	UQ$_6$		Lester and Crane (1959)
Candida utilis	UQ$_7$ + UQ$_9$	0·59	Lester and Crane (1959), Lester et al. (1958, 1959), Stevenson et al. (1962), McHale et al. (1962)

IV. Pathways of Electron Transport Alternative to the Phosphorylating Respiratory Chain

A. Cyanide-, Azide-, CO-, and Antimycin A- Insensitive Respiration

Many eukaryotic microorganisms show a residual respiration in the presence of CN^-, CO or azide, even when these inhibitors are present at concentrations sufficient to inhibit cytochrome c oxidase completely. The diversity of organisms showing this phenomenon is indicated in Table XVII. Organisms resistant to antimycin A inhibition are also included. The extent of the inhibitor-insensitive respiration observed, ranges from a small proportion of the total respiration in some species, to those cases in which hardly any sensitivity is observed.

Absorption bands of cytochrome a_3 have not been identified in *Chlamydomonas reinhardii* (Hiyama *et al.*, 1959), in some of the bloodstream forms of the trypanosomes (Ryley, 1956, 1962), in anaerobic protozoa such as *Trichomonas sp.* (Ryley, 1955), or in anaerobically-grown *Saccharomyces sp.* (Criddle and Schatz, 1969; Cartledge and Lloyd, 1972b). In these organisms complete insensitivity to the classical cytochrome c oxidase inhibitors is not unexpected. In all other species mentioned, electron transport proceeds to a greater or lesser extent via at least one pathway alternative to the main phosphorylating respiratory chain. In very few microorganisms have these alternative pathways been located to isolated mitochondrial fractions, and until further subcellular fractionation studies are attempted, it must be considered possible that many of the alternative routes of electron transport are extramitochondrial (e.g. in peroxisomes, endoplasmic reticulum or cytosol). The present discussion will therefore be limited to those cases in which CN^--insensitive respiration has been observed in isolated mitochondrial fractions.

The phenomenon of CN^--resistant respiration has been studied extensively in plant mitochondria, and the five different hypotheses proposed to explain its mechanism have been reviewed by Bendall and Bonner (1971):

1. Oxidations are mediated via a flavoprotein oxidase characterized by a low affinity for O_2.
2. An auto-oxidizable b-type cytochrome is present which accepts electrons from the respiratory chain on the substrate side of cytochrome c.
3. Incomplete inhibition of cytochrome c oxidase.
4. A second oxidase of the a-type as part of the respiratory chain.
5. A non-haem iron mediated pathway of electron transport to oxygen.

Bendall and Bonner (1971) have presented experimental evidence against hypotheses 1–4, and provide data which strongly suggest that a non-haem iron compound is associated with the cyanide-resistant pathway in plant mitochondria. It has also been shown that one site of oxidative phosphorylation is associated with electron transport occurring by way of the alternative pathway in plant mitochondria (Wilson, 1970).

Table XVII. Organisms showing respiration insensitive to inhibitors of cytochrome c oxidase.

Organism	Reference	Comments
A. ALGAE		
Prototheca zopfii	Webster and Hackett (1965), Epel and Butler (1970a)	Endogenous respiration resistant to CN^- azide and CO; CN^- and azide stimulate, possibly due to increased rate of auto-oxidation of cytochrome b_{564}
Chlorella pyrenoidosa	Warburg (1919), Emerson (1927), Genevois (1927), Tang and French (1933), Syrett (1951), Sargent and Taylor (1972)	Endogenous respiration stimulated by 0·1 mM-CN^-. Alternative oxidase not inhibited by CO, 0·1 mM-thiocyanate or 1 mM-8-hydroxyquinoline. $Km(O_2) = 6·7\ \mu M$
Chlamydomonas reinhardii	Hommersand and Thimann (1965)	Respiration of Zygospores CN^- and CO-sensitive; that of vegetative organisms resistant
Euglena gracilis (bleached strain)	Sharpless and Butow (1970b)	Alternative mitochondrial oxidase, CN^--insensitive induced by growth with antimycin A
Euglena gracilis (mutant W_3)	Perini et al. (1964)	Succinoxidase of sonicated cells only 15% inhibited by 95% CO: 5% O_2 gas mixture
Astasia klebsii	von Dach (1942)	50% inhibition of endogenous respiration by 10 mM-CN^-

B. PROTOZOA

Organism	References	Notes
Acanthamoeba sp.	Neff *et al.* (1958)	
Amoeba proteus	Clark (1945)	
Mayorella palestinensis	Reich (1955)	Low sensitivity to CN^- and azide
Tetrahymena pyriformis	Baker and Baumberger (1941), Hall (1941), McCashland (1956), McCashland *et al.* (1957), Ryley (1952), Turner *et al.* (1971)	Varying reports on CO and CN^- sensitivity (from 0–80 % inhibition by 1 mM-CN^-) depending on strain tested. Decreased CN^- sensitivity after growth with this inhibitor. Mitochondrial oxidations CN^- sensitive in strain ST
Paramecium caudatum	Clark (1945), Humphrey and Humphrey (1948), Pace (1945)	Only 50 % inhibition by 1 mM-CN^- or azide.
Paramecium aurelia	Pace (1945), Simonson and Van Wagendonk (1952)	Portion of respiration sensitive to CO, inhibition reversible by white light
Paramecium calkinski	Boell (1945, 1946)	
Trypanosoma vivax	Ryley (1956), Fulton and Spooner (1959)	CN^- and CO-resistant respiration
Trypanosoma congolense	von Brandt (1951), Fulton and Spooner (1959)	CN^-, Azide-, and CO-resistant respiration
Trypanosoma rhodesiense	Ryley (1956, 1962)	Respiration of bloodstream form CN^--insensitive culture form sensitive
Trypanosoma gambiense, evansi, equinum, equiperdum	Ryley (1956), Fulton and Spooner (1959)	Respiration insensitive to CN^-, azide, and CO
Trypanosoma brucei	Evans and Brown (1972, 1973)	Diphenylamine and CLAM inhibit CN^--insensitive respiration of bloodstream and newly established culture forms
Trypanosoma mega	Ray and Cross (1972)	
Trichomonas vaginalis	Ninomiya and Suzuoki (1952), Read and Rothman (1955)	Respiration entirely insensitive to CN^- and azide
Trichomonas foetus	Doran (1957, 1958), Lindblom (1961), Ryley (1955), Suzuoki and Suzuoki (1951a, b)	

Table XVII. (*continued*)

Organism	Reference	Comments
C. FUNGI		
Mucor genevensis (anaerobically-grown)	Clark-Walker (1972)	Respiration insensitive to CN^-, antimycin A
Ustilago sphaerogena	Grimm and Allen (1954)	Respiration partly CN^--insensitive
Myrothecium verrucaria	Kidder and Goddard (1965)	Mitochondrial oxidations quite insensitive to CN^- CO and azide
Aspergillus oryzae	Kawakita (1970a, b, 1971)	Mycelia grown in presence of antimycin A have mitochondria with CN^- and antimycin A-resistant respiration
Neurospora crassa (*poky* mutants)	Tissieres *et al.* (1953), Lambowitz and Slayman (1971), Lambowitz *et al.* (1972a, b)	65–85 % total respiration mediated by a mitochondrial alternative pathway which is CN^- and antimycin A-resistant but inhibited
Several other species	Sherald and Sisler (1972)	by SHAM
D. YEASTS		
Candida albicans	Ward and Nickerson (1958)	
Rhodotorula glutinis	Matsunaka *et al.* (1966a, b)	CN^- and antimycin A-resistant respiration mediated by oxidase with Km (O_2) = 23 μM
Candida utilis	Haddock and Garland (1971), Downie and Garland (1973a)	Alternative mitochondrial oxidase inducible in SO_4-limited chemostat cultures. Mutant containing an alternative oxidase (and having lost cytochrome *c* oxidase) selected by Cu-limited growth
Saccharomyces cerevisiae (anaerobically-grown)	Ephrussi and Slonimski (1950a, b), Criddle and Schatz (1969)	
Saccharomyces carlsbergensis (anaerobically-grown)	Cartledge and Lloyd (1972b, 1973)	

Some progress has also been made in elucidating the mechanisms of alternative routes of electron transport in mitochondria of eukaryotic microorganisms, especially where these pathways account for a high proportion of the total electron flux.

1. Euglena gracilis

When *E. gracilis* (bleached strain Z) is grown with succinate or ethanol as carbon source in the presence of antimycin A, a novel succinoxidase activity is induced (Sharpless and Butow, 1970b). The growth yield of the organism is decreased by two-thirds at concentrations of the inhibitor greater than 0·3 μg/ml; this result suggests that the efficiency of oxidative phosphorylation is drastically impaired. Whereas the succinoxidase activity of mitochondria from cells grown in the absence of antimycin A is almost completely sensitive to this antibiotic, growth in its presence produces a mitochondrial succinoxidase which is antimycin A-insensitive and stimulated by AMP. This activity was further stimulated by 1 mM-CN^-, but was completely blocked by an equimolar concentration of malonate. Cells grown on either succinate or on ethanol, in the absence of antimycin A, give mitochondria which have the novel succinoxidase as well as the normal CN^--sensitive activity, whereas mitochondria from cells grown on malate + glutamate show no trace of the AMP-stimulated oxidase pathway. Diphenylamine (1·7 mM) serves as a potent inhibitor of both pathways of electron transport.

The novel oxidase is more sensitive to 3·3 mM-TTFA than the conventional one. Stimulation by AMP of NADH and lactate oxidations (both of which are also further stimulated by CN^-) has also been observed in mitochondria possessing the alternative succinoxidase. Stimulation by GMP, IMP, ADP or ATP has also been observed. However this acceleration of respiration differs from the normal acceptor control (respiratory control) in that there is no coupled phosphorylation during the oxidation of succinate via the inducible oxidase; CCCP does not mimic the effect of addition of nucleotides. Competition between the AMP stimulated oxidase and PMS may be demonstrated, as AMP inhibits dye reduction in the PMS-DCPIP assay of succinate dehydrogenase in mitochondria from ethanol-grown cells. This suggests that AMP acts not at the level of substrate oxidation but at a stage of a common intermediate of the two alternative pathways. Free peroxide was not detected as an intermediate. The Km for oxygen was 3·7 μM (much higher than those values determined for cytochrome *c* oxidase which are in the range 0·02 to 0·05 μM (Chance, 1965). The cytochromes in mitochondria which possess the alternative oxidase appear spectrally identical with those of mitochondria which do not. Steady-state spectra showed that the redox state of *c*- and *a*-type cytochromes was almost entirely dependent on the activity of cytochrome c oxidase and that the cytochrome most affected by the addition of AMP was identical with that which became more reduced on the addition of antimycin A, i.e. a b-type cytochrome with an absorption maximum at 565 nm. This suggests that the alternative oxidase pathway branches from the main respiratory chain at the level of cytochrome b_{565}. The relatively

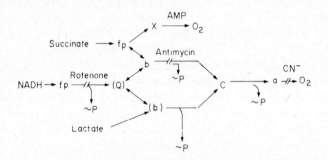

FIG. 52. Alternate pathways of electron transport in *Euglena* mitochondria. Since the relationships of the various spectroscopically distinguishable b- and c-type cytochromes are unknown, no attempt has been made to depict them in detail; however, the b cyto-chrome shown as closely associated with the antimycin-sensitive site and with the AMP-stimulated alternate oxidase almost certainly represents the 565 nm component seen in difference spectra. The alternate oxidase represented by X is active only in cells grown under inducing conditions. (Reproduced with permission from Sharpless and Butow, 1970b.)

small changes in redox state produced on adding AMP do not seem large enough to accommodate the 5–8 fold increase in the rate of electron flow. The nature of the alternative terminal oxidase remains obscure. The existence of a CO-binding cytochrome b (which may also react with CN^-) is not unique to those mitochondria possessing the novel succinoxidase, and hence the specific participation of a "cytochrome o" as an alternative oxidase seems unlikely. Sharpless and Butow (1970b) propose that one of the components of the succinate dehydrogenase may be altered to fulfil the role of a terminal oxidase. A scheme of possible alternative electron transport pathways in *E. gracilis* is shown in Fig. 52.

2. *Trypanosoma brucei*

Trypanosomes of the *T. brucei* sub-group undergo major changes in respiratory metabolism when they transform from the stages found in the bloodstream of the mammalian host to those which develop in the gut of the tsetse fly (Vickerman, 1971). Terminal respiration of the bloodstream forms is cyanide-insensitive and largely mediated by L-α-glycerophosphate oxidase systems (Grant and Sargent, 1960), one of which may be located extra-mitochondrically (Bayne *et al.*, 1969a, b). Although it has been reported that long established culture forms (believed to correspond to the mid-gut stage in the tsetse fly) have a terminal respiration which is sensitive to inhibition by CN^- (Ryley, 1962), it has recently been shown that organisms in newly established cultures, like the bloodstream forms, have respiration which is insensitive to CN^-, antimycin A and rotenone (Evans and Brown, 1971). The extent of CN^--insensitivity also depends upon the exact culture conditions employed. "Particulate preparations" isolated from culture forms under conditions which would be expected to give a mitochondrial-rich fraction, oxidized succinate, DL-α-glycerophosphate, NADH and ascorbate + TMPD; diphenylamine (0·3mM) gave 50%, 80%, 20% and 10% inhibition of the oxidation of these substrates respectively (Evans and Brown, 1972). Similar extents of inhibition of substrate oxidations were produced with whole homogenates of the primary culture form. With whole homogenates if the bloodstream forms, glycerophosphate oxidation was 75% inhibited and NADH oxidation was 60% inhibited by 0·3 mM-diphenylamine. The site of action of this inhibitor, which has been used to block CN^--resistant respiration in plants (Baker, 1963), probably lies between the flavoprotein and cytochrome components of the respiratory chain. The CN^--insensitive respiration of L-α-glycerophosphate by the pleomorphic bloodstream form of *T. brucei* and the monomorphic bloodstream form of *T. rhodesiense* was almost completely inhibited by 0·1 mM–CLAM (Evans and Brown, 1973). This compound has been shown to be a specific inhibitor of CN^--resistant respiration in plants (Schonbaum *et al.*, 1971). In the case of homogenates of culture forms of *T. brucei*, the residual CN^--insensitive respiration of succinate, L-α-glycerophosphate or L-proline (accounting for between 30 and 64% of the uninhibited respiration with these substrates) was completely sensitive to 0·1 mM–CLAM. This inhibitor (unlike diphenylamine) did not affect the rates of reduction of externally added mammalian cytochrome *c*. Further studies necessitate: (1) a more rigorous identification of the cytochromes present in the culture forms, together with an assessment of their functional role in cell respiration; (2) the elucidation of the site of action of CLAM; and (3) analytical subcellular fractionation to determine the location of the components of the L-α-glycerophosphate oxidase systems.

3. *Trypanosoma mega*

Experiments with mechanically-disrupted suspensions of *T. mega* (which had lost their cytosolic enzymes but retained "intact" organelles) indicate the operation of a pathway of electron transport alternative to the conventional respiratory chain (Ray and Cross, 1972). Inhibition of substrate oxidations (α-ketoglutarate or proline) by CN^-, azide, antimycin A or SHAM gave biphasic Dixon plots. SHAM specifically inhibited the cyanide-insensitive pathway. Although SHAM and CN^- used together were completely inhibitory, when used independently the sum of their inhibitions was not 100%. This strongly suggests the presence of a single branched pathway of electron transport. It was shown that the CN^--sensitive branch was comprised of cytochrome c and $a + a_3$. One of two b-type cytochromes remained oxidized in the aerobic steady-state in the presence of any of the inhibitors, placing the branch point at or before the position of cytochrome b, and suggesting that a b-type cytochrome is a component of the cyanide-insensitive branch.

4. Myrothecium verrucaria

The respiration of intact mycelia of *M. verrucaria* was only 20% inhibited in the presence of 0·46 mM–CN^- (Darby and Goddard, 1950; Kidder and Goddard, 1965). A 9:1 ratio of $CO:O_2$ gave no inhibition, although the inhibition produced by much higher partial pressures of CO was light reversible. Low concentrations of azide stimulated respiration, and 10 mM–azide only produced 50% inhibition of oxygen uptake. The succinoxidase activity of isolated mitochondria was only 43% inhibited by 0·46 mM–CN^-, although 70% of the cytochrome c oxidase activity was blocked at this concentration. Antimycin A (2·5 µg/ml) gave 95% inhibition of succinoxidase. Even in the presence of 4:1 $CO:O_2$ mixture succinoxidase activity was not inhibited, whereas a 19:1 mixture $CO:O_2$ gave a light reversible inhibition of cytochrome c oxidase. The affinity for O_2 of the alternative inhibitor-resistant oxidase was less than 1 µM. Spectral studies showed the presence of a normal cytochrome chain composed of 1 c-type, possible 3 b-type cytochromes, and cytochromes $a + a_3$. The possibility that phenol oxidases or ascorbate oxidase contribute to the alternative pathway was thus excluded, as these oxidases are characterized by a low affinity for O_2; neither did the evidence support the existance of an autoxidizable component at the level of cytochrome b. The "excess oxidase" hypothesis did not fit the spectral evidence either, as it was calculated that something like 100-fold excess cytochrome $a + a_3$ would be required to explain the data obtained. Thus the identity of the alternative oxidase in these mitochondria remains obscure.

5. Aspergillus oryzae

When *A. oryzae* is grown in liquid culture, mitochondria subsequently isolated show the classical pattern of inhibitor sensitivities (Kawakita, 1970a, b). Growth is rather less rapid in the presence of 10 µM–antimycin A (Kawakita, 1971), and mitochondria from mycelium grown with the inhibitor showed lower P/O ratios than those from normally grown cultures (1·2–1·5 for α-ketoglutarate, 0·9–1·2 for citrate, and 0·7–0·9 for succinate). The P/O ratio for the oxidation of TMPD + ascorbate (0·9) was not affected. Antimycin A (2·5 μM) failed to give any inhibition of oxygen uptake with α-ketoglutarate, citrate, NADH, succinate or ascorbate + TMPD; the oxidation of TMPD + ascorbate was virtually completely inhibited by 1 mM–cyanide, that of NADH was only 10% inhibited, whereas oxidation of the other three substrates tested was cyanide-insensitive. Spectral data confirmed the presence of a conventional cytochrome c oxidase (cytochromes $a + a_3$) and gave no indication that the formation of novel cytochrome component had occurred. Cytochrome b was extensively reduced, even in the aerobic steady state, and was only slowly oxidized by ferricyanide; Kawakita suggests that this component may be altered and rendered non-functional by growth in the presence of antimycin A. This explanation might account for the observation that electrons from TCA cycle intermediates do not reach the normal cytochrome c oxidase which is still present. A major difference in the ascorbate + TMPD-reducibility of a pigment absorbing at 565 nm (flavoprotein or non-haem iron protein) was noted between the two types of mitochondria; more than twice the absorption change detected in normal mitochondria was observed in the antimycin A-resistant organelles. Dual wavelength measurements suggest that even in the normal mitochondria (which also possess the antimycin-resistant pathway as a minor electron transport route) a component (measured at 465–500 nm) and located before cytochrome b in the chain is autooxidizable. This system provides a very clear case of an inducible alternative pathway and requires further investigation before the identity of its key components may be considered to be established.

6. Neurospora crassa

(a) Mutants mi-1 (poky) and mi-3. Wild-type N. crassa possesses mitochondria which contain flavoproteins, two b-type cytochromes, two c-type cytochromes and cytochrome $a + a_3$. The b- and c-type cytochromes of poky(mi-1) mitochondria are spectrally identical with those of the wild-type, but the room temperature absorption minimum of the poky iron-flavoprotein trough is shifted from 460 nm (the position in the wild-type) to 451 nm, and the position of the α-band (at 77°K) of poky cytochrome $a + a_3$ is shifted from 601 to 591 nm (Lambowitz et al., 1972a). The concentrations of the electron carriers are markedly different in poky mutants; poky mitochondria contain flavoproteins and cytochrome c (largely cytochrome c_{545}) at 30% higher concentrations than wild type-mitochondria but are grossly deficient in b- and a-type cytochromes, and have a reduced content of pyridine nucleotides. Cytochrome b_{562} (b_T) can be assayed by titration with antimycin A; cytochrome b_{556} reactions indicate it is a b_K. The low content of ubiquinone in the wild-type is increased 4 to 8-fold in *mi-1* (Drabikowska and Kruszewska, 1972; von Jagow *et al.*, 1973). The concentration ratio of cytochrome $a + a_3$ to cytochrome b is 0·3 compared with 1·0 in the wild-type mitochondria; this probably accounts for relatively high steady-state levels of reduction of cytochrome c in *poky* mitochondria. The molar ratios of the cytochromes $a:c:c_1:b_{562}:b_{556}$ in the wild-type are 1·0: 2·9: 1·0: 0·9: 0·9 and in *mi-1* are 0·05: 3·6: 1·0: 0·05: 0·3 (von Jagow *et al.*, 1973). As much as 66% of the total cell respiration of *poky* mycelium is mediated by an alternative oxidase system which is not sensitive to antimycin A or CN^- and which does not involve any of the cytochrome components (Lambowitz and Slayman, 1971; Lambowitz *et al.*, 1972a, b). This pathway can be inhibited by SHAM. It is present as a minor route of electron transport in wild-type, and the proportion of the total respiration for which it is responsible may be increased as much as 20-fold by growing wild-type in the presence of antimycin A, cyanide, or chloramphenicol. Confirmation that none of the cytochromes are involved in the CN^-- and antimycin A-insensitive pathway was obtained (Lambowitz *et al.*, 1972a) by showing that in a succinate $+$ antimycin A spectrum, the b-type cytochromes of *poky* were nearly completely reduced, and the c- and a-type cytochromes were nearly completely oxidized. When SHAM (180 μg/ml) was added to completely inhibit all electron flux through the alternate oxidase of *poky* mitochondria in the aerobic steady state of respiration, no increased reduction of cytochrome c was observed. CN^- (1·0 mM) added under these conditions, on the other hand, gave more than 90% reduction of the cytochrome c. This indicates that cytochrome c oxidation is mediated by a CN^--sensitive oxidase in *poky* which is presumably cytochrome $a + a_3$. Previous investigators (Haskins *et al.*, 1953; Eakin and Mitchell, 1970) had suggested a connection between cytochrome c and the alternative oxidase, in view of the large excess of this cytochrome in *poky*, but the latest studies invalidate this hypothesis. Tissières *et al.* (1953) suggested that the alternative oxidase may be a flavoprotein, and this suggestion is strengthened by the finding of a shift in the iron-flavoprotein trough in *poky* mitochondria (Lambowitz *et al.*, 1972a). A specific flavoprotein component of the hydroxamic acid-sensitive alternative oxidase system of higher plant mitochondria has recently been identified (Erecińska and Storey, 1970). The proposed routes of electron transport in *poky* mitochondria (Fig. 53) bear a marked similarity with those proposed for higher plant mitochondria.

Two lines of evidence suggest that, in *poky* the cytochrome chain and the alternate oxidase compete for reducing equivalent on the oxygen side of the dehydrogenases (Lambowitz *et al.*, 1972b): (a) total State 3 respiration rates are significantly lower than would be predicted from the sum of the two oxidases measured separately; and (b) a rapid, energy-dependent reduction of the alternate oxidase by TMPD + ascorbate via cytochrome *c*. *Poky* mitochondria show little respiratory control and low P/O ratios under most experimental conditions tested (Lambowitz *et al.*, 1972c).

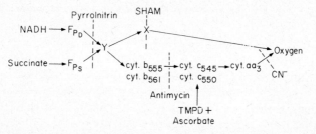

Fig. 53. Possible model for the respiratory system in the *poky* mutant of *Neurospora crassa*. X depicts the cyanide-resistant oxidase and Y an unspecified component which can transfer electrons from the flavin region to the cyanide-resistant oxidase or to the *b*-type cytochromes. Inhibition sites are indicated by dashed lines. (Reproduced by permission from Lambowitz *et al.*, 1972b.)

In these mitochondria essentially all the oxidative phosphorylation is inhibited by CN^- or antimycin A and this suggests that energy conservation is associated with the residual cytochrome system rather than with the alternate oxidase. Oxidative phosphorylation at Site I was not detectable in mitochondria from *poky*. The residual cytochrome system operates at maximal activity while the non-phosphorylating alternate oxidase is used only to accommodate surplus electron flux (Lambowitz *et al.*, 1972c). Somewhat different results have been reported by von Jagow *et al.* (1973) who found that oxidation of endogenous NADH by the cyanide- and antimycin A-insensitive pathway is coupled at one energy-conserving site. Two phosphorylation sites are involved in the oxidation of NADH, NADPH or succinate by the cyanide-sensitive pathway whereas there is no phosphorylation site for these substrate oxidations via the cyanide-insensitive route. Ubiquinone participates in both pathways. Mutant *mi-3* also shows cyanide and antimycin A- insensitive respiration which is inhibited by SHAM. Its cytochrome ratios (a: c: c_1: b_{562}: b_{556}) are 0·3: 3·2: 1·0: 0·9: 0·8. Thus it differs from *mi-1* in that only the content of its cytochrome $a + a_3$ is diminished (von Jagow *et al.*, 1973); the content of the reference cytochrome (c_1) was practically the same as in the wild-type. The UQ content was five times that of the parent strain.

Cyanide-insensitive respiration has also been studied in mutants *cni-1*, *cni-2* and *resp-1* (Edwards *et al.*, 1973).

(*b*) *Cu-depleted* N. *crassa*. Cyanide-resistant respiration also occurs in *N. crassa* after growth in Cu-deficient medium (Schwab, 1973, 1974) in cells which contained no detectable cytochromes $a + a_3$ (<0.02 μmol/g mitochondrial protein). The respiration of externally added NADH in these mitochondria is insensitive to CN^- and to antimycin A but is inhibited by SHAM.

7. Ustilago maydis, Ceratocystis ulmi, Neurospora sitophila and Saprolegnia sp.

Sporidia of *U. maydis* have an antimycin A- and azide-resistant electron transport pathway which has a low affinity for O_2 (Sherald and Sisler, 1972). Respiration by the normal pathway is only slightly affected by 1·5 mM–benzhydroxamic acid or 0·5 mM–8-hydroxyquinoline, whereas the alternate pathway is inhibted by 84% and 92% at these concentrations. Other inhibitors giving selective inhibition of the alternative pathway include 2-pyridinethiol-1-oxide, 2,2′-bipyridil, carboxin and diphenylamine. A similar pathway has also been found in conidia of *C. ulmi* and in *N. sitophila* (Sherald and Sisler, 1970). In *N. sitophila* and in *Saprolegnia sp.* (Unestam and Gleason, 1968), inclusion of antimycin A in the growth medium gives inhibition of respiration until the inducible alternative pathway is produced. A gene mutation eliminating the antimycin A-tolerant electron transport ability of *U. maydis* has been identified (Georgopoulous and Sisler, 1970).

8. Rhodotorula glutinis and Rhodotorula mucilaginosa

Inhibitor-tolerant respiration and low O_2-affinity have been studied in *R. glutinis* by Matsunaka *et al.* (1966a, b). In *R. mucilaginosa* the respiratory activity of whole cells is enhanced during the transition from the lag to the logarithmic phase of growth in the presence of 2% glucose. Stationary phase cultures yield mitochondrial fractions which show high levels of antimycin A- and azide-resistant respiration (Kitsutani *et al.*, 1970).

9. Candida lipolytica

Stationary phase cultures of *C. lipolytica* yield mitochondria which show CN^--insensitive respiration (Nyns and Hamaide, 1972).

10. Candida utilis

(*a*) *SO4-limited cells*. When *C. utilis* cells which have been grown under conditions of sulphate limitation in a chemostat are incubated with sulphate at concentrations higher than 250 μM, an alternative mechanism for cyanide- and antimycin A- (but not piericidin A-) insensitive respiration is induced (Haddock and Garland, 1971). Mitochondria isolated from these organisms oxidized pyruvate $+$ malate, DL-glycerol-3-phosphate, and α-ketoglutarate, and oxidation of these substrates was relatively insensitive to either antimycin A (2 nmol/mg protein) or KCN (2 mM). Although the segment of the respiratory chain from intra-mitochondrial NADH to cytochrome b was normal in that both piericidin A sensitivity and Site I energy conservation were present, the remainder of the respiratory chain (between the antimycin A-sensitive site and O_2) was probably being by-passed by a non-phosphorylating alternative route of electron transport. In the absence of inhibitors both routes were functional. In the presence of antimycin A or CN^- the P/O ratios were not significantly greater than unity, suggesting that under these conditions most of the ADP phosphorylation occurred at Site I. In the absence of these inhibitors the presence of energy conservation mechanisms at Site II and/or III in addition to an intact Site I was demonstrated by the energy-dependent reduction of NAD(P) by glycerol-3-phosphate oxidation. An energy-dependent electron flow from cytochrome c to cytochrome b (and thence to O_2 via the alternative route) results in the oxidation of cytochrome c in the presence of CN^-. A high State 3 reduction of cytochrome a indicates that the cytochrome oxidase is partially inhibited even in the absence of experimentally added inhibitors. No new carriers could be identified in difference spectra. The following inhibitors did not block the alternative pathway: CO (saturated), azide (3 mM), sulphide (1 mM), PCMB (1 mM) Mersalyl (1 mM), 1,10-phenanthroline (2 mM), 2,2'-bipyridyl (0·4 mM). No evidence for the presence of cytochrome o was seen in CO-difference spectra. The apparent Km for O_2 of the alternative oxidase was about 0·1 μM or less. The nature of the component(s) responsible for the alternative route of electron transport is unknown, but the critical dependence on SO_4 concentrations for the formation of the novel oxidase implicates some aspect of sulphur metabolism in the process. The alternative route accepts reducing equivalents at the level of cytochrome b (Fig. 54). Electron transport particles prepared from these mitochondria

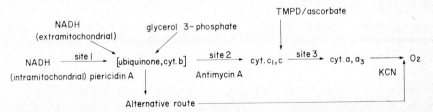

FIG. 54. Normal and alternative routes of electron transport in mitochondria from *Candida utilis*. Organisms were grown under sulphate-limited conditions then allowed to recover in the presence of 1 mM–sulphate. (Reproduced with permission from Haddock and Garland, 1971.)

contain elevated concentrations of acid-labile sulphur and non-haem iron compared to those from mitochondria not possessing the novel oxidase, and this implies that iron-sulphur protein(s) may play a role in the alternative pathway. This pathway cannot, however, be demonstrated in electron transport particles, a finding which suggests that the functional alternative oxidase is lost when the integrity of the mitochondria is destroyed.

(b) A mutant selected by Cu-limited growth. A mutant of *C. utilis* selected by Cu-limited growth with glycerol as carbon and energy source had less than 1% of the wild-type content of cytochrome $a + a_3$ although respiration rates were normal (Downie and Garland, 1973a). Mitochondrial respiration was insensitive to CN^- or antimycin A but not to rotenone or piericidin A. P/O ratios for isolated mitochondria were 1·0 for NAD-linked substrates and zero for others. The normal antimycin A-sensitive portion of the respiratory chain from cytochrome b to cytochrome c was present but was non-functional due to the absence of cytochrome c oxidase. The apparent K_m for O_2 of the mitochondrial oxidase was less than 10^{-6} M. The alternative oxidase communicated with the respiratory chain at about the level of cytochrome b. Growth of the mutant with Cu^{2+} for 20 generations resulted in the return of the cytochrome c oxidase (which was not restored in the presence of Cu^{2+} under non-growing conditions as is the case in the wild-type; Light, 1972a) but not the loss of the alternative oxidase which thus appeared to have become constitutive. The alternative oxidase was inhibited by a variety of substituted benz-hydroxamic acids, its identity is unknown; neither visible nor EPR spectroscopy revealed any unusual features.

11. Anaerobically-grown Saccharomyces sp.

When *S. cerevisiae* is grown under strictly anaerobic conditions it possesses none of the mitochondrial cytochromes and its respiration is insensitive to 1 mM–KCN (Ephrussi and Slonimski, 1950a, b; Slonimski, 1953a; Criddle and Schatz, 1969). The rate of oxygen uptake declines progressively as the concentration of O_2 falls below 0·3 μatom/ml (Criddle and Schatz, 1969) showing that the terminal oxidase has rather a low affinity for O_2. Similar results have been obtained with anaerobically grown *S. carlsbergensis* (Cartledge and Lloyd, 1973). The cyanide-resistant NADH oxidase activity is mediated by a non-mitochondrial membrane-bound flavin enzyme which is sensitive to CLAM and SHAM (Tustanoff and Ainsworth, 1973).

12. Aerobically grown Saccharomyces cerevisiae

Aerobic growth on lactate involves utilization of both D- and L-lactate dehydrogenases which feed electrons into the respiratory chain at the level of cytochrome c. Growth is thus antimycin-insensitive but CN^--sensitive (Pajot and Claisse, 1973).

B. The Alternative Oxidase of *Tetrahymena pyriformis*

The work of Turner *et al.* (1971) indicated that the haemoprotein with a broad absorption band at around 620 nm in mitochondrial suspensions prepared from *T. pyriformis* undergoes reduction when anaerobiosis is attained. The reoxidation of this cytochrome a_{620} by O_2 is more rapid than that of the b- and c-type cytochromes, and thus it reacts fast enough to be a terminal oxidase. On the basis of spectral changes in the Soret region on reaction with CO it is evident that these mitochondria also have a conventional cytochrome $a + a_3$ (Lloyd and Chance, 1972, see p. 111). Cytochrome a_{620} reacts with CO and CN^- and is unstable; its characteristic α-absorption band is lost from spectra of mitochondria which

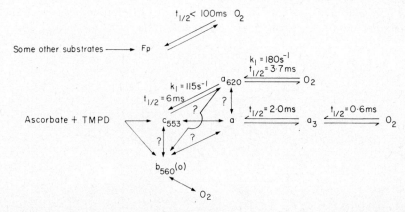

FIG. 55. Possible pathways of terminal oxidation in mitochondria of *Tetrahymena pyriformis*. Reaction half-times and pseudo-first-order kinetic constants for cytochromes a_{620} and c_{553} and flavoprotein oxidase are from the results of Turner *et al.* (1971). All kinetic results refer to oxidation rates at 24°C and 17 μM–O_2: calculations of kinetic constants assume irreversible reactions. (Reproduced with permission from Lloyd and Chance, 1972.)

have been aged or treated with ultrasound, detergents or organic solvents. Spectra of aged mitochondria also show a CO- reacting b-type cytochrome (Lloyd and Chance, 1972) which has been termed cytochrome o by Perlish and Eichel (1971). There is however no evidence that this haemoprotein is a functional oxidase, but it is included in the hypothetical scheme of electron transport (Fig. 55) as a possible second alternative oxidase.

C. Haemoproteins other than Cytochrome a_3 Reacting with Carbon Monoxide in Eukaryotic Microorganisms

Cytochrome a_3 reacts with CO, and a CO-difference spectrum (substrate or dithionite reduced + CO-substrate or dithionite reduced) may be used for the quantitative estimation of this haemoprotein (Chance, 1953). Under these conditions the peak to trough absorption difference (429·5 nm–445 nm) has a value for ϵ of 91×10^3 cm^{-1} M^{-1}. The α-region is characterized by an absorption minimum at around 605 nm and a maximum at 590 nm. The photochemical action spectrum for the relief of CO-inhibition of respiration is similar to the absorption spectrum of the CO-cytochrome a_3 complex, and action spectra of *Candida utilis* (Warburg and Negelein, 1929; Kubowitz and Hass, 1932), and of *Saccharomyces cerevisiae* (Melnick, 1942; Castor and Chance, 1955) confirm that cytochrome a_3 is the major terminal oxidase in these species (Fig. 56).

A wide variety of terminal oxidases have been shown to be present in various prokaryotes (see review of Kamen and Horio, 1970). Of particular interest is cytochrome o which shows α, β, and γ bands at 557–567, 532–537 and 415–429 nm respectively in CO-difference spectra; the functional significance of these pigments in bacteria and blue-green

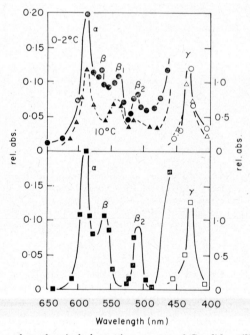

Fig. 56. (a) Relative photochemical absorption spectra of *Candida utilis* at 0·2°C and at 10°C from the data of Kubowitz and Hass (1932). (b) Relative photochemical spectrum of baker's yeast replotted from the data of Melnick (1942). (Reproduced with permission from Keilin and Hartree, 1953.)

algae has been demonstrated by action spectra based on light relief of CO inhibition of O_2 uptake (Castor and Chance 1959; Webster and Hackett, 1966).

Haemoproteins with spectral characteristics similar to those of bacterial cytochrome o have also been detected in many eukaryotic microorganisms (Table XVIII). No case has been recorded of a ferroprotein that combines with CO but does not react with O_2 (Keilin, 1966; Keilin and Hartree, 1939), and thus many of these pigments may be regarded as being possible candidates for the role of terminal oxidases. However it must also be recognized that any haemoprotein containing protohaem in a high-spin state would give CO-difference spectra with the characteristics of cytochrome o (e.g. haemoglobin, peroxidases). Furthermore there are very few cases listed where the CO-reacting haemoprotein has been located even in a crude mitochondrial fraction. In these cases further purification of mitochondria on density gradients might reveal an extra-mitochondrial location, as was shown with the CO-reacting peroxidase of plant mitochondrial fractions (Plesnicar et al., 1967).

The claim that "cytochrome o" acts as a terminal oxidase in *Trypanosoma mega* (Ray and Cross, 1972; p. 144) and also in a variety of other species (Hill and Cross, 1973) requires substantiation from action spectra. A similar proposal for the function of CO-binding pigments of *Crithidia fasciculata* (Hill, 1972) is invalidated by the work of Kusel and Storey (1973b) and Edwards and Lloyd (1973) who were unable to confirm any contribution from this so-called cytochrome o to photochemical action spectra for the relief of CO-inhibited respiration (Fig. 57). The data of Kusel and Storey (1973b) on the respiration of isolated mitochondrial fractions clearly reveals that cytochrome a_3 is the only functional oxidase present. Growth of the organism in the presence of 10 μM-euflavin leads to the production of cells containing less than 10% of the normal amount of spectrophotometrically-detectable cytochrome $a + a_3$. These cells still show no contribution from "cytochrome o" in photochemical action spectra. The respiration of whole cells may however proceed via more than one oxidase, as Edwards and Lloyd (1973) have evidence for a contribution from an unidentified oxidase with a low affinity for O_2.

It has been proposed that the cytochrome c peroxidase detected in both aerobically and (at low levels) in anaerobically-grown *Saccharomyces cerevisiae* (Lindenmayer and Smith, 1964; Yonetani and Ohnishi, 1966; Avers, 1967; Ishidate et al., 1969; Kawaguchi et al., 1969) may serve as an electron acceptor alternative to cytochrome $a + a_3$ under some conditions, and quantitative evidence for its importance in this role is now available. This enzyme is at least partly located as a freely soluble component of the intermembrane space, and over 90% of the mitochondrial cytochrome c can be oxidized by it (Erecińska et al., 1973); the H_2O_2

Table XVIII. Haemoproteins other than cytochrome a_3 reacting with CO in eukaryotic microorganisms.

Organism	Haemoprotein	Characteristics of CO-difference spectrum (position of absorption maxima and minima in nm)	Comments	Reference
A. IN ALGAE				
Eugena gracilis (bleached strain Z)	cytochrome b_{558} or b_{561}?	λmax 421–427 543 570; λmin 559	crude mitochondrial fraction	Sharpless and Butow (1970b)
Polytoma uvella		λmax 414 540 570; λmin 557; λmax 414	acetone-extracted whole cells; crude mitochondrial fraction	Webster and Hackett (1965)
Polytomella agilis Arago		λmax ~420 ~520 540 570; λmin ~560	whole cells	Webster and Hackett (1965)
Polytomella caeca Pringsheim	cytochromes P-416, P-450	λmax 416 450 570; ~560	microsomal fraction	Cooper and Lloyd (1972)
Astasia longa Pringsheim		λmax 418 470 520 540 568; λmin 425 557	crude mitochondrial fraction	Webster and Hackett (1965)
B. IN PROTOZOA				
Tetrahymena pyriformis	cytochrome b_{560}; cytochromes P-422, P-450	λmax 417 470 537 572; λmin 434 529 560; λmax 422 450	purified protein, also reacts with CN⁻; microsomal	Yamanaka et al. (1968); Poole et al. (1971a)
	cytochrome a_{620}; cytochrome b_{560}	λmax 597; λmax 418 470 522 542 560 572; λmin. 620; λmin	crude mitochondrial fraction	Turner et al. (1971); Lloyd and Chance (1972)
	"cytochrome o"	λmax 418 537 572; λmin 437 470 520 558; (oxygenated 545, 582; +CO, 574)	crude mitochondrial fraction	Perlish and Eichel (1971)
	haemoglobin		whole cells	Keilin and Ryley (1953)

Organism	Component	Absorption maxima	Material	References
Paramecium caudatum	haemoglobin	(oxygenated 545, 582; + CO 539, 574)	whole cells	Sato and Tamiya (1937), Keilin and Ryley (1953), Smith *et al.* (1962), Smith (1963),
Crithidia fasciculata	"cytochrome *o*"	λmax 419 540 570	whole cells, crude mitochondrial fraction	Hill and White (1968), Hill and Anderson (1970)
		λmax 410 540 570	whole cells	Hill (1972), Edwards and Lloyd (1973) Kusel and Storey (1973a, b)
Crithidia oncopelti	"cytochrome *o*"	λmax 418 540 570	crude mitochondrial fraction	Srivastava (1971)
Trypanosoma mega	"cytochrome *o*"			Ray and Cross (1972)
C. IN FUNGI				
Neurospora crassa / *Penicillium*	haemoglobin	(oxygenated 545, 583; + CO 539, 574 prominent in *poky* mutant of *N. crassa*)	whole cells, soluble in crude *N. crassa* extracts	Keilin and Tissieres (1953), Boulter and Derbyshire (1957)
Saprolegnia / *Aphanomyces*	"cytochrome *o*" or peroxidase?	λmax 417 554 573; λmin 561		Unestam and Gleason (1968)
D. IN YEAST				
Saccharomyces cerevisiae (grown aerobically)	haemoglobin	(oxygenated 583; + CO, 575) high concentration in *petite* mutant		Warburg and Hass (1934), Keilin (1953), Keilin and Tissieres (1954), Yčas (1956)
	haemoglobin	λmax 419 532 568 (oxygenated 610) 412 575	non-sedimentable fraction	Mok *et al.* (1969)
	"cytochrome *o*"	λmax 408 540 572	particulate fraction	

Table XVIII. (*continued*)

Organism	Haemoprotein	Characteristics of CO-difference spectrum (position of absorption maxima and minima in nm)	Comments	Reference
(grown anaerobically)	cytochrome P-450	λmax 449-450	whole cells P-450 microsomal?	Lindenmayer and Smith (1964)
	cytochrome P-420	λmax 418-420 540-542 570-575	P-420 soluble?	
	cytochrome P-450	λmax 450 (416 in reduced-oxidized)	microsomal?	Ishidate et al. (1969)
	cytochrome P-420	λmax 420	contribution from cytochrome c peroxidase	
(grown aerobically or anaerobically)	cytochrome c peroxidase	λmax 423 542 570		Yonetani and Ray (1965)
	haemoglobin	λmin 438 561 590		Ishidate et al. (1969) Mok et al. (1969)
Saccharomyces carlsbergensis (grown aerobically)	"cytochrome o" cytochrome c peroxidase		crude mitochondrial fraction	Yonetani and Ohnishi (1966)
(grown anaerobically)	cytochrome P-450	λmax 450		Cartledge and Lloyd (1972b)
	cytochrome P-420?	λmax 414-420		Cartledge et al. (1972)
Candida utilis (grown aerobically)	haemoglobin	λmax 408 (oxygenated 414 532 540 568 575) (610)	"non-sedimentable fraction", only after growth with high concentrations of glucose	Mok et al. (1969)
Candida mycoderma	"cytochrome o"	λmax 407 537 571	"particulate fraction"	Oshino et al. (1971, 1972a, b, 1973)
	haemoglobin	λmax 420 540 570 (oxygenated 415 542 577)	"non-sedimentable fraction" tightly bound to reductase system containing flavin and non-haem Fe	

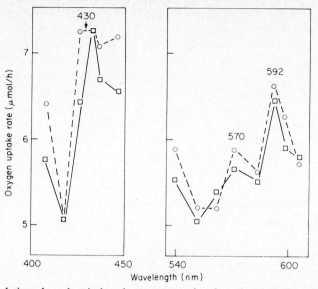

FIG. 57. Relative photochemical action spectrum for the release of CO-inhibition of respiration of *Crithidia fasciculata*. Oxygen uptake was measured manometrically under light of different wavelengths in the presence (- - - ○ - - -) and absence (—□—) of 60 mM-glycerol. No correction was made for the variation of energy content through the spectrum. (Reproduced with permission from Edwards and Lloyd, 1973.)

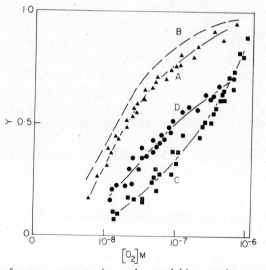

FIG. 58. Effects of oxygen concentration on haemoglobin, cytochrome *a* and cytochrome *a₃* in intact cells of *Candida mycoderma*. Curve A, oxygenation and deoxygenation of yeast haemoglobin in the presence of KCN (2 mM); Curve B, oxygenation curve of the haemoglobin which was obtained by subtraction as described in the text; Curve C, the redox change of cytochrome *a* ; Curve D, redox changes of cytochrome *a* + *a₃*. (Reproduced with permission from Oshino *et al.*, 1972b.)

necessary is generated intramitochondrially by the respiratory chain. The apparent rate of the cytochrome c peroxidase reaction is slower than that catalysed by cytochrome c oxidase. Thus cytochrome c peroxidase provides a system whereby a toxic by-product of respiration is eliminated and energy conservation is by-passed only at Site III. This enzyme is also present in *C. fasciculata* (Kusel *et al.*, 1973; Edwards and Lloyd, 1973).

The haemoglobin of *S. cerevisiae* has an even higher affinity for O_2 (half-oxygenation at 2×10^{-8} M–O_2) than does cytochrome a_3 (50% reduction at 3×10^{-8} M–O_2; Fig. 58). This observation (Oshino *et al.*, 1972a) suggests a strong competition by haemoglobin with the oxidase and this makes the physiological function of the haemoglobin difficult to explain. It appears that it may function in an oxidation-reduction reaction such as the decomposition of H_2O_2 (Oshino *et al.*, 1972b, 1973). Haemoglobin is found in many eukaryotic microorganisms (see Table XVIII).

D. Peroxide Formation and Superoxide Dismutases

Escherichia coli contains two distinct superoxide dismutase enzymes, a manganoprotein and an iron-containing protein (Yost and Fridowich, 1973). Mitochondria from chicken liver and pig heart also contain mango enzymes, whereas the cytosolic dismutases are cupro-zinc enzymes (Welsiger and Fridowich, 1973a). However Tyler (1973) has found a cupro-zinc enzyme in rat liver mitochondria, and Welsiger and Fridovich (1973b) have shown the presence of both types of superoxide dismutase in yeast mitochondria. The cupro-zinc enzyme is located in the inter-membrane space, and the mangano dismutase is a matrix enzyme. No evidence for membrane-associated dismutase was obtained. Both types of enzyme were also present in a ρ^0 yeast mutant. Tyler (1973) has suggested that these enzymes protect NADH dehydrogenase from superoxide radical anions generated during respiratory chain activity. The interaction of an energy-dependent component of the respiratory chain at the level of cytochrome b is probably responsible for H_2O_2 production during the respiration of mammalian mitochondria (Boveris and Chance, 1973; Loschen *et al.*, 1973) and a non-haem iron of the cytochrome b–c complex is a possible candidate for superoxide anion generation. H_2O_2 production also has been shown to accompany respiration of mitochondria isolated from *Crithidia fasciculata* (Kusel *et al.*, 1973).

4

Environmental Modifications of Mitochondrial Composition and Activities

I. Introduction

The ease with which the environment of microorganisms can be controlled has facilitated the study of alterations of their structure, composition and activity produced as a response to changes in culture conditions. Thus it is possible to alter the concentration of any nutrient in the medium and to observe the effect this has on the phenotype of organisms produced. The limitations of batch-culture methods are well recognized (Tempest, 1970), and in order to obtain reproducibility of cell composition, chemostat cultures are to be recommended. However, much of the work carried out on mitochondrial modifications has been performed using batch cultures; provided adequate precautions are taken to use standard inocula, to monitor growth, and to measure the concentrations of key nutrients and cell populations at time of harvesting, reproducibility can be achieved. Unfortunately, strict attention is not always paid to these growth parameters. Technical problems may in some cases make continuous culture of organisms difficult. For instance, fragile protozoa cannot be grown to high cell densities, as they are disrupted by vigorous aeration or stirring, and their oxygen demand cannot be satisfied. A possible disadvantage of continuous culture also arises from the need to collect culture effluent for some time to obtain sufficient cells for subsequent experimentation; many organisms change markedly during storage, even if the suspensions are kept cold and anaerobic. However, these disadvantages have not hindered the use of continuous culture in the study of the effects of specific nutrient limitation on yeast mitochondria (Garland, 1970).

Modifications of mitochondrial structure, composition and function during growth under conditions of environmental stress range from the massive reorganization possible in the facultatively anaerobic yeasts to the more specific changes produced by limitation of certain nutrients (e.g. iron). These alterations are of interest from at least three points of view:

(a) from the intrinsic interest of the extent of variability possible; (b) modifications provide insight into the roles of individual mitochondrial components in mitochondrial function and enable mechanisms of mitochondrial processes to be investigated; and (c) they enable mechanisms of assembly of functional macromolecular complexes and membranes to be examined. In short, these systems provide material for the investigation of (often reversible) processes of assembly or the deletion of specific mitochondrial components *in vivo*, even in the absence of net cell growth i.e. in non-proliferating cell suspensions.

Environmental stimuli may also trigger complex changes of structure and function involved in cellular differentiation; modification of mitochondrial activity usually accompanies those changes.

Finally the only method of achieving almost complete homogeneity of organisms in a culture so that the whole population approximates that of a single organism (and possibly so that the whole population of mitochondria acts as a single organelle) is to set up cultures of synchronously dividing cells. Only in these cultures can the temporal sequence of events involved in the elaboration of complex multienzyme-containing organelle membranes, during the normal growth processes, be adequately investigated.

II. Modifications of Mitochondria Produced by Environmental Factors

A. Effects on Mitochondria of Growth at Different Oxygen Tensions

1. Anaerobic growth of yeasts

(a) *Cytochromes.* The most extensive modification of the respiratory apparatus which is possible in yeast is produced by growth in the absence of oxygen. Differences between the spectra of aerobically- and anaerobically-grown yeast were first reported by Euler and Fink (1927) and by Euler *et al.* (1927). The cyanide-sensitive respiratory chain, composed of the classical cytochrome components ($a + a_3$, b, c_1 and c) is not present in cells grown under strictly-anaerobic conditions (Chin, 1950; Ephrussi and Slonimski, 1950a, b; Slonimski, 1953a, b), but anaerobically grown cells do contain pigments characterized at room temperature by α-absorption bands at 558 and 590 nm which have been designated cytochromes "b_1" and "a_1" respectively. The cytochrome b_1 absorption maximum is split at liquid nitrogen temperature to give two maxima (at 558 nm and 551 nm, Lindenmayer and Estabrook, 1958). These authors considered that this absorption had no contribution from cytochrome b_2, as lactate dehydrogenase was not detectable in extracts. The absorption maxima in the Soret region for cytochromes b_1 and a_1, at room temperature occur at 430 nm and 445 nm respectively (Lindenmayer and Smith, 1964). Depending on the degree of catabolite repression and availability of lipids, additional absorption bands are sometimes observed at 502, 575 and 630 nm (Chaix, 1961; Linnane, 1965; Wallace *et al.*, 1968). Ishidate *et al.* (1969) have suggested that the absorption band at 590 nm may have a contribution from cytochrome *c* peroxidase. Cytochrome b_1 has been purified, and it resembles the cytochrome b_5 of liver microsomes in its properties (Yoshida and Kumaoka, 1969).

Two CO-binding haemoproteins have been demonstrated in anaerobically-grown *S. cerevisiae*, the 450–CO pigment and the 420–CO pigment (Lindenmayer and Smith, 1964). The pigment responsible for the 450 nm maximum in CO-difference spectra was shown to be very similar to the cytochrome *P*-450 of liver microsomes (Ishidate *et al.*, 1969). The 420 nm maximum in these spectra is a composite band, arising from unidentified haemoproteins and cytochromes *c* peroxidase (Ishidate *et al.*, 1969). The 420–CO pigment also shows absorption maxima in CO-difference spectra at 570–575 nm and 540–542 nm, features also characteristic of protohaem (Lemberg and Legge, 1949), yeast haemoglobin (Keilin, 1953), "cytochrome *o*" which is sometimes present in aerobically-grown yeast (Mok *et al.*, 1969) and a degradation product of cytochrome *P*-450 (Omura and Sato, 1964).

Of these pigments, only cytochromes b_1 and *P*-450 have been shown to be membrane-associated (Hebb and Slebodnik, 1958; Chaix, 1961;

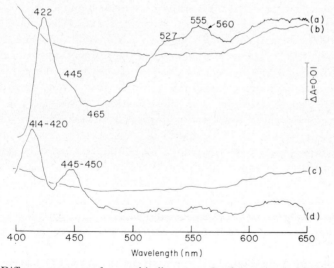

FIG. 59. Difference spectra of anaerobically-grown *Saccharomyces carlsbergensis*. Cells were grown to a density of 2×10^8/ml, harvested in an atmosphere of nitrogen, and resuspended at a density of $6 \cdot 7 \times 10^8$ cells/ml. Spectra were recorded at room temperature. (a) Oxidized-oxidized; (b) dithionite-reduced-oxidized; (c) dithionite-reduced-dithionite-reduced; (d) dithionite reduced + CO-dithionite-reduced. (Reproduced with permission from Cartledge and Lloyd, 1972b.)

Lindenmayer and Smith, 1964; Clark-Walker and Linnane, 1967;
Ishidate *et al.*, 1969; Criddle and Schatz, 1969; Watson *et al.*, 1970,
1971); both pigments can be reduced by NADH or NADPH (Ishidate
et al., 1969).

Glucose-repressed anaerobically-grown *S. carlsbergensis* show spectra
similar to those of *S. cerevisiae* except that cytochrome a_1 is not detectable
(Fig. 59; Cartledge and Lloyd, 1972b; Cartledge *et al.*, 1972). Cytochrome b_1 was detected in the mitochondrial fraction from these cells and
was reducible by NADH or NADPH (Cartledge and Lloyd, 1972b).

(*b*) *Mitochondrial enzyme activities in homogenates of anaerobically-*
grown yeasts. The literature abounds with controversies over the levels
of activities of respiratory enzymes found in homogenates of anaerobically-grown yeasts. Somlo and Fukuhara (1965) have demonstrated that,
contrary to earlier reports (e.g. Tustanoff and Bartley, 1964a, b; Lindenmayer and Smith, 1964), cytochrome *c* oxidase is not detectable in cells
grown with 5% glucose or 5% galactose when rigorous precautions are
taken, (a) to remove traces of oxygen from the commercially available
nitrogen used to sparge cultures and, (b) to inhibit respiratory adaptation
during harvesting. The formation of enzymes under conditions of strict
anaerobiosis is also influenced by the lipid composition of the medium,
and to some extent by catabolite repression. There is now general agreement that succinate dehydrogenase is present in homogenates of anaerobically-grown yeasts (Hebb *et al.*, 1959; Schatz, 1965; Criddle and
Schatz 1969; Watson *et al.*, 1970) as well as fumarate reductase, (Hauber
and Singer, 1967). The level of activity of the primary dehydrogenase (as
measured using PMS as electron acceptor, Hebb *et al.*, 1959), is higher
after lipid-supplemented anaerobic growth than after lipid-depleted
anaerobic growth (Criddle and Schatz 1969; Watson *et al.*, 1970).
Succinate-ferricyanide reductase and succinate-cytochrome *c* reductase
activities are so low as to be hardly detectable after strictly anaerobic
growth (Slonimski, 1953a, b; Somlo and Fukuhara, 1965; Criddle and
Schatz, 1969; Cartledge and Lloyd, 1972b). Both (D)- and (L)-lactate-
cytochrome *c* oxidoreductases are repressed to low levels of activity
(Somlo and Fukuhara, 1965); the enzyme catalysing the oxidation of the
L-isomer has been reported to be undetectable in anaerobically-grown
glucose-repressed *S. carlsbergensis* (Cartledge and Lloyd, 1972b). NADH-
and NADPH-cytochrome *c* oxidoreductases of anaerobically-grown yeasts
are antimycin A-insensitive; the activity of the NADPH-linked enzyme
complex is as high in homogenates of anaerobically-grown glucose-repressed cells as in those from glucose-derepressed aerobically grown cells
(Cartledge and Lloyd, 1972a, b). The activities of some tricarboxylic acid
cycle enzymes (citrate synthase, malate dehydrogenase, fumarase and

aconitase) are repressed after anaerobic growth (Duntze et al., 1969). The level of activity of particulate malate dehydrogenase is also influenced by catabolite repression and lipid-supplementation during anaerobic growth (Watson et al., 1970). Both the specific activity of ATPase and its sensitivity to inhibition by oligomycin is lower in homogenates of anaerobically-grown cells than in those from aerobically-grown cells (Schatz, 1965; Somlo, 1968; Criddle and Schatz, 1969). Cytochrome c peroxidase is detectable in anaerobically-grown S. cerevisiae (Lindenmayer and Smith, 1964; Avers, 1967; Kawaguchi et al., 1969) despite an earlier report to the contrary (Chantrenne, 1955).

(c) Biochemical properties of fractions containing mitochondria-like organelles. Subcellular fractionation of homogenates of anaerobically-grown S. cerevisiae obtained by mechanical disruption revealed the presence of a number of different particulate populations (Schatz, 1965). By flotation of the crude particulate fraction through a Urografin gradient, resolution into four zones was achieved. The most prominent band at $\rho = 1\cdot15$ g/cm^3 accounted for 35–55% of the protein, 80–100% of the succinate-PMS oxidoreductase, 50–70% of the NADPH-ferricyanide oxidoreductase and 10–30% of the NADH-ferricyanide oxidoreductase of the crude particulate fraction. This band also contained ferrochelatase and oligomycin-sensitive ATPase. On a sucrose gradient (containing 20 mM–tris–SO$_4$ at pH 7·4, and 2 mM–EDTA) this band sedimented to an equilibrium density of $1\cdot16$–$1\cdot18$ g/cm^3 (Criddle and Schatz, 1969). The identity of the ATPase of these particles was confirmed as a mitochondrial F$_1$–ATPase, as it was sensitive to the naturally occurring F$_1$–ATPase inhibitor of Pullman and Monroy (1963) and to a specific antiserum against purified F$_1$–ATPase from aerobic yeast mitochondria (Schatz et al., 1967). The ATPase associated with the "promitochondria" of wild-type yeast is cold stable and inhibited by concentrations of oligomycin similar to those blocking ATPase of aerobic yeast mitochondria. The total amount of F$_1$–ATPase per mg of homogenate protein was one-third to one-half of that found in homogenates of aerobically-grown cells for cells grown anaerobically on 10% glucose in the presence of lipid supplements: homogenates of cells grown in the absence of lipid supplements showed about a sixth of the F$_1$–ATPase of the homogenates of aerobically-grown cells. Lipid depletion during anaerobic growth decreases the succinate-PMS reductase activity of the "promitochondria" by about ten fold. Criddle and Schatz (1969) concluded that, regardless of the composition of the growth medium, the "promitochondria" lacked an integrated electron transport chain as they were incapable of oxidizing succinate with cytochrome c or oxygen as acceptors. Whereas in the mitochondria of aerobically-grown cells the activity of succinate-ferricyanide reductase

is almost as high as that of succinate dehydrogenase, and is antimycin A-sensitive, the "promitochondria" showed a barely-detectable level of activity of succinate-ferricyanide reductase. After anaerobic growth for 8 generations on a synthetic medium containing 10% glucose, unsaturated lipids and sterols, no membrane bound succinate-PMS oxidoreductase was detectable ($<5 \times 10^{-9}$ mol/mg protein/min) and no membrane-bound flavoproteins were found ($<10^{-11}$ mol/mg protein) (Goffeau, 1969). Loss of the atractyloside binding capacity and ADP binding characteristic of the mitochondrial adenine-nucleotide transporter found in the mito-chondria obtained from aerobically-grown cells has also been observed for the particles from anaerobically-grown cells (Lauquin and Vignais, 1973).

Spectral examination of fractions containing "promitochondria" indi-cated the presence of cytochrome b_1 which was dithionite- and NADH-reducible, but not reducible on the addition of succinate. A flavoprotein (+ non haem iron) absorption at 450 nm was produced on reduction with succinate or NADH. The percentage of the total protein of the homogenates represented by promitochondrial protein was similar to that accounted for by the mitochondria of aerobically-grown cells under similar conditions of glucose repression, i.e. 3–4% after growth with 10% glucose and lipid supplements, about 10% after growth with 0·8% glucose plus lipids, and 2–3% after growth with 10% glucose in the absence of lipid supplements. Thus Criddle and Schatz (1969) were able to discount the possibility that the "promitochondria" represent degener-ate remnants of mitochondria of the aerobic inoculum. The "promito-chondria" also contained "structural" protein which was immunologically and electrophoretically similar to that of the "structural" protein of mitochondria of aerobically-grown yeast. Analysis of the promitochondrial DNA indicated the presence of about 5 µg of DNA per mg of particle protein (a similar amount to that present in the mitochondrial fraction of aerobically-grown cells).

The "promitochondria" differ from mitochondria of aerobically-grown yeast in their lipid complement (Table XIX; Paltauf and Schatz, 1969). They are particularly low in ergosterol and have an unusual fatty acid composition (Table XX). In promitochondrial fractions from cells grown in the presence of Tween 80 and ergosterol, oleic acid accounts for 90% of the total unsaturated fatty acids. Promitochondrial fractions from cells grown in the absence of a lipid supplement exhibit an extremely low degree of unsaturation and a high content of short chain ($<C_{14}$) saturated fatty acids. Regardless of the lipid composition of the growth medium, promitochondria contain all of the major phospholipid components typical of aerobic yeast mitochondria, although the relative proportion of phosphatidyl inositol and phosphatidyl ethanolamine are significantly

different. Thus although the lipid moieties of promitochondria and mitochondria are quite dissimilar, they share some important common features. The sterol content of the membranes of the anaerobic mito- chondria-like particles can be manipulated, and results indicate that membrane-bound enzymes are profoundly influenced by membrane lipid composition (Cobon and Haslam, 1973). The temperature at which a transition in the Arrhenius plots for ATPase occurs increased by up to $7°C$ as sterol content was decreased.

Subcellular fractionation of gently disrupted sphaeroplasts of glucose- repressed anaerobically-grown *S. carlsbergensis* indicates the presence of a diversity of particles containing NADH- and NADPH-cytochrome c oxidoreductases and ATPases (Fig. 60) (Cartledge and Lloyd, 1972b). As it is not possible to achieve complete separation of these different populations on an equilibrium density basis on sucrose gradients, it is evident that the promitochondrial fractions used for biochemical investi- gations of the properties of these mitochondrial counterparts are heavily contaminated by other membraneous organelles. These results are in agreement with those of Goffeau (1969). The presence of 2 mM–$MgCl_2$ during cell breakage and density-gradient centrifugation helps to maintain the integrity of these organelles.

The promitochondria of anaerobically-grown yeast as seen in freeze- etched preparations resemble the mitochondria of aerobically-grown yeast in that they have both outer and inner membranes regardless of lipid- depletion or lipid-supplementation during anaerobic growth (Plattner and Schatz, 1969). However Watson *et al.* (1970, 1971) have evidence that the structural organization of the inner membrane is less complex in the promitochondria of lipid-depleted cells than in those from cells grown with Tween 80 and ergosterol.

Although promitochondria lack the components of the respiratory chain (cytochromes $a + a_3$, b, c_1, c, UQ and iron-sulphur proteins with a $g = 1·94$ signal at $77°K$) they do possess a mitochondrial energy- transfer system (Groot *et al.*, 1971). Thus they catalyse a Pi–ATP exchange reaction at about 25% of the rate of mitochondria. This reaction in promitochondria is inhibited by the uncoupler 1799 as well as by ruta- mycin, atractyloside and by valinomycin $+ K^+$, indicating that the promitochondria possess an adenine nucleotide translocator and a K^+ pump. As no UQ was detectable in these promitochondria, the possibility of the participation of a phosphorylated ubiquinone compound in ATP synthesis (Parson and Rudney, 1966) was excluded. An $[^{18}O]$ H_2O–Pi exchange, which was inhibited by rutamycin and the uncoupler 1799, showed an activity of about 20% of that seen in mitochondria from aerobically-grown yeast. Groot *et al.* (1971) were also able to demonstrate an energy-linked decrease in the fluorescence of ANS bound to the

G

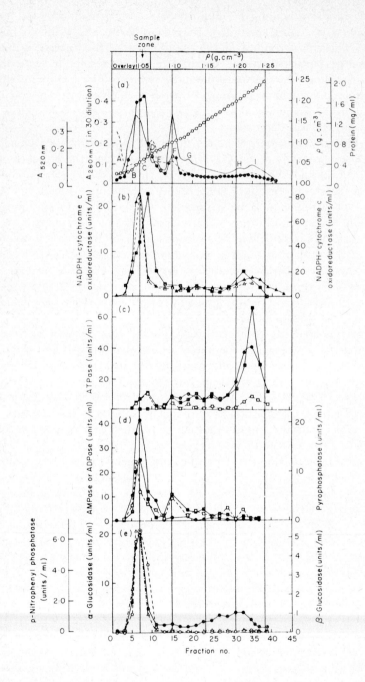

promitochondrial membranes on the addition of ATP, and its reversal by uncoupler (Fig. 61). The promitochondrial membranes are impermeable to protons and K^+ ions. Thus, at below pH 6·0, an acid pulse gave only slow equilibration of H^+ and addition of uncoupler produced little effect (Fig. 62). Fast proton equilibration was observed only after the membrane was made permeable to K^+ by the addition of valinomycin. The low K^+ permeability of promitochondria was also shown by their slow swelling in isotonic potassium acetate in the absence of valinomycin.

2. Effects of low oxygen tensions on mitochondria in organisms other than yeasts

(a) Effects of low oxygen tensions on Aspergillus oryzae. Changes in the cytochrome absorption bands of *A. oryzae* under various growth conditions were noted by Tamiya (1928). Mycelia produced by surface growth for 3–4 days possessed the full complement of mitochondrial cytochromes; the absorption bands were less distinct after 7 days growth, and were not detectable after 30 days. When grown in submerged culture, disappearance of the cytochromes was complete within 10 h; this loss of the cytochromes was reversible. Aerobic culture for 1 day was sufficient to restore the cytochromes to hyphae grown for 5 days under submerged culture conditions. There was a strict parallelism between the respiratory activity and intensity of the cytochrome absorption bands in mycelia produced under different growth regimes; the anaerobic organisms showed absorption bands at 555–564 nm and 530–535 nm on reduction, which were attributed to haemochromogens.

FIG. 60. Fractionation of a whole homogenate of anaerobically-grown *Saccharomyces carlsbergensis*. Whole homogenate (22·5 ml) containing 131 mg of protein was loaded on to the gradient. The volumes of homogenate (diluted 1 in 20) and of fractions taken for assay were as follows: catalase, cytochrome *c* oxidase and succinate-, D(−) lactate-, L(+)-lactate-, NADH- and NADPH-cytochrome *c* oxidoreductases, all 0·1 ml; acid *p*-nitrophenyl phosphatase acid α- and β-glycosidases, ATPase, AMPase, ADPase and pyrophosphatase all 0·2 ml. Centrifugation was at 35,000 rev/min for 165 min (6 × 10⁶ g-min at the sample zone; $\int_0^t \omega^2 . dt = 1·45 \times 10^{11}$ rad²s⁻¹). Fractions (10 ml) were collected at 5°C. (a) Sucrose density gradient (○), light-scattering at 520 nm (- - - -), 260 nm (——) and protein (●); (b) activities of NADH-cytochrome *c* oxidoreductase (■) NADPH-cytochrome *c* oxidoreductase (▲) and antimycin A-insensitive NADPH cytochrome *c* oxidoreductase (△); (c) activities of ATPase (■) oligomycin-insensitive ATPase (□) and ouabain-insensitive ATPase (●); (d) activities of AMPase (●) ADP (■) and pyrophosphatase (□); (e) activities of acid *p*-nitrophenyl phosphatase ●), α-glucosidase (○) and β-glucosidase (△). Specific activities of enzymes in the whole homogenate were as follows (recoveries in parentheses); protein (97%), NADH-cytochrome *c* oxidoreductase (137%), NADPH-cytochrome *c* oxidoreductase 59·5 (77%), ATPase 46·9 (75%) AMPase 14 (88%), ADPase 15·3 (81%), pyrophosphatase 8·25 (67%) acid *p*-nitrophenyl phosphate 2·4 (81%), acid α-glucosidase 6·8 (98%) and acid β-glucosidase 1·16 (129%). (Reproduced with permission from Cartledge and Lloyd, 1972b.)

Table XIX. Lipid and phospholipid composition of mitochondria and "promitochondria" of *S. cerevisiae*. (Reproduced with permission from Paltauf and Schatz, 1969.)

A. Lipids

Particle preparation	Neutral lipid[a]	Phospholipid (mg/mg of protein[b])	Ergosterol (μg/mg of protein)
Mitochondria (from cells grown in presence of added lipids)	65	0·23	30
Promitochondria (from cells grown in presence of added lipids)	63	0·34	8
Promitochondria (from cells grown without added lipids)	60	0·24	<5

[a] Expressed as micrograms of fatty acids per milligram of protein bound to neutral lipids. [b] Determined as phospholipid phosphorus and calculated on the basis of 40 μg of phosphorus/mg of phospholipids.

B. Phospholipids

Particle preparation	% of total phospholipid phosphorus						
	Cardiolipin	Phosphatidyl-ethanolamine	Phosphatidyl-serine	Phosphatidyl-inositol	Lysophosphatidyl-ethanolamine + Lysophosphatidyl-serin	Lysophosphatidyl-choline	Phosphatidyl-choline
Mitochondria (from cells grown in presence of lipids)	10·9	30·6	4·2	8·1	4·5	3·2	38·5
Promitochondria (from cells grown in presence of lipids)	6·2	19·3	10·0	12·6	4·1	0·3	47·5
Promitochondria (from cells grown without added lipids)	8·9	17·9	3·9	26·0	8·2	0·8	34·3

a Not determined.

Table XX. Fatty acid composition of total phospholipids and neutral lipids of mitochondria and "promitochondria" of *S. cerevisiae*. (Reproduced with permission from Paltauf and Schatz, 1969.)

A. Phospholipids

Particle preparation	Wt % of total fatty acids							
	C_{10}	C_{12}	C_{14}	C_{16}	$C_{16:1}$	C_{18}	$C_{18:1}$	$C_{20:1}$
Mitochondria (from cells grown in presence of added lipids)	tr	tr	0·6	17·9	43·7	3·6	34·2	tr
Promitochondria (from cells grown in presence of added lipids)	tr	tr	4·5	20·5	6·5	3·9	61·5	3·1
Promitochondria (from cells grown without added lipids)	14·3	8·9	10·4	33·7	12·0	13·7	7·0	tr

B. Neutral Lipids

Particle preparation	Wt % of total fatty acids							
	C_{10}	C_{12}	C_{14}	C_{16}	$C_{16:1}$	C_{18}	$C_{18:1}$	$C_{20:1}$
Mitochondria (from cells grown in presence of lipids)	tr	tr	1·2	16·1	43·1	3·6	36·0	tr
Promitochondria (from cells grown in presence of lipids)	tr	tr	4·9	18·0	2·1	10·3	63·0	1·7
Promitochondria (from cells grown without added lipids)	8·5	10·5	14·4	34·0	7·7	17·5	6·4	1·0

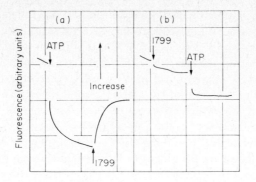

Fig. 61. Effect of ATP and the uncoupler 1799 on the fluorescence of ANS in the presence of a "promitochondrial" fraction of (strain DT-X11) of *Saccharomyces cerevisiae*. At the arrows ATP and 1799 were added at a final concentration of 2 mM and 10 μM, respectively. (Reproduced with permission from Groot *et al.*, 1971.)

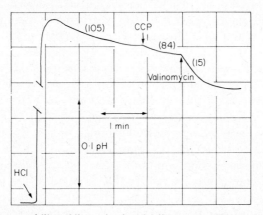

Fig. 62. Proton impermeability of "promitochondria" of strain DT-X11 of *Saccharomyces cerevisiae*. A suspension containing 5 mg of promitochondrial protein in 400 mM–mannitol, 150 mM–KCl, and 3·3 mM–glycylglycine (pH 6·5) was incubated for 10 min. The final volume was 2·5 ml, the final pH 5·9. At the first arrow, 825 nmol of HCl was added and the change of the pH was monitored. During the passive proton equilibration the uncoupler CCCP (4 μM final concentration) and valinomycin (1 μg/ml) were added at the arrows. The figures between parentheses are the half-times in seconds for the passive equilibration of protons. (Reproduced with permission from Groot *et al.*, 1971.)

(*b*) *Anaerobic growth of* Mucor sp. The ability to grow under strictly anaerobic conditions is a property peculiar to only a few fungi other than yeasts (Tabak and Cooke, 1968). Some species of *Mucor* are facultative anaerobes which exhibit dimorphism, and exhibit a yeast-like morphology when grown anaerobically or in the presence of 2-phenyl ethanol (Terenzi and Storck, 1969a, b). The respiration of *M. genevensis* has been studied during aerobic and anaerobic growth, and during the transition produced on aeration of anaerobic cultures (Clark-Walker, 1972). The respiration of cells from a culture grown aerobically in the presence of 1% glucose was completely sensitive to inhibition by 1 mM–KCN or 1 μg of antimycin A/ml; glucose exhaustion from the medium was not accompanied by any increase in respiration rate. Anaerobically-grown cells respired at about one-fifth of the rate of aerobically-grown organisms and the respiration was insensitive to KCN or antimycin A. Respiratory adaptation occurred on aeration of anaerobically-grown cells at similar rates in growth media containing 0·1% or 10% glucose or 2% glycerol. Thus, in this organism, the glucose-repression effect seen in yeasts (p. 196) is not pronounced. During adaptation in either of the three media, growth of the yeast-like cells into hyphae occurred, but the change in morphology was retarded by 10% glucose. Absorption spectra of the anaerobically-grown cells showed the presence of pigments absorbing at 623, 596, 560, 530 and 499 nm. Respiratory adaptation for 4 h produced organisms with a spectrum similar to that of aerobic mycelia, which showed typical mitochondrial cytochromes. Electron microscopic examination indicated that the mitochondria of the anaerobic *M. genevensis* still contain highly organized cristae. Clark-Walker (1972) suggested that, whereas in various species of yeast, mitochondrial structure and function is repressed by both anaerobiosis and glucose, in the mould lack of oxygen is a much more significant factor than glucose concentration.

(*c*) *Facultative anaerobiosis in* Neurospora crassa. An examination of twelve wild-type strains of *N. crassa* showed that none were capable of growth under anaerobic conditions, but it it possible to induce mutants which are facultatively anaerobic (Howell *et al.*, 1971). Genetic analysis suggests single nuclear gene inheritance of this property, and the mutants were indistinguishable from the wild-type in morphology and growth rate under aerobic conditions. Anaerobic growth (at oxygen concentrations less than 0·2% air saturation) required undefined nutritional supplements (other than ergosterol and Tween 80) which were not required for aerobic growth. Anaerobiosis, or the addition of chloramphenicol to the growth medium, both affected mitochondrial enzyme activities and morphology. Anaerobiosis led to reduction in both cytochrome *c* oxidase and malate dehydrogenase activities and produced enlarged mitochondria with few cristae.

(*d*) *Effects of decreased aeration on the electron transport system of* Tetrahymena pyriformis. Growth of poorly-aerated cultures of *T. pyriformis* results in an increased ratio of cytochromes *b* and *c* to *a* (Shrago *et al.*, 1971). This mitochondrial modification has not been studied in isolated organelles.

3. Effects of growth at high partial pressures of oxygen

Exposure of *Astasia longa* to $O_2 + CO_2$ (19:1) at atmospheric pressure leads to an inhibition of growth rate and of respiration (Bégin-Heick and Blum, 1967). Growth resumes at the normal rate as soon as oxygenation is discontinued, but respiration recovers more slowly. Mitochondria prepared from cells exposed to high partial pressures of O_2 during growth have considerably decreased activities of succinate-cytochrome *c* oxidoreductase, NADH-cytochrome *c* oxidoreductase, succinate dehydrogenase and succinate oxidase activities as compared with mitochondria isolated from cells exposed to air $+ CO_2$ (19:1). The cytochrome *c* oxidase activity of the mitochondria is not appreciably inhibited by exposure of the cells to 95% O_2. The mitochondrial content of ergosterol-containing compounds and rhodoquinone is increased by exposure to 95% O_2, whereas the UQ_9 content is not changed. Similar inhibition of respiration after growth at high O_2 concentrations is observed in cultures of streptomycin-bleached *E. gracilis*, and in this case phosphate-deprivation produced a sensitizing effect (Blum and Bégin-Heick, 1967).

B. Respiratory Adaptation in Yeasts

1. Development of respiratory activities; changes in cytochrome spectra and enzyme activities

Oxygen itself is known to take part in the synthesis of haem (Jacobs *et al.*, 1967) as well as that of ergosterol and the unsaturated fatty acids (Kováč *et al.*, 1967b; Jollow *et al.*, 1968; Paltauf and Schatz, 1969) which are needed for the formation of mitochondria membranes (Wallace *et al.*, 1968; Plattner and Schatz, 1969). It has been suggested that oxygen itself may also be an inducer of respiratory enzyme synthesis (Ephrussi and Slonimski, 1950a, b; Lindenmayer and Estabrook, 1958; Somlo and Fukuhara, 1965).

Respiratory adaptation in yeast was demonstrated by Fink and Berwald (1933) and Fugita and Kodama (1934) who showed that repeated subculture of brewer's yeast under highly aerobic conditions led to a modification of the absorption spectrum. These experiments did not exclude the possibility of a gradual selection of mutant organisms, but Chin (1950) was able to show a transformation of the cytochrome spectrum of anaerobically-grown brewer's yeast into that typical of aerobic yeasts during the aeration of non-growing cell suspensions. Similar results were obtained by Ephrussi and Slonimski (1950a, b) with baker's yeast. Using cells which had been grown in anaerobic cultures without the addition of ergosterol and oleic acid, Slonimski (1953a, b; 1956) showed that the cytochrome *c* content of the cells, as well as the cytochrome *c* oxidase activity and rate of oxygen uptake increased linearly from the beginning of aeration. Lindenmayer and Estabrook (1958), using cells harvested from cultures grown anaerobically in the presence of lipid-supplements, found that stationary-phase cells gave a similarly linear adaptation, whereas cells from the start of aeration, during which oxygen-uptake rates changed very little during adaptation in phosphate buffer containing 3% glucose. A comparison of spectra of cells (recorded at the temperature of liquid nitrogen) from the anaerobic cultures with those of cells at different stages of adaptation enabled Lindenmayer and Estabrook (1958) to study the kinetics of cytochrome changes during the process. Whereas unadapted cells contained cytochrome "b_1" (λ_{max} 557·5 nm and 551 nm) and cytochrome "a_1" (585–590 nm), progressive development of cytochrome b (559 nm), c (548 nm) and c_1 (554 nm) was observed over 150 min of respiratory adaptation of cells harvested from stationary-phase; cells from exponentially-growing cultures again showed a lag in the development of cytochromes of the aerobic pattern. The kinetics of the process suggested a possible precursor-product relationship between cytochrome b_1 and the mitochondrial cytochromes. The influence of growth conditions on subsequent appearance of succinate dehydrogenase and cytochrome c oxidase during respiratory adaptation was also studied by Hebb and Slebodnik (1958). It has been suggested that a component (which may bind with molecular oxygen) necessary for induction of cytochrome synthesis may accumulate in the stationary-phase of growth and make possible immediate adaptation on subsequent aeration (Lindenmayer and Estabrook, 1958). Cells grown aerobically with galactose adapt more quickly than those grown with glucose (Linnane, 1965; Tustanoff and Bartley, 1964a, b).

These results with non-proliferating cell suspensions show some similarities with the development of cytochromes in growing cultures, where the composition of mitochondrial membranes is influenced by the conditions of aeration. In *S. cerevisiae* the molar ratios of cytochromes a_3: a: b: $c + c_1$ under conditions of "low" aeration were 0·7: 1·0: 4·1: 2·8, whereas conditions of "high" aeration gave ratios of 0·9: 1·0: 1·0: 1·0 (Biggs and Linnane, 1963).

Linear rates of induction of both cytochrome c oxidase and haem a were obtained over the first 120 min of respiratory adaptation of *S. carlsbergensis* after anaerobic growth with 3% galactose followed by an anaerobic "step-down period" in the adaptation medium which contained galactose, glycerol and phosphate buffer (Chen and Charalampous, 1969). Under these conditions linear adaptation was observed for cells from either the exponential or stationary phases of growth. The synthesis of haem a was completed about 60 min before maximum activity of cytochrome c oxidase was attained. The levels attained were only about one-third of those seen in aerobic cells.

The respiration of glucose-derepressed anaerobically-grown *S. carls-bergensis* is antimycin A-insensitive and only 30% inhibited by 1 mM–KCN (Cartledge and Lloyd, 1973). Cytochrome *c* oxidase activity was not de-tectable in homogenates prepared from these cells ($<0·05$ nmol/mg protein/min). Respiratory adaptation for 1 h led to the development of respiration which was more than 80% inhibited by antimycin A or KCN and the appearance of cytochrome *c* oxidase and antimycin A-sensitive NADH-cytochrome *c* oxidoreductase activities (Fig. 63).

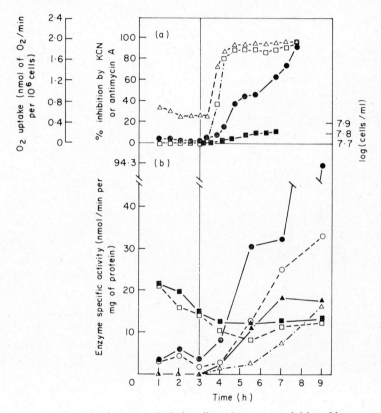

Fig. 63. Changes in respiration rates of whole cells and enzyme activities of homogenates during glucose-derepression and respiratory adaptation of anaerobically-grown *Sacc-haromyces carlsbergensis*. The period (0–1 h) indicates the time taken to harvest (at 6×10^7 cells/ml) and resuspend cells in adaptation medium; 1–3 h is the period of anaerobic derepression and 3–9 h is the 6 h of adaptation. (a) Conc. of cells (log (cells/ml), ■), O₂ uptake (nmol of O₂/min per 10⁶ cells, ●), and % inhibition of respiration in the presence of 2 μg of antimycin A/ml (□) of 1 mM–KCN (△); (b) specific activities of cytochrome *c* oxidase (▲), catalase (△), NADH-cytochrome *c* oxidoreductase (◉), antimycin A-insensitive NADH-cytochrome *c* oxidoreductase (○), NADPH-cytochrome *c* oxidoreductase (■) and antimycin A-insensitive NADPH cytochrome *c* oxidoreductase (□). (Reproduced with permission from Cartledge and Lloyd, 1973.)

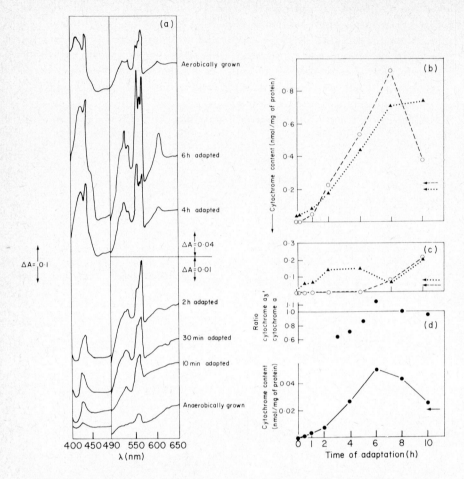

FIG. 64. Cytochromes of mitochondrial fractions and supernatant fractions of homogenates prepared from sphaeroplasts of *Saccharomyces carlsbergensis* at stages during respiratory adaptation. (a) Reduction was achieved by adding excess dithionite to the sample cuvette contents. Oxidation of the reference suspensions was by aeration; the cuvettes were then immersed in liquid N_2 and spectra recorded at 77°K. The path length throughout was 2 mm and spectral band width 1 nm. Protein concentrations were 15 mg/ml throughout, apart from that for the mitochondrial suspension from anaerobically-grown cells, which was 10 mg/ml. (b) Total dithionite reducible cytochrome *b* (▲) and ascorbate + TMPD reducible cytochrome *c* (○) of the mitochondrial fractions. (c) Total dithionite-reducible cytochromes *b* ▲ and *c* (○) of the supernatant fractions. (d) Cytochrome *a* + a_3 content and ratio cytochrome a_3/cytochrome *a* of mitochondrial fractions. Cytochrome content was calculated from changes in E observed at 445–458 nm in flow-flash photolysis experiments. Ratios cytochrome a_3/cytochrome *a* were obtained by measuring changes in absorption in anaerobic photolysis (contribution of cytochrome a_3) and photolysis in the presence of 17 μm–O_2 (contribution of cytochromes *a* + a_3). Arrows indicate values for mitochondrial fractions from cells grown aerobically to the phase of glucose-derepression. (Reproduced with permission from Cartledge *et al.*, 1972.)

Mitochondrial fractions isolated at various stages over the first 4 h of respiratory adaptation contained all the c- and a-type cytochromes of the lysed sphaeroplasts, but at later times in adaptation it was more difficult to extract intact mitochondria (Cartledge et al., 1972). Increases in a-, b- and c-type cytochromes occur at different rates so as to produce mitochondria with different ratios of these cytochromes at different stages in adaptation (Fig. 64).

After 1 h the proportions of total a-: b-: c-type cytochromes in mitochondrial fractions was 1:29:19, whereas after 6 h adaptation these proportions were 1:14:19. These ratios may be compared to those of mitochondrial fractions from aerobically-grown glucose-derepressed cells (1:8:8). Despite these marked changes in composition, no evidence was obtained from measurements of kinetics of cytochrome oxidation (Table XXI) that extensive cytochrome dislocation had occurred or that pools of non-functional cytochromes had accumulated. Thus the formation of fully-functional respiratory chains seems to occur as individual components are synthesized. The low mitochondrial content of cytochromes $a + a_3$ after 30 min adaptation made the resolution of these two components impossible; the half-time for oxidation after aerobic photolysis of the CO-cytochrome a_3 complex was 50 ms. After 4 h adaptation this reaction was resolved into a rapid ($t_\frac{1}{2} = 0·3$ ms) and slower phase ($t_\frac{1}{2} = 4$ ms), corresponding to the oxidation of cytochromes a_3 and a respectively. Over a similar period of adaptation the half-time for the oxidation of cytochrome c decreased from 100 ms to 25 ms (Table XXI; Cartledge et al., 1972).

Changes in CO-binding haemoproteins other than cytochrome a_3 have also been observed in yeasts during respiratory adaptation. Thus the haemoprotein P-450 of anaerobically-grown S. cerevisiae decreases until after 1 h of aeration it becomes undetectable, (Ishidate et al., 1969). A poorly defined haemoprotein of unknown function, "cytochrome P-420", showed an increased concentration in mitochondrial fractions over an 8 h period of adaptation of S. carlsbergensis (Cartledge et al., 1972).

The development of iron-sulphur proteins detectable by EPR spectroscopy occurs during respiratory adaptation of S. carlsbergensis (T. Ohnishi, T. G. Cartledge and D. Lloyd, unpublished results). SMPs from anaerobically-grown, glucose-derepressed cells contained no detectable iron-sulphur protein with a "$g = 1·94$" signal. On respiratory adaptation, succinate dehydrogenase iron-sulphur protein (Fe-S_{s-1}), (FeS_{s-2}, $g = 2·03$, 1·94 and 1·92), Centre 5 ($g = 2·09$, 1·89) and Rieske's iron-sulphur centre ($g = 2·03$, 1·90, 1·80) all become evident after 2 h aeration, but are hardly detectable at 7°K after 30 min of adaptation.

2. Inhibition of respiratory adaptation by protein synthesis inhibitors

The respiration rate of anaerobically-grown yeast does not increase if

Table XXI. Kinetics of cytochrome reactions in mitochondria from *S. carlsbergensis* at different stages in respiratory adaptation. (Reproduced with permission from Cartledge et al. (1972).)

Cytochrome	Half-times (ms) for oxidation in mitochondrial suspensions									Cells grown aerobically
Adaptation time of cells...	30 min	1 h	2 h	3 h	4 h	5 h	6 h	7 h	8 h	
a_3	~50	~50	~20	0·5	0·3	0·5	(<7)	0·35	0·25	0·3 (0·3)
a	~100	<100	30	20	4	6	(15)	3·7	3	5 (15)
c		<100		20	25	6·5	(50)	10	<15	(<100)
c_1		200		45	25		(70)		30	
b_{560}		No oxidation		150		100	Reduction			
b_{566}				500		>200	90			

Respiratory substrates were 10 mM-ascorbate + 20 μM-TMPD in experiments with mitochondria from cells adapted for 30 min, 1 h and 6 h and with those from aerobically grown organisms. In all other cases substrate was 6 mM-sodium succinate + 3 mM-sodium malonate. Temperature of incubation was 24°C except for values shown in parentheses, where experiments were at −3°C in the presence of 20% (v/v) ethanediol.

chloramphenicol or acriflavine is present throughout the period of aeration; cycloheximide also inhibits respiratory adaptation and sometimes leads to a decrease in oxygen-uptake rates (Kováč *et al.*, 1967b). The synthesis of cytochrome *c* oxidase is inhibited when acriflavine is added to non-growing cell suspensions during respiratory adaptation (Slonimski, 1953a, b; Nagao and Sugimura, 1965); actinomycin D inhibits only if added 2 h before the commencement of aeration (Asano, 1972). The effects of protein synthesis inhibitors on the formation of haemoprotein *a* and cytochrome *c* oxidase activity have been investigated by Chen and Charalampous (1969). In this study it was shown that cells adapted for 25 min continue to form cytochrome *c* oxidase when aeration is stopped. Cycloheximide added before the commencement of aeration prevents the expression of this ability, whereas chloramphenicol or acriflavine are without effect. Separate addition of one of these inhibitors prior to aeration allows the formation of different amounts of haemoprotein *a* and cytochrome *c* oxidase activity, and under these conditions the amount of inhibitor-resistant synthesis of haemoprotein *a* is greater than that of cytochrome *c* oxidase activity. When cycloheximide is added less than 1 h after the start of aeration, cytochrome *c* oxidase development is not completely blocked, but if chloramphenicol is also added no further increase in enzyme activity occurs. Cells adapted for periods longer than 80 min can however form additional cytochrome *c* oxidase when further incubated in the presence of chloramphenicol, and this capacity is cycloheximide-sensitive. In these experiments a difference in the kinetics of appearance of haemoprotein *a* and cytochrome *c* oxidase suggested the occurrence of several possible steps in the expression of this activity, i.e. the asynchronous production of cytochromes *a* and a_3, or the requirement for polymerisation or integration into mitochondrial membranes. An increase in the ratio of cytochrome a_3 to cytochrome *a* as adaptation proceeds has been noted by Cartledge *et al.* (1972; Fig. 64) and confirmed by Chen and Charalampous (1973). The results of Chen and Charalampous (1969) strongly suggest that the development of cytochrome *c* oxidase activity during the respiratory adaptation of non-proliferating cells requires the participation of both mitochondrial and extra-mitochondrial protein synthesis systems. Similar results were reported for growing cultures undergoing respiratory adaptation by Yu *et al.* (1968). That accumulation of intermediates occurs during respiratory adaptation carried out in the presence of cycloheximide or chloramphenicol has also been confirmed by Rouslin and Schatz (1969). These results indicated that the promitochondrial synthesis of certain intermediates required for respiratory adaptation could proceed in the absence of cytoplasmic protein synthesis. These intermediates, presumably proteins, had a half-life of about 2 h, but could only be used for adaptation when products of the cytoplasmic protein synthesis were present. One product of the latter system did not accumulate when mitochondrial protein synthesis was blocked.

Benzimidazole (9 mM) almost completely inhibits respiratory adaptation of *S. cerevisiae*, but only inhibits protein synthesis by about 60% (Sels and Verhulst, 1971). The production of cytochrome *c* peroxidase and iso-2-cytochrome *c* occurs during adaptation by insertion of haem groups into the respective apoenzymes which are present in the anaerobically-grown cells (Fukuhara, 1966; Sels and Cocriamount, 1968; Fukuhara and Sels, 1966). Benzimidazole does not interfere with the development of these holoenzymes during adaptation (Sels and Verhulst, 1971). The *de novo* synthesis (Yčas and Drabkin, 1957) of the major species of cytochrome *c* (iso-1-cytochrome *c*) is mediated by the extra-mitochondrial protein synthesis system (Fukuhara, 1966) and the structural information is encoded by a nuclear gene (Sherman *et al.*, 1968): iso-2-cytochrome *c* acts as a repressor of iso-1-cytochrome *c* synthesis (Slonimski *et al.*, 1965). This synthesis is also insensitive to benzimidazole (Sels and Verhulst, 1971). The effect of benzimidazole appears to be particularly on cytochrome *c* oxidase synthesis, which is more than 95% inhibited. Removal of benzimidazole is followed by an increased rate of respiratory adaptation (Slonimski, 1956) and, under these conditions, insensitivity of adaptation to cycloheximide suggests the accumulation of inactive precursors during benzimidazole-inhibition (Sels, 1969). The mechanism of the inhibition of expression of cytochrome *c* oxidase activity is not clear; Sels and Verhulst (1971) were not able to distinguish between the three possible stages at which this could occur i.e. synthesis of the apoenzyme, association of the apoenzyme with its specific prosthetic group, or integration of the holoenzyme into mitochondrial membranes. Activation on removal of benzimidazole is insensitive to inhibitors of translation and is energy-requiring (Sels and Verhulst, 1972). It occurs in two steps: (1) formation of spectroscopically-detectable cytochromes $a + a_3$, and (2) activation to give functional assemblies, and it proceeds at the expense of the double set of cytoplasmic and mitochondrially-synthesized products accumulated in the presence of the inhibitor (Schmitt-Verhulst *et al.*, 1973). Thus benzimidazole acts as a specific inhibitor of cytochrome $a + a_3$ assembly.

Inhibition of respiratory adaptation by CN^-, azide DNP *p*-fluorophenylalanine and 2-phenyl ethanol was demonstrated by Bartley and Tustanoff (1966). In these experiments iodoacetate and NaF were not inhibitory.

Parallel induction of ATPase(s) and respiratory enzymes occurs during the respiratory adaptation of a non-proliferating cell suspension of *S. cerevisiae* (Somlo, 1968). The three-fold increase in specific activity of ATPase(s) which occurred over a 17 h adaptation period did not occur in the presence of cycloheximide, and it was concluded that a *de novo* synthesis (rather than oxygen-activation of pre-existing ATPase(s) without protein synthesis) was involved. Partial restoration of ADP- and carboxy-tractyloside-binding capacity of mitochondria also occurs (Lauquin and Vignais, 1973). As is the case with cytochrome *c* oxidase, L- and D-lactate cytochrome *c* oxidoreductases (Kattermann and Slonimski, 1960; Fukuhara 1965; Somlo 1965, 1966) are also synthesized *de novo* during respiratory adaptation.

The formation of malate dehydrogenase, succinate dehydrogenase and fumarase during respiratory adaptation in cultures of *S. cerevisiae* which had been grown anaerobically with galactose and lipid supplements was sensitive to cycloheximide but not to chloramphenicol (Vary *et al.*, 1969). Lipid-depleted cells aerated for 4 h before addition of chloramphenicol or erythromycin also show the development of all three enzymes, whereas addition of these inhibitors of mitochondrial protein synthesis at the onset of aeration completely blocks the O_2-induced syntheses (Vary *et al.*, 1970). Cytochrome *c* oxidase production was inhibited by erythromycin or chloramphenicol irrespective of the history of the cells prior to antibiotic addition. Cycloheximide inhibited induction of all four enzymes in all cases. A catabolite-derepressed anaerobically-grown *petite* mutant did not form significant amounts of the O_2-inducible enzymes. These results emphasize the complex interrelationships between the biosynthetic processes proceeding in the mitochondria and cytoplasm during respiratory adaptation and indicate that the mitochondrial protein synthesis system (absent from the lipid-depleted anaerobes) is induced first; this system then gives products necessary for the increased cytoplasmic syntheses of the three TCA cycle enzymes.

3. Changes in mitochondrial lipids during respiratory adaptation

The UQ content of mitochondrial fractions obtained from *S. cerevisiae* increased from <0.02 to 2.7 μg/mg protein over a period of 8 h respiratory adaptation, irrespective of the lipid-status of the anaerobically-grown cells (Gordon and Stewart, 1971). During the adaptation of lipid-supplemented cells, synthesis of UQ was barely affected by chloramphenicol, but was inhibited by about 50% by cycloheximide. In lipid-depleted cells both antibiotics gave $60–70\%$ inhibition of UQ synthesis.

The sterol content of yeast is increased during aeration of washed cells in the presence of glucose (Maguigan and Walker, 1940; Klein *et al.*, 1954; Starr and Parks, 1962; Alexander *et al.*, 1965). The kinetics of changes in the content of sterols, total fatty acids and phospholipids in *S. cerevisiae* undergoing respiratory adaptation were followed alongside the development of respiration (Kováč *et al.*, 1967*b*). In these experiments the lipid content typical of anaerobically-grown cells changed gradually to the content characteristic of aerobically-grown yeast; the phospholipid content increased sharply in the first 2 h of adaptation, whilst the total fatty acids and sterols increased in parallel with the respiratory capacity over an 8 h period. In the presence of chloramphenicol or acriflavine the phospholipid and fatty acid contents of the cells still increased during respiratory adaptation but sterol synthesis was considerably inhibited.

Cycloheximide produced no detectable inhibition of phospholipid synthesis, whereas fatty acid synthesis was partially inhibited and sterols showed no increase in the presence of the inhibitor. This correlation between sterol synthesis and the formation of the respiratory apparatus was also demonstrated by Parks and Starr (1963). Adaptation in the presence of DNP, azide or CN^- stimulate sterol production, and the effects of acriflavine or chloramphenicol suggest that a component of sterol formation is a product of protein synthesis on mitochondrial ribosomes which is induced in the presence of O_2 (Adams and Parks, 1969). Changes in the lipid composition of mitochondria isolated from cells at stages in respiratory adaptation have also been observed by Gordon and Stewart (1971). During aeration of lipid-depleted cells, the increase in mitochondrial unsaturated fatty acid content of the mitochondrial fraction occurred concurrently with the decline of saturated fatty acids. The greater part of this increase represents a net increase in total fatty acids, rather than desaturation and elongation of pre-existing saturated fatty acids (Table XXII). Both chloramphenicol and cycloheximide inhibited ergosterol and unsaturated fatty acid synthesis in the mitochondria of lipid-depleted cells by about 60% (Table XXIII). The stability of the lipids of the membranes of mitochondrial structures during aeration was assessed by growing the cells anaerobically with linoleic acid and cholesterol and measuring the rate of replacement of these by normal unsaturated fatty acids (palmitoleic and oleic acids) and ergosterol (Fig. 65). These experiments indicated that a substantial fraction of the unsaturated fatty acids of mitochondrial membranes is conserved during mitochondrial development, whereas a rapid and almost complete replacement of cholesterol by ergosterol occurred in the early stages of adaptation (Gordon and Stewart, 1971).

In a further study (Gordon et al., 1972a) have investigated the coordination between the synthesis of lipids and enzymes during respiratory adaptation. The production of mitochondrial enzymes on exposure to O_2 of lipid-depleted cells is inhibited by D(−)-*threo* chloramphenicol or erythromycin; 50% inhibition was observed at <1 mM, which is similar to the concentrations required to inhibit protein synthesis *in situ*. The synthesis of unsaturated fatty acids, ergosterol and phospholipids induced by aeration was inhibited by high concentrations of D(−)-*threo* chloramphenicol (12 mM) but unaffected by erythromycin. L(+)-*threo* chlor-

Table XXII. Effect of anaerobic growth and aeration conditions on fatty acid composition of the mitochondrial fraction. (Reproduced with permission from Gordon and Stewart, 1971.)

Conditions of growth and aeration	Fatty acid*:								
	8:0	10:0	12:0	14:0	14:1	16:0	16:1	18:0	18:1
LIPID DEPLETED									
Anaerobe	0·6	1·4	7·2	13·8	0·4	54·7	5·7	9·7	6·5
Aerated 8 h	0·1	0·4	1·3	2·4	0·4	18·1	43·5	3·0	30·8
Aerated 8 h + CAP†	0·2	0·8	1·9	3·8	0·9	22·6	40·1	2·8	26·9
Aerated 8 h + CYC†	0·3	1·5	2·7	9·4	0·4	43·0	26·6	8·1	8·0
LIPID SUPPLEMENTED									
Anaerobe	tr‡	2·1	2·5	3·7	0·0	29·4	8·5	18·5	35·3
Aerated 8 h	1·3	0·3	0·6	1·5	0·8	12·9	37·6	3·0	42·1
ANAEROBIC BATCH GROWN	0·7	2·0	2·2	2·5	1·0	15·7	36·8	11·1	28·0

* Fatty acids are denoted by the convention — number of carbon atoms : number of double bonds.
† Chloramphenicol (CAP), 9 mM, or cycloheximide (CYC), 20 µM, were added at beginning of aeration.
‡ tr indicates that the particular fatty acid represented less than 0·1% of the total fatty acid.

Table XXIII. Effect of antibiotics on the lipid composition of mitochondria from anaerobic and aerated cells. (Reproduced with permission from Gordon and Stewart, 1971.)

Conditions of growth and aeration	Fatty acids (µg/mg protein):				Ergosterol (µg/mg protein)
	Short chain (8:0, 10:0, 12:0, 14:0)	Long chain saturated (16:0, 18:0)	Long chain unsaturated (16:1, 18:1)	Total	
LIPID DEPLETED					
Anaerobe	40	113	22	175	<1
Aerated 8 h	11	55	194	260	29
Aerated 8 h + CAP*	13	48	129	190	32
Aerated 8 h + CYC*	10	37	25	72	5
LIPID SUPPLEMENTED					
Anaerobe	3	17	158	178	6
Aerated 8 h	8	33	196	237	18
AEROBIC BATCH GROWN	16	75	159	250	32

Cells were grown anaerobically and aerated in the presence of chloramphenicol or cycloheximide. Mitochondrial fractions were obtained from cell homogenates prepared by the glass-bead method.
* Chloramphenicol (CAP), 9 mM, or cycloheximide (CYC), 20 µM, was added at the beginning of aeration.

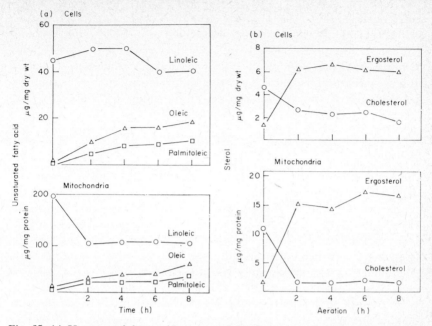

Fig. 65. (a) Unsaturated fatty acid content of *Saccharomyces cerevisiae* cells and mito-chondria from these cells grown anaerobically on linoleic acid and cholesterol and aerated in the absence of these lipids. Cells were grown anaerobically with linoleic acid (600 μg/ml) and cholesterol (20 μg/ml) replacing the usual lipid supplements (Tween 80 and ergosterol). After harvesting and extensive washing, the cells were aerated in media containing no added lipid. Samples were removed at the times indicated and mitochondrial fractions prepared. Fatty acid and sterol analysis was carried out on whole cells and on the isolated mitochondria. Cholesterol and ergosterol accounted for more than 90% of the total sterol in each case and linoleic, oleic, and palmitoleic acids for more than 98% of the unsaturated fatty acids extracted. (b) Sterol content of cells, and of mitochondria from these cells, grown anaerobically on linoleic acid and cholesterol and aerated in the absence of these lipids. (Reproduced with permission from Gordon and Stewart, 1971.)

amphenicol affected neither lipid synthesis nor enzyme synthesis, and had no inhibitory effect on protein synthesis *in vivo*. All three compounds inhibited the oxidative activity of isolated mitochondria, the chloram-phenicol isomers also inhibited phosphorylation.

Aeration of a euflavine-derived mutant (which lacked mitochondrial protein synthesis and had impaired respiration) led to development of lipids but not to the synthesis of fumarase, malate dehydrogenase or succinate dehydrogenase. Lipid synthesis as in these cells is not inhibited by D(−)-*threo* chloramphenicol.

The extent of development of respiration induced on aeration of an anaerobically grown unsaturated fatty acid auxotroph of *S. cerevisiae* is determined by the availability of endogenous or exogenously-supplied UFA, and the synthesis of mitochondrial and cytoplasmic enzymes (malate dehydrogenase, fumarase, succinate dehydrogenase cytochrome c oxidase and catalase) during aeration, appears to have a similar basis of regulation (Gordon et al., 1972b). Levels of UFA that permit synthesis of mitochondrial enzymes also result in a substantial stimulation of cellular protein synthesis. Thus there may be a specific dependency of the synthesis of mitochondrial enzymes on synthesis of mitochondrial lipids.

4. Changes in the physical properties of mitochondria during respiratory adaptation

The mitochondria-like organelles of anaerobically-grown yeasts (see Chapter 1) are often regarded as incomplete mitochondria which differentiate during respiratory adaptation to yield fully-functional organelles (Flavell, 1971). However, the evidence for the insertion of newly-synthesized lipids and enzymes into the pre-existing framework of the promitochondrion is very incomplete, and the possibility that the membranes of the anaerobic structures are completely disassembled in the early stages of respiratory adaptation must still be considered (Cartledge and Lloyd, 1973). In an attempt to demonstrate the physical continuity between promitochondria and the respiring organelles produced on respiratory adaptation, Plattner et al. (1970) selectivity labelled the promitochondria in vivo by incubating cells with (^{3}H)-Leu in the presence of cycloheximide. After washing away the inhibitor, the cells were allowed to undergo respiratory adaptation in the presence of unlabelled Leu. Mitochondria isolated after adaptation were found to be radioactive, whereas iso-1-cytochrome c which is synthesized de novo during the process showed very little labelling. However, Plattner et al. (1970) found that about 50% of the promitochondrial label was lost during the adaptation to O_2, and suggested that this loss might be explained by one of three different mechanisms: (a) differentiation of promitochondria, followed by multiplication of respiring mitochondria in the unlabelled medium; (b) degradation of (labelled) promitochondrial membranes and simultaneous de novo formation of (unlabelled) respiring mitochondria; or (c) differentiation of promitochondria and concomitant turnover of (pro)mitochondrial proteins. Alternative (a) was thought to be unlikely, as the specific activity of F_1–ATPase in homogenates did not increase during respiratory adaptation. Alternative (b) also seemed unlikely, as cells incubated under nitrogen after the initial labelling period, also showed a loss of the promitochondrial label. Quantitative analysis of the distribution of promitochondrial label by electron microscope autoradiography in vesicles stained for cytochrome c oxidase indicated an association of the radioactivity with the stained vesicles. Plattner et al. (1970) propose that these results favour the concept of the promitochondrion as a direct mitochondrial precursor, but in the opinion of the author the evidence presented is far from compelling. The changes in enzyme distributions in subcellular fractions obtained from sphaeroplasts of *S. carlsbergensis* prepared at different stages during the process of respiratory adaptation strongly suggest that mitochondrial assembly may involve more than the simple insertion of newly-synthesized molecular components into the existing promitochondrial framework (Cartledge and Lloyd, 1973). Cells were grown anaerobically in the presence of 10% glucose and lipid supplements, subjected to anaerobic glucose derepression for 2 h in a medium containing 0·4% glucose, and finally aerated in this medium. Sphaeroplasts were prepared at various stages in this process, and after gentle mechanical breakage, the resulting extracts were fractionated by high speed zonal centrifugation through sucrose gradients containing Mg^{2+}. Anaerobic glucose-derepressed cells contained no detectable cytochrome *c* oxidase activity, and during derepression the total ATPase activity of whole homogenates decreased to a value one-half that found in homogenates of glucose-repressed cells. Of the ATPase in the repressed cells, 58% was inhibited by oligomycin whereas only 7% inhibition was found after anaerobic glucose derepression. The density of the major sedimentable zone containing oligomycin-sensitive ATPase activity (previously used as a marker enzyme for promitochondrial membranes (Criddle and Schatz, 1969)) was 1·27 g/cm³ in the homogenates of glucose-derepressed cells; oligomycin-insensitive ATPase was found between $\rho = 1·17$ and 1·26 g/cm³. After only 10 min of respiratory adaptation, cytochrome *c* oxidase was present in functional electron transport chains and was found in particles large enough to be completely sedimented in 0·25 M–sucrose after centrifugation for 10^5 *g* min (Cartledge *et al.*, 1972). These particles showed a bimodal distribution after density-gradient centrifugation, and the major sedimentable zone showed an extremely heterogeneous density distribution (between $\rho = 1·15$ and 1·30 g/cm³) as did ATPase activity, which was in this case 95% inhibited by oligomycin. Highly organized mitochondria were already present throughout this zone; these were completely different in ultrastructure from the organelles of the anaerobically-grown cells that occurred in fractions containing oligomycin-sensitive ATPase and are usually called promitochondria (Criddle and Schatz, 1969) or mitochondrial precursors (Watson *et al.*, 1970) (see Fig. 28, p. 46). Enzyme distributions and electron micrographs indicate that in fact several populations of particles overlap in their distributions through the gradient and cannot be separated by equlibrium density centrifugation (Cartledge and Lloyd, 1973). Partial separation of these populations has been achieved on iso-osmotic gradients (T. G. Cartledge and D. Lloyd, unpublished results). Highly osmophilic particles of about 0·1 µm diameter, which had not been detected in anaerobically- or aerobically-grown cells (Cartledge and Lloyd 1972a, b) were of frequent occurrence throughout the fractions containing newly-assembled mitochondria. The high E_{260nm}/protein ratio of these fractions suggested the presence of ribosomal aggregates involved in enzyme and membrane synthesis, and that these are often intimately associated with membranes was confirmed by electron microscopy. A remarkable alteration of the subcellular distribution of acid p-nitrophenyl phosphatase activity suggested an involvement of acid-hydrolase-containing organelles in the turnover and reorganization of membrane distribution within the cells (Cartledge and Lloyd, 1973).

Further changes in density distributions of cytochrome c oxidase and oligomycin-sensitive ATPase were observed after 30 min of respiratory adaptation; at this stage the major sedimentable zone of particles containing these enzymes had shifted to $\rho = 1\cdot25$ g/cm^3. After 3 h adaptation these mitochondrial marker enzymes were almost entirely confined to a zone at $\rho = 1\cdot23$ g/cm^3. which corresponds to the density of mitochondria from aerobically grown, glucose-derepressed cells (Cartledge and Lloyd, 1972a). A summary of these changes in density distributions is given in Fig. 66. Whilst it is impossible to discount the possibility that newly assembled mitochondria in the early stages of adaptation may be more susceptible to damage, the finding that all the particles are large enough to be sedimented at 10^5g min and have a wide range of densities suggests that they are not produced by membrane comminution. Cartledge and Lloyd (1973) suggest that the marked changes in density distributions are compatible with the hypothesis of massive reorganization of membrane composition during the transition from the anaerobic to the aerobic phenotype. Thus particles of low buoyant density may represent a steady-state population of lipid-rich enzyme-containing vesicles, which fuse with particles of high buoyant densities, to give finally a population of intermediate density. Thus the assembly of fully-functional mitochondria may involve fusion of large membrane units of differing enzymic composition, which do not prexist in anaerobically grown cells, rather than the direct conversion of the promitochondrion by the insertion of new components of molecular dimensions.

Further evidence for the complexity of the mechanism of mitochondrial assembly during respiratory adaptation has been obtained by a study of transition points in Arrhenius plots of cytochrome c oxidase activity (Ainsworth and Tustanoff, 1972). These transition points are characteristic of the major fatty acids found in the mitochondrial membrane. By growing *S. cerevisiae* anaerobically with linoleic acid, it was possible to produce cells in which this unsaturated fatty acid became a major membrane component. The production of new membranes during subsequent respiratory adaptation in the presence of oleic acid was then monitored by the characteristic change in the Arrhenius profile for cytochrome c oxidase (Fig. 67). The results obtained imply that the cytochrome c oxidase

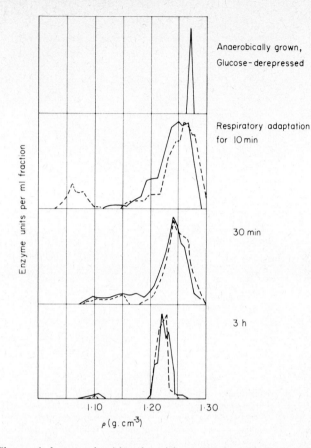

FIG. 66. Changes in buoyant densities of particles containing oligomycin-sensitive ATPase and cytochrome *c* oxidase during the respiratory adaptation of *Saccharomyces carlsbergensis*. Solid line, oligomycin-sensitive ATPase; dashed line, cytochrome *c* oxidase. (Replotted from data of Cartledge and Lloyd, 1973.)

activity which develops immediately after O_2 induction is associated with a linoleic-dominated membrane which had been synthesized under anaerobic conditions, but that the activity developed between 0·5 and 1·0 h of adaptation is associated with newly-synthesized membrane showing oleic-type transition points. There are at least two possible interpretations of these results: (a) that cytochrome *c* oxidase synthesized immediately after induction is integrated into anaerobically synthesized membrane until all the available sites become saturated, and then integration into newly-synthesized membranes occurs; or (b) that some precursors of cytochrome *c* oxidase are synthesized under anaerobic conditions and integrated into the anaerobic membranes. The first-formed enzyme activity then results

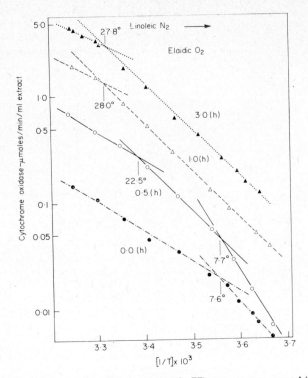

FIG. 67. *Saccharomyces cerevisiae* (laboratory strain 77) were grown anaerobically to their late exponential phase in a medium containing 3% glucose, 0·5% casein hydrolysate, 0·1% Triton X-100, 1·2% yeast extract, 0·002% ergosterol and 0·01% linoleic acid. Cells were harvested and after washing in fresh medium containing elaidic acid *in lieu* of linoleic acid, were grown aerobically in this new medium with vigorous aeration. Cells were removed at various times during this adaptation period and after washing in the presence of cycloheximide and D-chloroamphenicol were subjected to French pressure cell homogenization. The resulting extracts were monitored spectrophotometrically for cytochrome *c* oxidase activity at temperature ranges from 0°C to 40°C, using a immersible thermistor to follow the reaction temperature. The break at 7·6°C for anaerobic cells indicates membrane whereas those seen after 0·5, 1 and 3 h adaptation reflect an elaidic membrane. (Reproduced with permission of P. J. Ainsworth and E. R. Tustanoff.)

from the activation of these precursors. Subsequently the newly-synthesized cytochrome *c* oxidase is integrated into newly-formed membranes.

The transition temperatures of some electron transport enzymes and of F_1–ATPase determined from Arrhenius plots changed from 24°–27°C (characteristic of membranes from anaerobically-grown cells) to 11–14°C (characteristic of fully functional mitochondria) during respiratory adaptation (Watson *et al.*, 1973a, b). These changes are correlated with the changes in lipid/protein ratio and in the fluidity of the mitochondrial membrane.

5. Effect of anaerobiosis on aerobically-grown yeasts

Ephrussi and Slonimski (1950a, b) showed that the reversal of respiratory adaptation (i.e. the loss of cytochrome components characteristic of aerobically-grown yeast) did not occur rapidly in the absence of cell growth. Some changes in mitochondria of aerobically-grown *S. cerevisiae* are produced when nonproliferating cell suspensions are sparged with argon (Luzikov *et al.*, 1971). Thus removal of O_2 leads to a decrease (over a period of 4 h) of the complete respiratory system as measured by the NADH oxidase or succcinoxidase of mitochondrial fractions isolated at intervals (Fig. 68). The activities of segments of the respiratory chain

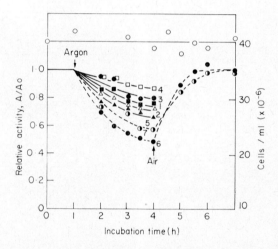

FIG. 68. Changes in activities of certain mitochondrial enzymes upon deaeration and subsequent aeration of yeast cells. Yeast cells were subjected to anaerobiosis at the end of the exponential growth phase. At the times indicated in the figure the required amounts of the cell suspension were drawn off, mitochondria were isolated and the activities indicated below were assayed. A/A_0 is the ratio of current to initial specific activity. Curve 1, NADH-ferricyanide oxidoreductase; 2 succinate/ferricyanide reductase; 3, NADH-cytochrome *c* oxidoreductase; 4, succinate-cytochrome *c* oxidoreductase; 5 cytochrome *c* oxidase; 6, NADH oxidase; 7, succinate oxidase. (Reproduced with permission from Luzikov *et al.*, 1971.)

(NADH-ferricyanide oxidoreductase, succinate-ferricyanide oxidoreductase, NADH-cytochrome *c* oxidoreductase, succinate-cytochrome *c* oxidoreductase and cytochrome *c* oxidase) are also reduced. Addition of Tween 80 or ergosterol to the medium prior to deaeration does not prevent inactivation, and on re-aeration the activities are restored to their initial levels. Decreases in mitochondrial content of cytochrome *b* and cytoch⁻omes

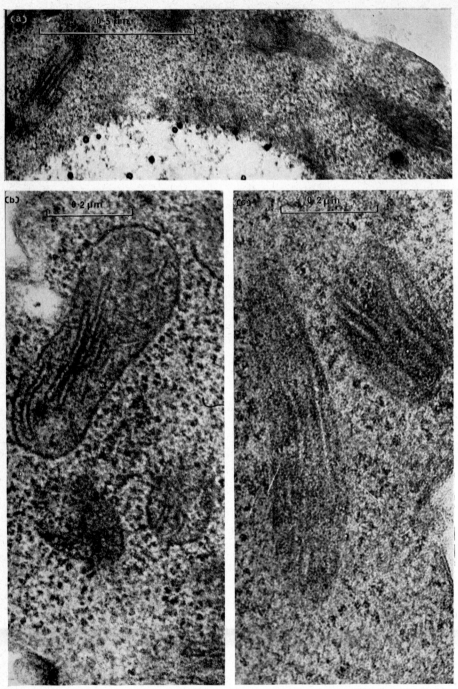

FIG. 69. *Saccharomyces cerevisiae* grown aerobically for 13 h and fixed with glutaraldehyde (a, b). Mitochondria of non-deaerated yeast cells and (c) those of cells de-aerated for 30 min. In all micrographs mitochondria are in orthodox conformation. (Reproduced with permission from Luzikov, 1973.)

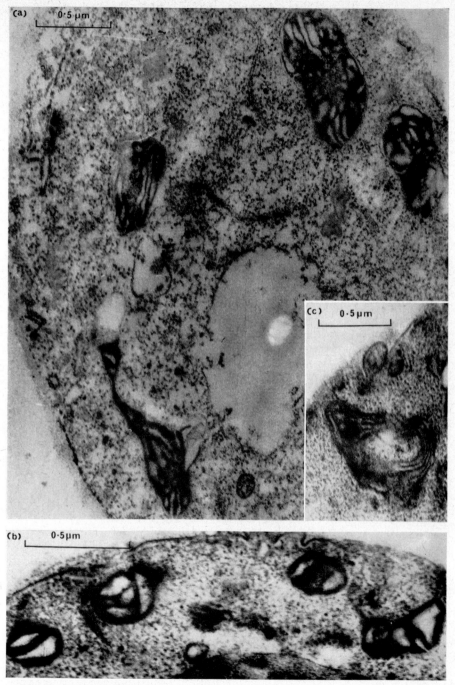

FIG. 70. Morphological changes in mitochondria of de-aerated *Saccharomyces cerevisiae*. The culture was grown for 13 h (the end of the exponential phase) and then de-aerated for 3 h. (a, b, c) Mitochondria in the condensed conformation. (Reproduced with permission from Luzikov, 1973.)

$a + a_3$ were also observed after deaeration, and again these changes were reversible. The total cytochrome content of whole cells was not altered under these conditions, and the authors suggest that the effect is not due to a destruction of cytochromes but rather to their loss from mitochondrial membranes. A reversible disorganization of the cristae membranes was also evident (Figs. 69, 70). Chloramphenicol did not inhibit the restoration of NADH oxidase activity. Cycloheximide promoted a rapid restoration of NADH oxidase and cytochrome c oxidase activities, so that these were restored within 1 h of reaeration to their former levels, but these initial rapid phases of restoration were followed by dramatic decreases of activity. The reappearance of mitochondrial activities was in all cases correlated with the appearance of organelles showing typically complex mitochondrial ultrastructure. Luzikov *et al.* (1971) concluded from these results that the functions of O_2 in the formation of mitochondria are not restricted to the induction of protein synthesis, and that it has a "direct" effect on the assembly and stability of the respiratory system and mitochondrial membranes in circumstances where all the components are already present.

6. Degradation of the mitochondrial respiratory system in the presence of respiratory inhibitors and uncouplers

Addition of the respiratory inhibitors or uncouplers to an aerobically grown culture of *S. cerevisiae* at the end of logarithmic growth produces a decreased respiratory activity of mitochondrial suspensions subsequently prepared at stages over a 4 h period (Luzikov *et al.*, 1970). Thus when

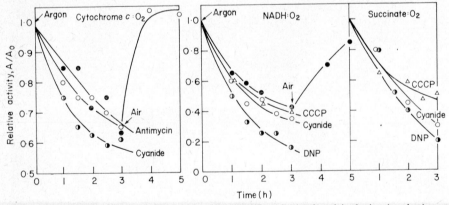

FIG. 71. (a) Decrease in cytochrome c oxidase activity of mitochondria during incubation of aerobically-grown yeast cells under anaerobic conditions and in the presence of CN^- or antimycin A. A_0: specific activity of mitochondria isolated from yeast cells immediately after introducing argon or inhibitors. (b) Decrease in NADH oxidase and succinate oxidase activities of mitochondria upon incubation of aerobically-grown yeast cells in anaerobic conditions and in the presence of CN^-, DNP, or CCCP. (Reproduced with permission from Luzikov *et al.*, 1970.)

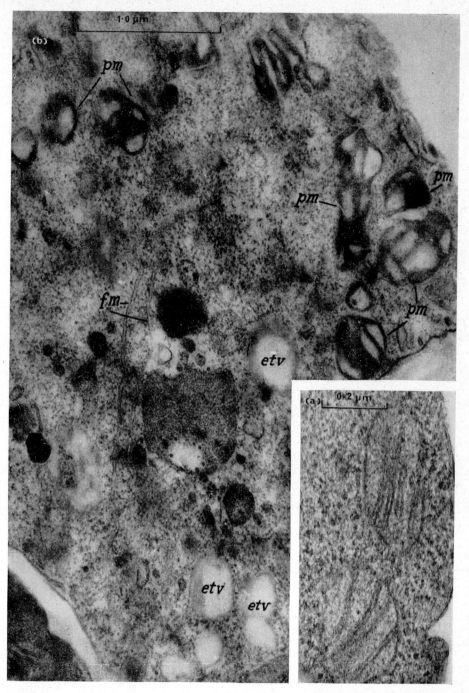

(see next page for caption)

0·11 µg/ml antimycin A, CN⁻ (3 mM) or DNP (0·1 mM) were added to the culture they appeared to lead to a degradation of the respiratory system *in vivo* in a manner analogous with the effects produced on deaeration of the culture (Figs. 71, 72). A 3 h incubation of the cells with CN⁻ decreased the NADH oxidase activity by more than 50%, whereas the content of cytochromes $a + a_3$ in the mitochondrial fraction decreased only by 10–15%.

It was concluded that the organization of the mitochondrial membrane is stabilized when involved in carrying out oxidative phosphorylation, but that the changes observed are not a direct result of de-energization of the mitochondria.

C. Catabolite Repression and Derepression

1. Catabolite repression

The cellular numbers, morphology and enzymatic activities of yeast mitochondria are all influenced by the nature and concentration of carbon source in the growth medium. High concentrations of glucose or other rapidly fermentable sugars inhibit the synthesis of mitochondrial components (Ephrussi *et al.*, 1956), and aerobic growth under these conditions leads to the production of cells which show a fermentative pattern of metabolism. This impairment of respiration resulting from catabolite repression (Slonimski, 1956) can be accounted for by (a) the "reverse Pasteur effect" (inhibition of biosynthesis of respiratory enzymes), or (b) the "Crabtree effect" (inhibition of the activity of respiratory enzymes, Crabtree, 1929). The low Q_{O_2} of intact cells grown with high concentrations of glucose is accompanied by a reduction in cytochrome content (Ephrussi *et al.*, 1956) and reduced levels of activity of enzymes of the tricarboxylic acid cycle (MacQuillen and Halvorson, 1962; Polakis *et al.*, 1964, 1965; Polakis and Bartley, 1965). Other enzymes subject to repression by glucose include alcohol dehydrogenase (Witt *et al.*, 1966), NADH oxidase and NADH dehydrogenase (Linnane, 1965), D-lactate oxidase, L-lactate oxidase, L-α-glycerophosphate oxidase (Utter *et al.*, 1968) and oligomycin-sensitive ATPase (Somlo, 1968). The rate of mitochondrial ADP transport is decreased 10 fold when 3% ethanol is replaced by 10% glucose in the growth medium (Lauquin and Vignais, 1973). The number of atractyloside-sensitive or (^{35}S)-carboxyatractlyoside binding sites is not

---

FIG. 72. Morphological changes in the mitochondria of *Saccharomyces cerevisiae* in cells incubated with 1 µM CCCP. The culture was grown aerobically to the end of the exponential growth phase and the uncoupler was added. (a) After incubation for 30 min with the uncoupler the cells still possess mitochondria which show the orthodox conformation. (b) Incubation for 3 h with the uncoupler yields cells containing "free membranes" (*fm*), "electron transport vesicles" (*etv*) and aberrant mitochondria (*pm*). (Reproduced with permission from Luzikov, 1973.)

modified, the affinity for ADP is not altered and the endogenous nucleotide pool is not decreased in size. UQ (Gordon and Stewart, 1969), phospholipids and palmitoleic acid (Lukins *et al.*, 1968) are all catabolite-repressible mitochondrial components. Glucose is a more effective repressor of tricarboxylic acid cycle enzymes than is galactose (Polakis and Bartley, 1965), melibiose or raffinose (Reilly and Sherman, 1965). In *S. cerevisiae* the threshold for inhibition of biosynthesis of respiratory enzymes is 6 mM–glucose (Slonimski, 1956); for *Shizosaccharomyces pombe* cytochrome *c* oxidase and malate dehydrogenase are derepressed below 12 mM–glucose, NADH and succinate-cytochrome *c* oxidoreductases and succinate dehydrogenase are derepressed below 30 mM–glucose (Poole and Lloyd, 1973) whereas in *S. carlsbergensis* derepression occurs below 70–100 mM–glucose (Cartledge and Lloyd, 1972a). The concentration of glucose present in the medium, has practically no influence on the respiratory activity of *Kluyveromyces fragilis* (Chassang-Douillet *et al.*, 1973). Under conditions of aerobic glucose repression the mitochondria are few in number, large, irregular and have poorly developed cristae (Yotsuyanagi, 1962a; Linnane, 1965; Lukins *et al.*, 1968; Neal *et al.*, 1970). Glucosamine and 2-deoxyglucose, two sugars which are not metabolized beyond the stage of formation of their phosphate esters, also act as catabolite repressors in yeast; Holtzer (1968) has based a suggestion on this evidence that glucose itself, glucose-6-phosphate, ATP, ADP or Pi may be the metabolites directly responsible as effectors of glucose repression. Glucose at very high concentrations (30%) does not block completely the formation of the cytochromes (Reilly and Sherman, 1965). Differences in rates of consumption of various sugars by fermentation or by respiration can be correlated with the extent of repression of enzyme synthesis in yeasts (De Deken, 1966a; Beck and von Meyenburg, 1968; Görts, 1971). The phenomenon of glucose repression in yeast is reminiscent of catabolite repression in bacteria, where it has been shown that high concentrations of glucose in the growth medium leads to a depletion of intracellular 3' 5' cyclic AMP (Makman and Sutherland, 1965). Evidence is available that this nucleotide is required in bacterial systems for the expression of catabolite repressible operons (Perlman and Pastan, 1968a, b). Fang and Butow (1970) found that protoplasts prepared from anaerobically grown cultures of yeast are capable of respiratory adaptation on aeration in a medium containing 0·1% glucose. In media containing 5% or 10% glucose respiratory adaptation was inhibited, but this inhibition could be overcome by the addition of 3', 5' cyclic AMP. Other nucleotides, notably 5' AMP also produced this effect, and the nucleotide-stimulated respiratory adaptation was sensitive to cycloheximide and chloramphenicol. The intracellular level of cyclic AMP is low after growth with glucose and increases during ethanol or maltose utilization (T. A. Tuan and A. Goffeau,

H

personal communication; van Wijk and Konijn, 1971). It has been suggested that control of phosphodiesterase activity by ATP, pyrophosphate and polyphosphate levels, regulates the rate of degradation of cyclic AMP in *S. carlsbergensis* (Speziali and van Wijk, 1971).

When aerobically-grown yeast containing high levels of activity of tricarboxylic acid cycle, glyoxylate cycle and electron transport enzymes is transferred to medium containing 10% glucose, there is a lag phase which lasts 30 min, followed by exponential growth (Chapman and Bartley, 1968). During the lag phase 45% of the malate dehydrogenase, 17% of the isocitrate dehydrogenase, 27% of the succinate-cytochrome c oxidoreductase and 40% of the NADH-cytochrome c oxidoreductase activities are lost; the investigators suggest a process of active removal of these enzymes. The loss of other enzymes during the subsequent growth of cells (aconitase, fumarase, glucose-6-phosphate dehydrogenase, and α-ketoglutarate dehydrogenase) could be accounted for by dilution by cell division.

The loss of respiratory activity of yeast cells during glucose repression and its restoration on derepression has been correlated with changes in the activities of dehydrogenases involved in the oxidation of NADH, succinate, ethanol and L-lactate (Jayaraman *et al.*, 1966). Particulate malate dehydrogenase was not sensitive to repression. The corresponding oxidase activities of mitochondria isolated at various stages also showed decreased activities during repression, and increased activities during derepression. The highest respiratory control ratios were obtained in the mitochondria from the derepressed cells. The cytochrome composition of the particles changed continuously during derepression, not only with respect to absolute amounts, but also relative amounts of individual cytochromes.

In a study of the lipid and enzyme composition of mitochondria isolated from cells grown under different conditions of catabolite repression (Lukins *et al.*, 1966, 1968), it was shown that the syntheses of both membrane bound and soluble enzymes of the mitochondria are influenced by the extent of repression. Membrane bound cytochromes and dehydrogenases are affected to different extents. Thus the concentration of a-type cytochromes of lactate-grown cells may be as high as 1·11 nmoles/mg protein, and in mitochondria from 5% glucose-grown cells, as low as 0·07 nmoles/mg protein, a sixteen-fold change. The corresponding changes in succinate and NADH dehydrogenase activities are only about eight fold and three fold respectively.

The individual cytochromes are also affected to a variable extent; the ratios of cytochromes $a + a_3$: b: $c + c_1$, in lactate grown cells is 1:1:1 whereas a ratio of 1:2:2·5 is observed with mitochondria isolated from cells grown with 5% glucose. Both the total amount of phospholipid, and fatty acid composition, are markedly affected by catabolite repression. Mitochondria isolated from repressed cells have less total phospholipid

and the fatty acids of this fraction are more saturated than those of derepressed cells. The neutral lipid fatty acid composition is almost unaffected
by varying catabolite repression. Phospholipase D activity is increased
under conditions of glucose repression (Dharmalingam and Jayaraman,
1971; Grossman *et al.*, 1973). No qualitative differences in membrane
proteins were seen on gel electrophoresis of mitochondrial extracts prepared from glucose-repressed and derepressed *S. carlsbergensis* (Kaplan
and Criddle, 1970). However the amount of one component was greatly
increased relative to all others on derepression.

2. Catabolite derepression

Growth of a batch culture of yeast with glucose as the carbon source is a
biphasic process. Throughout the early stages, where the glucose concentration is high, the cells are subject to glucose repression and have a highly
fermentative pattern of metabolism (Fig. 73). Exhaustion of the glucose
is accompanied by changes in internal nucleotide concentrations (Ball and
Tustanoff, 1968), the derepression of respiratory activities, and oxidation
of accumulated ethanol (Lemoigne *et al.*, 1954; Slonimski, 1956; Ephrussi
et al., 1956; Ball and Tustanoff, 1971). The growth characteristics of some
commonly used strains of *S. cerevisiae* with respect to some of those
properties have been compared by Ball *et al.* (1971). Asynchrony in the
development of various mitochondrial activities has been observed during
glucose derepression (Lenaz *et al.*, 1969). Thus increases in activity of
tricarboxylic acid cycle enzymes occur more rapidly than that of ATPase
or the observed increases in efficiency of oxidative phosphorylation.
During glucose derepression the mitochondria become smaller and show
development of organization of the cristae (Yotsuyanagi, 1962a; Linnane,
1965; Lukins *et al.*, 1968; Neal *et al.*, 1970) as well as a decreased density
in sorbitol gradients (Neal *et al.*, 1971). The processes of derepression
may also be followed in sphaeroplasts, and this provides a convenient
system for the study of changes in mitochondria (Neal *et al.*, 1970).
Analysis of the mitochondrial populations obtained from sphaeroplasts at
various times during derepression by rate and isopycnic zonal centrifugation on sorbitol gradients suggests that smaller, less dense ($\rho = 1\cdot194$
g/cm^3) mitochondria replace the large, dense ($\rho = 1\cdot211$ g/cm^3) mitochondria characteristic of the repressed yeast. These results were consistent
with the idea that repressed mitochondria serve as precursors of derepressed
mitochondria. Matile and Bahr (1968) have presented quantitative electron
microscopic evidence for the existence of two populations of mitochondria
in a homogenate prepared from derepressed yeast by mechanical methods.
Isolated organelles gave two bands in Urografin gradients corresponding
to equilibrium densities of $1\cdot122$ and $1\cdot100$ g/cm^3; the two populations
showed differences in ratios of specific activities of respiratory enzymes.

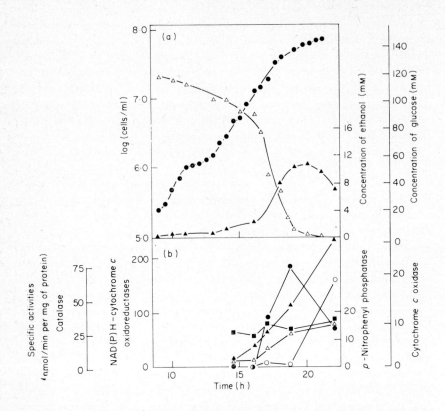

FIG. 73. Changes in cell population, glucose and ethanol concentration in the medium and specific activities of marker enzymes of homogenates from *Saccharomyces carlsbergensis* during aerobic growth. (a) Shows the cell concentration (log(cells/ml)●), glucose concentration (△) and ethanol enconcentration (▲). (b) Shows the activities of cytochrome *c* oxidase (●), catalase (○), *p*-nitrophenyl phosphatase (■), NADH-cytochrome *c* oxidoreductase (▲) and NADH-cytochrome *c* oxidoreductase (△). (Reproduced with permission from Cartledge and Lloyd 1972a.)

Cartledge and Lloyd (1972a) have also demonstrated the presence of two populations of mitochondria in sphaeroplast lysates prepared from cells of *S. carlsbergensis* during glucose derepression (Fig. 74). The mitochondria banding at equilibrium density $\rho = 1 \cdot 21$ g/cm³ after zonal centrifugation showed a mean diameter of $0 \cdot 38$ μm in electron micrographs whereas those at $\rho = 1 \cdot 235$ g/cm³ had a diameter of $0 \cdot 62$ μm. Both populations showed a variety of conformational states; the lighter mitochondria were predominantly in the "condensed" form, whereas the heavier mitochondria showed approximately equal proportions of "condensed" and "orthodox" forms. The ratios of specific activities of various enzymes (cytochrome *c*

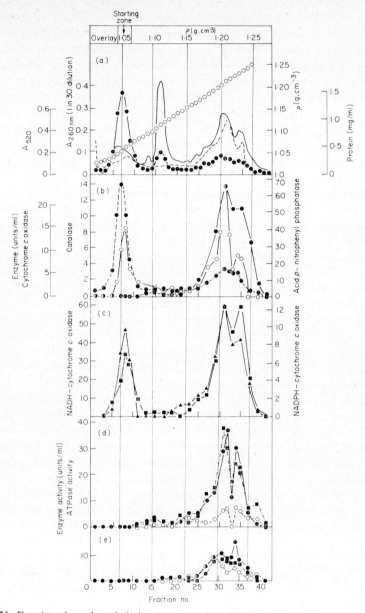

FIG. 74. Fractionation of a whole homogenate of aerobically-grown glucose-derepressed *Saccharomyces carlsbergensis* by zonal centrifugation. Whole homogenate (20 ml), containing 114 mg of protein, was loaded on the gradient. The volumes of homogenate (diluted 1 in 20) and of fractions taken for assay were as follows: catalase, cytochrome *c* oxidase, NADH- and NADPH-cytochrome *c* oxidoreductases, all 0·1 ml, acid *p*-nitrophenyl phosphate and ATPases 0·2 ml. Centrifugation was at 35000 rev/min for 165 min (6 ×

(*continued on next page*)

oxidase, NADH- and NADPH-cytochrome c oxidoreductases and ATPases) were different in the two populations, but electron micrographs indicated that mitochondria were not the only membraneous structures present. Cartledge and Lloyd (1972a) concluded that it was not possible to decide on the basis of these results whether the two populations represent physiologically distinct types of mitochondria present in the cells, or whether changes in mitochondrial permeability to sucrose during isolation and centrifugation give rise to the observed heterogeneity. It also seems likely that all cells of the cultures are not derepressed synchronously, and so the experiment used a mixed population of repressed and derepressed cells (see p. 74).

During the release from glucose repression the synthesis of both cytoplasmic and mitochondrial enzymes is sensitive to inhibition by cycloheximide (Henson et al., 1968a, b). The synthesis of inner mitochondrial membrane enzymes are preferentially inhibited by chloramphenicol in short-term experiments; the inhibition of derepression of the NADH-cytochrome c oxidoreductase is greater than that of cytochrome c oxidase. Cytoplasmic enzyme synthesis was not affected. Over long periods (6 h exposure to chloramphenicol) both cytoplasmic and mitochondrial enzymes were affected. Similar results have been reported by Görts and Hasilík (1972a, b) with a strain of S. cerevisiae sensitive to 0·5 μg/ml chloramphenicol. Evidence was produced for interference with the ultrastructural integrity of complex multienzyme systems in mitochondria during glucose derepression in the presence of the antibiotic; again the primary lesion appeared to be inactivation of assembly of cytochrome c oxidase. No net synthesis of phospholipids takes place during glucose derepression, and the turnover of phospholipids is drastically reduced during this process, (Jayaraman and Sastry, 1971). This suggests that many of the structural phospholipid elements for mitochondrial development are already available in the cells. The changes in phospholipid composition of the mitochondrial membranes during derepression are not directly responsible for any of the major alterations in observed enzyme activities (Bertoli et al., 1971).

$10^6 g$ min at the sample zone; $\int_0^t \omega^2 \, dt = 1 \cdot 45 \times 10^{11} \mathrm{rad}^2 \mathrm{s}^{-1}$). (a) Sucrose density gradient (○), light scattering at 520 nm (– – – –) 260 nm (———) and protein (●). (b) Cytochrome c oxidase (●————●), catalase (○), and acid p-nitrophenyl phosphatase (●– – – –●). (c) NADH-cytochrome c oxidoreductase (■) and NADPH-cytochrome c oxidoreductase (■). (d) ATPase (●) oligomycin-insensitive ATPase (○) and ouabain-insensitive ATPase (■) assayed after storage for 18 h at 4°C. (e) ATPase (●), oligomycin insensitive ATPase (○) and ouabain-insensitive ATPase (■) assayed after storage for 18 h at 18°C. Specific activities of enzymes in the whole homogenate were as follows (recoveries in parentheses): protein (96·7%) catalase 7·1 (123%), cytochrome c oxidase 18·1 (104%), NADH-cytochrome c oxidoreductase 74·2 (76%) NADPH-cytochrome c oxidoreductase 15·2 (75%) and acid p-nitrophenyl phosphatase 9·7 (99%). (Reproduced with permission from Cartledge and Lloyd, 1972a.)

3. Anaerobic glucose derepression

Cell suspensions of *S. carlsbergensis* grown anaerobically in the presence of 10% glucose and lipid supplements were harvested, resuspended and maintained anaerobically in the presence of 0·4% glucose (Cartledge and Lloyd, 1973). Throughout this period of anaerobic glucose derepression these cells showed a respiration which was insensitive to antimycin A and inhibited only 30% by 1 mM–KCN. In extracts prepared at various times over a two hour period no detectable cytochrome *c* oxidase activity was observed and the oligomycin sensitivity of ATPase declined from 58% to 7%.

Lowden *et al.* (1972) have compared mitochondrial activities in *S. cerevisiae* after growth anaerobically in batch culture (glucose-repressed) with those of cells grown anaerobically in chemostat cultures with glucose as the limiting substrate (glucose-derepressed). Both types of cell showed similar low levels of cyanide-sensitive respiration, malate dehydrogenase, succinate dehydrogenase and fumarase; levels of activity of cytochrome *c* oxidase and UQ contents were so low as to be not detectable. Thus decreased mitochondrial function in anaerobiosis occurs independently of the effects of glucose repression.

The Dio-9 sensitive ATPase activity of anaerobically-grown *S. cerevisiae* (in lipid-supplemented, chemically defined medium) was about one-third that of cells grown under aerobic conditions in an identical medium. The degree of repression was further enhanced at high glucose concentrations, but was less in media containing yeast extract than in the defined medium (A. Goffeau, E. Mrena and A. Claude, personal communication).

4. Resolution of the effects of oxygen tension and catabolite repression

The response of *Candida utilis* to changes in O_2 tension and glucose concentration have been studied in continuous culture (Moss *et al.*, 1969). This organism is highly aerophilic and grows only very slowly under anaerobic conditions. Independent variations of these two parameters indicate that there is an inverse relationship between dissolved O_2 and total cytochrome content and also between glucose concentration and total cytochrome content. When the feed-rate of glucose was limiting, suppression of cytochrome development occurs at high levels of dissolved O_2 and the concentrations of *a*-, *b*- and *c*-type cytochromes attained maximal values when dissolved O_2 was present at concentrations less than 0·1 μM.

In the highly fermentative yeast *S. carlsbergensis* under similar conditions of growth a different response is noted (Moss *et al.*, 1971). When glucose was limited and O_2 abundant, oxidative catabolism predominated over ethanolic fermentation; cytochrome $a + a_3$ content of the cells was highest under these conditions. At high levels of glucose, ethanolic fermentation is the principal metabolic route, and cytochrome $a + a_3$ content is lower. As was the case with *C. utilis* a step change from high to low O_2 was followed by a significant increase in total cytochrome content. Of the cytochromes, $a + a_3$ show the most marked changes. A summary of the effects of O_2 and glucose on cytochrome concentrations in these two yeasts and in *S. cerevisiae* in batch and continuous cultures is presented by Rickard *et al.* (1971).

The process of respiratory adaptation of cells after growth under glucose-limited and strictly anaerobic conditions was studied by Faures and Fukuhara (1965). The influence of O_2 tension and concentrations of glucose and galactose on the physiology of *S. cerevisiae* and other yeasts has also been investigated in continuous cultures by Brown and Johnson (1970, 1971) and by Johnson *et al.* (1972).

The formation of visible profiles of mitochondria (in sections of per-manganate-fixed cells) and of detectable mitochondrial cytochromes in a glucose-limited chemostat culture of *S. cerevisiae* occurred at 3 μM–O_2 (Ferdouse *et al.*, 1972). The definition and numbers of profiles were enhanced by addition of ergosterol + Tween 80 during induction, by glucose limitation, or by increased O_2 tensions. Mitochondria in cells produced by aeration in the absence of lipid supplements were enriched in cytochromes compared to those produced in the presence of ergosterol and Tween 80. Elevation of O_2 from 3 μM to 52 μM led to increased definition of mitochondrial membranes, and to the formation of larger mitochondria, but not to increased numbers of profiles per cell. These results suggest that mitochondrial development and cytochrome synthesis can proceed to an independently variable extent under different growth conditions. Half maximal levels of membrane-bound cytochromes ($a + a_3$, b and c_1), unsaturated fatty acids and sterol were reached in cells of *S. cerevisiae* at O_2 concentrations around 0·1 μM (Rogers and Stewart, 1973a), whereas the synthesis of ubiquinone and cytochrome c and the increase in fumarase activity were all essentially linear functions of dis-solved O_2 concentration up to 3·5 μM–O_2. The synthesis of succinate dehydrogenase, succinate-cytochrome c oxidoreductase and cytochrome c oxidase showed different responses to changes in O_2 levels in the medium. Both cytochrome P-450 and cyanide-insensitive respiration were maximal at 0·25 μM–O_2 and declined at more anaerobic or aerobic conditions. Cytochrome c peroxidase and catalase were high in all but the most strictly anaerobic cells. Synthesis of peroxisomal enzymes, including those of the glyoxylate cycle, occurred at significantly higher dissolved O_2 con-centrations than that of the mitochondrial enzymes (Rogers and Stewart, 1973b).

D. Effects of Specific Nutrient Limitation

1. Iron or sulphate limitation

Mitochondria prepared from *Candida utilis* grown in a chemostat with iron-limited growth (at an iron concentration of 0·2 µg atom/l) lack energy-conservation but not electron flow in that segment of the respiratory chain leading from intramitochondrial NADH to the cytochromes (Light *et al.*, 1968; Light and Garland, 1971). Growth of this yeast under conditions of limitation by glycerol, ammonium or magnesium did not lead to loss of Site I energy conservation, and phosphate-limited cells yielded mitochondria which did not exhibit respiratory control. The oxygen uptake rates and P/O ratios of mitochondria from cells grown under these different conditions of nutrient limitation are shown in Table XXIV. The oxidation of all the substrates listed was sensitive to both antimycin A and KCN. The oxidation of NAD-linked substrates by mitochondria from cells grown under conditions of iron limitation was insensitive to inhibition by rotenone or piericidin A, whereas sensitivity was observed in mitochondria prepared from glycerol-, ammonium-, magnesium- or phosphate-limited cells. The mitochondrial concentration of the piericidin A-sensitive component was calculated to be about 75 pmol/mg of protein.

The absence of Site I energy conservation in mitochondria from iron-limited cells was confirmed by the observation that no energy-dependent reduction of intramitochondrial NAD(P) or of low potential fluorescent flavoprotein occurred on addition of glycerol-3-phosphate. These reactions did occur in mitochondria from glycerol-, ammonium- and magnesium-limited cells.

Iron limitation decreased the concentrations of cytochromes about five fold when compared with glycerol limitation (Table XXV). The decrease was similar in both whole cells and mitochondria, and therefore iron limitation did not affect the proportion of cell dry weight accounted for by mitochondrial protein. No marked differential effects were observed for the different cytochromes.

The effects of iron-limited growth are not due to the selection within the chemostat of a new genotype lacking the first energy-conservation site as the aeration of iron-limited cells for 12 h under non-growing conditions results in the recovery of Site 1 energy conservation and sensitivity to piericidin A (Light *et al.*, 1968; Clegg and Garland, 1971). When this recovery procedure was carried out in the presence of cycloheximide (100 µg/ml), restoration of piericidin A or rotenone sensitivity, but not Site I was prevented. Chloramphenicol, erythromycin and tetracycline were without effect on the recovery of Site 1 and piericidin A sensitivity (Table XXVI). Iron-limited growth of *C. utilis* lowers the concentration of both non-haem iron and acid-labile sulphide of SMPs by over 20 fold

Table XXIV. Oxygen-uptake rates and P/O ratios of mitochondria prepared from cells of *C. utilis* after growth with different growth-limiting nutrients. (Reproduced with permission from Light and Garland, 1971.)

Limited by	Glycerol		Iron		Ammonium		Phosphate		Magnesium	
Substrate and conc. (mM)	O_2 uptake	P/O ratio	O_2 uptake	P/O ratio	O_2 uptake	P/O ratio	O_2 uptake	P/O ratio	O_2 uptake	P/O ratio
Endogenous	40–90	2·18 ± 0·33 (5)	30–70	1·67 ± 0·4 (4)	225	3·6	85	—	210	3·69
2-oxoglutarate (5)	120–140	3·83 ± 0·42 (3)	125–270	2·9 ± 0·2 (4)	225	2·8	82	—	180	2·04
Pyruvate (2) and L-malate (2)	145–205	3·0 ± 0·21 (5)	166–275	1·95 ± 0·2 (4)						
Citrate (5)	65–115	2·4 ± 0·28 (2)	N.T.	N.T.	N.T.	N.T.	N.T.	N.T.	N.T.	N.T.
Ethanol (1)	135–160	2·6 ± 0·16 (4)	N.T.	N.T.	N.T.	N.T.	N.T.	N.T.	N.T.	N.T.
Succinate (5)	40–90	—	N.T.	N.T.	N.T.	N.T.	N.T.	N.T.	N.T.	N.T.
DL-glycerol 3-phosphate	390–480	1·77 ± 0·15 (3)	332–480	1·97 ± 0·17 (4)	90	1·2	254	—	420	1·89
NADH	370–420	1·65 ± 0·16 (4)	280–350	1·6 ± 0·3 (4)	590	1·92	286	—	N.T.	N.T.

The oxygen-uptake rates (ng-atoms of O_2/min per mg of protein) are not corrected for the endogenous rates but are the observed rates in State 3 (Chance and Williams, 1956). N.T., not tested.

Table XXV. Effects of iron-limited growth on the concentrations of cytochromes in whole cells of *C. utilis* and their mitochondria. (Reproduced with permission form Light and Garland, 1971.)

| | Conc. of cytochromes in | | | |
| | Mitochondria (nmol/mg of protein) | | Cells (nmol/g dry wt) | |
	Glycerol-limited	Iron-limited	Glycerol-limited	Iron-limited
Cytochrome $a + a_3$	0·21	0·038	26·8	5·6
Cytochrome b	0·22	0·053	21·6	7·0
Cytochrome $c + c_1$	0·50	0·073	47·3	7·2

The results given are for dual-wavelength spectrophotometric assays on two preparations and are similar to those obtained by wavelength-scanning spectroscopy on any other preparation derived from iron-limited growth without added iron or glycerol-limited growth with 100 μM added iron. In all cases, spectroscopic measurements were made on sufficient material to give a final concentration of cytochrome $(a + a_3)$ in the cuvette of about 0·1 μM. For whole cells, the suspension in air-saturated 50 mM-potassium phosphate buffer, pH 7·2, was allowed to become anoxic while the extinction difference at an appropriate wavelength pair was measured. A similar technique was used for mitochondria except that air-saturated incubation medium was used containing 0·25 μM-FCCP and the reduced-minus-oxidized extinction change at a given wavelength pair attributable to a cytochrome was calculated from the sum of changes occurring on (i) the addition of 5 mM-DL-glycerol 3-phosphate, and (ii) the exhaustion of oxygen.

Table XXVI. Effects of inhibitors of protein synthesis on the transition from F- to FR-cells of *C. utilis*. (Reproduced with permission from Clegg and Garland, 1971.)

| Additions to recovery medium | | Mitochondrial properties | | | | Inhibition of pyruvate and malate oxidation by piericidin A (%) |
| | | Oxygen uptake rate (ng-atoms/min per mg of protein) | | P/O ratio | | |
Inhibitor	FeSO$_4$ (μM)	Pyruvate + malate	DL-glycerol 3-phosphate	Pyruvate + malate	DL-glycerol 3-phosphate	
None	200	128	235	3·06	1·84	95
	200	130	260	3·4	2·14	80
	200	270	510	2·6	2·0	65
	200	44	148	2·4	1·75	30
	1	74	140	2·6	1·5	70
	1	113	246	2·7	1·6	58
	1	86	230	2·4	1·6	81
	1	34	61	3·0	1·8	80
Cycloheximide (100 μg/ml)	200	150	135	2·3	1·7	0
	200	35	73	2·5	1·4	0
	200	120	135	2·2	1·7	0
Erythromycin (2·0 mg/ml)	200	50	100	No respiratory control		
Cycloheximide (100 μg/ml)	1	121	149	2·2	1·7	0
Chloramphenicol (2·5 mg/ml)	200	50	180	—	—	10
	200	72	242	2·3	1·8	40
	200	93	140	2·8	1·9	90
Tetracycline (2 mg/ml)	100	235	240	2·9	1·9	60
	1	129	224	2·9	1·5	60

The recovery procedure was carried out in the presence of FeSO$_4$ and inhibitors of protein synthesis as indicated. Mitochondria were then prepared and assayed for respiration rates, inhibitor sensitivities and P/O ratios as described by Light and Garland (1971). Each line of the table represents a separate mitochondrial preparation.

Table XXVII. Properties of mitochondria from F- and FG-cells of *C. utilis*. (Reproduced with permission from Clegg and Garland, 1971.)

Type of cells	No. of preparations	Oxygen uptake rate (ng-atoms of O_2/min per mg of protein)		P/O ratio		Inhibition of pyruvate and malate oxidation (%)	
		Pyruvate + malate	DL-glycerol 3-phosphate	Pyruvate + malate	DL-glycerol 3-phosphate	Piericidin A	Antimycin A
F	7	170 ± 40	260 ± 27	1·83 ± 0·28	1·76 ± 0·17	0	>90
FG	4	182 ± 41	150 ± 82	2·5 ± 0·35	1·7 ± 0·23	0	>90

FG-cells were obtained in continuous culture by poising growth between iron and glycerol limitation. The methods were otherwise as described by Light and Garland (1971). The values are the means ± S.E.M.

compared with those prepared from glycerol-limited cells. Increases in the non-haem iron and acid-labile sulphide concentrations of SMP's accompany the recovery of Site 1 and piericidin sensitivity and these increases are halved by the presence of cycloheximide.

The production of cells poised between iron and glycerol limitation ("FG cells") results in organisms in which mitochondria possess Site 1 energy conservation but lack piericidin A sensitivity (Table XXVII;

Table XXVIII. Concentrations of non-haem iron, acid-labile sulphide and cytochromes in ETPm preparations from various cell types of *C. utilis*. (Reproduced with permission from Clegg and Garland, 1971.)

Cell type	Non-haem iron	Acid-labile sulphide	Cytochromes		
			$a + a_3$	b	$c + c_1$
G (growth on 2 μM-iron)	0·82	0·61	0·28	0·33	0·21
F	0·11	0·03	0·09	0·187	0·07
FR_1	0·46	0·47	0·09	0·10	0·13
FR_0	0·186	0·164	0·10	0·10	0·12
FR_0 (cycloheximide)	0·119	0·065	0·08	0·08	0·07
FR_1 (cycloheximide)	0·174	0·34	0·12	0·13	0·12

All concentrations are expressed as nmol/mg of protein. Non-haem iron was assayed by the radioisotopic method except for G-cells where the spectrophotometric procedure was used.

Clegg *et al.*, 1969; Clegg and Garland, 1971), and these results indicate that energy conservation at Site 1 does not require electron flow to proceed through a piericidin A- or rotenone-sensitive route. Restriction of the iron supplied to chemostat cultures of *C. utilis* to a concentration just above that required for growth limitation demonstrates that a 10- to 20-fold decrease of non-haem iron of both cells and mitochondria does not affect the growth yield per unit carbon source. The non-haem iron content of SMPs prepared from these functionally normal cells is about 0·5–0·8 ng-atom/mg of protein, a value close to the concentrations of cytochromes and flavoproteins. A summary of the concentrations of non-haem iron, acid-labile sulphide and cytochromes of electron transport particles obtained from cells of various phenotypes is presented in Table XXVIII.

In a study of sulphate-limited chemostat cultures of *C. utilis*, Haddock and Garland (1971) showed loss of piericidin A sensitivity and Site 1 energy conservation (Table XXIX). Aerobic incubation for 8 h of sulphate-limited cells with low concentrations of sulphate (50 μM or less) led to the recovery of both these mitochondrial properties. The use of higher concentrations of sulphate (250 μM or more) in the recovery process also led to the appearance of an alternative pathway of electron

Table XXIX. Respiration rates, respiratory control ratios and P/O ratios for mitochondria from sulphate-limited cells (S-cells) of *C. utilis*. (Reproduced with permission from Haddock and Garland, 1971.)

Substrate	Respiration rate (ng-atoms of oxygen/min per mg of protein)	Respiratory-control ratio	P/O ratio
Pyruvate (5 mM) + L-malate (5 mM)	134 ± 33 (6)	2·6 ± 0·3 (3)	1·9 ± 0·1 (6)
DL-glycerol-3-phosphate (5 mM)	196 ± 81 (6)	2·3 ± 0·3 (6)	1·9 ± 0·2 (6)
2-oxoglutarate (5 mM)	38 ± 7 (4)	2·1 ± 0·3 (4)	2·1 ± 0·2 (4)
NADH (0·5 mM)	195 ± 14 (2)	2·4 ± 0·2 (2)	1·8 ± 0·1 (2)
Endogenous	7 ± 4 (6)	1·0 (6)	—

The respiration rates listed are the rates observed in the presence of ADP and substrate and are uncorrected for any contribution by endogenous substrates. Respiratory-control ratios and P/O ratios were calculated as described by Light and Garland (1971). The experiments were repeated with several different preparations of mitochondria, and the results are expressed as the mean ±S.E.M., with the number of observations in parentheses.

Table XXX. Respiration rates, respiratory-control ratios and P/O ratios for mitochondria from SR_0 and SR_0 (cycloheximide)-cells of *C. utilis*. (Reproduced with permission from Haddock and Garland, 1971.)

Substrate	Mitochondria from SR_0-cells			Mitochondria from SR_0(cycloheximide)-cells		
	Respiration rate (ng-atoms of oxygen/min per mg of protein)	Respiratory-control ratio	P/O ratio	Respiration rate (ng-atoms of oxygen/min per mg of protein)	Respiratory-control ratio	P/O ratio
Pyruvate (5 mM) + L-malate (5 mM)	234 ± 26 (4)	3·4 ± 0·4 (4)	2·6 ± 0·2 (4)	220 ± 30 (2)	2·5 ± 0·3 (2)	2·7 ± 0·1 (2)
DL-glycerol 3-phosphate (5 mM)	349 ± 52 (4)	2·7 ± 0·3 (4)	1·9 ± 0·2 (4)	283 ± 37 (2)	2·7 ± 0·3 (2)	1·7 ± 0·1 (2)
2-oxoglutarate (5 mM)	134 ± 31 (4)	2·9 ± 0·3 (4)	2·9 ± 0·4 (4)	89 ± 11 (2)	2·2 ± 0·2 (2)	2·9 ± 0·3 (2)
NADH (0·5 mM)	340 ± 25 (2)	2·0 ± 0·3 (2)	1·9 ± 0·2 (2)	276 ± 31 (2)	2·0 ± 0·2 (2)	1·8 ± 0·2 (2)
Endogenous	107 ± 34 (4)	2·3 ± 0·5 (4)	2·8 ± 0·5 (4)	46 ± 23 (2)	1·0	—

transport (see p. 148). The inclusion of cycloheximide in the recovery medium resulted in the recovery by mitochondria of Site 1, but not of piericidin sensitivity (Table XXX). Measurements of the cytochrome contents of sulphate-limited cells and their mitochondria indicated that the concentrations of the cytochromes were similar in mitochondria from sulphate-limited cells and in glycerol-limited cells, although the concentration of mitochondria within the cells was decreased by one-third in the sulphate-limited organisms (Table XXXI). Sensitivity to antimycin A or

Table XXXI. Cytochrome contents of glycerol- and sulphate-limited cells of *C. utilis* and their mitochondria. (Reproduced with permission from Haddock and Garland, 1971.)

	Whole cells (nmol/g dry wt)		Isolated mitochondria (nmol/mg of protein)	
	Glycerol-limited	Sulphate-limited	Glycerol-limited	Sulphate-limited
Cytochrome $a + a_3$	34	20	0·10	0·09
Cytochrome b	25	15	0·13	0·11
Cytochrome $c + c_1$	63	25	0·11	0·10

Spectrophotometric measurements of cytochrome concentrations were made as described by Light and Garland (1971).

CN^- and the second and third energy-conservation sites were unaffected by sulphate-limited growth. The non-haem iron concentration of SMPs was unaffected by sulphate limitation, whereas the acid-labile sulphide concentration was lowered ten fold; the recovery procedure resulted in marked increases in the acid-labile sulphide concentration. These effects are similar to those found in mitochondria from iron-limited cells (Light *et al.*, 1968; Light and Garland, 1971), and provide independent support for the conclusion that non-haem iron proteins are involved in piericidin A sensitivity and in the mechanism of energy conservation at Site 1. EPR spectroscopy at 98°K demonstrates that the $g = 1·94$ signal associated with the NADH dehydrogenase system need not be present in detectable amounts for the occurrence of Site 1 energy conservation (Clegg *et al.*, 1969), although a close correlation between this signal and piericidin A sensitivity does occur. This correlation is not so clearly demonstrable in iron-limited batch cultures (Ohnishi *et al.*, 1969) which cannot give uniform cell populations of precisely-controlled composition. A parallel increase of all species of EPR signals detectable at 22°K has been observed when the iron concentration was raised in the culture medium of batch cultures (Ohnishi *et al.*, 1971). EPR studies at 19°K (Ohnishi *et al.*, 1972b) have shown that when Site 1 phosphorylation is induced *in vivo* by aerating *C. utilis* cells grown under iron-deficient conditions, the induction

Table XXXII. Phenotypic modifications of *C. utilis*. (Reproduced with permission from Ragan and Garland, 1971.)

Abbreviation	Growth or other conditions	Properties of derived mitochondria	
		Site 1 energy conservation	Piericidin A sensitivity
G*	Glycerol-limited	+	+
F†	Iron-limited	−	−
S†	Sulphate-limited	−	−
FR*	F-cells recovered with iron	+	+
FR(cycloheximide)‡	F-cells recovered with iron and cycloheximide	+	−
FR(anaerobic)‡	F-cells recovered with iron anaerobically	−	−
SR₀†	S-cells recovered without sulphate	+	+
SR₀(cycloheximide)†	S-cells recovered without sulphate in the presence of cycloheximide	+	−
SR₁†	S-cells recovered with 1 mM-sulphate	+	+

(+) and (−) indicate presence and absence of the property listed. References are: * Light and Garland (1971); † Haddock and Garland (1971); ‡ Clegg and Garland (1971).

of iron-sulphur proteins and piericidin A sensitivity also occurs. The development of both these properties is inhibited by cycloheximide. Unlike the results of Clegg et al. (1969) those of Ohnishi et al. (1972b) suggest that EPR detectable iron-sulphur proteins (at least Centres 1 and 2) may play a role in Site 1 energy conservation. A summary of the properties of mitochondria derived from the various phenotypic modifications of C. utilis (Ragan and Garland, 1971) are shown in Table XXXII; spectro-scopic resolution of flavoproteins and non-haem iron proteins of these mitochondria has been achieved (see p. 128).

Iron-deficiency in C. utilis leads to an eight-fold reduction in the NADH-ferricyanide reductase activity of SMPs (Clegg and Skyrme, 1973). The piericidin-sensitive reduction of UQ, and duroquinone by NADH is completely lost and the rates of reduction of substituted naph-thoquinones is about halved, as is the rate of NADH oxidation by O_2. Polyacrylamide gel electrophoretic analyses of mitochondrial membrane proteins revealed that iron-deficiency results in the loss of at least two polypeptides from the mitochondrial membrane; neither of these poly-peptides is part of a cytochrome, but one of them is part of the NADH dehydrogenase. The non-haem iron proteins of mitochondrial membranes cannot be identified electrophoretically in solutions containing SDS as the prosthetic group is released. This can be avoided by use of non-ionic detergents, but satisfactory conditions for electrophoretic separation have not yet been discovered. The observation that iron deficiency gives a 5–10-fold reduction in cytochrome content with few concomitant changes in electrophoretically-separated membrane proteins suggests that the synthesis of cytochrome apoproteins and their integration into the mito-chondrial membranes is not controlled by, or coordinated with, synthesis of the haem prosthetic groups.

2. Copper limitation

The concentration of copper in the growth medium influences the cyto-chrome c oxidase activity of yeast (Elvehjem, 1931), and the growth of S. cerevisiae is diminished in media containing less than 0·15 μM copper (Giorgio et al., 1963). Growth of S. cerevisiae in a medium containing between 0·04 μM and 0·08 μM copper led to a decrease of 70% in the concentration of cytochrome $a + a_3$ (Wohlrab and Jacobs, 1967). Keyhani and Chance (1971) have studied the synthesis of cytochromes which occurs when C. utilis cells, grown in a copper-deficient medium are returned to copper-sufficient media. Under these conditions a rapid synthesis of cytochromes $a + a_3$ occurs over the first 10 h of growth accompanied by small fluctuations in cellular content of cytochromes c, b_T and b_K. Hardly any alteration in the ratio cytochrome $a + a_3$ to cyto-chrome a was observed. The total amount of cytochrome c oxidase produced increased proportionally with the amount of copper present up to 200 μg/l. Synthesis of cytochrome c oxidase did not proceed in the presence of copper if net cell growth was prevented by omitting the carbon source. A mixture of cycloheximide (10 μg/ml) and chloramphenicol (4 mg/ml) in the growth medium inhibited the restoration of cytochrome c oxidase during copper supplementation.

The effects of copper limitation have also been studied using continuous cultures of $C.$ $utilis$ (Light, 1972a, b). The minimal cell content of cytochrome $a + a_3$ under copper-limited growth conditions was about 5 nmole/g dry wt of cells which represented a ten-fold reduction from the normal content of glycerol-limited copper-supplemented cells. The concentrations of cytochromes b and $c_1 + c$ of the copper-deficient cells were increased such that the total haem content of the cells was virtually unchanged; the ratio cytochrome b: cytochrome $c_1 + c$: cytochrome $a + a_3$ was 1: 2: 0·2, whereas that in cells from carbon-limited cultures was 1: 2: 2. Oxygen-uptake by the copper-deficient cells was about one-third that of cells from carbon-limited cultures and was not sensitive to inhibition by rotenone. Mitochondria isolated from the copper-limited cells showed decreased State 3 respiration rates with pyruvate + malate or ascorbate + TMPD as substrates; these results strongly suggest that cytochrome c oxidase had become rate-limiting for respiration. The P/O ratios obtained with L-glycerol-3-phosphate (1·8) or NADH (1·65) were similar in the two types of mitochondria; those for pyruvate + malate were somewhat reduced for the mitochondria from copper-limited cells (2·45), and it was evident that all three energy conservation sites associated with the respiratory chain of this yeast were still functional. Sensitivity to antimycin-A and CN^- were not altered. Rotenone-insensitivity of pyruvate + malate oxidation by mitochondria isolated from copper-deficient cells was an unexpected finding, as this property has been specifically correlated with alteration of the mitochondrial content of iron-sulphur protein (Garland, 1970). It was suggested that the drastic changes in relative cytochrome concentrations of the mitochondrial membranes produced during copper-limited growth may lead to an alteration of the packing of respiratory components, and this might be correlated with a diminished sensitivity to rotenone. Aerobic incubation of copper-deficient cells in 40 μM–copper sulphate solution overnight at 30°C resulted in a five-fold increase in the cell concentration of cytochrome $a + a_3$, an increase in mitochondrial respiration rate and a return of rotenone sensitivity. This copper-recovery system provides a non-growing system in which cytochrome $a + a_3$ is restored, presumably to a pre-existing membrane; recovery was inhibited when cycloheximide (0·1 mg/ml) was present. Some synthesis of cytochrome $a + a_3$ did occur in the presence of chloramphenicol, but this may have reflected impermeability of the

Table XXXIII. Cytochrome contents in mitochondria from *P. polycephalum* grown under normal and low-haem conditions. (Reproduced with permission from Barnes *et al.*, 1973.)

Methods of reduction	Growth conditions	Conc. of cytochrome $a + a_3$ (nmol/mg of protein)	Conc. of cytochrome b (nmol/mg of protein)	Conc. of cytochrome c (nmol/mg of protein)
Succinate (anaerobic)	Normal	0·0713 ± 0·01	0·061 ± 0·01	0·165 ± 0·045
Succinate (anaerobic)	Low haem	0·036 ± 0·012	0·031 ± 0·012	0·091 ± 0·05
Dithionite	Normal	0·072 ± 0·013	0·215 ± 0·021	0·260 ± 0·008
Dithionite	Low haem	0·055 ± 0·022	0·102 ± 0·06	0·123 ± 0·06
NADH (anaerobic)	Normal	0·0622 ± 0·005	0·036 ± 0·009	0·177 ± 0·054

After reduction with substrates or dithionite, the concentrations of cytochromes a, b and c were calculated from the extinction changes at 605 minus 630 nm, 560 minus 570 nm, or 550 minus 540 nm, by using ϵ mM; of 16, 20 and 20 respectively (Chance, 1964).

intact organisms to the antibiotic (Light, 1972b). Selection of mutants through their ability to by-pass a copper-requiring process which was previously rate-limiting for growth has been observed in copper-limited continuous culture of *C. utilis*, Downie and Garland (1973b) see p. 150. Copper-limited *Neurospora crassa* is deficient in cytochrome *c* oxidase, has a SHAM-sensitive alternative electron transport system and retains a functional mitochondrial system of protein synthesis (Schwab, 1973, 1974).

3. Effect of haem limitation

Physarum polycephalum requires protohaem for growth and cytochrome synthesis (Colleran and Jones, 1973); the following enzymes of haem synthesis were not detectable in extracts: δ-aminolaevulinate synthetase, conversion of δ-aminolaevulinate into porphyrinogens and porphyrins, ferrochelatase (Barnes *et al.*, 1973). Haem-limited cultures were slow growing and the cell yield was only 10–20% of normal cultures. The concentration of cytochromes in mitochondrial fractions was about 50% that of normal mitochondria (Table XXXIII).

4. Effect of pantothenic acid deficiency

The respiration rate of a pantothenic acid-requiring strain of *S. cerevisiae* was reduced to about 6% of the normal rate after growth in a defined medium deficient in this growth factor (Hosono *et al.*, 1972). This decreased respiratory ability was quantitatively a more pronounced effect than that produced by growing this strain in the presence of high glucose concentrations (5%). Pantothenic acid-deficient yeast accumulated hydrogen sulphide in the medium. The cellular content of cytochromes $a + a_3$ and b was decreased and the cytochrome c oxidase activity of extracts was reduced one thousand fold; cytochrome c peroxidase activity was halved. The specific activities of malate dehydrogenase, fumarase and succinate dehydrogenase were also lowered in extracts of pantothenic acid-deficient cells, whilst those of the enzymes of haem-synthesis were not appreciably altered. Hosono *et al.* (1972) concluded that the decreased respiratory activities produced by pantothenate deficiency may be a consequence of the rather specific effect on cytochrome c oxidase content of the cells.

5. Effect of inositol deficiency

When cultures of the inositol-deficient yeast *S. carlsbergensis* are grown in the absence of inositol the respiration rates and cytochrome complement of cells are not markedly altered (Paltauf and Johnson, 1970; Ghosh *et al.*, 1960; Ghosh and Bhattacharyya, 1967), and no gross damage to mitochondrial morphology is evident. Inositol-deficient cells of *Shizosaccharo-myces pombe* show an enhanced regulatory control of fermentative rates by O_2 (Dohi *et al.*, 1973).

6. Effect of choline deficiency

The concentrations of the cytochromes and the specific activities of malate and succinate dehydrogenases are the same in mitochondria isolated from choline-deficient as in those from choline-supplemented cultures of a choline-requiring strain of *Neurospora crassa*, but the activity of cytochrome c oxidase is reduced (Luck 1965a). The phospholipid: protein ratio is halved in the mitochondria isolated from cells grown at low concentrations of choline as compared with those from organisms grown at high concentrations.

7. Effect of carbon starvation

Withdrawal of carbon sources leads to a succession of ultrastructura changes and ultimately to increased autophagy of *Euglena* mitochondria (Brandes, *et al.*, 1964; Leedale and Buetow, 1970). Respiratory activity of organisms declines during starvation e.g. in *Prototheca zopfii* both the ability to respire added acetate and the endogenous respiration decrease (with little loss of cell viability) over a period of forty days. Although cells starved for this period still show recognizable mitochondrial profiles in electron micrographs, examination of spectra of whole cell suspensions at $77°K$ indicates complete loss of a- and c-type cytochromes. Dithionite reduced-oxidized spectra reveal the presence of absorption maxima at 559, 553 and 428 nm (D. Lloyd, unpublished results).

In aerated stationary phase cultures of yeast the respiratory activity declines in stepwise fashion (Miura and Yanagita, 1972). The first loss is attributed to a rise in NADPH-dependent lipid peroxidation. The second phase of decreased respiration occurred after 14 days and was accompanied by loss of viability. Inactivation of succinate-cytochrome c oxido-reductase and cytochrome c oxidase commenced at about 7 days.

E. Effects of Growth with Different Carbon Sources

1. Formation of inducible mitochondrial enzymes

It has been shown that the respiratory systems of algae can be extensively modified by changes of carbon and energy sources in their growth media. Adaptation to growth on a different substrate results in the induced synthesis of enzymes of the appropriate energy-yielding pathway. Adaptation of acetate-grown *Polytoma uvella* to growth on butyrate (Cirillo,

1956, 1957) is an example of this process, and in this case the presence of a nitrogen source is obligatory. Adaptation can also occur in the absence of net cell growth. Non-proliferating suspensions of the colourless alga *Prototheca zopfii* (grown on acetate) oxidize propionate only after a lag of 20–30 min, and the adaptation involves induction of the enzymes for β-oxidation of the new energy source (Lloyd and Venables, 1967). These enzymes are partly located in mitochondria, and organelles isolated from propionate-grown cells oxidized propionyl-CoA, β-hydroxypropionate and malonic semialdehyde. Rates of oxidation in the presence of Pi were ADP dependent, and the P/O ratios determined for these substrates were 2·2, 2·9 and 2·7 respectively. Mitochondria from acetate-grown cells did not possess the key enzymes of the β-oxidation pathway and intermediates were not oxidized. The difficulty of disintegration of cells of *P. zopfii* makes critical subcellular fractionation studies impossible, and further studies were carried out with the more easily disrupted flagellate, *Polytomella caeca* (Lloyd *et al.*, 1968). Adaptation of this organism to propionate after growth on acetate as sole carbon source also involves the formation of β-oxidation enzymes. The β-hydroxypropionate dehydrogenase and malonic semialdehyde dehydrogenase activities of both propionate-grown and propionate-adapted cells are located in mitochondrial fractions. Mitochondria isolated from propionate-grown cells, and from acetate-grown cells fully adapted to propionate, oxidize succinate, α-ketoglutarate, β-hydroxypropionate and malonic semialdehyde, and show respiratory control with these substrates. Mitochondria from acetate-grown cells exhibit ADP-dependent oxidation of succinate and α-ketoglutarate but do not oxidize β-hydroxypropionate or malonic semialdehyde (Table XXXIV). Mitochondria isolated from acetate-grown cells adapted to propionate for 5 h slowly oxidize β-hydroxypropionate and malonic semialdehyde, but no tightly coupled phosphorlyation is detectable (Fig. 75). The two key enzymes of propionate oxidation are located within the NAD-impermeable barrier and appear to be membrane bound. The formation of these inducible enzymes during propionate adaptation is inhibited by cycloheximide and actinomycin D, but not by chloramphenicol. These results suggest that this adaptation in the absence of net cell growth involves extramitochondrial enzyme synthesis followed by a progressive integration of the enzymes of the new pathway into the inner mitochondrial membranes of pre-existing mitochondria.

Thr adaptation of starved baker's yeast to acetate results in increasing respiration rates of intact cells and an increased activity of isocitrate lyase (Gosling and Duggan, 1971). No marked changes in cytochrome content of these cells were detected and subcellular fractionation of homogenates was not reported.

Some of the enzymes synthesized on transferring *Neurospora* mycelium

Table XXXIV. Effect of adaptation of *Polytomella caeca* on respiratory activity of isolated mitochondria. (Reproduced with permission from Lloyd *et al.*, 1968.)

| Source of mitochondria | | Acetate-grown cells | | | Acetate-grown cells adapted to propionate | | | | | |
| | | | | | 5 h adaptation | | | 18 h adaptation | | |
Substrate	Final conc. (mM)	Uptake of O₂ (nmoles/mg of protein/min.) (state 3)	Respiratory control ratio	P/O ratio	Uptake of O₂ (nmoles/mg of protein/min.) (state 3)	Respiratory control ratio	P/O ratio	Uptake of O₂ (nmoles/mg of protein/min.) (state 3)	Respiratory control ratio	P/O ratio
α-oxoglutarate	10·0	11·4	3·5	3·1	14·2	3·1	3·3	11·4	3·5	3·1
β-hydroxy-propionate	10·0	0	—	—	7·5	1·0	—	14·0	2·6	2·4
Malonic semialdehyde	1·0	0	—	—	3·2	1·0	—	4·1	1·0	—

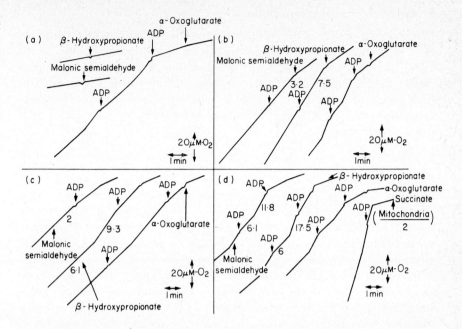

Fig. 75. Polarographic traces of oxygen uptake by mitochondrial suspensions from *Polyto-mella caeca*. Mitochondria (1–1·5 mg of protein) were added to 0·9 ml of buffer, pH 7·4 (0·25 M–sucrose 5 mM–EDTA–15 mM–KH₂PO₄–bovine serum albumin (0·15%, w/v) that had been equilibrated with air at 25C°. Additions of substrates (10 μmoles, except for malonic semialdehyde, 1 μmole) and ADP (80 nmoles) were made as indicated. The temperature of incubation was 25°C. The numbers refer to oxygen uptake rates in nmoles/mg, of protein/min above the endogenous values; ADP did not accelerate the endogenous uptake rates. Mitochondria were: (a) from cells grown with acetate; (b) from non-proliferating acetate-grown cells adapted to propionate for 5 h; (c) from acetate-grown cells adapted to propionate for 18 h; (d) from propionate grown cells. (Reproduced with permission from Lloyd, *et al.*, 1968.)

from a medium containing sucrose to one containing acetate may be selectively inhibited by procedures which inhibit mitochondrial respiration (Flavell and Woodward, 1971). This system also provides evidence for the energy-dependent transference of newly-synthesized proteins from cytoplasmic ribosomes into mitochondria.

Growth of *Candida utilis* in chemostat cultures under conditions of acetate-limitation results in the loss from mitochondria of the external rotenone-insensitive NADH dehydrogenase (Fukami *et al.*, 1970). Growth on acetate is completely inhibited by rotenone or piericidin A, whereas these inhibitors do not prevent growth on glycerol. The results of Ohnishi (1972), confirm this observation, but she presents an alternative explanation for rotenone sensitivity of acetate grown cells, i.e. loss from the inner membrane of a by-pass of the rotenone sensitive site.

2. Alteration of lipid composition

Unusual fatty acids can be incorporated into the mitochondrial membranes of *C. lipolytica* by growth of the organisms on *n*-alkanes (Skipton *et al.*, 1973). Supplementation of a defined medium with specific fatty acids results in an enrichment of the resulting mitochondrial lipids with the fatty acid added (Janki *et al.*, 1974). The activities and transition temperatures in Arrhenius plots of ATPase and kynurenine hydroxylase respond to such modifications.

F. Growth in the Presence of Inhibitors of Electron Transport

Phenotypic modifications of mitochondria may also be produced by inclusion of inhibitors of electron transport in the growth medium. Thus growth of *Euglena gracilis* in the presence of antimycin A leads to the development of a pathway of electron transport alternative to the respiratory chain (Sharpless and Butow, 1970b). An alteration of mitochondrial electron transport pathways has also been observed after growth with antimycin A in mycelia of *Aspergillus oryzae* (Kawakita, 1970a, b), *Neurospora crassa* (Lambowitz and Slayman, 1971) and also in many other fungi (for review see Sherald and Sisler, 1972). The alternative pathways involved have been described in Chapter 3. The cellular content of haemoglobin in yeast is increased after growth with antimycin A (Yčas, 1956).

Growth of *Tetrahymena pyriformis* in the presence of increasing concentrations of CN^- results in a decreased sensitivity to this inhibitor (McCashland, 1956); the mechanism of this phenomenon has not been investigated.

G. Growth in the Presence of Inhibitors of Mitochondrial Macromolecular Synthesis

1. Growth with inhibitors of mitochondrial protein synthesis

Chloramphenicol, a specific inhibitor of bacterial protein synthesis, also inhibits amino acid incorporation by intact mitochondria from *Tetrahymena pyriformis* (Mager, 1960) and mammalian tissues (Kroon, 1963), but does not inhibit protein synthesis by isolated cytoplasmic ribosomes from eukaryotic cells (von Ehrenstein and Lipmann, 1961; So and Davie, 1963; Bretthauer *et al.*, 1963). Growth of *S. cerevisiae* in the presence of 4 mg/ml of chloramphenicol yields cells deficient in cytochromes *b*, c_1 and $a + a_3$ (Huang *et al.*, 1966). Both the cellular respiration rates and the activities of NADH and succinate dehydrogenases in extracts were diminished after chloramphenicol-inhibited growth, a five-fold reduction in the rate of ADP translocation in isolated mitochondria also has been observed (Lauquin and Vignais, 1973). Similar changes in cytochrome contents were also seen in chloramphenicol-inhibited cultures of the obligately aerobic yeast *Candida parapsilosis* (Linnane *et al.*, 1968b; Kellerman *et al.*, 1969), and in continuous cultures of *S. carlsbergensis* (Gray and Rogers, 1971a, b). Oscillations in cytochrome content of both *S. carlsbergensis* and *C. utilis* are produced following the addition and the removal of chloramphenicol. The period of these oscillations is a function of growth rate.

The effects produced by chloramphenicol were reversible on washing the cells and allowing them to grow in the absence of the antibiotic; thus the modification is a phenotypic one and does not involve the selection of a mutant population.

The primary effect of chloramphenicol in both facultatively anaerobic *petite*-forming yeasts (e.g. *S. cerevisiae*) and in non-*petite*-forming yeasts (e.g. *Hansenula wingei* and *C. utilis*) under identical physiological conditions (in the absence of catabolite repression) has been shown to be in the elaboration of cytochrome *c* oxidase activity and the formation of spectoscopically detectable cytochromes $a + a_3$ (Mahler and Perlman, 1971). The properties of mitochondrial ATPase are also modified during chloramphenicol-inhibited growth of *S. cerevisiae* (Schatz, 1968). Decreased sensitivity to oligomycin and cold lability of this enzyme suggest an alteration of a component of the ATP synthetase complex, analogous to that seen in some cytoplasmic *petite* mutants (see p. 311). It is clear that these effects are correlated with the finding that chloramphenicol inhibits protein synthesis by isolated yeast mitochondria (Wintersberger, 1964a, b; Linnane *et al.*, 1968b; Lamb *et al.*, 1968). The similarity of the mitochondrial protein synthesis system to that of bacteria was confirmed by the finding that several other antibiotics known to inhibit bacterial protein synthesis (tetracycline, oxytetracycline, erythromycin, carbomycin, spiramycin, oleandomycin and lincomycin) also cause a total or partial inhibition of cytochrome $a + a_3$ formation in growing cultures of *S. cerevisiae* (Clark-Walker and Linnane, 1966). Erythromycin gives similar results for the inhibition of respiratory activities in glycerol—but not glucose-grown *Shizosaccharomyces pombe* (Michel *et al.*, 1971). Similar results have also been obtained with neomycin B, neomycin C and paromomycin (Davey *et al.*, 1970). In all these experiments cytochrome *c* synthesis was unaffected; indeed elevation of cellular levels of cytochrome *c* sometimes is observed. In glucose-repressed cells chloramphenicol had little effect on the synthesis of malate dehydrogenase or fumarase. The activities of both these enzymes was however markedly decreased in extracts of cells from chloramphenicol-inhibited, glucose-derepressed cultures (Clark-Walker and Linnane, 1967). Growth with the antibiotic did not affect the synthesis of the cytochromes or of malate dehydrogenase or fumarase in a cytoplasmic respiratory deficient mutant.

The high concentrations of chloramphenicol used in these experiments may produce effects other than those specific to the mitochondrial protein synthesis system. The degree of sensitivity to antibiotics is genetically determined, and strains of *S. cerevisiae* sensitive to 50 μg/ml chloramphenicol or 10 μg/ml erythromycin may be selected (Wilkie *et al.*, 1967). Use of these strains can minimize the possibilities of side-effects when investigating the effects of antibotics as specific inhibitors of mitochondrial protein synthesis. Chloramphenicol at high concentrations is known to block the respiration of heart mitochondria in the region of Site 1 (Freeman and Haldar, 1968) and may have a similar effect in microorganisms which possess this coupling site. Neither the D(−) nor the L(+)-*threo*-isomers affect substrate oxidation by coupled mitochondria from *S. cerevisiae* (Ball and Tustanoff, 1970); D(−)-chloramphenicol is the isomer responsible for the inhibition of the development of the respiratory phase of growth of yeast.

Mycelium of *Pythium ultimum* grown with 100 μg/ml chloramphenicol contained no detectable cytochromes $a + a_3$ or b, but showed an increased content of cytochrome c (Marchant and Smith, 1968a). The only morphological lesion observable in electron micrographs was an alteration in the inner-membrane subunits; these could not be demonstrated in negatively-stained preparations from the chloramphenicol-inhibited hyphae. Growth of a facultatively anaerobic mutant of *Neurospora crassa* with 2 mg/ml chloramphenicol led to a marked decrease in the cytochrome c oxidase activity of extracts as compared with those from control cultures not containing the antibiotic (Howell *et al.*, 1971); abnormal mitochondrial morphology was also evident.

When wild-type *N. crassa* is grown in the presence of chloramphenicol (2 mg/ml) the development of cyanide-resistant respiration occurs (Lambowitz and Slayman, 1971). Mitochondria isolated from these hyphae (like those from *poky*, Lambowitz *et al.*, 1972a) contain both a cytochrome system which is sensitive to CN^- and an alternate oxidase resistant to CN^- but sensitive to SHAM (see p. 145). These mitochondria showed little respiratory control and low P/O ratios; essentially all oxidative phosphorylation is inhibited by CN^- or antimycin A.

Growth of *N. crassa* in the presence of increasing concentrations of chloramphenicol leads to a gradual alteration of the respiratory chain components and a production of phenotypes approaching those of the *mi-1* mutant (von Jagow and Klingenberg, 1972). The effect on cytochromes $a + a_3$ and cytochrome b_T is most marked (Fig. 76, Table XXXV); the cytochrome b_T content is finally lowered to one-sixteenth whereas that of cytochrome b_K only to about one-half. A close correlation between the antimycin A titre and the content of cytochrome b_T supports

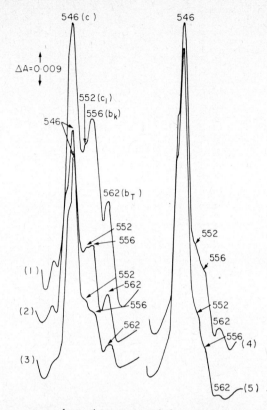

Fig. 76. Low temperature absorption spectra of the mitochondrial b-type and c-type cytochromes from different cultures of *Neurospora crassa* containing increasing concentrations of chloramphenicol. The concentrations of chloramphenicol in the different cultures (1–5) are given in the second column of Table XXXV. The protein concentration in the cuvette (mg/ml) of the preparations 1–5 was, 4·0, 2·7, 3·0, 3·5, and 2·7 respectively. The spectra were obtained with a light path of 5 mm; reduction of the measuring sample was with dithionite; the reference sample was kept aerobic. (Reproduced with permission from von Jagow and Klingenberg, 1972.)

the suggestion that the binding of the inhibitor is closely related to the b cytochrome associated with energy transfer.

Spectral studies on mycelia of *Aspergillus oryzae* grown in the presence of 2 mg/ml chloramphenicol indicate that a reduction in the content of the a-type and all three b-type cytochromes (λ_{max} (77°K) at 556, 559·5 and 564 nm) is produced (Wakiyama and Ogura, 1972).

The succinoxidase and NADH oxidase activities of mitochondria isolated from the flagellate *Polytomella caeca* after chloramphenicol-inhibited growth were decreased by 43% and 72% respectively (Lloyd *et al.*, 1970b). The content of cytochromes $a + a_3$ was decreased by 40%

Table XXXV. Alteration of the contents of respiratory chain components and of sensitivity of NADH-respiration to KCN as a function of chloramphenicol concentration in growth medium. (Reproduced with permission from von Jagow and Klingenberg, 1972.)

No. of culture	CAP	UQ	Cytochromes					Inhibition of NADH-respiration by KCN
			b_T	b_K	c_1	c	aa_3	
	(g/l)		(μmoles/g protein)					(%)
1	0	0·40	0·32	0·39	0·36	1·30	0·30	100
2	0·25	0·45	0·23	0·37	0·37	1·30	0·26	90
3	0·50	0·62	0·15	0·24	0·44	1·42	0·22	78
4	1·00	1·10	0·07	0·26	0·42	1·47	0·13	32
5	5·00	3·21	0·02	0·23	0·51	2·07	0·01	24

and the cytochrome c oxidase activity was halved. No alteration in the activities of rotenone-sensitive NADH-cytochrome c oxidoreductase or succinate-cytochrome c oxidoreductase was detected. However the rate of reduction of the individual electron transport components, during the aerobic-anaerobic transition of mitochondria respiring succinate, suggested that supply of electrons to the respiratory chain was rate-limiting, and that a chloramphenicol-induced lesion at the dehydrogenase level might be more important than the decreased cytochrome c oxidase activity in the overall respiratory deficiency of the organisms. No major cytochrome dislocation was evident from observation of steady-state levels of reduction of these components, suggesting that the respiratory chain was itself intact (Table XXXVI). No mitochondrial morphological changes were seen at the level of resolution possible in electron micrographs. The presence of chloramphenicol does produce changes in mitochondrial ultrastructure of the Chrysophyte alga, *Ochromonas danica* (Smith-Johannsen and Gibbs, 1972).

When chloramphenicol (500 μg/ml) is added to an exponentially growing culture of *Tetrahymena pyriformis*, growth is completely inhibited within a few generations (Turner and Lloyd, 1971). Organisms from chloramphenicol-inhibited cultures have a greater number of mito-chondria than normal cells; growth and division of mitochondria is apparently uncoupled from the growth and division of the organisms (see p. 456). Growth with chloramphenicol did not lead to any significant alteration in the cytochrome content per cell or per mg of mitochondrial protein of isolated mitochondria; it is possible that a specific effect on cytochromes $a + a_3$ in this organism was masked by cytochrome a_{620}. Mitochondria from cells grown with chloramphenicol show de-creased oxygen uptake rates with α-ketoglutarate or succinate as respiratory substrates and increased sensitivity to the Site I inhibitors, rotenone and piericidin A. Respiratory control was not detectable and the oligomycin-sensitivity of the F_1–ATPase is reduced in mitochondria from cells grown with the antibiotic (Table XXXVII). After inoculation into medium not containing chloramphenicol, these respiratory deficient organisms grow normally after a lag of 24 h. Impaired coupling of oxida-tive phosphorylation is not seen in mitochondria from organisms starved for 24 h in the presence of chloramphenicol: this suggests that the turnover of enzymes involved in phosphorylating electron transport proceeds only very slowly in the absence of net cell growth when chloramphenicol is present. Decreased content of cytochromes has been observed in mito-chondria from chloramphenicol-inhibited *Tetrahymena pyriformis* by Rohatgi and Krawiec (1973) and by Mason *et al.* (1970).

In sensitive strains of *Paramecium aurelia* growth with chloramphenicol or erythromycin leads to an elongation of the mitochondria, a reduction

Table XXXVI. Steady-state levels of reduction of electron-transport components (substrate 6 mM-succinate), and half-times for their reduction in the aerobic–anaerobic transition (State 4 → State 5) in mitochondria from *Polytomella caeca* grown in the absence and presence of chloramphenicol. (Reproduced with permission from Lloyd et al., 1970b.)

Electron-transport component	Cytochrome				Flavo-protein (+iron-S proteins)
	$a(+a_3)$	$a_3(+a)$	c	b	
Wavelength pair (nm)	605–630	445–455	550–540	430–410	565–510
Without chloramphenicol					
$t_{\frac{1}{2}}$ red^n State 4 → State 5 (s)	2·0	2·5	1·0	1·3	1·3
% red^n State 4	5·0	—	21·0	48·0	26·0
% not reduced in State 5 but reduced by dithionite	—	—	—	33·0	39·0
With choramphenicol					
$t_{\frac{1}{2}}$ red^n State 4 → State 5 (s)	7·5	7·5	5·0	7·0	7·0
% red^n State 4	—	6·4	14·0	44·3	23·2
% not reduced in State 5 but reduced by dithionite	—	—	—	37·0	49·0

100% reduction of a component taken as anaerobic reduction level minus level in absence of exogenous substrate (State 5–State 2), (Chance and Williams, 1956). Absence of a figure indicates that if any reduction occurred it was below the limit of detection. Figures representative of results obtained with eight different mitochondrial suspensions. Temperature of incubation 20°C.

Table XXXVII. Oxidative phosphorylation in mitochondria isolated from *Tetrahymena pyriformis* after: (A) growth at various chloramphenicol concentrations, and (B) after starvation in the absence and presence of chloramphenicol. (Reproduced with permission from Turner and Lloyd, 1971.)

A.

Chloramphenicol (μg/ml)	Time of Growth (h)	Substrate					
		9 mM-sodium succinate			9 mM-sodium α-ketoglutarate		
		O₂ uptake (nmoles/min/mg protein) (State 3)	Respiratory control ratio	P/O ratio	O₂ uptake (nmoles/min/mg protein) (State 3)	Respiratory control ratio	P/O ratio
0	24	99 ± 10 (3)	1·1–16	0·8–1·5	51 ± 6 (3)	1·3–2·3	1·2–2·2
100	24	34 ± 8 (3)	2·6–2·9	0·7–1·5	43 ± 2 (3)	1·1–1·4	1·1–1·2
500	48	16 ± 2 (3)	—	—	7·5 ± 1 (3)	—	—

B. Organisms were harvested and resuspended in the inorganic medium of Hamburger and Zeuthen (1957); starved cultures were force-aerated.

(a) Starvation in the absence of chloramphenicol

		O₂ uptake (State 3)	Respiratory control ratio	P/O ratio	O₂ uptake (State 3)	Respiratory control ratio	P/O ratio
for 4 h		74 ± 6 (4)	1·3–1·4	0·6–0·8	34 ± 3 (3)	1·2–1·3	0·6–1·0
for 24 h		64 ± 5 (3)	1·3–1·4	0·8	30 ± 5 (3)	—	—

(b) Starvation in the presence of 500 μg/ml chloramphenicol

		O₂ uptake (State 3)	Respiratory control ratio	P/O ratio	O₂ uptake (State 3)	Respiratory control ratio	P/O ratio
for 4 h		94 ± 9 (7)	1·4–2·1	0·7–1·0	48 ± 4 (6)	1·6–2·9	1·1–1·7
for 24 h		91 ± 7 (5)	1·7–2·6	1·0–1·7	50 ± 6 (4)	1·6–2·5	1·5–2·1

Absence of a figure indicates failure to detect respiratory control

I

in the number of cristae and the appearance of abnormal lamellar cristae and periodic structures (Adoutte *et al.*, 1972).

Canavanine, a structural analogue of arginine, selectively inhibits mitochondrial protein synthesis in yeast mitochondria *in vivo* in certain strains (Wilkie, 1970a, b, c), and a similar effect has been noted with *p*-fluorophenylalanine although there was no correlation in activity between the two analogues. In sensitive strains, a marked reduction in growth rate of cells on non-fermentable (but not fermentable) medium was accompanied by inhibition of synthesis of *a*- and *b*-type cytochromes and reduced respiratory activity. Up to one hundred-fold difference in sensitivity between the mitochondrial and cytoplasmic protein synthesis systems has been demonstrated.

Lampren (an imino phenazine) completely inhibits the growth of *S. cerevisiae* on non-fermentable medium when present at 0·5 μg/ml (Rhodes and Wilkie, 1973) but only leads to incomplete inhibition of glucose-supported growth even at 10 μg/ml. Protein and RNA synthesis is depressed by 50% by 5 μg/ml of the drug in cells growing on glycerol-containing medium, but 20 μg/ml are required to produce an equivalent effect in glucose medium. All the cytochromes are significantly decreased in inhibited cells. Lampren also acts as an artificial electron acceptor and restores oxygen-uptake in CN^--inhibited cells.

2. Growth in the presence of agents which react with mitochondrial DNA

(a) *Phenotypic changes produced by growth with acriflavine.* Acriflavine inhibits the synthesis of respiratory enzymes during the respiratory adaptation of cells of *S. cerevisiae* (Slonimski, 1953a, b), this effect is apparently independent of the formation of *petite* mutants by this dye (Ephrussi, 1953). De Deken (1961) and Bulder (1964b) have shown that whereas all species of *Saccharomyces* are induced to form *petites* by acriflavin (*petite* positive) most other species of yeasts are "*petite* negative" and do not give rise to *petites*. Although there is this clear division into two groups, the synthesis of respiratory enzymes is inhibited to a similar extent in all the yeast species studied. Bulder has suggested therefore that the primary action of the dye is on the inhibition of synthesis of respiratory enzymes, and that the mutagenic effect of acriflavin is a secondary one in *petite*-positive species. The dye is known to intercalate with the DNA double helix *in vitro* (Lerman, 1963) and with mt DNA *in vivo* (Tewari *et al.*, 1966a, b). Euflavine present in the growth medium of the obligately aerobic yeast *Candida parapsilosis* leads to the production of mitochondria with markedly disorganized cristae deficient in cytochromes $a + a_3$, b, c_1 and succinate dehydrogenase (Kellerman *et al.*, 1969).

Acriflavine induces loss of the DNA-containing material from the kinetoplast of trypanosomatids (Trager and Rudzinska, 1964; Mühlpfordt, 1963a, b; Kusel *et al.*, 1967). After growth of *Crithidia fasciculata* for 3–4 days with 5 μM–acriflavine, the organisms were cytochrome deficient (Hill and Anderson, 1969). There was a five-fold decrease in the cytochrome $a + a_3$ of the mitochondria of these cells, and a decrease in the mitochondrial enzyme activities of NADH-succinate- and L-α-glycerophosphate oxidases and succinate- and L-α-glycerophosphate dehydrogenases was also evident. Acriflavine-resistant strains of *Leishmania tarentolae* have been isolated (Strauss, 1972).

(*b*) *Effects of growth with nalidixic acid.* Nalidixic acid, a specific inhibitor of DNA synthesis in bacteria (Goss *et al.*, 1965) also acts specifically on mitochondrial DNA synthesis in the yeast *Kluyveromyces lactis* (Luha *et al.*, 1971). In experiments with *S. cerevisiae*, Mounolou and Perrodin (1968) showed that nalidixic acid inhibited total cell DNA synthesis during respiratory adaptation. In the presence of 1% glucose, inhibition of growth and DNA synthesis was only temporary: the duration of the transition phase from the utilization of the fermentable substrate (glucose) to its oxidizable products in batch cultures is increased in the presence of nalidixic acid (Kaplan and Criddle, 1970; Gross and Smith, 1972). The *petite* frequency was also increased by a small but significant amount with increasing concentrations of the drug. The pattern of effects on the phenotypic expression of mitochondrial enzyme activities produced by addition of nalidixic acid to growing cultures of *S. cerevisiae* is similar to that produced by chloramphenicol (Mahler *et al.*, 1968).

(*c*) *Effects of growth with ethidium bromide.* Ethidium bromide is a phenanthridine dye which intercalates between the bases of double-stranded DNA (Waring, 1965, 1968). This effect potentially leads to interference with the replication and transcription of the genome in mitochondria (Goldring *et al.*, 1970a, b; Nass, 1970) and kinetoplasts (Riou, 1967). The drug also preferentially inhibits mitochondrial DNA polymerase *in vitro* as compared with the nuclear enzyme (Meyer and Simpson, 1969) and the synthesis of RNA and protein present in mitochondria (Zylber *et al.*, 1969; Zylber and Penman, 1969; Knight, 1969; Fukuhara and Kajawa, 1970; Perlman and Mahler, 1971a, b). Ethidium bromide treatment leads to loss of both the kinetoplast and some mitochondrial enzymes (succinate dehydrogenase, α-glycerophosphatase and ATPase) (Thirion and Laub, 1972b) in trypanosomes (Riou, 1967) and is an extremely efficient cytoplasmic mutagen in facultatively anaerobic yeasts (Slonimski *et al.*, 1968). The rate of mutation increases from the spontaneous rate of 1% up to 100% within two hours of treatment. In

obligately aerobic yeasts the synthesis of cytochromes $a + a_3$, b_1 and c_1
are inhibited by ethidium bromide, although the *petite* mutation is not
induced in this case (Kellerman *et al.*, 1969). Reversible induction of
respiratory deficiency involving loss of cytochromes $a + a_3$ and b has
also been shown in *Shizosaccharomyces pombe* (Schwab *et al.*, 1971) and
the kinetics of respiratory degeneration and regeneration have been
followed (Luha and Whittaker, 1972). Ethidium bromide does not inhibit
the growth of *S. cerevisiae* on a fermentable carbon source, but in the
presence of 3% lactate growth continues for two generations and then
stops (Mahler and Perlman, 1971). After this time effects due to the
irreparable alterations in mt DNA become manifest, but during these
first two generations there is a total inhibition of cytochrome $a + a_3$

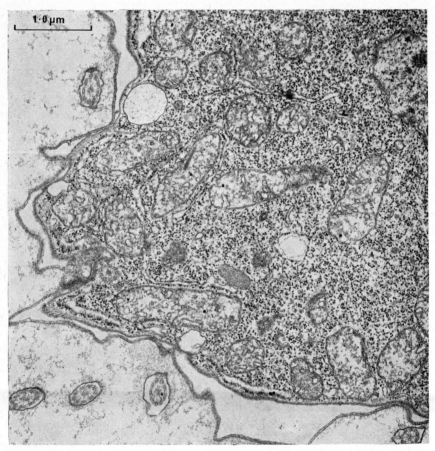

FIG. 77. Loss of mitochondrial cristae in *Tetrahymena pyriformis* after growth for 12 h in
the presence of 20 μg/ml ethidium bromide. (Reproduced with permission of R. Charret.)

synthesis and a continued synthesis of cytochromes $c + c_1$ as measured both in growing cells and isolated mitochondrial fractions. Thus ethidium bromide gives rise to alterations in phenotype qualitatively similar to those produced by chloramphenicol. Mitochondrial membranes so produced still contain material which can cross-react with antiserum to cytochrome c oxidase (Shakespeare and Mahler, 1972). In the *petite negative yeast). C. utilis* cytochrome c oxidase synthesis was also inhibited by the inclusion of ethidium bromide in the growth medium.

Progressive disorganization of the ultrastructure of cristae occurs when cultures of *Tetrahymena pyriformis* are grown in the presence of 20 µg/ml of ethidium bromide (Fig. 77; Meyer *et al.*, 1972; Charret, 1972). Reduction in mitochondrial cytochrome content accompanies this disorganization (Rohatgi and Krawiec, 1973). Decreased respiratory activity and cytochrome c oxidase levels accompany the disorganization of mitochondrial ultrastructure produced on treatment of *Acetabularia mediterranea* with ethidium bromide (Hielporn and Limbosch, 1971). Cytochrome c oxidase synthesis is also blocked in amoebae of *Dictyostelium discoideum* by ethidium bromide (Stuchell *et al.*, 1973).

H. Effect of Light upon Mitochondrial Functions

1. Interaction between chloroplasts and mitochondria

A hypothesis for the interaction of chloroplasts and mitochondria was developed to explain observations on the pale green mutant of *Chlamydomonas reinhardii* (Hiyama *et al.*, 1959). Thus the light-induced oxidation of cytochrome b_{563} is eliminated by its O_2-induced oxidation, whilst O_2-induced oxidation is greatly diminished under illumination. Antimycin A diminished the O_2-induced oxidation of cytochrome b_{563} but did not affect the light-induced oxidation. Thus it was proposed that mitochondrial reducing equivalents either pass to O_2 down the electron transport chain, or proceed by way of cytochrome b_{563} to oxygen or to cytochrome f. The possibility that the apparent derepression of respiratory oxygen uptake which accompanies activation of Photosystem I in *C. reinhardii* (Kok effect) involves diversion of a respiratory reductant has been discussed by Hewley and Myers (1971).

The alterations of mitochondrial structure and function which accompany chloroplast development in higher plants (Borque and Naylor, 1972) have not been studied in algae.

2. Effect of light on respiration and the respiratory apparatus of nonphotosynthetic eukaryotes

Blue light enhances the endogenous respiration of a starved chlorophyll-free mutant of *Chlorella vulgaris*; saturation was achieved at low intensities (500 ergs cm^{-2} s^{-1}) to give a three-fold stimulation of oxygen uptake

(Kowallik, 1967). The action spectra reveal two maxima (at 450 nm and 375 nm) which implicate either a flavin or a carotenoid (Kowallik, 1968), but further experiments to eliminate one of these candidates were not conclusive. It has been suggested (Pickett, 1967) that a similar effect on *Chlorella pyrenoidosa* may involve the CN^--sensitive portion of dark respiration.

The inhibitory effects of high light intensities (2.9×10^4 ergs cm^{-2} s^{-1}) on the growth of *Prototheca zopfii* (Epel and Krauss, 1966) have been traced to photodestruction of cytochromes $c(551)$ $b(559)$ and a_3 (Fig. 78; Epel and Butler, 1969, 1970b). O_2 is required during irradiation

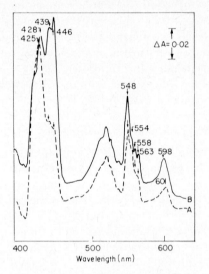

FIG. 78. Photodestruction of cytochromes in *Prototheca zopofii*. Low temperature difference spectra (8.7×10^6 cells/ml, with added light scatter agent, 0.33 g of $CaCO_3$/ml). Curve A: Cells irradiated 2 h (AH-6 super pressure mercury lamp, Corning filters 5433 + 3850, $I = 2.5 \times 10^6$ ergs cm^{-2} s^{-1}); dithionite reduced versus substrate-(0.25% ethanol) respiring aerated cells; curve B; dark control cells; dithionite versus substrate-respiring aerated cells. (Reproduced with permission from Epel and Butler, 1969.)

for cytochrome loss; CN^- protects cytochrome a_3 from damage. Visible light, at sub-lethal doses, also inhibits the formation of oxygen-induced respiratory enzymes of yeast during respiratory adaptation (Guerin and Sulkowski, 1966). Recovery of adaptation ability occurs when cells are illuminated and then left in the dark. The most light-sensitive period of adaptation occurs during the first hour of aeration and the sensitivity of the cells depends on the phase of growth at which the cells are harvested. The recovery of the respiratory capacity of irradiated starved yeast cells on incubation in the dark for 24 h is an energy-dependent process which may be prevented by the addition of uncoupling agents such as DNP (Ninnemann, 1972). The recovery, which involves the synthesis of cyto-chrome c oxidase is also inhibited by cycloheximide and by chloram-phenicol.

III. Modifications of Mitochondria During Cellular Differentiation

The study of processes of cellular differentiation involved during the life-cycles of many eukaryotic microorganisms constitutes a major field of biochemical activity. Yet, with a few exceptions, there has been little attempt to investigate the extent of changes in mitochondrial composition and activity or the control mechanisms implicated in the striking varia-tions in respiration observed at the whole cell level. Morphological changes in mitochondria observed during cellular differentiation have been outlined in Chapter 1. Systems amenable to study include differenti-ation in the water mould *Blastocladiella emersonii* (Cantino, 1966), in the acellular slime mould *Physarum polycephalum* (Sauer, 1973) and in the cellular slime mould *Dictyostelium discoideum* (Garrod and Ashworth, 1973). Changes in the activities of succinate dehydrogenase and cyto-chrome c oxidase and alterations in their distributions between different cell types produced during the differentiation of *D. mucorales* have been reported (Takeuchi, 1960).

The complex life-cycles of many algal species also provide interesting systems for the study of modified mitochondrial structure and function. When zygospores of *Chlamydomonas reinhardii* (Dang) germinate, there is a remarkable change in the characteristics of respiration (Hommersand and Thimann, 1965). The oxygen-uptake of the zygospores is cyanide-sensitive and is inhibited by carbon monoxide: inhibition by CO is reversible by blue light. The respiration of the vegetative cells is resistant to CN^- (1·0 mM) and CO (99%). The transition from the sensitive to the resistant state occurs between the eighth and twelfth hour of germination, i.e. at the same time as photosynthesis begins in the light. The terminal oxidase of the zygospores has a higher affinity for O_2 than that present in vegetative cells, and these observations suggest that a classical inhibitor-sensitive electron transport chain present in the zygospores is not the major route of electron transport during the greatly increased respiratory activity of the vegetative organisms.

Studies at the level of isolated mitochondria have been commenced with three types of differentiating system: the encystment of protozoa, the bloodstream and culture forms of trypanosomatids, and the sporu-lation of fungi.

A. Changes in Mitochondria During the Encystment of Protozoa

Mitochondria isolated from encysting cells of *Hartmannella castellanii* showed considerable differences from those isolated from vegetative amoebae (Griffiths *et al.*, 1967; Lloyd and Griffiths 1968). Respiration rates with succinate, malate, pyruvate + malate, NADH, α-ketoglutarate, 3-hydroxybutyrate and glutamate were higher in mitochondria isolated from amoebae after 4 h of encystment, but after 24 h encystment, respiration with these substrates was decreased. These changes were accompanied by a progressive impairment of the phosphorylating ability and loss of cytochromes (Fig. 79). Isolation of intact mitochondria from

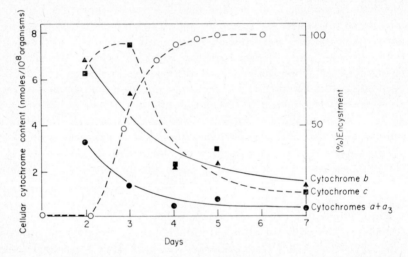

FIG. 79. Loss of mitochondrial cytochromes during the encystment of *Hartmannella castellanii*. The organisms were grown in proteose-peptone containing 20 mM–MgCl₂. Encystment commenced after 2 days growth and was complete after 4·5 days. Progress of encystment (○) measured by estimation of cellulose. Cellular content of cytochromes measured in different spectra (dithionite reduced—ferricyanide oxidized). (Reproduced with permission of A. H. Chagla and A. J. Griffiths.)

organisms at later stages of encystment (2–6 days) was not possible, as harsh mechanical treatment was necessary for cell breakage. Mitochondria isolated from encysting cells of *Acanthamoeba rhysodes* however, show similar respiratory metabolism to those of vegetative cells (Band and Mohrlok, 1969). Both types of mitochondria oxidize succinate, α-ketoglutarate and pyruvate + malate, show respiratory control and carry out oxidative phosphorylation. P/O ratios obtained for these substrates were 1·5–1·8 2·2–3·1 and 2·5–3·2 respectively. However, complete encystment leads to the development of mitochondria which oxidized these substrates only very slowly.

The presence of phospholipase activity in *Colpoda steinii* interferes with the study of mitochondrial properties during the encystment of this ciliate (Tibbs and Marshall 1969). Succinate-PMS oxidoreductase decreases by only 30% when this organism encysts, suggesting that the process involves no major degradation of mitochondria. This view is also supported by the rapidity of restoration of respiration of whole cells during excystment.

B. Changes in Mitochondrial Function During the Life-cycle of Trypanosomatids

The processes of differentiation during the life cycles of Trypanosomatidae have been reviewed by Newton *et al.* (1973b; Fig. 80). Studies of the bloodstream and culture forms of *T. brucei* have shown that the morphological changes occurring during the life-cycles of these organisms are accompanied by alternating development and regression of mitochondrial structures, TCA cycle enzymes and cytochromes. Of particular interest is the switch from the cyanide-insensitive respiration of the bloodstream forms to the predominantly cyanide-sensitive electron transport system of culture forms (Evans and Brown, 1971; Srivastava and Bowman, 1972; Bowman *et al.*, 1972; Brown *et al.*, 1973). The proportion of respiration mediated by either of the two alternative pathways in culture forms (see p. 142) depends upon the culture conditons (Evans and Brown, 1971; Ray and Cross, 1972).

C. Mitochondrial Activities During the Development of Fungal Spores

Changes of mitochondrial activities during the germination of conidia to produce hyphae in wild-type *Neurospora crassa* have been studied by several workers. Zalokar (1959) reported a ten-fold increase in succinate dehydrogenase during the first 8 h of growth, beginning with ungerminated conidia, whilst Turian (1962) reported a reciprocal correlation between succinate dehydrogenase activity and the conidiogenic capabilities of hyphae. Results in variance with these reports were obtained by Weiss (1965) who showed no change in the activity of this enzyme as assayed by the PMS method. A ten-fold increase in the activity of cytochrome c oxidase accompanied conidial germination. No alterations in the P/O ratios obtained with succinate as substrate were noted in mitochondria isolated at different stages of development suggesting no difference in phosphorylating capacity of *Neurospora* mitochondria between conidia and germinated conidia. The development of the cytochrome system of *Phycomyces* during the first 8 h of spore germination parallels the increased oxygen-uptake observed during this process (Keyhani *et al.*, 1972a). The presence of cytochromes $a + a_3$, b, c and c_1 was evident in ungerminated

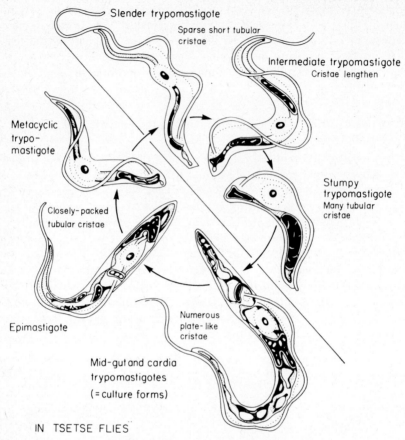

FIG. 80. The life-cycle of *Trypanosoma brucei* showing changes in the morphology of the mitochondria. (Reproduced with permission from Vickerman, 1971.)

spores, and all the cytochromes, increased in parallel over the period of germination studied, to give a 15-fold increase on a dry weight basis over 8 h. Inhibitors of mitochondrial macromolecular syntheses inhibit different steps in this differentiation, e.g. chloramphenicol inhibits germ tube formation, kanamycin allows formation of only a few germ tubes, and celesticetin, while allowing the growth of a mycelium prevents sporangiophore formation (Keyhani *et al.*, 1972b). Further studies on mitochondria isolated at different stages of fungal differentiation would be of great interest; the microcycle conidiation technique which enables examination of conidiation of *Aspergillus* spp. in submerged culture and eliminates the vegetative growth phase (Anderson and Smith, 1971) appears to provide an ideal experimental system for this purpose.

Induction of competence to form flocs, a prerequisite for conjugation in *Shizosaccharomyces pombe* requires both mitochondrial function and cytoplasmic protein synthesis (Calleja, 1973).

There is some evidence that mitochondrial protein synthesis is required during yeast sporulation. Puglisi and Zennaro (1971) observed that erythromycin prevents sporulation in erythromycin-sensitive strains but not in erthromycin-resistant strains. The antibiotic had no effect on the respiration of either strain. The development of respiration during a synchronously germinating ascospore population of *S. cerevisiae* has been followed (Rousseau and Halvorson, 1973).

Erythromycin also inhibits morphogenesis (development of mycelial forms) of the dimorphic yeast, *Endomycopsis capsularis* (Puglisi and Algeri, 1974).

IV. Changes in Mitochondrial Activity During the Cell-Cycle

Variations of mitochondrial activity during the cell-cycle, that is during the sequence of temporally-organized events occurring between successive cell divisions, have been studied in single cells and in cultures synchronized either by induction synchrony or cell selection techniques (Mitchison, 1971). The methods of choice for obtaining synchronous cultures (in which metabolic disturbances due to experimental procedures are kept at a minimum) are those based on cell selection procedures which do not involve temperature or nutritional shocks. The events occurring in the culture may ideally be interpreted as approximating to the changes occurring in individual cells, and this interpretation is clearly valid in cases where the culture obeys the criteria of "balanced" growth (Campbell, 1957). Thus parameters such as dry weight, total protein and respiration should double in magnitude over one cell-cycle, the sequence of events should be repeated in successive cycles, and the length of a cycle should be identical with the mean generation time of an exponential culture growing under similar conditions.

A. Changes in Respiration of Whole Cells During the Cell-cycle

1. Protozoa and algae

Measurements of respiration on single cells of *Tetrahymena pyriformis* indicate that the rate of oxygen consumption per cell increases linearly throughout the cell-cycle until just before cell division when the rate stops increasing (Zeuthen, 1953). Variations on this pattern were found in different experiments; some cells showed a uniformly linear increase throughout the entire cycle, whereas in other cells the increase was an exponential one (Lövlie, 1963). The respiration rate of *Tetrahymena* (expressed per cell) falls as cell division occurs in cultures synchronized by temperature shocks (Padilla *et al.*, 1966), and similar cultures of *Astasia longa* also show this phenomenon (Wilson and James, 1966).

Synchronous cultures of *Crithidia fasciculata* show complex changes in respiration rates during the cell-cycle (C. Edwards and D. Lloyd, unpublished results). A complex pattern of oxygen uptake changes also occurs during the cell-cycle of *Chlamydomonas reinhardii* (Osafune *et al.*, 1972a). The reversible formation of giant mitochondria by the fusion of many smaller mitochondria coincides with a decreased oxygen uptake of cell suspensions. However, changes in the highly branched mitochondria of this organism cannot be easily deduced without serial sectioning (Arnold *et al.*, 1972).

Studies with cultures of *Chlorella* indicate that the rates of endogenous respiration and of oxygen-uptake in the presence of added respiratory substrates show variations through the growth cycle (Talbert and Sorokin, 1971). However the exact course of changes is much in dispute, and is very dependent on the methods of synchronization used. Observations on cells separated into age groups by a centrifugation technique indicates that small cells possess higher respiratory activity than large cells approaching the division stage (Talbert and Sorokin, 1971).

Activities of succinate dehydrogenase and cytochrome *c* oxidase both increase in single step during the last third of the cell-cycle in *Chlorella pyrenoidosa* (Forde and John, 1973).

A sharp doubling of oxygen consumption per cell was observed at an intermediate stage in the cell-cycle of cultures of *Polytomella agilis* subjected to temperature-induced synchrony (Cantor and Klotz, 1971). The oxygen uptake pattern in a culture of *Polytomella caeca* synchronized by a process involving thiamine starvation is more complex. The addition of chloramphenicol (1 mg/ml) or ethidium bromide (20 μg/ml) does not prevent a doubling of cell numbers in this culture, whereas in an exponential culture either of the inhibitors allows only a 20% increase in cell count during a normal mean generation time. It is suggested that either, (a) a mitochondrial product required for cell division is produced early in the cell cycle, or (b) the result reflects the close interlock between the mitochondrial and cytoplasmic systems of macromolecular synthesis. Cycloheximide (100 μg/ml) allows only a 6% increase in cell numbers over one mean generation time in both exponential and synchronous cultures (M. Cantor and D. Lloyd, unpublished result).

2. Yeasts

Rates of oxygen uptake which increase in a series of abrupt steps, remaining constant in the intervening periods were reported for a synchronous culture of *Saccharomyces cerevisiae* prepared by a starvation method (Scopes and Williamson, 1964). The steps were less than doublings, and the time interval between steps was less than the length of the cell-cycle: total cell mass increased continuously through three doublings in cell numbers. Similar results were also obtained by Greksák *et al.* (1971), whereas a uniformly exponential increase in oxygen uptake was observed in synchronized yeast cultures by Cottrell and Avers (1970, 1971).

Oscillations of respiratory activity with one maximum occurring per cell-cycle have been observed in *S. cerevisiae* in synchronous cultures prepared by a size-selection method, irrespective of whether growth was in a defined medium (Küenzi and Fletcher, 1969; Nosoh and Takamiya, 1962) or in a complex medium (von Meyenburg, 1969; Wiemken *et al.*, 1970). Oscillations observed in respiratory activity in glucose-grown cultures were not evident when maltose was the carbon source (Dharmalingam and Jayaraman, 1973).

Single stepwise increases in oxygen uptake rate per ml of culture were observed during the cell-cycle of *Shizosaccharomyces pombe* synchronized by a nutritional induction method in a complex growth medium (Osumi and Sando, 1969; Osumi *et al.*, 1971); these authors claimed to find a doubling of mitochondrial numbers per cell coinciding with the abrupt increases in respiration rate. However the difficulties of estimation of mitochondrial numbers in sectioned organisms necessitate a more thorough investigation before this claim is confirmed, and it seems unlikely that respiration rate is proportional to mitochondrial numbers rather than mass. A more complex progress curve for oxygen uptake was obtained in this species when grown synchronously in a defined medium (containing 1% glucose) after a density-gradient centrifugation size selection procedure (Poole *et al.*, 1973). In this case both total cell protein and respiration increase exponentially overall so as to double during the cell-cycle, but an oscillation in oxygen uptake also occurred so as to produce two maxima per cycle (Fig. 81). The first maximum was observed during the phase of cell elongation prior to cell division and the second maximum coincided with the doubling in cell numbers. A doubling of the respiratory rate of samples withdrawn from the culture at a respiratory maximum was produced when 16 μM–CCCP was added; organisms taken at the minima showed a less marked response to the uncoupler. Heat production of the culture increased uniformly through the cell-cycle. Addition of concentrations of antimycin A or CN^- sufficient to give about 50% inhibition produced no attenuation of the respiratory oscillation, and on the basis of these results Poole *et al.* (1973) concluded that the respiration of the synchronous culture has two components. One of these increases exponentially at all times in the cycle, cannot be easily uncoupled and is preferen-

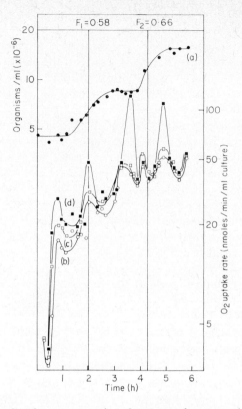

Fig. 81. Oxygen-uptake of yeast suspensions from a synchronous culture of *Schizosac-charomyces pombe* growing in the presence of 1% glucose and the effect of CCCP. F_1 and F_2 denote synchrony indices of the first and second doublings in cell numbers (a) respectively. O_2-uptake measurements on culture samples removed at frequent intervals from a synchronous culture were made in the absence (b), or the presence of 8·1 μM–CCCP (c), or 16·2 μm–CCCP (d). (Reproduced with permission from Poole *et al.*, 1973.)

tially blocked by inhibitors of electron transport acting at phosphorylation Sites II or III. The second component consists of easily-uncoupled electron transport chains which are only inhibited by high concentrations of antimycin A or CN^-. The activity of this component oscillates with a periodicity equal to 0·5 of a cell cycle, presumably due to the periodic expression of activity of a rate-limiting entity.

In a culture growing synchronously in the presence of 1% glycerol, oxygen-uptake shows a step pattern (Fig. 82). The two abrupt increases which occur during the cell cycle appear to correspond temporally with the rises leading to overshoot previously observed in glucose-grown *S. pombe* (Poole and Lloyd, 1974).

In a culture of *Candida utilis* (defined medium containing 1% glucose)

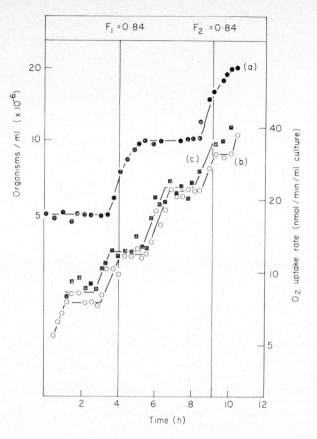

FIG. 82. Oxygen-uptake of yeast suspension from a synchronous culture of *Schizosaccharomyces pombe* growing in the presence of 1 % glycerol and the effect of CCCP. F_1 and F_2 denote the synchrony indices of the first and second doubling in cell numbers (a) respectively. O_2-uptake measurements on culture samples removed at frequent intervals were made in the absence (b) or the presence (c) of 0.3 μM–CCCP. (Reproduced with permission from Poole and Lloyd, 1974.)

made synchronous by means of a size-selection technique, oxygen-uptake increased to maxima three times in the cell cycle (R. K. Poole and D. Lloyd, unpublished results). In this system addition of the optimal concentration of uncoupler (1.1 μM–CCCP) to samples withdrawn from the culture produced a preferential stimulation of respiration at the minima so as to attenuate the oscillation, (Fig. 83). Evidently the control mechanisms involved in the production of variations in oxygen-uptake rates in the presence of 1% glucose are quite different in *C. utilis* from those observed in *S. pombe*.

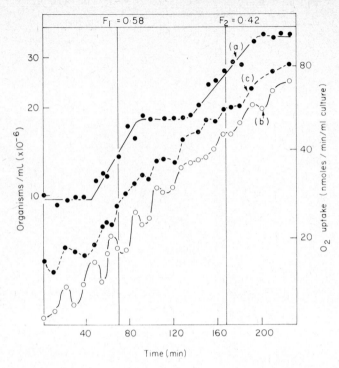

Fig. 83. Oxygen-uptake of yeast suspension from a synchronous culture of *Candida utilis* growing in the presence of 1 % glucose and the effect of CCCP. F_1 and F_2 denote the synchrony indices of the first and second doubling in cell numbers (a) respectively. O_2-uptake measurements on culture samples removed at frequent intervals from a synchronous culture were made in the absence (b), or the presence of 1·12 μM–CCCP (c). (Experiment of R. K. Poole.)

B. Changes in Mitochondrial Enzyme Activities during the Cell-cycle

Although many workers have mapped the changes in enzyme activities through the cell-cycle in synchronized cultures of eukaryotic organisms (Mitchison, 1971), little information is available on enzymes specifically associated with different classes of organelles. In *Saccharomyces cerevisiae* both cytochrome *c* oxidase and malate dehydrogenase activities exhibit single stepwise increases during each cell doubling; these increases coincided with the onset of budding (Cottrell and Avers, 1970, 1971). No attempt was made to distinguish between mitochondrial and extramitochondrial malate dehydrogenase. Similar stepwise increases were also reported for succinate dehydrogenase by Greksák *et al.* (1971). Periodic increases in total amounts of several mitochondrial enzymes were observed in glucose-repressed *Shizosaccharomyces pombe* by Poole and Lloyd (1973). In this case "peak" patterns

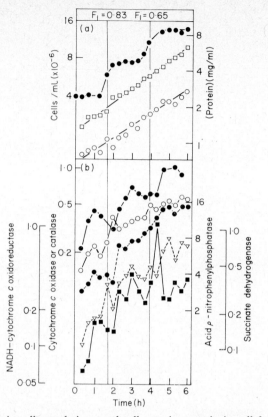

Fig. 84. Changes in cell population, total cell protein, protein in cell-free extracts and amounts of enzymes during growth in the presence of 1% glucose of a synchronously dividing culture of *Schizosaccharomyces pombe*. (a) Shows the synchrony (F_1 and F_2) of the first and second doublings in cell numbers respectively (●), and total (□) and released (○) protein of cell suspensions used to prepare cell-free extracts. Vertical lines indicate the midpoints of doublings in cell numbers. (b) Shows amounts of cytochrome *c* oxidase (●), catalase (○) acid *p*-nitrophenylphosphatase (●– – –●), NADH-cytochrome *c* oxidoreductase (■) and succinate dehydrogenase (▽). Amounts of enzymes are expressed as units/ml of culture. (Reproduced with permission from Poole and Lloyd, 1973.)

rather than steps were obtained (Fig. 84). Whereas the total cytochrome *c* oxidase rose to a maximum in the culture once in the cell-cycle, succinate-dehydrogenase showed two maxima. Other enzymes such as NADH-cytochrome *c* oxidoreductase, malate dehydrogenase and isocitrate dehydrogenase (NADP linked) also gave two maxima in the cycle. The timing of appearance of maximum activities in the division cycle was checked by assaying the enzymes in homogenates of cells which had been separated into classes by virtue of, (a) their size or, (b) their density, properties which vary uniformly through most of the cell-cycle; the

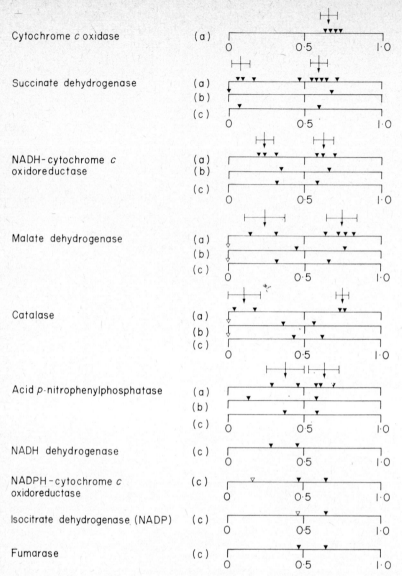

FIG. 85. Timings of maxima of enzyme amounts during the cell-cycle of glucose-repressed *Shizosaccharomyces pombe*. In (a) the abscissa represents (as a linear scale from 0–1·0) the cell-cycle taken as the period between the mid-points of the first and second doublings in cell numbers in synchronous cultures. (b) Shows the results of analysis of the cell-cycle by rate-zonal centrifugation, assuming the cell-cycle to be represented linearly with distance from the rotor centre, and by using the single maximum of cytochrome *c* oxidase at 0·67 of a cell-cycle as a reference point for timings of maxima of other enzymes. (c) Shows the results of analyses of the cell-cycle by isopycnic-zonal centrifugations, assuming the cell-cycle to be represented linearly as a reciprocal function of distance from the rotor centre and again normalizing the positions of maxima of enzyme amounts with respect to the position of the maximum of cytochrome *c* oxidase. ▼, indicate the possible occurrence of maxima of enzyme amounts. Arrows with cross bars are mean values with S.D. of timings of maxima determined from results obtained with synchronously-dividing cultures. (Reproduced with permission from Poole and Lloyd, 1973.)

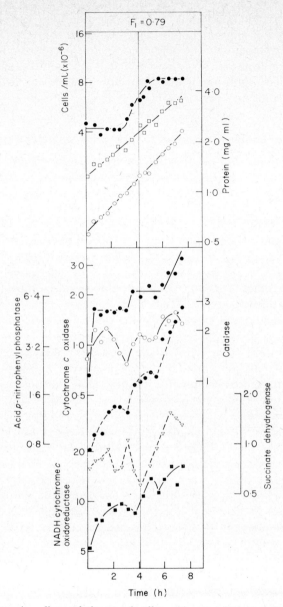

FIG. 86. Changes in cell population, total cell protein, protein in cell-free extracts and amounts of enzymes during growth of a glycerol-containing synchronously-dividing culture of *Shizosacchomyces pombe*. (a) Shows the synchrony index (F_1) of the first doubling in cell numbers (●) and total (–□–) and released (–○–) protein of cell suspensions used to prepare cell-free extracts. The vertical line indicates the mid-point in doubling of cell numbers. (b) Shows amounts of cytochrome *c* oxidase (–●–), catalase (–○–), acid *p*-nitro-phenyl phosphatase (– –●– –), NADH-cytochrome *c* oxidoreductase (–■–) and succinate dehydrogenase (– –▽– –). Amounts of enzymes are expressed as units/ml culture. (Reproduced with permission from Poole and Lloyd, 1974).

three methods of assessment gave good agreement (Fig. 85). In glycerol-grown cells cytochrome c oxidase showed two step-wise increases in amount, one during cell division and second at 0·5 of the cell-cycle (Fig. 86). All other enzymes assayed (including succinate dehydrogenate) exhibited two maxima per cell-cycle.

C. Changes in Cellular Content of Cytochromes During the Cell-cycle

In a synchronous culture of S. *pombe* growing in the presence of 1% glucose the time course of synthesis of cytochromes $a + a_3$ is similar to that of b_{563} (b_T); both cytochromes reach two maxima per cell-cycle (Poole *et al.*, 1973, Fig. 87). These cytochromes (the synthesis of which are partially dependent on the mitochondrial protein synthesis system, see pp. 422 and 427), show a completely different pattern of development from cytochromes c_{548}, b_{554} and b_{560}, all of which increase linearly over the first part of the cycle, then plateau off. Thus the ratio of individual cytochrome components of the inner mitochondrial membrane varies considerably throughout the cell-cycle. Comparison of the spectroscopically detectable cytochromes $a + a_3$ with the cytochrome c oxidase activities of extracts reveals that the increase in glucose-reducible a-type cytochrome over the first third of the cell-cycle does not result in a detectable increase in enzyme activity. The trough of enzyme activity represents a real disappearance in spectrophotometrically-detectable haemoprotein, and the subsequent maximum in the haemoprotein is also reflected in cytochrome c oxidase activity. The control mechanisms involved in this complex pattern of changes remains to be elucidated. Variation in the ratio cytochrome a_3: cytochrome a over the range 0·4–1·3 was observed during the cell-cycle indicating that the syntheses of the two components of the cytochrome c oxidase complex do not occur together. The time course for the development of cytochrome a is also different from that of cytochrome a_3 during the respiratory adaptation of S. *carlsbergensis* (Cartledge *et al.*, 1972; Chen and Charalampous, 1973). The relationship between cytochrome a and cytochrome a_3 to the seven subunits of the cytochrome c oxidase complex (see p. 424) is not yet clear; further studies on these systems may provide some insight into this relationship. Cytochrome P-450 shows a gradual increase through the cell-cycle whereas cytochrome P-420 declines and therefore shows the behaviour expected of a cytochrome precursor. The subcellular location of these two CO-reacting haemoproteins has not been elucidated in S. *pombe*.

The progress of synthesis of cytochrome $a + a_3$ also differs from that of cytochromes c_{548}, b_{554} and b_{560} in the cell-cycle of S. *pombe* growing in the presence of 1% glycerol (Fig. 88; Poole and Lloyd, 1974). In this case stepwise patterns of synthesis were observed, and whereas

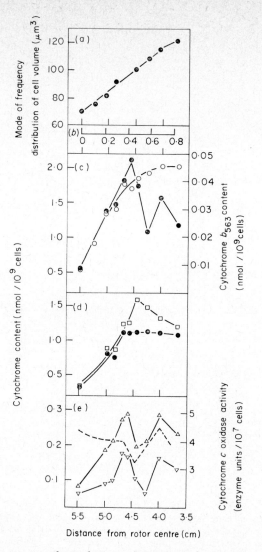

FIG. 87. Cellular contents of cytochromes after analysis of the cell-cycle by isopycnic zonal centrifugation on a dextran gradient of cells from a glucose-repressed exponentially-growing culture of *Schizosaccharomyces pombe*. 30 ml of cell suspension (containing 15·8 g wet wt cells) was loaded on the dextran gradients in a Beckman Ti-14(B29) zonal rotor. Centrifugation was at 35,000 rev/min for 50 min ($2 \times 10^6 \, g$ min at the sample zone: $\int_0^t \omega^2 \, dt = 4 \cdot 53 \times 10^{10} \, rad^2 \, s^{-1}$. Difference spectra (reduced-oxidized) of intact cells were recorded in successive fractions removed from the rotor; reduction was in the presence of 1 % glucose, oxidation was in the presence of 0·1 % H_2O_2. Spectra were recorded in a path length of 2 mm with a spectral bandwidth 2 nm, at 77°K on cell suspensions at 3×10^9 cells/ml. (a) Modes of frequency distribution of cell volumes in successive fractions removed from the rotor. (b) Resolved portion of the cell-cycle, represented on a linear scale, and normalized with respect to a cell volume of 115 μm^3 at 0·67 of a cell cycle. (c) c_{548} (–○–) and b_{563} (–●–). (d) b_{554} (–□–) and b_{560} (–●–) (e) $a + a_3$ (measured at 445–458 nm (–△–) and at 600–630 nm (–▽–) respectively, together with cytochrome *c* oxidase activity plotted as a function of the cell-cycle, as determined in a separate experiment (– – –). (Reproduced with permission from Poole *et al.*, 1974.)

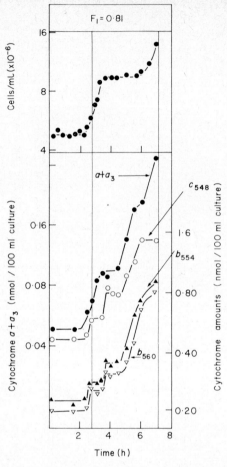

FIG. 88. Changes in cellular content of cytochrome during the growth of a glycerol-containing synchronously-dividing culture of *Schizosacchormyces pombe*. F_1 denotes the synchrony index of the first doubling in cell numbers (a). (b) Shows the amount of cytochrome $a + a_3$ (–●–), c_{548} (–○–), b_{554} (–▲–) and b_{560} (–▽–). Amounts of cytochromes are expressed as nmol/100 ml of culture. (Reproduced with permission from Poole and Lloyd, 1974.)

cytochrome $a + a_3$ shows a single step per cell-cycle, the other three cytochromes (2 b-type and 1 c-type) show a double step pattern.

Variations in the cellular content of cytochromes 556 and a_{608} during the cell-cycle of a synchronous culture of *Euglena gracilis* have also been observed (Calvayrac and Claisse, 1973). The highest content of cytochrome 556 appears to coincide with the non-dividing phase in which a mitochondrial network is observed.

PART II

BIOGENESIS

5

Mitochondrial DNA in Eukaryotic Microorganisms

I. Introduction

The biosynthesis of a functional mitochondrion requires the cooperation of two distinct genetic systems, that of the nucleus and that of the mito-chondrion itself. The physiochemical properties of mitochondrial DNA, its replication and possible genetic functions have been reviewed by Nass *et al.* (1965); Nass (1967); Borst *et al.* (1967); Borst and Kroon (1969); Nass (1969b) and Borst 1972, 1974a). In summary the mt DNA of animal cells has been characterized as a homogeneous population of covalently closed circular molecules, about 5 μm in contour length (corres-ponding to a mol. wt of 10^7) and comprising less than 1% of the total cell DNA. Plant mt DNA is larger than animal mt DNA, and lengths up to 62 μm have been reported (Wolstenholme and Gross, 1968). No obvious relation exists between the densities of mitochondrial and nuclear DNAs present in the same cell, and the values show "the pattern that might be expected of two distinct genetic systems each undergoing gradual but essentially independent changes in composition with evolu-tion" (Rabinowitz and Swift, 1970). The information content of mt DNAs studied by quantitative renaturation experiments suggests that the mole-cular weights calculated from renaturation rates are identical with those obtained from length measurements. Thus large-scale heterogeneity or gene repetitions are absent, and it is likely, but not proven, that all mt DNA molecules from a cell have the same base sequence; the possibility of microheterogeneity for mitochondrial tRNA genes cannot however be excluded (Borst, 1972). The question whether the nucleus contains one or more "master copies" of mt DNA (Wilkie, 1963) is not settled for all organisms, although the absence of an integrated copy of mt DNA in the nuclear genome of *Tetrahymena pyriformis* has been indicated in the studies of Flavell and Trampé (1973). Early conclusions that *extensive* base sequence homology occurs between mt DNA and nuclear DNA are incorrect (Borst, 1972).

II. Properties of mt DNA

A. Mt DNA in Wild-type Yeasts

The mt DNA of yeasts has been extensively investigated following the early work of Schatz et al. (1964) and Tewari et al. (1965).

Lysis of yeast mitochondria releases mt DNA as a homogeneous population of twisted circles 25 μm in circumference (Fig. 89); this length is equivalent to a mol. wt of 50×10^6 (Hollenberg et al., 1969, 1970a, b). Attempts to isolate these circles intact have been unsuccessful; heterogeneous linear DNA with an upper size limit of 26 μm has been reported by several investigators (Avers et al., 1968; Guerineau et al., 1968; Van Bruggen et al., 1968; Billheimer and Avers, 1969); many preparations have yielded DNA sedimenting at about 30S indicating mol. wts of the order of 20×10^6 (Goldring et al., 1970a, b; Hollenberg et al., 1970a). Sedimentation behaviour compatible with a mol. wt of 46×10^6 has recently been reported for DNA obtained from isolated yeast mitochondria (Blamire et al., 1972), a value which agrees well with size estimates by electron microscopy and reassociation studies. Another class of DNA has also been prepared from yeast; this consists of a pure population of covalently closed twisted molecules, homogeneous in size (monomers 2·2 μm, multiple length oligomers of $n \times 2·2$ μm), (Guerineau et al., 1971). A mitochondrial origin for this DNA has not been ruled out, but it has a similar density to that of nuclear DNA and is not the same as the heavy satellite DNA (γ-DNA) described by Moustacchi and Williamson (1966). This o-DNA is present in *petite* mutants which completely lack mt DNA (Clark-Walker, 1973).

Measurement of the kinetic complexity of yeast mt DNA in renaturation experiments indicates a value of the order of $50–100 \times 10^6$, which is in good agreement with the physical size data (Borst 1970; Hollenberg et al., 1970a, b). The uncertainty in this determination arises because the calculated kinetic complexity varies with the fragment size of the mt DNA used in the reaction, and this finding suggests the presence of repeated sequences in the mt DNA (Christiansen et al., 1971).

The base composition of *S. cerevisiae* mt DNA is 17% (G + C) (Bernardi et al., 1970; Mehrotra and Mahler, 1968). The discrepancy between this analytical value and the value of 24% (G + C) determined from a buoyant density of 1·683 g/cm^3 in CsCl (Bak et al., 1969; Fig. 90) or the value of 13% (G + C) calculated from a Tm in 0·15 M–NaCl of 74°C, may result from the presence of regions rich in A and T. This suggestion is reinforced by the characteristics of the CD and ORD spectra (Bernardi and Timasheff, 1970). These anomalous properties of yeast mt DNA, indicating an intramolecular heterogeneity at a size level of $1·5 \times 10^6$ daltons, correlate with the compositional heterogeneity of fragments of

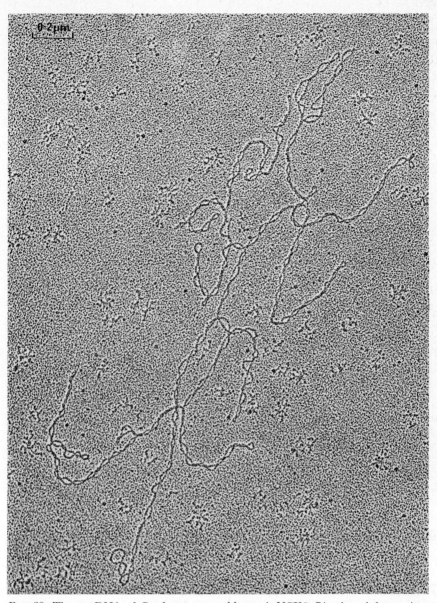

FIG. 89. The mt DNA of *Saccharomyces carlsbergensis* NCYC 74 released from mitochondria by osmotic shock and prepared by the Kleinschmidt technique. (Reproduced with permission of C. Hollenberg and W. Van der Vegte.)

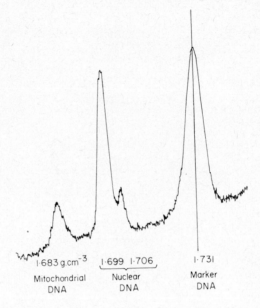

FIG. 90. Denistometer tracing of an equilibrium CsCl density gradient preparation of DNA from a respiratory competent diploid yeast. (Reproduced with permission from Williamson, 1970.)

mt DNA separated by hydroxyapatite chromatography (Bernardi *et al.*, 1972). Similar methods applied to enzymically produced fragments show that a large fraction, and possibly all of the yeast mitochondrial genome is formed by (G + C)-rich and (A + T)-rich stretches which have mol. wts of the order of 10^5 to 10^6 and which are intermingled (Piperno *et al.*, 1972). Whilst the former may correspond to mitochondrial genes (like those coding for ribosomal RNA) the biological significance of the (A + T)-rich stretches is not clear; it has been suggested that they may act as "spacer" sequences which are not transcribed (but see Borst, 1972). Long stretches are absent and estimates of the possible existence of repeating nucleotide sequences and amounts of alternating and non-alternating dA,T sequences have been made (Ehrlich *et al.*, 1972). This work suggests that the (A + T)-rich stretches are basically formed by intermingled short alternating and non-alternating dA,T sequences.

Williamson *et al.* (1971b), have measured the mt DNA contents of extracts of yeasts and in most cases mt DNA comprises between 15% and 23% of the total DNA of the cells. The haploid yeast cell (which contains approximately half as much mt DNA as a diploid yeast cell) contains about 0·005 pg (3×10^9 daltons) of mt DNA (Williamson, 1970), and accepting a figure of 26 μm for its length, then the haploid cell contains 50 mt DNA molecules each of 50×10^6 daltons. The number of copies of mt DNA per mitochondrion cannot yet be stated with any precision, as estimates of the number of mitochondria present in a yeast cell vary from as many as 25 (Avers, et al., 1964) to as few as 7 (Mahler, 1974) or even one (Hoffmann and Avers, 1973) for a haploid organism.

Recent experiments suggest that the idea of a "master-copy" of mt DNA present in the nucleus may be incorrect, but the matter is not finally settled. Thus attempts to reveal sequence homologies between mt DNA and nuclear DNA have not produced any convincing evidence for extensive homologies. Although RNA from yeast mitochondria hybridizes with both mt DNA and nuclear DNA, purification of the mt RNA by prehybridization with mt DNA eliminates its ability to hybridize with nuclear DNA (Faures and Genin, 1969; Cohen et al., 1970; Fukuhara, 1970). Even so, if the nucleus contained only one master copy, it could be barely detected in hybridization experiments. The amino acid coding capacity of yeast mt DNA so rich in A and T has been discussed by Grossman et al. (1971); much of this DNA cannot specify proteins. Even so Borst (1972) finds "no compelling reason to assume that the five-fold difference in size between the mt DNAs of fungi and animals is not the reflection of a much larger number of genes coding for mitochondrial proteins on the fungal mt DNAs".

The mt DNA of *petite*-negative yeasts (Bulder, 1964a) has been characterized in CsCl gradients. In *S. lactis* its buoyant density is about 1·685 g/cm³ (Smith et al., 1968; Luha et al., 1971). In a haploid strain of *Shizosaccharomyces pombe* (NCYC 132) mt DNA accounts for 6% of the total cell DNA in cultures growing exponentially in the defined medium containing 1% glucose, and this figure rises to 14% in the stationary phase of growth (Bostock, 1969). In this organism the similarities in densities of nuclear and mt DNAs ($\rho = 1·695$ and $1·689$ g/cm³ respectively) make exact measurements of relative amounts difficult.

B. Mt DNA in Fungi

1. Neurospora crassa

Many of the earlier preparations of mt DNA from *N. crassa* consisted of very heterogeneous linear molecules in the size range 2–25 μm (Luck and Reich, 1964; Reich and Luck, 1966; Wood and Luck, 1969) but Schäfer et al. (1971) have more recently isolated linear molecules mainly of 25 μm length. From renaturation data, Wood and Luck (1969) have calculated a minimal mol. wt of 66×10^6 for *Neurospora* mt DNA; its base composition is 40% (G + C) (Richter and Lipmann, 1970). Anomalous renaturation kinetics have led Wood and Luck (1969) to conclude that repeating sequences compose a large fraction of the DNA, but this conclusion has been questioned (see Borst, 1972). Disruption of isolated mitochondria by osmotic shock releases mt DNA which appears to consist for a large part of circular molecules; most of which show the supertwisting characteristic of closed circular duplex DNA (Fig. 91; Agsteribbe *et al.*, 1972). In this study the contour length of many of the molecules was 19 μm but small circles ranging in size from 0·5 to 7 μm were also present and a circular molecule of 39 μm (probably a dimer) was also observed. Rapid extraction of DNA from a cell wall—less strain under conditions which avoided both harsh disruption of the cells and prolonged exposure to endogenous nucleases, also permits the demonstration of a population of closed-circular mt DNAs with a mean contour length of 20 μm banding in a CsCl-ethidium bromide gradient (Clayton and Brambl, 1972).

2. Mt DNA in other fungi

Buoyant density data for mt DNA from *Aspergillus nidulans* has been obtained by Edelman *et al.* (1970), and for fourteen other species of fungi by Villa and Stork (1968) who found that nuclear and mt DNAs were of the same densities in three Mucorales, whereas the two classes of DNA could be separated by equilibrium density centrifugation in nine species of Ascomycetes and in two Basidiomycetes. (G + C) contents for these mt DNAs were also presented. Circular DNA (14 μm contour length) probably represents the mt DNA of *Saprolegnia* (Clark-Walker and Gleason, 1973).

C. Mt DNA in Protozoa and Algae

1. Mt DNA of Tetrahymena pyriformis

CsCl density-gradient equilibrium centrifugation analyses of DNAs from whole-cells and mitochondrial fractions of several different strains of *T. pyriformis* (Syngens 4, 9 and GL) indicated that the buoyant densities (g/cm^3) for mt DNAs and nuclear DNAs were: 4 (1·686, 1·692); 9 (1·684, 1·690); GL (1·684, 1·688) (Suyama and Preer, 1965; Suyama, 1966). The purified mt DNA from strain ST analysed on a sucrose gradient by the sedimentation velocity method, showed a major sedimenting component of $s^0_{20,w} = 41·6S$ corresponding to a molecular weight of 40×10^6, and this appeared to represent the molecular size in its native state. The size of the major component has been independently confirmed by Brunk and Hanawalt (1969). A second component with a mol. wt of 1×10^6 was presumed to represent an enzymically degraded product of the major component. Examination of *Tetrahymena* mt DNA by the Kleinschmidt spreading technique indicated that the 41S purified DNA consisted of linear filaments of mean length 17·6 μm corresponding to $3·4 \times 10^7$

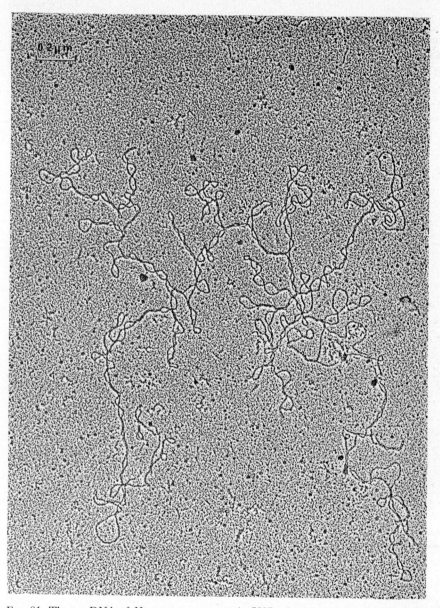

FIG. 91. The mt DNA of *Neurospora crassa* strain 5297 released from mitochondria by osmotic shock and prepared by the Kleinschmidt technique. (Reproduced with permission of E. Agsteribbe and C. Hollenberg.)

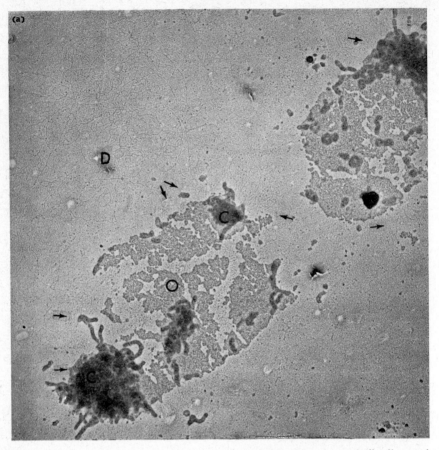

FIG. 92. (a) The mt DNA of *Tetrahymena pyriformis* released from osmotically-disrupted mitochondria. Mitochondrial outer membrane (O), the shape of which is well preserved, and clumps of cristae (C) are seen. DNA filaments (D) radiate from many points of disrupted mitochondria.

daltons (Suyama and Miura, 1968; Suyama, 1969). Osmotically-disrupted mitochondria each released a cluster of seven or more strands of this length (Fig. 92). These results have been confirmed for other strains of *T. pyriformis* (Schutgens, 1971; Flavell and Jones, 1970). Renaturation kinetics indicates a complexity of approximately 3×10^7 daltons and the base-pairing in the renaturation product is precise, as its T_m (80·5°C) is indistinguishable from that of native DNA. From these results Flavell and Jones (1970) concluded, in agreement with Suyama and Miura (1968), that the mitochondrion of *T. pyriformis* contains about eight molecules that are at least of extremely similar base sequence. Band splitting of *Tetrahymena* mt DNA ($\Delta\rho = 6$ mg/cm³) in an alkaline CsCl density

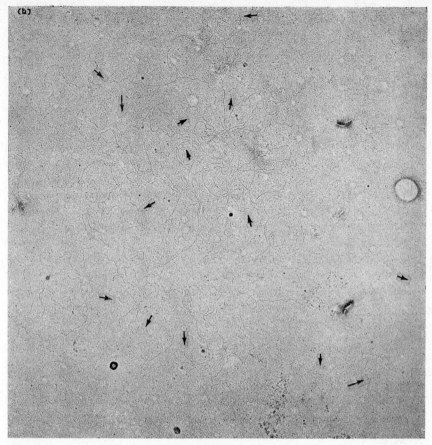

FIG. 92. (b) Points of association of DNA to cristae are indicated by arrows. A cluster of mt DNA released from osmotically-disrupted mitochondria. The total DNA mass corresponds to about 110 μm with 14 ends (arrows) which are connected to the DNA mass in the centre. (Reproduced with permission from Suyama and Miura, 1968.)

gradient suggests separation of the complimentary strands (Flavell and Jones, 1970). Strand separation in neutral CsCl gradients was achieved by differential binding of poly(U) or poly(U, G) to the denatured DNA (Schutgens et al., 1973). All attempts to demonstrate mt DNA circles in purified mt DNA, ethidium-CsCl gradients of whole cell lysates, or in osmotically-lysed mitochondria have failed. Borst (1972) has suggested that the molecules may be circular in situ, and that the isolation of a population of linear full-length molecules of unique base sequence could be due to the presence of a specific nuclease that might be part of the replication mechanism. Recent work (Borst et al., 1973b) has shown that the isolated DNA consists of a permuted collection of molecules without repetitive

K

ends, arising by single random ds breaks during extraction. The base composition is about 26% (G + C), but there appears to be a few long stretches of regions very rich in A and T (Flavell and Jones, 1971a). The possibility of the existence of a "master-copy" of mt DNA in the nuclear DNA of *T. pyriformis* has been examined by hybridization of complementary RNA made on mt DNA with nuclear DNA extracted from the macronucleus (Flavell and Trampé, 1973). Results indicated that no integrated "master-copy" can occur, as less than 0·04 copies per haploid nuclear genome were found.

2. Mt DNA of Paramecium

Paramecium mt DNA (ρ = 1·702 g/cm^3, Tm = 70°C, Suyama and Preer, 1965) has a base composition of 40% (G + C) and a kinetic complexity indicating a size of 35 × 10^6 daltons (Flavell and Jones, 1971b). No sequence homology was found between the mt DNAs of *Tetrahymena* and *Paramecium*. In some strains of *Paramecium* the presence of endo-symbionts provides another location for extranuclear DNA (Gibson *et al.*, 1971).

3. Mt DNA of Physarum polycephalum

Mt DNA in *P. polycephalum* has a base composition of 26% (G + C) (Guttes *et al.*, 1967; Evans and Suskind, 1971). Heterogeneous linear mt DNA up to 28 μm long has been isolated from this species (Sonnenshein and Holt, 1968) and observed by electron microscopy (Kessler, 1969): the minimal size of the mt DNA is 40–60 × 10^6 daltons.

The first evidence for the presence of methylated bases in mt DNA from any source was provided by Evans and Evans (1970) who reported the presence of about 2% 5-methyl-C in the total mt DNA.

4. Mt DNA of Dictyostelium discoideum

Mt DNA in *D. discoideum* has a base composition of 28% (G + C) a buoyant density of 1·688 g/cm^3, and a thermal transition normal for double-stranded DNA (Tm = 80°C) (Sussman and Rayner, 1971). No evidence was obtained for the circularity of this mt DNA.

5. Mt DNA of Euglena gracilis

Mt DNA from streptomycin-bleached cells of *E. gracilis* var. bacillaris has a buoyant density of 1·691 g/cm^3 in CsCl (31–32% G + C) and is double stranded (Edelman *et al.*, 1966). Mitochondria purified from bleached *E. gracilis* strain Z contain only DNA with a density of 1·688 g/cm^3 (Krawiec and Eisenstadt, 1970a, b). Autotrophically-grown

Table XXXVIII. Summary of buoyant density and denaturation data of nuclear and satellite DNA species from kinetoplastidae. (Reproduced with permission from Simpson, 1972.)

Species	Density in caesium chloride (gm/cm³)[a,b,c]	Percent of total cell DNA	T_M(°C)[d,e]	Hyper-chromicity (%)	$\Delta \rho$ on denaturation (gm/cm³)[f]	Percent GC by analysis	Reference[g]
Crithidia fasciculata Culture	N 1·716 (57 %)	—	—	—	—	—	14
	S 1·688 (29 %)	—	—	—	—	—	13
	S 1·701 (42 %)	—	—	—	—	—	14
	S 1·698 (39 %)	—	—	—	—	—	10
Crithidia oncopelti Culture	N 1·712 (53 %)	—	—	—	—	—	14
	N 1·709 (50 %)	—	—	—	—	—	9
K-DNA	S₁ 1·702 (43 %)	—	—	—	—	—	14
	S₁ 1·699 (40 %)	—	—	—	—	—	9
	S₁ 1·701 (42 %)	—	—	—	—	—	10
Bipolar body DNA	S₂ 1·691 (32 %)	—	—	—	—	—	9
	S₂ 1·693 (34 %)	—	—	—	—	—	10
Leishmania enrietti Culture	N 1·721 (62 %)	—	93 (57 %)	25	—	—	5
	S 1·702 (43 %)	—	83 (33 %)	8·4	—	—	5
Leishmania tarentolae Culture	N 1·716±0·0006[h] (9 runs) (57 %)	80–83	92·6±0·7[h] (3 runs) (57 %)	38·3±0·9[h] (3 runs)	0·014	—	7
	2 1·703±0·0003[h] (12 runs) (44 %)	17–20	Component 1 80·2 ±0·3[h] (3 runs)	2·79 ±3·2[h] (7 runs)	0·003–0·016	—	7
			Component 2 84·1±0·8[h] (3 runs)	37.[i]		—	7
			Component 3 86·3±0·8[h] (3 runs)	—		—	7

Table XXXVIII. (continued)

Species	Density in caesium chloride (gm/cm³)[a,b,c]	Percent of total cell DNA	T_M(°C)[d,e]	Hyperchromicity (%)	Δρ on denaturation (gm/cm³)[f]	Percent GC by analysis	Reference[g]
Trypanosoma mega Culture	N 1·703 (44 %)	—	—	—	—	—	12
	S 1·693 (34 %)	—	—	—	—	—	12
Trypanosoma lewisi Bloodstream strain	N 1·719 (60 %)	—	—	—	—	—	10
	N 1·707 (48 %)	82	—	—	—	—	15
	S₁ 1·699 (40 %)	9	—	—	—	—	15
	S₂ 1·721 (62 %)	9	—	—	—	—	15
Trypanosoma lewisi Culture strain	N 1·711 (52 %)	81	91·1 (53 %)	30	0·018	—	15
	S 1·699 (40 %)	19	85–87	1[k]	0·004	—	15
	S 1·699 (40 %)	19	86·5 (42 %)	24[l]	—	—	15
Trypanosoma cruzi Culture forms	N 1·710 (51 %)	—	91 (53 %)	36	0·017	52	3
	N 1·710 (51 %)	—	—	40	—	—	6
	N 1·709 ± 0·0003 (10 runs) (50 %)	—	—	—	—	—	7
	S₁ 1·699 (40 %)	4	86 (40 %)	31	0·012	40	3
	S₁ 1·699 (40 %)	15–20	—	—	—	—	4,6
	S₁ 1·696 ± 0·0001 (7 runs) (37 %)	—	—	—	—	—	7
	S₂ 1·686 (26 %)	10	83 and 73	33	0·023	—	3
	S₂ 1·688 (29 %)	0·5–1·0	—	—	—	—	8
Trypanosoma equiperdum Bloodstream	N 1·707 (48 %)	—	89 (48 %)	40	0·018	—	1
	N 1·707 (48 %)	90–92	—	48	—	46	6
	S 1·701 (42 %)	6	86 (42 %)	42	—	—	1
	S 1·701 (42 %)	8–10	—	42	—	42	6

Trypanosoma gambiense
Bloodstream

N 1·707 (48 %)	—	89 (48 %)	38	—	1
N 1·707 (48 %)	81–84	—	—	46	6
S₁ 1·701 (42 %)	11	86 (42 %)	36	—	1
S₁ 1·701 (42 %)	8–10	—	—	41	6
S₂ 1·690 (31 %)	—	81 (30 %)	32	—	1
S₂ 1·690 (31 %)	5–6	—	—	—	6

[a] N, Nuclear DNA; S, satellite DNA.

[b] Assuming *Escherichia coli* DNA has the density of 1·710 gm/cm³.

[c] Equivalent percent GC is shown in parentheses as calculated with the formula: $\rho = 0.098 \, (GC) + 1.660$ (Schildkraut et al., 1962b).

[d] T_M, Midpoint of the melting curve (Marmur and Doty, 1962).

[e] Equivalent percent is shown in parentheses as calculated with the formula: $T_M = 69.3°C + 0.41 \, (GC)$ (Marmur and Doty, 1962).

[f] Change in equilibrium buoyant density in caesium chloride after denaturation by heat or pH, and quick-cooling or neutralization.

[g] Key to references.

[h] Average ρ plus or minus the standard deviation.

[i] Purified by preparative caesium chloride centrifugation; untreated prior to melting.

[j] Purified as in footnote *i*. Sonicated prior to melting to linear fragments with an average molecular weight of 0.25×10^6.

[k] The lower band of an ethidium bromide–caesium chloride gradient of total cell DNA.

[l] DNA isolated as in footnote *k*, but sonicated prior to melting.

REFERENCES

1. Riou et al. (1966).
2. Riou and Pautrizel (1967).
3. Riou and Paoletti (1967).
4. Riou and Delain (1969a).
5. DuBuy et al. (1966).
6. Riou and Pautrizel (1969).
7. Simpson and da Silva (1971).
8. Riou et al. (1970).
9. Marmur et al. (1963).
10. Schildkraut et al. (1962a).
11. Du Buy et al. (1965).
12. Steinert and Van Assel (1967a).
13. Newton (1967).
14. Newton (1968).
15. Renger and Wolstenholme (1970).
16. da Silva and Simpson, unpublished results.

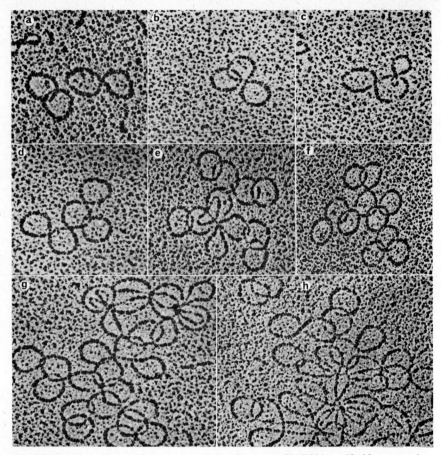

FIG. 93. Electron micrographs of various molecular types of k-DNA purified by successive CsCl equilibrium centrifugations. Each minicircle is 0·29 μm in circumference. (a–c) Monomeric minicircles and fused dimers. (d) Circular oligomer. (e–h) Small catenanes. (i) Electronmicrographs of a single k-DNA association. (Reproduced with permission from Simpson, 1973.)

E. gracilis (Klebs strain z) cells yield mt DNA of density 1·690 g/cm³ (Manning *et al.*, 1971). Lysates of mitochondrial fractions from these cells contained linear molecules ranging in length from 1 to 19 μm with a mean length of 1·3 μm. Only linear molecules were found in both these mitochondria and in those from strains lacking chloroplasts. These studies confirm the early reports of Ray and Hanawalt (1964, 1965) and the recent finding of Schori *et al.* (1970), that *Euglena* mt DNA appears to be smaller than that isolated from other eukaryotes. Further work is necessary to confirm these size determinations by renaturation studies and electrophoresis of denatured DNA in acrylamide gels.

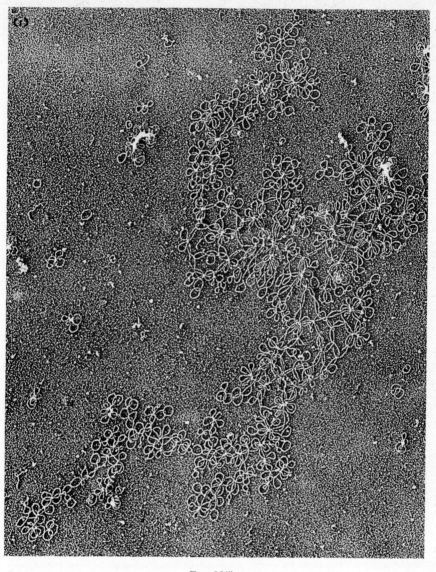

FIG. 93(i).

D. Kinetoplast DNA in Haemoflagellates

The existence of DNA satellite bands present in DNA extracts from *Crithidia* on equilibrium density centrifugation in CsCl gradients (Schild-kraut *et al.*, 1962a, b; Marmur *et al.*, 1963) suggested that the DNA known to be present in the kinetoplast was distinct from nuclear DNA in physical

Table XXXIX. Buoyant densities of main band and satellite DNAs and contour lengths of circular DNA molecules of eight strains of *crithidia*. (Reproduced with permission from Renger and Wolstenholme, 1972.)

Strain	Buoyant density (g/cm³)		Mean contour length (μm)
	Main band	Satellite band	
C. acanthocephali	1·717 ± 0·0001 (5)	1·702 ± 0·0001 (8)	0·80 ± 0·002 (95)
C. fasciculata (*Culex pipiens*) (Nöller strain)	1·717 ± 0·0001 (2)	1·701 ± 0·0001 (2)	0·73 ± 0·007 (20)
C. fasciculata (*Culex pipiens*)	1·717 ± 0·0002 (2)	1·701 ± 0·0002 (2)	0·72 ± 0·013 (20)
C. luciliae	1·717 ± 0·0001 (3)	1·705 ± 0·0002 (3)	0·75 ± 0·011 (20)
C. rileyi	1·717 ± 0·0002 (2)	1·703 ± 0·0001 (2)	0·72 ± 0·016 (20)
Crithidia sp. (*Arilus*)	1·717 ± 0·0002 (2)	1·702 ± 0·0004 (2)	0·80 ± 0·004 (20)
Crithidia sp. (*E. davisi*)	1·717 ± 0·0002 (2)	1·701 ± 0·0003 (2)	0·73 ± 0·006 (20)
Crithidia sp. (Syrphid)	1·717 ± 0·0002 (2)	1·701 ± 0·0003 (2)	0·69 ± 0·012 (20)

For each of the buoyant density values the standard error and the number of runs are given. For each of the mean contour lengths, the standard error and the number of molecules examined are given. The insect of origin is shown in parentheses for some of the strains.

properties, The first conclusive evidence that a satellite band was attributable to kinetoplast DNA was provided by DuBuy *et al.* (1965); it accounts for between 6% and 28% of total cell DNA in various species (Newton and Burnett, 1972). A summary of the properties of DNAs from Kinetoplastidae is given in Table XXXVIII (Simpson, 1972).

In the electron microscope, purified kinetoplast DNA from *T. cruzi* (Riou and Delain, 1969a) *T. lewisi* (Renger and Wolstenholme, 1970, 1971) and *L. tarentolae* (Simpson and da Silva, 1971) was found to consist of the following classes: (a) large associations in which molecular structure was not resolved but in which long DNA molecules with free ends could be seen; (b) small associations consisting of minicircles catenated with each other and "figure eight"-type molecules; (c) small catenates (2–10 minicircles) and (d) free minicircles (Fig. 93). Evidence for the existence of a large proportion of long DNA molecules in kinetoplast-DNA was provided in the case of *T. mega* by Laurent and Steinert (1970); the mol. wt of the longest linear molecule was 43×10^6. The minicircles from *L. tarentolae* form a homogeneous population with a mol. wt of 0.56×10^6 ($\equiv$ contour length 0.29 μm), those from *T. cruzi* have a mol. wt of 0.94×10^6 ($\equiv 0.49$ μm) and those from *C. fasciculata* (Table XXXIX) have a mol. wt of 1.49×10^6 ($\equiv 0.79$ μm). The information content of these minicircles is very small; at most one small protein could be encoded. The relative proportions of minicircles and linear molecules has not been ascertained, except in the case of *L. tarentolae* where it has been estimated that the linear molecules account for 33% of the total (Simpson and daSilva, 1971). Multiphasic melting behaviour of fragmented satellite DNA from *L. tarentolae* (sonicated to linear fragments and open minicircles) suggests the presence of reiterative sequences; the supercoil density is lower than in other mt DNAs examined (Simpson, 1972). Whereas no detectable hybridization occurs between nuclear and kinetoplast DNAs in *L. tarentolae*, the technique is not sensitive enough to enable us to conclude that no part of the kinetoplast DNA is represented in the nucleus (Simpson and daSilva, 1971). Hybridization between *L. tarentolae* and *T. cruzi* kinetoplast DNAs suggests at least 2–4% homology; no cross hybridization between kinetoplast DNAs of *C. luciliae* and *T. mega* can be demonstrated (Steinert *et al.*, 1973). As many as 1.3×10^4 minicircles may occur per kinetoplast in *L. tarentolae* and 2.4×10^4 minicircles per kinetoplast in *T. cruzi* (Simpson and daSilva, 1971; Riou and Pautrizel, 1969). If the minicircles in each species are identical in base sequence (and renaturation kinetics indicates the presence of only one or two classes) this represents on enormous gene amplification of the order of 10^4 (Simpson, 1973; Wesley and Simpson 1973, a, b, c). In the kinetoplast of *C. acanthocephali*, Renger and Wolstenholme (1972) showed the presence of large DNA masses of associations of mainly covalently closed 8 μm circular molecules (mol.

wt $= 1.54 \times 10^6$). The circles are held together in rosettes of up to 46 by the topological interlocking of each circle with many others of the group, and associations comprise on average about 2.7×10^4 circles (total mol. wt $\simeq 41 \times 10^9$). This may represent the total kinetoplast DNA complement of a single cell.

Treatment with ethidium bromide or acriflavine (but not proflavine or 5-aminoacridine) induces dyskinetoplasty in *Leishmania tropica* (Morales *et al.*, 1972); growth for several generations with either of these two dyes leads to loss of kinetoplast DNA. This report also reviews previous work on dye-induced dyskinetoplasty in various trypanosomatids. Other agents leading to rearrangement of kinetoplast DNA structure include dauno-mycine, and the nonintercalating trypanocides, hydroxystilbamidine and berenil (Delain *et al.*, 1972).

III. Synthesis of mt DNA in Eukaryotic Microorganisms

A. Possible Replicative Intermediates

Analysis of the mechanism of mt DNA replication at the molecular level in eukaryotic microorganisms has not proceeded as far as comparable studies with the mt DNA of higher organisms (Arnberg *et al.*, 1973a). The branched circles described by Kirschner *et al.* (1968) and the displacement loops (D-loops) discovered by Kasamatsu *et al.* (1971) are likely to represent intermediates in DNA replication in higher animals. Multiple D-loops have been observed in *Tetrahymena* mt DNA molecules (Arnberg *et al.*, 1972). Up to six clearly separated loops (mean length 0.22 μm) are located apparently at random on a single linear molecule. Two molecules with a double-stranded form were also identified. These investigators (Arnberg *et al.*, 1973b) and Borst and Grivell (1973) suggest that the results provide evidence for the circularity of *Tetrahymena* mt DNA *in situ* and that there are multiple initiation sites for mt DNA replication, as schematic circula-rization of the multiple D-loops showed them clustered in eight unique positions.

In the kinetoplast DNA of exponentially growing *Trypanosoma cruzi*, replicating double-branched circular molecules are very rare (Brack *et al.*, 1972b) as the possibility of extracting the very small kinetoplast DNA molecules during replication is rather small. Treatment of the organisms with ethidium bromide (Riou and Delain, 1969b), or the trypanocides Berenil (4,4'-diazoamino dibenzamidine diaceturate $4H_2O$) or hydroxystilb-amidine increased the proportion of replicating molecules by a factor of 10^3. These results indicate that each of the several thousand minicircles present in the kinetoplast is able to replicate independently.

B. Synthesis of mt DNA *in vitro*

Mitochondria isolated from a range of organisms synthesize mt DNA as shown by isotope incorporation experiments *in vitro*; evidence has been obtained for the occurrence of both replicative and repair processes (Karol and Simpson, 1968; Ter Schegget and Borst 1971a, b). The presence of specific enzymes for the synthesis of deoxythymidine triphosphate in mitochondria of *N. crassa* has been confirmed by Rossi and Woodward (1973). The maximum net DNA synthesis observed *in vitro* does not exceed 0·5%, and in chick liver the process is not inhibited by phenyl ethanol and to only a limited extent by high concentrations of nalidixic acid. Ethidium bromide (12 μM) gives complete inhibition of DNA synthesis. Up to 70% of the product cosediments with closed circular duplex mt DNA in sucrose or CsCl gradients containing ethidium. Treatment of the product (designated I) with DNAase gives molecules with the sedimentation properties of open circular mt DNA.

The presence of a DNA-polymerase in isolated yeast mitochondria was reported by Wintersberger (1966a). Incorporation of (^{3}H)-dATP into acid insoluble product by isolated mitochondria required the simultaneous presence of dGTP, dCTP, dTTP, and Mg^{2+}. Actinomycin and mitomycin C are inhibitory. The purified product was digested by pancreatic DNAase and by snake venom phosphodiesterase.

DNA synthesis has also been demonstrated in mitochondria isolated from *Physarum polycephalum* by Brewer *et al.* (1967), who observed the incorporation of (^{3}H)-dATP into an acid insoluble product with the same density as *Physarum* mt DNA. The reaction required all four deoxyribonucleoside triphosphates and Mg^{2+}, was unaffected by exogenous DNA, DNAase or ATP, and was inhibited by Ca^{2+} and spermine.

Mitochondrial-DNA polymerase as distinct from that of the nuclear-DNA polymerase, has been purified eighty fold from *S. cerevisiae* (Wintersberger and Wintersberger, 1970a, b). The purified mitochondrial enzyme (mol. wt 150,000) requires all four deoxyriboside triphosphates and Mg^{2+} and different DNAs, in native or denatured form, serve as templates and primers. A preference for denatured over native DNAs was indicated, whilst mt DNA promoted mt DNA synthesis. Mitochondria from a cytoplasmic *petite* also contained DNA polymerase with properties indistinguishable from those of the enzyme from wild-type yeast. Both the mitochondrial enzymes and the polymerase from mitochondrial-free extracts were inhibited by acriflavine. Close correlation between DNA and RNA polymerases at the transcriptive level is indicated by the finding that 65% inhibition of (^{3}H)-dTTP incorporation occurs at 25 μg/ml rifampicin (R. S. Criddle, unpublished resul.). Iwashiwa and Rabinowitz (1969) have also purified DNA polymerases from yeast. Unlike Wintersberger and Wintersberger (1970a, b) they report similar properties for the mitochondrial and extramitochondrial enzymes, but did find a difference in their mol. wts.

A mitochondrial DNA-polymerase has also been purified from *T. pyriformis* (Westergaard and Pearlman, 1969; Westergaard, 1970; Westergaard *et al.*, 1970; Keiding and Westergaard, 1971). In *Tetrahymena*, damage of the DNA by thymine starvation, ethidium bromide treatment, or irradiation with ultraviolet light or electrons, leads to an increased activity of the mt DNA polymerase activity up to fifty fold. This process requires RNA synthesis and is cycloheximide- but not chloramphenicol-sensitive; these results suggest that damage to mt DNA leads to transcription of a nuclear gene for mt DNA polymerase which is translated extramitochondrially. This inducible polymerase is clearly a repair rather than a replicative enzyme. A similar specific repair mechanism has also been demonstrated in yeast (Moustacchi, 1969, 1971; Moustacchi and Enteric, 1970).

C. Synthesis of mt DNA *in vivo*

The mechanism of replication and inheritance of mt DNA was studied in cultures of *Neurospora crassa* (Reich and Luck, 1966) using the density-labelling technique first applied to *E. coli* by Meselson and Stahl (1964). Cultures were grown using $^{15}NH_4$$^{15}NO_3$ as the sole N source, and the buoyant densitities of DNA from purified mitochondrial and nuclear fractions were analysed after resuspending cells in a medium containing only ^{14}N-nitrogen sources. Density measurements indicated that most of the DNA synthesized during the first mass-doubling cycle consisted of strands containing wholly or largely undiluted ^{15}N. During the second doubling cycle following transfer to ^{14}N medium, an increasing proportion of ^{14}N was incorporated into mt DNA; on denaturation slightly over half of the DNA consisted of strands containing undiluted ^{15}N. By the end of the third doubling cycle, the ^{15}N content of the DNA was 40%, but most of the DNA-^{15}N was diluted with ^{14}N, since most of the DNA banded after denaturation at a position intermediate between pure DNA-^{14}N and DNA-^{15}N. From these results, Reich and Luck (1966) concluded that the synthesis of mt DNA in *N. crassa* is associated with the conservation of intact pre-existing polynucleotide. The nitrogenous precursors for mt DNA synthesis are derived from a pool which is large in relation to mt DNA, turns over slowly relative to the kinetics of mt DNA synthesis, and hence resists dilution by exogenous nitrogen sources. Results for nuclear DNA synthesis suggest that the contrary is so in this case, and therefore that the two processes are metabolically independent. As "hybrid" DNA molecules were not observed at the anticipated times, these experiments did not establish unequivocally that the mechanism of replication of mt DNA is semiconservative, but the investigators suggest that the appearance of a distinct band with near-hybrid density in native mt DNA is strongly suggestive of such a mechanism. These conclusions have been questioned by Borst and Kroon (1969) who are extremely critical of the suggestions of Reich and Luck (1966) regarding the identity of the "store" of ^{15}N compounds upon which mitochondria draw for DNA replication. A satisfactory explanation is still not available for the continuing incorporation of ^{15}N nucleotides into *Neurospora* mt DNA after the shift to ^{14}N medium, and the conclusions of Reich and Luck (1966) require confirmation.

Corneo *et al.* (1966) have data from ^{15}N transfer experiments in yeast which are similar to the results obtained with *N. crassa*. Thus in *S. carlsbergensis* heavy ($^{15}N–^{15}N$) mt DNA could still be detected after nuclear DNA had undergone one complete replication to give a hybrid ($^{15}N–^{14}N$) form. These results suggest that mt DNA replication does not occur synchronously with nuclear DNA replication, and that the turnover rate of mt DNA is low in relation to the rate of duplication in rapidly dividing cells. A process of turnover of mt DNA has been established in non-dividing or slowly-dividing cells, e.g. in animal tissues (Neubert *et al.*, 1968; Gross *et al.*, 1968). It has been suggested that this may reflect degradation of mitochondria as a unit as a consequence of the engulfment of mitochondria by lysosomes; autophagosomes have been observed to contain mitochondria undergoing digestion (deDuve and Baudhuin, 1966). Studies of turnover rates of mt DNA in microorganisms under various conditions of restricted growth should prove interesting.

D. Timing of mt DNA Synthesis in the Cell-cycle

1. Mt DNA synthesis in Physarum polycephalum

The relative timing of mt DNA and nuclear DNA syntheses in the cell-cycle has been studied (a), by audioradiographic analysis of pulse-labelled cells and (b), by measurement of the rate of incorporation of precursors into mitochondrial and nuclear fractions obtained at different stages of the cell-cycle of synchronous cultures. Incorporation of (^3H)-thymidine into the mitochondria of *P. polycephalum* was observed autoradiographically by Guttes and Guttes (1964); the labelled product was insoluble in acid, was not digested by RNAase, but was removed by DNAase. Evidence that the cytoplasmic DNA differed in density from nuclear DNA and represented about 5% of the total cellular DNA was presented by Evans (1966) who was also able to show that the synthesis of this mt DNA proceeded at an approximately constant rate through the cell-cycle, whereas the synthesis of nuclear DNA is confined to a short S phase. In isolated mitochondria the activity of DNA polymerase is independent of the stage in the mitotic cycle at which the mitochondria are extracted (Brewer *et al.*, 1967). The combined approaches of autoradiography and CsCl-density sedimentation analysis of labelled DNAs led Guttes *et al.* (1967) to confirm that the incorporation of (^{3}H)-thymidine into mt DNA occurs at all times in the mitotic cycle of *P. polycephalum*. Similar results have also been reported by Holt and Gurney (1969), who took into account the presence of a nuclear satellite DNA in this organism. Thus whereas the most of the nuclear DNA is replicated in the S phase, both the mitochondrial and nuclear satellite DNAs are synthesized at constant rates through the cell-cycle (Braun and Evans, 1969); this constant rate of mt DNA synthesis applies to the total mitochondrial population and the question of whether individual mitochondria have specific S periods cannot be decided by these experiments.

Ethidium bromide (10 µg/ml) inhibits mt DNA synthesis by over 90% and has no measurable effect on the synthesis of principal nuclear DNA or nuclear satellite DNA (Horwitz and Holt, 1971). According to Cummins and Rusch (1966) and Muldoon *et al.* (1971) the nuclear DNA of *P. polycephalum* is synthesized in successive rounds of replication, each step being regulated by specific proteins which are only synthesized after preceding rounds have been completed. Cycloheximide (50 µM), added during the S-phase, strongly suppresses the labelling of the major nuclear component of *P. polycephalum*, but only slightly affects (^{3}H)-thymidine incorporation into mt DNA during the G$_2$ phase (Werry and Wanka, 1972). These results suggest that the hypothetical control mechanism involved in nuclear DNA replication are not paralleled in mt DNA replication. However the validity of conclusions from similar experiments on bacterial DNA replication has recently been questioned (Cooper and Wuethoff, 1971).

2. Mt DNA synthesis in Tetrahymena pyriformis

Practically all the cytoplasmic radioactivity of (^{3}H)-thymidine-labelled *T. pyriformis* has been shown by electron microscope autoradiography to be associated with the mitochondria (Fig. 94; Stone *et al.*, 1964). Phase-contrast autoradiography has indicated continuous incorporation of (^{3}H)-thymidine into the mitochondria, and the identity of the product as DNA was confirmed by enzymic digestion (Parsons, 1965). Pulse labelling indicated that mitochondrial incorporation of the isotope occurred at periods in the cell-cycle when nuclei are not synthesizing DNA. These findings were in large part confirmed by Cameron (1966) who also concluded that mt DNA synthesis, unlike nuclear DNA synthesis, occurs during most if not all of the cell-cycle. However in Cameron's experiments an increased rate of cytoplasmic labelling during micronuclear DNA synthesis and at the start of macronuclear DNA synthesis may reflect either, an increased rate of mt DNA synthesis, or a change in size of DNA precursor pools at this

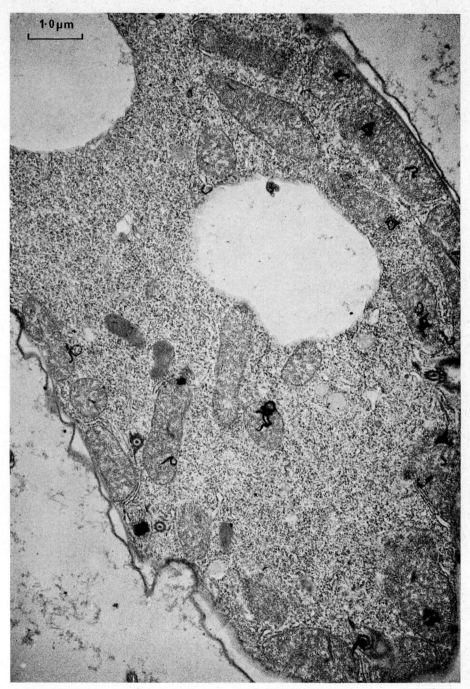

Fig. 94. Electron microscope radioautograph of *Tetrahymena pyriformis* grown in the presence of (^{3}H)-thymidine. (Reproduced with permission from Charret and André, 1968.)

time in the-cell cycle. Further work by Parsons and Rustad (1968) confirmed that, although some labelling of *T. pyriformis* mt DNA occurred at all stages of the cell-cycle, a significant increase in labelling was found during the period of macronuclear DNA synthesis, suggesting the possibility of a common trigger for both mt DNA and macronuclear DNA syntheses. All mitochondria incorporated (^{3}H)-thymidine within one population-doubling time, and the retention of label and its random distribution amongst all the mitochondria (through four generations of daughter mitochondria) satisfy two of the prerequisites for a genetic material and suggest the production of mitochondria by growth and division of pre-existing mitochondria. Charret and André (1968) also report a persistance of mitochondrial incorporation of (^{3}H)-thymidine in the absence of nuclear incorporation at some stages of the cell-cycle in exponentially growing cultures, but point out that the apparent continuity of DNA synthesis in the entire mitochondrial population may or may not represent a series of asynchronous discontinuities. Ethidium bromide (20 µg/ml) inhibits mt DNA synthesis but does not affect nuclear DNA synthesis (Charret, 1972). There has been no report of the time course of mt DNA synthesis in synchronized cultures of *T. pyriformis*.

3. Kinetoplast-DNA synthesis in Haemoflagellates

Electron microscope autoradiography of (^{3}H)-thymidine incorporation into the culture forms of *T. lewisi* (Burton and Dusanic, 1968), and into *C. fasciculata* (Anderson and Hill, 1969) suggest that this method could give information on the mechanism of replication of the kinetoplast DNA nucleoid if quantitative data and statistical analyses were available. The results of Simpson and daSilva (1971) on the distribution of totally-labelled kinetoplast DNA among progeny kinetoplasts of *L. tarentolae* during several generations in unlabelled medium suggest that there are more than 14 separate DNA units that segregate independently at kinetoplast division. Simpson (1972) points out that "the replication of the highly redundant minicircular complement of the kinetoplast DNA within the complex quaternary structure of the kinetoplast nucleoid represents an interesting and unsolved problem".

In all species of kinetoplastidae examined, close temporal connection of nuclear and kinetoplast DNA syntheses have been found (Fig. 95). Division of the kinetoplast occurs just before nuclear division in *C. fasciculata* (Steinert and Steinert, 1962) and in *L. tarentolae* (Simpson and Braly, 1970), whereas in *C. luciliae* it occurs just after nuclear division (Steinert and Van Assel 1967a, b). Initiation of kinetoplast DNA synthesis occurs just before or just after initiation of nuclear DNA synthesis in *C. fasciculata*, depending on the temperature of growth (Cosgrove and Skeen, 1970). In *C. luciliae* the initiation and completion of the synthesis of

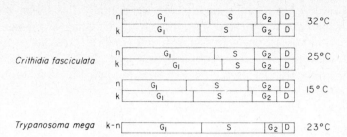

FIG. 95. Diagram of the relative portions of the cell cycle occupied by G_1, S, G_2 and D (the period from the time of complete separation of the daughter kinetoplasts to the complete separation of the daughter cells) for *Crithidia fasciculata* grown at three different temperatures, and for *Trypanosoma mega* at 23°C. n, nucleus; k, kinetoplast. (Reproduced with permission from Simpson, 1972.)

nuclear and kinetoplast DNAs are simultaneous (Steinert and Van Assel, 1967a, b). The mechanism of regulation of this tight coupling between nuclear and kinetoplast DNA synthesis is not clear, although it has been suggested that common periodicity may be conferred by the sharing between nucleus and kinetoplast of a common pool of deoxyribonucleotide precursors which are synthesized at one time in the cell-cycle (Cosgrove and Skeen, 1970). It has been demonstrated that hydroxyurea treatment leads to synchronization of cultures of *C. luciliae* (Steinert, 1969) and *L. tarentolae* (Simpson and Braly, 1970). In the latter case, release from inhibition was followed by a commencement of both nuclear and kinetoplast DNA synthesis, followed by mitosis and then by a second cycle of these sequential events. Simultaneous labelling of kinetoplast and nuclear DNAs has also been demonstrated by quantitative autoradiographic studies of (^{3}H)-thymidine incorporation into cells of a culture of *C. luciliae* synchronized by hydroxyurea treatment (Van Assel and Steinert, 1971).

4. Mt DNA synthesis in Euglena gracillis

Dark-grown cells of *E. gracilis* were synchronized in a defined medium and the time course of incorporation of (^{3}H)- and (^{14}C)-adenine into DNA was followed in whole cells and their mitochondria (Calvayrac *et al.*, 1972). Preferential incorporation occurred at the end of the non-dividing phase and at the beginning of cell division, into both a major peak assigned to nuclear DNA, and a satellite peak which appeared to correspond to mt DNA. Essentially similar results were obtained both for etiolated *Euglena* containing mitochondria and proplastids, and for mutants lacking plastid DNA. Fragmentation of the giant reticulate mitochondrial complex was observed in electron micrographs at the start of nuclear division. Richards *et al.* (1971) have shown that cycloheximide (15 μg/ml) significantly reduced the synthesis of nuclear DNA but allowed the continued synthesis of mt DNA, and similar results were obtained by Calvayrac *et al.* (1972). Chloramphenicol had no effect on the labelling or on the distribution of chloroplast or mt DNAs relative to nuclear DNA for at least 5·5 generations, and thus the continued synthesis of extra-nuclear DNAs suggests either that all proteins necessary are made on cytoplasmic ribosomes, or that these proteins turn over only very slowly.

5. Mt DNA synthesis in yeasts

It has been claimed that replication of mt DNA is periodic during the synchronous growth of cultures of *Saccharomyces lactis* (Smith *et al.*, 1968) and *S. cerevisiae* (Cottrell and Avers, 1970), and occurs at a time different from that at which nuclear DNA is synthesized. However it has recently been shown that difficulties in quantitative extraction lead to the observed variations in the ratio of mt DNA to nuclear DNA in these two reports. Careful analyses by Williamson and Moustacchi (1971) by analytical iso-pycnic centrifugation in CsCl, as well as by pulse labelling with (^{3}H)-adenine, confirms that the synthesis of mt DNA in *S. cerevisiae* growing synchronously in the presence of 2% glucose is not closely coupled to the replication of chromosomal DNA (Fig. 96) nor is it triggered by any unique condition occurring at only one stage of the cell-cycle as suggested by Smith *et al.* (1968). Rather it seems that conditions favouring synthesis of mt DNA exist at all times in the cell-cycle, and the increasing rate of synthesis observed reflects the influence of flexible regulatory influences (Fig. 97). Williamson (1969) has pointed out that these results tell us nothing of the pattern of DNA synthesis in the individual mitochondrion which may be either continuous or discontinuous with respect to the cell-cycle. However they clearly indicate that the replication of mt DNA under these conditions of synchronous growth (produced by an induction method, see p. 239) is not initiated by the same event that triggers the replication of nuclear DNA.

The observation that the frequency of induction of mitochondrial mutants by N-methyl-N'-nitro-N-nitrosoguanidine is greatest late in the cell-cycle has led Dawes and Carter (1974) to suggest that mt DNA replication occurs at this time, and at a different time from nuclear DNA replication. This conclusion depends on the assumption that the mutagen induces mutations selectively at the replication point of the DNAs.

However, the synthesis of mt DNA in yeast, as in *Euglena* (Richards *et al.*, 1971) can be dissociated from nuclear DNA synthesis under conditions which inhibit protein synthesis (Grossman *et al.*, 1969). Thus mt DNA is the major species synthesized in the presence of cycloheximide, on amino acid starvation, or under suitable growth conditions in temperature-sensitive mutants. Synthesis of mt DNA proceeds for 4–6 h after the cessation of protein synthesis, and during this period the relative amount of mt DNA doubles and can be labelled to contain 90% of the total DNA radioactivity. Similar results were also obtained with a cytoplasmic *petite*. Chloramphenicol inhibition did not affect the relative amount of mt DNA

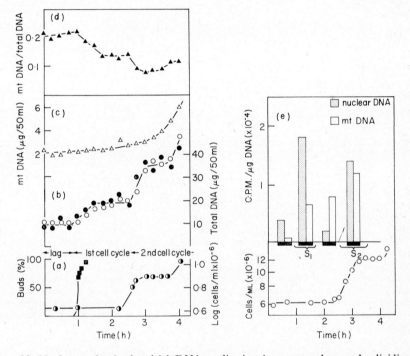

Fig. 96. Nuclear and mitochondrial DNA replication in two synchronously dividing cultures. (a) ■——■, per cent young buds; ◑—◑, cell counts; (b) ●—●, total DNA by diphenylamine analysis; ○—○, total DNA obtained by adding separate data for nuclear and mt DNA provided by the analytical CsCl method (c) △—△, mt DNA; (d) ▲—▲, mt DNA/total DNA. In (d) the ultracentrifuge data for total DNA have been normalized by adjusting the average of the first three points to fit that of the first three diphenylamine points. (e) Specific radio-activities of nuclear and mt DNA obtained on pulse labelling with (^{3}H)-adenine at different times in a synchronously dividing culture. The black bars indicate the durations of the pulses, and the periods occupied by the first and second S periods (determined by separate diphenylamine analyses) are indicated. (Reproduced with permission from Williamson and Moustacchi, 1971.)

present in the organisms. The mechanism of the cycloheximide effect is not clear, but Grossman *et al.* (1969) suggest that cytoplasmic protein synthesis may be involved in nuclear histone production, and that a cessation of histone synthesis may lead to inhibition of nuclear DNA synthesis.

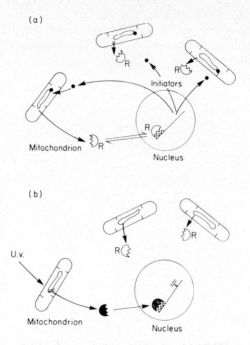

FIG. 97. Schematic model for the regulation of mitochondrial DNA synthesis and for the involvement of the regulatory system in the induction of the cytoplasmic *petite* mutation. In (a) the replication of mt DNA is under "feedback" control; production of essential "initiator" molecules by a repressible nuclear operon being regulated by repressor molecules produce by the mitochondria. In (b), UV irradiation damage to a mitochondrion causes production of an altered repressor which irreversibly blocks initiator production and thereby inhibits mt DNA synthesis throughout the cell. R, Repressor. (Reproduced with permission from Williamson, 1970.)

E. Control of Mt DNA Synthesis by Environmental Factors and During Differentiation

1. Glucose repression

Reports of extensive repression of the mt DNA content of *S. lactis* (Smith *et al.*, 1968) and of *S. cerevisiae* (Moustacchi and Williamson, 1966; Cottrell and Avers, 1970) when grown aerobically at high glucose concentrations have recently been placed in doubt. An improved procedure for extraction and estimation of the separated DNA components, which gives high mol. wt species with no selective losses of any component, has recently been developed (Williamson *et al.*, 1971b). Using this technique, Williamson (1970) has obtained results which suggest that the drop in mt DNA content in the presence of high concentrations of glucose is much less than was formerly supposed (Table XL). Thus in strain 74J growing in the presence of 8% glucose the mt DNA (as a % of total DNA)

Table XL. The amount of mitochondrial DNA in various respiratory competent strains of *Saccharomyces cerevisiae*. (Reproduced with permission from Williamson, 1970.)

Growth conditions and status of respiratory system	Strain	Ploidy	Phase of culture	mtDNA (% of total DNA)
Aerobic shaken culture with 2% glucose "non-repressed"	74HD3	Haploid	Stationary	19
	74HD6	Haploid	Stationary	20
	74HD7	Haploid	Stationary	23
	239	Diploid	Log	14
	239	Diploid	Stationary	19
	351	Haploid	Stationary	26
	S-42-11	Haploid	Stationary	22
	Z65	Diploid	Log	16
	745	Diploid	Stationary	22
	745	Diploid	Log	13
Anaerobic liquid culture, 4% glucose "anaerobically repressed"	681-B	Haploid	Stationary	23
	74J	Diploid	?	28
Aerobic shaken culture, 8% glucose "glucose repressed"	Z65	Diploid	Log	15
	DV147	Haploid	Log	11
	DV147	Haploid	Stationary	23
	878	Haploid	Log	13
	733	Haploid	Log	13
	745	Diploid	Log	12
	745	Diploid	Stationary	20
Aerobic shaken culture, 2% glucose + 3 mg/ml erythromycin "erythromycin repressed"	74J	Diploid	Stationary	13
	Z65	Diploid	Stationary	27
	239	Diploid	Stationary	13
Aerobic shaken culture, 4% glycerol or 2% ethanol "fully derepressed"	681-B	Haploid	Stationary	20
	74J	Diploid	Stationary	25
	239	Diploid	Stationary	20
	DV147	Haploid	Log	20

only underwent a reduction from 15% to 10%. Comparable small reductions have been observed in cells growing in the presence of lower glucose concentrations, and substantial amounts of mt DNA were observed in chemostat cultures in the presence of 8% glucose. Fukuhara (1969) also finds a constant mt DNA/nDNA ratio under a wide range of growth conditions.

Chemostat cultures of *S. carlsbergensis* have also been used to investigate this problem (Bleeg *et al.*, 1972), and the results differ somewhat from those obtained with *S. cerevisiae*. The proportion of mt DNA varied from 11·5% of nuclear DNA in nonrepressed to 3·3% in repressed cells in this species, and that numbers of mitochondria per cell (counted as Janus green-staining particles) showed a similar decrease from 11 to 4. Thus the ratio between the number of mitochondrial genomes and the number of mitochondria is almost constant, and calculations revealed that this figure works out at 3–4 genomes per mitochondrion.

2. Anaerobiosis

A number of workers have observed that anaerobiosis in the presence of excess lipid supplements does not affect the synthesis of mt DNA (Mounolou, 1967; Criddle and Schatz, 1969; Fukuhara, 1969; Williamson, 1970). Anaerobic growth of cells in limiting concentrations of essential lipids does appear however to reduce the cellular content of mt DNA to about a half. On aeration of anaerobically-grown lipid supplemented cells a burst of mt DNA synthesis occurs, leading to an increase of 10–35% of the mt DNA (Fig. 98; Mounolou *et al.*, 1968; Rabinowitz *et al.*, 1969). In similar circumstances lipid-depleted cells give a three-fold increase in cellular content of mt DNA. The mechanism of this O_2-induced DNA synthesis requires further investigation.

3. Antibiotics

Mitochondrial repression by temporary exposure to erythromycin, even for several generations, does not lead to a drastic drop in mt DNA content (Williamson, 1970). Prolonged exposure leads to the irreversible loss of mt DNA in some strains, and hence to the production of cytoplasmic *petites* (Williamson *et al.*, 1971a). Ethidium bromide, acriflavine and nalidixate (but not chloramphenicol or erythromycin) considerably lower the mitochondrial DNA content of the *petite*-negative yeast, *Kluyveromyces lactis* (Luha *et al.*, 1971).

4. Mt DNA during differentiation

The base composition and physical properties of DNA isolated from germinated spores of the fungi *Botryodiplodia theobromae* and *Rhizopus*

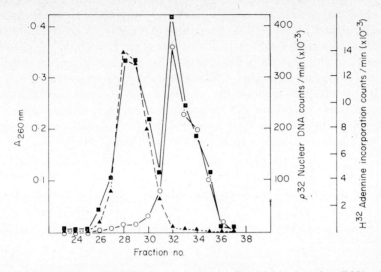

FIG. 98. Incorporation of (^{3}H)-adenine into mitochondrial and nuclear DNA of O_2-adapting yeast. CsCl density-gradient centrifugation of DNA samples isolated from yeast during respiratory adaptation. After a 2 h period of shaking in a nitrogen atmosphere, 1 mCi of (^{3}H)-adenine was added to the flask and aeration began. The yeast was collected in 1 h. The DNA was extracted and purified from protoplasts. ^{32}P-labelled nuclear DNA (4 × 10^3 counts/min, specific activity 4 × 10^4 counts/min per μg) previously purified by CsCl centrifugation was added to each gradient as a marker. Optical density at 260 nm, ■———■; (^{3}H)-adenine incorporation into DNA counts/min, ○———○; ^{32}P-labelled nuclear DNA marker counts/min, ▲- - - -▲. (Reproduced with permission from Rabinowitz et al., 1969.)

stolonifer were indistinguishable from those of DNA from ungerminated spores (Dunkle and Van Etten, 1972a, b). Experiments on the incorporation of (8-^{14}C) guanine by *B. theobromae* indicated that both nuclear and mt DNA become labelled at about the same time during germination, and that synthesis of both components continues throughout the process. Ethidium bromide inhibited synthesis of mt DNA without having a significant effect on nuclear DNA synthesis, on the rate or extent of spore germination, or on oxygen uptake (Dunkle and Van Etten, 1972a, b). Thus mt DNA synthesis is apparently not required for spore germination.

The observation that cytoplasmic *petite* yeast mutants cannot be induced to sporulate (Ephrussi, 1956) suggests, either that mt DNA or mitochondrial energy formation may play a key role in this differentiation process. However, completely derepressed cells can sporulate after ethidium bromide treatment (which virtually eliminated mitochondrial protein synthesis and left no mt DNA detectable by analytical ultracentrifugation in CsCl, Küenzi et al., 1974). Partially repressed cells, treated with ethidium bromide, showed no ascus formation after transfer to sporulation medium. These results suggest that once cells are fully derepressed, no mitochondrial genetic information has to be expressed during meiosis and ascus formation, although it is conceivable that the DNA analyses were not capable of detecting small residual amounts of mt DNA, and these studies did not test this possibility by applying the criterion of suppressiveness (see also Puglisi and Zennaro, 1971). These experiments provide a method for the production of haploid cytoplasmic *petite* mutants. Ethidium bromide does not inhibit the differentiation of *Dictyostelium discoideum* although the synthesis of both cytochrome *c* oxidase and succinate-cytochrome *c* oxidoreductase is inhibited (Stuchell *et al.*, 1973).

6
Extrachromosomal Mutations Affecting Mitochondrial Functions

I. Introduction

The existence of genes in the cytoplasm was first suggested by the observations of Correns (1909) and of Baur (1909) that the patterns of inheritance of leaf variegation did not obey the rules of Mendelian genetics. The first systematic study of cytoplasmic genetics in a microorganism was that on the green alga *Chlamydomonas*: this study had led to the discovery of a cytoplasmic linkage group which is probably located in chloroplast DNA (see Sager, 1972, for review). Early work on yeasts, *Neurospora* and other fungi, and on a variety of protozoan species, also clearly revealed the influence of extrachromosomal determinants on the inheritance of certain characteristics (Ephrussi, 1953; Wilkie, 1964). The discovery of mt DNA (Nass and Nass, 1963a, b; Bell and Mühlethaler, 1964) provided proof of the molecular identity of the cytoplasmic genes, and the rapid advances over the past ten years, both in the elucidation of the physico-chemical characteristics of mt DNA and in the isolation of mutants with altered mt DNAs, have seen molecular biologists and geneticists providing contributions in a joint venture. This chapter describes the alterations in mt DNA which are known to alter various aspects of mitochondrial functions in eukaryotic microorganisms.

II. Extrachromosomal Mutants of Yeasts
A. The Life-Cycle of Budding Yeasts

Investigations of the genetics of *Saccharomyces cerevisiae* began with the description of the sexual cycle of the organism (Kruis and Satava, 1918) and the development of analytical methods by Winge (1935) and Lindegren and Lindegren (1943). Haploid strains are distinguished by a pair of chromosomal alleles a and α which determine their mating type. Mixing of cells of opposite mating types gives rise to a diploid zygote which can divide mitotically to give a clone of diploid cells (Fig. 99). Single diploid cells under suitable conditions can be induced to sporulate; during this process meiosis occurs and results in the production of four ascospores.

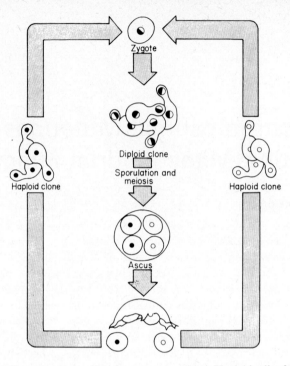

FIG. 99. Sexual life-cycle of yeast (*Saccharomyces cerevisiae*). Haploid cells of opposite mating types, a and α, fuse to form the diploid zygote, which grows by budding to form a diploid colony. Diploid cells can be induced to undergo meiosis and sporulation, producing four ascospores (two are mating type a and two are α), which are the four products of meiosis, in an unordered ascus. Each ascospore grows to form a haploid clone of cells which can fuse with cells of opposite mating type and repeat the cycle. (Reproduced with permission from Sager, 1972.)

Each ascospore then can give rise to a clone of haploid cells. Genetic analysis can be performed by, (1) separation of each of the four ascopores followed by their cloning and characterization (tetrad analysis) or, (2) by separation of individual buds from the diploid vegetative cell. Nuclear genes segretate in a 2:2 ratio in the ascopores; any deviation from this Mendelian ratio is indicative of a cytoplasmically-inherited character.

B. Isolation of Cytoplasmic *Petite* Mutants and their Biochemical Characteristics

Ephrussi and his coworkers observed that when a culture of baker's yeast, whether diploid or haploid, is plated on a solid medium containing standard nutrients and glucose as the carbon and energy source, 0·1–0·2% of the colonies which develop are distinctly smaller than average (see Ephrussi, 1953). Cells taken from the big colonies, when replated, give rise to some small colonies, whereas cells from small colonies give rise to small colonies only, even over thousands of cell generations. Thus the cells of the large colonies can, in the course of normal vegetative reproduction, constantly give rise not only to cells similar to themselves, but also to stable mutant organisms, the "vegetative *petites*". The observations that the *petites* grew more slowly aerobically on glucose than the wild-type organism, although both types of organism grew at similar rates anaerobically, suggested that the mutants have normal fermentation mechanisms, but have a defective respiratory system (Slonimski, 1949; Tavlitzki, 1949; Ephrussi, 1956). The mutants had in fact lost cytochromes $a + a_3$, c_1 and b, succinate dehydrogenase, cytochrome c oxidase, NADH-cytochrome c oxidoreductase and α-glycerophosphate dehydrogenase, and were no longer capable of giving a positive Nadi reaction (indophenol oxidase) (Slonimski, 1952; Slonimski and Ephrussi, 1949). The major characteristics of these mutants are: (a) the absence of cyanide-sensitive respiration; (b) the presence of an actively fermentative metabolism; (c) the presence of cyanide-insensitive respiration; and (d) their inability to grow on non-fermentable substrates.

Examination of enzyme activities in *petite* mutants led Slonimski and Hirsch (1952) to conclude that it was the sedimentable enzymes which were most affected. Lactate dehydrogenase, alcohol dehydrogenase and cytochrome c were all increased, whilst several TCA cycle enzymes were reduced in their activities (malate dehydrogenase and isocitrate dehydrogenase to 50%, fumarase to about 30%, and aconitase to about 15%). Mahler *et al.* (1964a) showed that the patterns of some primary dehydrogenases were similar in respiratory particles obtained from *petites* to those of the wild-type. Antibodies raised to these particles from *petites* gave different extents of inhibition of particulate enzyme's activities in the order: succinate dehydrogenase < NADH dehydrogenase ≃ cytochrome c oxidase < NADH oxidase < NADPH-cytochrome c oxidoreductase (Mahler *et al.* 1964b). Respiratory particles from *petites* contained cross-reacting material capable of binding or reacting with the anti-NADH oxidase cross-reacting material of sera prepared against particles for wild-type cells. This cross-reacting material was not identical with the primary dehydrogenase. Material cross-reacting with purified yeast cytochrome c oxidase is also present in crude mitochondrial fractions of *petite* cells (Kraml and Mahler, 1967). Evidence has also been presented for the presence in some *petites* of a protein identical to (or at least closely similar to) the apoprotein of cytochrome c oxidase (Tuppy and Birkmayer, 1969). Incubation of SMPs of the *petite* with cytohaemin gave 30% of the specific activity of cytochrome c oxidase found in SMPs from the wild-type yeast. Comparison of the primary structure of cytochrome c isolated from a *petite* with that from the parent strain indicated that the two were identical (Yaoi, 1967). Iso-1 and iso-2 cytochromes c both possess one residue of ϵN-trimethyl Lys in ρ^-

and ρ^+ strains (Verdière and Lederer, 1971). F_1–ATPase is also present in crude mitochondrial preparations from homogenates of *petite* cells (Kováč and Weissová, 1967; Schatz, 1968, see p. 311).

The similarity in phenotype of a wide variety of *petite* mutants with regard to activities of lactate dehydrogenases was pointed out by Roodyn and Wilkie (1967), who used automated multiple enzyme assays to compare activities in homogenates of the different mutants with wild-type yeast. In all cases a dramatic fall in the L(+)-lactate-ferricyanide oxidoreductase was accompanied by a rise in the D(—)-lactate-ferricyanide oxidoreductase.

Katoh and Sanukida (1965) reported that preparations of structural protein from wild-type and respiratory-deficient mutants of yeast were indistinguishable chemically or serologically, but Tuppy and Swetly (1968) reported diminished ATP binding by a preparation from a ρ^- mutant; at least one of the antigenic determinants was also missing (Tuppy *et al.*, 1968). The significance of these observations is doubtful owing to uncertainties of the molecular components of "structural protein" preparations (see p. 450).

The most detailed investigation of intact mitochondria from a haploid *petite* mutant with a high (A + T) content of mt DNA (see p. 290) is that by Perlman and Mahler (1970a, b) and Mahler *et al.* (1971a). Rate centrifugation through sucrose gradients was employed to remove contaminating particles from the mitochondria; throughout this procedure the mitochondria retained both outer- and (non-cristate) inner-membranes. They resembled mitochondria from the wild-type strains in many of their physical features (size, shape and buoyant density). Their content of L-malate dehydrogenase, NADP-dependent isocitrate dehydrogenase, and ATPase was comparable with that found in the wild-type mitochondria. The ATPase was cold-stable, oligomycin insensitive, Dio-9 sensitive, and susceptible to inhibition by the F_1-inhibitor from beef heart (see p. 311). They differed from wild-type mitochondria in that they possessed the pleiotropic lesions of the respiratory chain always associated with the ρ^- mutation.

Swelling of isolated *petite* mitochondria in ammonium salts solutions demonstrates that the carriers for phosphate, succinate, and L-malate are retained (Kolarov *et al.*, 1972a). The three distinct carriers for TCA cycle anions (one for L-malate and succinate, one for citrate, and a third for α-ketoglutarate) are unaffected in a *petite* mutant which has no detectable mt DNA (Perkins *et al.*, 1973). Whereas the adenine nucleotide transporter was apparently unaffected in the ρ^- mutant of Kolarov *et al.* (1972a), this enzyme is drastically modified in the ρ° mutant (see p. 291) studied by Haslam *et al.* (1973a). ATP-uptake became completely insensitive to atractyloside in this case, and the high affinity binding site was lost.

The mitochondria isolated from a ρ^- mutant are like those from the wild-type yeast in that they are impermeable to protons (Kováč et al., 1972). In both types of mitochondria, passive permeability for K^+ was enhanced by the addition of valinomycin; together with uncoupler valinomycin also induced proton permeability. However, ATP hydrolysis in the mutant mitochondria could not support efflux of H^+ or uptake of K^+; these mitochondria showed neither uncoupler-sensitive ^{32}Pi-ATP exchange, nor a decrease in ANS fluorescence an addition of ATP. It was concluded that the terminal segment of the energy transfer system is rendered non-functional by the cytoplasmic mutation.

Other changes which accompany the induction of the *petite* mutation, including alteration of mt DNA, RNAs and the protein synthesis machinery, are discussed in Section D. It is evident that the variety of cytoplasmic *petites* now recognized (with respect to the extent of alteration of mt DNA) necessitates the future re-examination of the range of resulting phenotypes in biochemical terms.

C. Induction of Cytoplasmic *Petite* Mutants

The simultaneous loss of a number of distinct enzymes (i.e. a pleiotropic effect) does not usually result from a mutation; single enzyme deficiencies are usually produced. Another unusual feature of the *petite* mutation is the high spontaneous mutation rate (10^{-2} to 10^{-3}) which is in contrast to the figure of 10^{-6} for the nuclear mutation rate. Although the process of mutation to give *petites* is thus frequent and unidirectional, yeast populations do not normally transform into populations of mutants because the high mutation rate is compensated for by the selective advantage of the normal cells, i.e. their rapid multiplication. However, if the yeast is grown in the presence of acridine dyes, highly active and specific mutagens which enormously increase the frequency of mutation without affecting the selection process, an almost total conversion of the population into *petites* is observed (Ephrussi *et al.*, 1949a, b; Ephrussi and Chimenes, 1949; Marcovitch, 1951). Acridines can also induce in the cell a long lasting "unstable state" so that the offspring from acridine-treated cells, even when grown in normal medium without the dye, go on producing normal and mutant buds alternately over several generations (Ephrussi, 1951). Other chemical mutagens inducing a high frequency of cytoplasmic *petite* mutations include euflavine (but not proflavin, Marcovitch, 1951), 5-fluorouracil (Lacroute, 1963; Marcovitch, 1951), ethidium bromide (Slonimski *et al.*, 1968; Mahler and Perlman, 1972a, b), 2,3,5-triphenyltetrazolium (Laskowski, 1954), phenyl ethanol (Wilkie and Maroudas, 1969), nitrosomethyl urethane (Schwaier *et al.*, 1968), dithranitol (Gillberg *et al.*, 1967), *N*-methyl *N'*-nitro-*N*-nitrosoguanidine (Nordstrom, 1967), Mn^{2+}, Co^{2+}, Cu^{2+} and Ni^{2+} (Lindegren *et al.*, 1958), SDS (Pinto da Costa and Bacila, 1973) and the non-intercalating mutagen berenil (Mahler and

Perlman, 1973a, b). The relationship between chemical structure and mutagen efficiency for phenanthridines and acridines has been explored by Mahler (1973). A further list of agents inducing respiratory deficiency is to be found in the reviews by Nagai *et al.* (1961) and Kraepelin (1967, 1972). Nitrous acid, X-rays, and anaerobiosis are not effective. Growth at elevated temperatures or heat shock at 54°C has also been found to induce *petites*, (Yčas, 1954; Sherman, 1959), as has starvation (Wallis *et al.*, 1972). *Petite* mutants show a greater sensitivity to elevated growth temperatures than their *grande* counterparts (Gause and Kusakova, 1970). The *petite* mutant has a selective advantage over wild-type cells in media containing Co^{2+} (Horn and Wilkie, 1966) or TTC (Bachofen *et al.*, 1972).

The continued presence of ATP inside mitochondria is necessary for the normal replication of mt DNA and for a function related to cellular multiplication (Sǔbik *et al.*, 1972a). Respiratory-deficient mutants are produced in aerobic cultures when respiration is inhibited using antimycin A or CN^-, when influx of ATP from cytosol to mitochondria is simultaneously prevented by bongkrekic acid. The susceptibility of yeast to UV mutagenesis with respect to the efficiency of *petite* induction is highest at the onset of budding (Kováčová *et al.*, 1969). Spontaneous mutants resistant to acriflavine (1–20 μg/ml) have been characterized (Thomas and Wilkie cited by Roodyn and Wilkie, 1968); resistance is controlled by a recessive nuclear gene. Similar mutants are also resistant to ethidium bromide, but in the case of this dye, resistance is controlled by a cytoplasmic factor which is presumably mt DNA (Gouhier and Mounolou, 1973). Cytoplasmic *petites* are also induced at high frequencies by extended exposures to high concentrations (1·3–3·0 mg/ml) of either erythromycin or chloramphenicol (Williamson *et al.*, 1971a).

Reversion of naturally arising cytoplasmically-inherited respiratory deficiency has been described (Rank and Pearson, 1969). Mutants which are less susceptible to the induction of the ρ^-mutant have been studied by Moustacchi (1973).

Rifampicin enhances the mutagenic effect of ethidium bromide treatment, but depresses the mutagenic effects of acriflavine, although it has no mutagenic effect alone (Whittaker and Wallis, 1971). Growth on galactose can delay the induction of *petite* mutants by ethidium bromide (Hammond *et al.*, 1974).

D. Alteration of mt DNA by Intercalating Dyes: mt DNAs of Cytoplasmic *Petite* Mutants

Cytoplasmic *petites* are induced with high efficiency by concentrations of intercalating dyes (such as acridines or ethidium bromide) which have no apparent effect on the nuclear DNA of yeast (Ephrussi, 1951; Slonimski,

1953a; Slonimski *et al.*, 1968). It has been shown that on treatment with these mutagens, synthesis of mt DNA ceases, transcription of mitochondrial DNA is inhibited (Fukuhara and Kujawa, 1970), and breakdown of pre-existing mt DNA occurs (Goldring *et al.*, 1970a, b; Perlman and Mahler, 1971a). A mitochondrial DNAase stimulated by ethidium bromide has been characterized (Paoletti *et al.*, 1972). In *petite*-negative yeasts (as in other obligately aerobic organisms such as *Paramecium*; A. Adoutte, personal communication) mt DNA replication is inhibited by ethidium or acridine treatment (Luha *et al.*, 1971) and mitochondrial transcription and translation is prevented (Heslot *et al.*, 1970b; Kellerman *et al.*, 1969; Schwab *et al.*, 1971; Mahler and Perlman, 1971), but these effects are fully reversible. Mutation by ethidium can be prevented by glucose repression (Hollenberg and Borst, 1971) under conditions where mt DNA replication proceeds normally. Mutagensis of wild-type yeast by acridines is inhibited by long-term pretreatment with protein synthesis inhibitors, and established *petites* are resistant to further mutagenesis by the dyes (Michaelis *et al.*, 1971).

An early event in the interaction of ethidium with mt DNA is the establishment of a premutational state, in which the potential mutagenesis is not expressed if the cells are subjected to heat shock (Perlman and Mahler, 1971a, b). These data suggest that mutagenesis is not a result of a primary effect on DNA replication, but rather interference with a process like DNA repair or recombination (Borst, 1972). A product of mitochondrial protein synthesis is involved and the initial damage is repairable. Thus it has been suggested that intercalation of dyes into mt DNA results in attack of the DNA by a mt DNA repair endonuclease. A *Tetrahymena* mt DNA polymerase is inducible in the presence of ethidium (Westergaard *et al.*, 1970).

Brief treatment with ethidium bromide leads to the induction of *petites* in which mt DNA is the same size as that of the *grande* parent strain. Longer treatment leads to fragmentation of the mt DNA (Goldring *et al.*, 1971). High concentrations of the mutagen applied for extended periods leads to the formation of cytoplasmic *petites* that contain no detectable mt DNA (Nagley and Linnane, 1970; Williamson *et al.*, 1971a; Michaelis *et al.*, 1971) and these *petites* behave as neutral *petites* in crosses with wild-type strains. Cytoplasmic *petites* lacking mt DNA (Nagai, 1969; Goldring *et al.*, 1970a, b; Michaelis *et al.*, 1971) have been denoted as ρ° (Nagley and Linnane, 1970, 1972a) and show no suppresiveness (see p. 296).

When dye treatment is stopped before all the mt DNA is degraded, the remaining DNA fragments may resume replication; the cytoplasmic *petites* thereby generated have altered mt DNA. A selection process giving clones with only one type of aberrant mt DNA may require as many as

65 generations (Hollenberg *et al.*, 1972a, b). The base composition may be altered from the wild-type 17% (G + C) to 4% (G + C) in extreme cases giving "low density" *petites* (Mounolou *et al.*, 1966; Carnevali *et al.*, 1966, 1969; Bernardi *et al.*, 1968, 1970; Mehrotra and Mahler, 1968; Perlman and Mahler, 1970a, b; Borst, 1970). Mt DNA may also be altered in size in some cytoplasmic *petites* (Tewari *et al.*, 1966a, b); Hollenberg *et al.*, 1969; Goldring *et al.*, 1970a, b). Loss of genetic complexity is revealed in the renaturation properties of altered mt DNA (Mehrotra and Mahler, 1968). The mt DNA (3% G + C) from a "low density" ethidium-induced cytoplasmic neutral *petite* (RD$_1$A) has been shown to consist of a perfect tandem repetition of a sequence of 150 nucleotides or less (Hollenberg *et al.*, 1972a, b; Van Kreije *et al.*, 1972; Borst, 1974b). The mt DNA has been replaced in this mutant by an equivalent amount of mutant mt DNA. This DNA ($\rho = 1.671$ g/cm^3) sediments in 3 M–alkalineCsCl as heterogenous linear DNA with a median mol. wt of 1.7×10^6, and electron microscopy revealed that the DNA was organized as large networks and occasional small circles.

Replication of sequences is accurate since no change was observed in the Tm of renatured DNA, or in the fingerprint of complementary RNA, after an additional 72 generations (Borst *et al.*, 1973a, b). All of the RD$_1$A mt DNA hybridizes with about 0.3% of the wild-type mt DNA, showing that it represents the amplification of a small AT-rich segment of the wild-type genome (Saunders *et al.*, 1973). RD$_1$A mt DNA does not cross-hybridize with the mt DNAs of the two other "low-density" *petites*, and this shows that the initiation site for DNA replication is not part of the repeating sequence. The RD$_1$A homoduplex has a Tm 6°C higher than that of the heteroduplex found between RD$_1$A and wild-type mt DNAs; this suggests some miscopying in the original mutagenic event. If copying errors occur in all highly repetitive *petite* mt DNAs this may limit attempts at gene purification by the isolation of *petite* mutants with appropriate genetic markers. However, the isolation, purification and physicochemical characteristics of yeast mt DNA segments conferring resistance either to chloramphenicol or erythromycin has been described by Fukuhara *et al.* (1974).

Deletions are compensated by reiteration of the non-deleted segments of the wild-type mt DNA leading to gene redundancy (Faye *et al.*, 1973). Periodic tandem repetitions of the sequence have been shown in long molecules of mt DNA by electron microscopy of half-denatured linear molecules and by the detection of circular molecules which occur in a series of multiples of a constant basic circumference. This repetitive DNA can be transcribed into repetitive RNA. With the help of suitable genetic markers stable ρ^- mutants can be selected in which different parts of the mt DNA are repeated and amplified in various combinations up to a hundred fold with practically no common sequences. Thus selective purification of different tRNA genes, rRNA genes and messenger RNA genes can be achieved. These purified mt DNA sequences have been characterized physico-chemically by both high resolution analysis of differential denaturation and by buoyant density measurements. It has been shown that the gene conferring erythromycin resistance to mitochondrial ribosomes has a higher G + C content than the gene conferring chloramphenicol resistance.

Intramolecular heterogeneity observed in fragments of *petite* mt DNA also suggests that the (G + C) content (comprising only 4% of the molecules in this case) could be localized in a limited region of the molecule, which could hence still carry genetic information (Carnevali and Leoni, 1972); for instance for mitochondrial $tRNA_{ser}$, although that for $tRNA_{phe}$ is lost (Carnevali *et al.*, 1973). Loss of antibiotic-resistance genes, originally present in the mt DNA may occur (Linnane *et al.*, 1968a,c; Gingold *et al.*, 1969; Rank, 1970a, b; Wilkie, 1970a; Coen *et al.*, 1970; Saunders *et al.*, 1971). In some cytoplasmic *petites* alterations in mt DNA are not detectable by gross physical methods, but can be shown by nearest-neighbour analysis of the RNA polymerase product of the altered DNA template (Grossman *et al.*, 1971) or by hybridization experiments using the mt DNA of wild-type and mutant (Fauman and Rabinowitz, 1972; Michaelis *et al.*, 1972; Gordon and Rabinowitz, 1973). The latter technique yields results which suggest deletion of *grande* mt DNA and generation of localized stretches of non-homogeneous base sequences. Loss of genes recognizable by hybridization with mitochondrial rRNA has also been reported (Mahler *et al.*, 1971b). Mt DNA homologous to certain mitochondrial tRNAs is retained in some *petites* and lost in others (Cohen and Rabinowitz, 1972). One *petite* cell may have a molecular population of mt DNAs which are heterogeneous with respect to information retention (Nagley and Linnane, 1972a). No cytoplasmic *petite* is known which has a completely functional system for protein synthesis (Borst, 1972). While ρ° cells lack detectable mt DNA, the cellular level of mt DNA in other classes of cytoplasmic *petites* (ρ^{-}) is very similar to the characteristic level maintained in ρ^{+} cells. If ρ^{-} cells contain mt DNA reduced in size as compared with ρ^{+} cells, problems arise as to the mechanism of control of mt DNA levels in yeast, since there would seem to be a large number of molecules of mt DNA in such ρ^{-} cells as compared to the wild-type (Nagley *et al.*, 1974). The regulatory processes controlling the cellular level of mt DNA in ρ^{-} cells must be provided for by cytoplasmic protein synthesis, and are probably under direct control of the nuclear genome.

The processes of ethidium bromide-induced disruption of the mitochondrial genomes in ρ^{+} and ρ^{-} cells are similar in principle. Therefore the mt DNA in ρ^{-} cells is replicated and maintained by similar processes

L

to those involved in ρ^+ cells even though the mt DNA fragments retained in ρ^- cells show considerable variation on their physical and genetic properties, representing different regions of the mt DNA of the ρ^+ parent cell. Those base sequences which are retained are considerably amplified in order to maintain the cellular level of mt DNA.

The mutagenic effect of ethidium bromide on *petite* strains varies in different clones and requires a higher concentration than in *grande* strains (Nagley *et al.*, 1973).

E. Genetic Properties of *Petite* Mutants

The behaviour of the *petite* mutant in sexual reproduction was studied by crossing two haploid strains, a normal yeast and a vegetative *petite* of opposite mating types (Ephrussi and Tavlitzki, 1949). Vegetative growth of the resulting diploids gave only respiratory competent cells. Sporulation of the diploids gave rise to four haploid ascospores, which were found on germination to be all respiratory competent (Fig. 100). In these experiments the nuclear (Mendelian) markers indicated that the nuclear behaviour of the hybrid was normal, as the markers showed a 2:2 segregation in each ascus, whereas the mutant character vanished on crossing and did not reappear in the spore progeny. Repeated backcrossing (i.e. crossing of the spores formed by the hybrid with the mutant parent) indicated that, in order to interpret the experimental results in Mendelian terms, it would have to have been assumed that the manifestation of the mutant character depended on the simultaneous presence of at least a dozen recessive genes, a very unlikely explanation. Ephrussi concluded that his working hypothesis (Ephrussi, 1949), that "the vegetative *petite* is the result of a cytoplasmic rather than of a nuclear mutation", was the most likely explanation. It was not possible to establish the behaviour of the mutant character through meiosis of diploid respiratory mutants as these do not sporulate.

Further confirmation that inheritance of cytoplasmic units (ρ particles, Sherman, 1964) is partly responsible for determination of the respiratory competence of yeast cells, came from the experiments of Wright and Lederberg (1957). After crossing strains of *S. cerevisiae* var. *ellipsoideus*, mixing of the two cytoplasms occurs, but the haploid nuclei remain separate and pass into the buds without fusing (heterokaryosis). Nuclear markers segregated independently of the ρ particles, and thus the respiratory competence of the progeny clearly depended on extra-nuclear inheritance.

It has been claimed that incubation of sphaeroplasts of a ρ^- yeast with wild-type mitochondria results in the transformation of some of the sphaeroplasts to ρ^+ (Tuppy and Wildner, 1965).

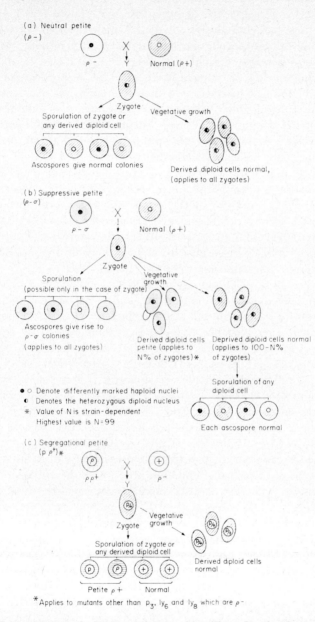

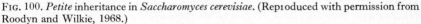

FIG. 100. *Petite* inheritance in *Saccharomyces cerevisiae*. (Reproduced with permission from Roodyn and Wilkie, 1968.)

F. The Phenomenon of Suppressiveness

Studies of a variety of *petites* showed that the straight-forward results described above were not always obtained, and that crosses between most *petites* and normal *grande* cells give rise to a percentage of both respiratory competent and deficient diploid cells (Fig. 100; Ephrussi *et al.*, 1955): this is the phenomenon of "suppressiveness". Thus whereas the first *petite* mutants to be studied behaved as if their mutation was a recessive one, other *petite* mutants behave as if their determinant is dominant; the respective *petite* strains are termed "neutral" (0% suppressive) and suppressive. The suppressive *petite* strains are divisible into those that enforce respiratory deficiency consistently on 99% of derived diploids from matings with normal strains, and those in which only a certain proportion of the resulting diploids are *petite*. This latter proportion can range from 0–99% depending on the strain. The degree of suppressiveness can be lowered by the action of mutagens. *Petites* of a given suppressiveness in general can give rise to progeny of similar characteristics (Ephrussi and Grandchamp, 1965), but also give rise to a number of organisms of both higher and lower suppressiveness (Ephrussi *et al.*, 1966).

Three types of zygotic clones are obtained following a cross. The zygote is initially unstable; some of the single zygotes give rise to both stable respiratory competent and stable respiratory deficient diploid cells, as well as to further unstable diploids. If zygotes formed by crossing a suppressive *petite* and a normal cell are immediately induced to sporulate and tetrad analysis performed, both 0:4 and 4:0 segregations for respiratory competence: respiratory deficiency are obtained. If however the induction of sporulation is delayed, the established respiratory deficient diploids do not sporulate, and only 4:0 respiratory competence: respiratory deficiency segregations are observed. A model accounting for some of these features has been proposed by Whittaker (1969).

Further clarification of the nature of suppressiveness came from a study of another type of cytoplasmic mutation, that which determines resistance to erythromycin (Thomas and Wilkie, 1968a; Linnane *et al.*, 1968c; Gingold *et al.*, 1969; Saunders *et al.*, 1971, see p. 299). *Petites* induced with acridine, euflavine or ethidium bromide (10 μg/ml) are nearly all of the genotype ρ^-ER° (which when crossed with a ρ^+ER^S cell gives only pure ρ^+ER^S diploid colonies, and when crossed with a ρ^+ER^R cell gives only pure ρ^+ER^R diploid colonies). *Petites* arising spontaneously are often of the class ρ^-ER^R or ρ^-ER^S, and only a proportion are ρ^-ER^O. This means that the mutation to cytoplasmic *petite* does not necessarily lead to loss of the erythromycin determinant, that cytoplasmic *petites* vary in their information content, and that the cytoplasmic *petite* mutation is not unitary (Saunders *et al.*, 1971). A correlation was noted between the degree of suppressiveness and the retention of cytoplasmic information.

A number of authors (Carnevali *et al.*, 1969; Rank, 1970a, b; Rank and Bech-Hansen 1972b) have suggested that a competition of replication rates between the mt DNAs of ρ^- and ρ^+ is involved. Saunders *et al.* (1971) supported this suggestion, and found that *petites* with the least altered mt DNA are of intermediate suppressiveness, and are thus able to compete equally with ρ^+ DNA in the zygotes. The change down towards neutrality or up to higher suppressiveness is a result of the DNA becoming less or better equipped to compete with the ρ^+ DNA. "Low density" *petites* which have grossly altered mt DNA (see p. 290) behaved like neutral *petites* with regard to their suppressiveness (Moustacchi, 1972). Neutral *petites* contain no detectable mt DNA and can be designated ρ° (Nagley and Linnane, 1970; Michaelis *et al.*, 1971).

Hypotheses involving recombination in suppressiveness have also been proposed (Carnevali *et al.*, 1969; Coen *et al.*, 1970; Shannon *et al.*, 1972; Michaelis *et al.*, 1973). Deutsch *et al.* (1974) envisage that the processes of *petite* mutation and suppressiveness may be due to the same basic mechanism, i.e. the spreading of errors in the mt DNA sequence by successive rounds of recombination. If this is so then the *petite* mutation can be regarded as a kind of "internal suppressiveness", and it would suffice to produce one modified mt DNA molecule in the cell to eventually suppress all the remaining normal molecules.

A further unexpected recent observation about suppressiveness has been reported by Bech-Hansen and Rank (1973) who have identified two neutral *petites* lacking detectable amounts of mt DNA as showing a suppressive phenotype when crossed with a certain ρ^+ strain (GR 25a) but which retained a neutral phenotype when crossed with other ρ^+ strains. This bivious suppressiveness could not be attributed to features of the ρ^- parents. Furthermore, crosses of two ρ^+ strains usually generates only ρ^+ zygote colonies, but crosses of GR 25a(ρ^+) with another ρ^+ strain produced a significant proportion of ρ^- zygotes. These observations suggest that strain G25a harbours a recessive nuclear mutation (q factor) in a gene (Q) which functions in the maintenance of the ρ^+ state. The proposed action of the q factor is similar to the effect of some segregational *petites* (Sherman and Ephrussi, 1962) that cause loss of the ρ^+ state, but unlike segregational *petite* mutations the q factor does not result in a respiratory-deficient phenotype.

G. The Number of Cytoplasmic Genomes

The number of mt DNA molecules per yeast cell varies under differing physiological conditions (see p. 280) but may be as many as 50–100 (Williamson, 1970). Studies on the induction of the *petite* mutation by growth at elevated temperatures (Sherman, 1959), or by intercalating agents (Sugimura *et al.*, 1966*b*; Slonimski *et al.*, 1968; Mahler *et al.*, 1971a) have suggested the presence of a small number (between 2 and 20) of cytoplasmic determinants which are altered in the cytoplasmic *petite*. Target analysis of the induction by UV of *petite* mutants in cells growing under anaerobic conditions led to the suggestion that a single "master template" may be present (Wilkie, 1963; Maroudas and Wilkie, 1966), but Allen and MacQuillan (1969) have interpreted the results obtained from similar experiments as indicative of the presence of at least three cytoplasmic determinants after anaerobic growth. These latter investigators found up to 20 targets in aerobically-grown cells. The work of Saunders *et al.* (1971) on the characteristics of the zygotes and zygotic buds which result from an $ER^R \times ER^S$ cross, showed that all clones obtained from single buds were mixed, and a large variation in the degree of mixedness was found. If a single master template did occur in aerobic cells, buds derived from the initial zygote would have been pure ER^R or ER^S. Considering the large variation in mixedness resulting from vegetative segregation it seems likely that at least more than two cytoplasmic determinants enter each bud.

The rate of formation of cytoplasmic *petites* by ethidium bromide is decreased in the presence of 10% glucose or nalidixic acid (500 µg/ml), but not in the presence of chloramphenicol (Hollenberg and Borst, 1971). These results are interpreted as indicating that the mt DNA can replicate normally in the presence of ethidium bromide and that mutagenesis may require another process, like DNA repair or recombination which can be repressed by glucose. A different mechanism for these effects of nalidixic acid is proposed by Vidová and Kováč (1972) who suggest that this agent interferes with nicking-closing cycles of mt DNA triggered by ethidium bromide. The normal replication of mt DNA in yeast is not inhibited by nalidixic acid in many strains of yeast. Whittaker *et al.* (1972) have shown that 50 µg/ml nalidixic acid is sufficient to prevent induction of cytoplasmic *petites* by ethidium bromide or 5-fluoro-uracil, although degradation of mt DNA does occur. It is suggested that at least a single copy of mt DNA (a "master copy"?) must be retained in order to maintain the respiratory-competent genotype in the presence of nalidixic acid. Similar results were obtained when very low concentrations of cyclo-heximide (1–100 µg/ml) were used instead of nalidixic acid (Whittaker and Wright, 1972). Although *petite* induction by ethidium bromide was prevented by cycloheximide, loss of mt DNA still occurred, and again these results were interpreted in terms of a "master copy" of mt DNA.

The discrepancy between the number of mt DNA molecules present and the number of targets for mutagens detected may thus be explained by the "master copy" hypothesis. More recent alternative hypotheses propose: (1) that a membrane bound complex of mt DNA is the actual target (Mahler and Perlman, 1972a) or, (2) a genetic recombination of mutated molecules with unaffected mt DNA molecules may lead to the spreading of initial errors amongst almost all the mt DNA molecules of the cell (Deutsch *et al.*, 1973).

H. Cytoplasmic Mutants Resistant to Antibiotic Inhibitors of Protein Synthesis

The discovery of a new class of cytoplasmically-inherited mutations in yeast, those conferring resistance to antibiotics (Wilkie *et al.*, 1967; Linnane *et al.*, 1968b, c; Thomas and Wilkie, 1968a, b) has had far-reaching consequences in the advance of mitochondrial genetics. These mutants also provide a promising approach to the identification of mitochondrial gene products. Prior to this, studies on extrachromosomally-induced lesions of mitochondrial functions in yeast had been limited to only one class of mutant, the cytoplasmic *petite*. The drastic effect of *petite* mutation on the mitochondrial respiratory system makes some genetic approaches impossible; e.g. respiratory deficient diploids cannot be analysed by tetrad analysis, as they cannot be induced to sporulate. Neither somatic recombination (extensively employed in studies of nuclear genetics, Roman, 1956; Wilkie and Lewis, 1963) nor complementation has been detected in many millions of diploid cells derived from crosses between different ρ^- strains, both spontaneous and variously induced (Roodyn and Wilkie, 1968).

Growth of *S. cerevisae* on glycerol (or on lactate) as the sole carbon and energy source eliminates energy production by fermentative pathways and makes obligatory the utilization of the respiratory system for the provision of energy for growth. Under these conditions, inhibitors having a specific effect on mitochondrial energy-yielding processes, or on the development of functional mitochondria will inhibit growth. The experimental procedure for exploiting this situation is to make plates of fermentable and nonfermentable media containing different concentrations of the inhibitor, and to lay down replicas of different strains. Strains sensitive to the mitochondrial inhibitor will grow only on the plates of medium containing fermentative substrate. Inhibitor-resistant mutants, which are selected for on the nonfermentative medium will show up as colonies on a background of inhibited cells (Fig. 101, Roodyn and Wilkie, 1968; Wilkie, 1970a, b, c). Resistance in such mutants may originate in an altered mitochondrial ribosome, an altered mitochondrial membrane, impermeability to the antibiotic, or reflect a type of detoxification mechanism. It has been suggested that antibiotic-resistant cells are due to intracellular selection of resistant mitochondrial genomes (Birky, 1973). High concentrations of antibiotics (e.g. 4 mg/ml chloramphenicol) were necessary to inhibit cytochrome formation in the original diploid strain of *S. cerevisiae* used in these studies (Clark-Walker and Linnane 1966, 1967; Huang *et al.*, 1966). Analysis of a number of haploid strains of this species indicated that the degree of sensitivity to each of the antibiotics tested was strain dependent, i.e. genetically determined (Wilkie *et al.*, 1967). Strains sensitive to as little as 50 μg/ml chloramphenicol were isolated, while sensitivity to 10 μg/ml erythromycin was seen in other strains. Sensitive strains gave rise spontaneously to stable resistant mutants, both of the recessive and dominant

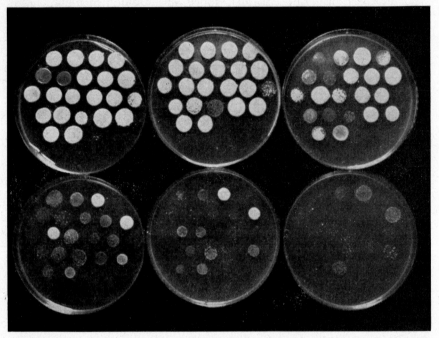

Fig. 101. A screening method for inhibitory effects of drugs on the mitochondrial system of a number of yeast strains. Replica drops from cell suspensions are put down in a series of plates containing non-fermentable substrate in the presence of increasing concentration of inhibitor (read from left to right). Resistant papillae can be seen against a background of inhibited cells. These effects would not be seen in parallel series containing fermentable substrate (Reproduced with permission from Wilkie, 1970b.)

types, and the resistance levels in some cases were so high as to correspond with the maximum solubility of the drug. Nearly all mutants showing spontaneous resistance to chloramphenicol had a simultaneous increase in tetracycline resistance, but no increase in erythromycin resistance. Spontaneous resistance to erythromycin had no striking effect on tolerance to chloramphenicol and tetracyline.

Increases in erythromycin resistance were often accompanied by increased resistance to other macrolide antibiotics (carbomycin, oleandomycin, spiramycin and lincomycin). Mutants with widely differing biochemical characteristics often map at identical genetic loci. None of the mutations are deleterious to mitochondrial ribosomal function; both the growth rates of the mutants and the activities of isolated ribosomes are similar to those of wild-type strains. Mutants selected for resistance to high levels of erythromycin have been shown to contain mitochondria having an amino acid-incorporating activity which is completely resistant to the drug (Linnane et al., 1968a, c). This change similarly affects the sensitivity of the mitochondrial protein synthesis system to other macrolides. On the other hand chloramphenicol resistance, in the strains then studied, was due to changes in cell permeability which extends to tetracycline and thereby accounts for the observed cross resistance shown by intact cells towards these two antibiotics. Inheritance of resistance to erythromycin did not segregate as a single chromosomal gene, but rather showed extrachromosomal inheritance (Linnane et al., 1968a, c; Fig. 102). The cytoplasmic determinant for erythromycin resistance, and the factor associated with the cytoplasmic petite mutation, appeared not to be identical. It was suggested that cytoplasmic coding for erythromycin resistance might mean the existence of an extrachromosomal determinant for a mitochondrial ribosomal protein. Thomas and Wilkie (1968a) showed that strains resistant to spiramycin, paromomycin and erythromycin when crossed gave recombinant clones. Whether the recombinant progeny contained copies of both parental mt DNAs or recombinant mt DNA molecules was not clear in these experiments, and a sporulation test which would have distinguished between these two possibilities was not applied. However, it was proposed that the mechanism of recombination might involve the pairing of mt DNA strands during meiosis and after mating in the zygote, resulting in genetic crossing over (Wilkie, 1970c).

Thomas and Wilkie (1968b) working with euflavine-induced petites reported that the loss of respiratory competence is always accompanied by the loss of the erythromycin determinant, and concluded that the two determinants are carried on the same factor. However, petites derived spontaneously from erythromycin-resistant strains frequently retained the erythromycin resistance determinant, indicating that the two cytoplasmic determinants for respiratory competence and erythromycin resistance are not closely linked (Gingold et al., 1969).

Biochemical analysis of this erythromycin-resistant mutant was initially held up by the difficulty of finding conditions that would allow isolated mitochondrial ribosomes to use a message other than poly-(U), as poly-(U)-directed poly-Phe synthesis is insensitive to many antibiotics. The modified fragment reaction (De Vries et al., 1971) provided a convenient way of overcoming this problem for erythromycin or chloramphenicol (Grivell et al., 1971a), and by this assay it was shown unequivocally that the mutation to erythromycin resistance in strain 6–81c led to a change in the

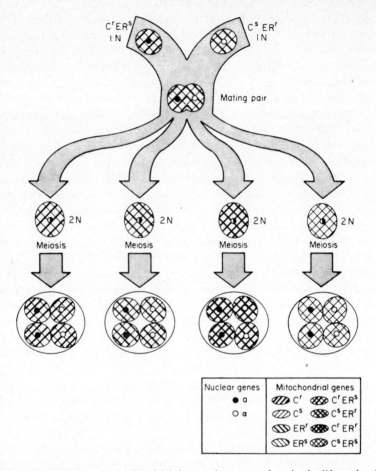

FIG. 102. Segregation of mitochondrial drug-resistance markers in the life-cycle of yeast. The haploid parental strains differ in the nuclear mating type alleles a and α, and in two pairs of mitochondrial genes: C^R/C^S and ER^R/ER^S. Zygote formation is followed by segregation of the parental and recombinant types during vegetative growth of diploid colonies. After one subculture each diploid cell gives rise to a uniform clone, either parental or recombinant. When these pure clones are induced to sporulate, they produce asci in which the four products segregate 4:0 for the mitochondrial genes and 2:2 for the mating type alleles a and α. (Reproduced with permission from Sager, 1972.)

mitochondrial ribosomes rather than in the membrane. These experiments did not enable an exact location of the lesion to be determined; it was suggested that this was either in a protein of the large ribosomal subunit or in the ribosomal RNA.

Similar studies with several other genetically distinguishable mutants of mt DNA (resistant to chloramphenicol, erythromycin, and/or spiramycin) also have altered mitochondrial ribosomes (Grivell *et al.*, 1973;

Molloy *et al.*, 1973). Acrylamide gel electrophoresis of the proteins of the large ribosomal subunit from wild-type or several mutant strains failed to reveal any differences; however it is possible that amino acid substitutions may occur without change in charge. It is considered that the antibiotic resistance of the mutants may result from alterations to rRNA. Mt DNA has only one cistron for each rRNA per molecule. If each molecule consti- tutes an independent unit for mutation and segregation, then a mito- chondrion with a full complement of resistant ribosomes can arise as a result of a single mutation, and under selective conditions replace the original antibiotic sensitive population.

Two other cytoplasmically-determined mutants of *S. cerevisae*, resistant to chloramphenicol and mikamycin have been isolated (Mitchell *et al.*, 1970; Bunn *et al.*, 1970). Amino acid incorporation into protein by mitochondria isolated from either cells resistant or sensitive to mikamycin or chloramphenicol was inhibited by these antibiotics. Although aerobically- grown resistant strains were not affected by mikamycin or chloramphenicol, it was found that the mitochondrial protein synthesis systems of anaerobi- cally-grown organisms was inhibited *in vivo*.

Cross resistance among the antibiotics chloramphenicol, mikamycin, erythromycin, lincomycin, carbomycin, and spiramycin was observed. All erythromycin-resistant mutants, on the other hand, were resistant to ery- thromycin both *in vivo* and *in vitro*. The simple explanation that holds in this case (i.e. that erythromycin resistance is a consequence of a specific alteration in a mitochondrial ribosomal protein) cannot easily explain the observations on mikamycin and chloramphenicol resistance. It was pro- posed that the mutation to mikamycin resistance leads to changes in the membrane, which render the mitochondria impermeable to mikamycin, chloramphenicol, lincomycin and carbomycin. Damage to mitochondrial membranes during isolation and assay would then allow penetration of the antibiotics *in vitro*. The altered membranes produced by anaerobic growth allow these effects *in vivo*. It was also proposed that conformational changes, either in membranes or in ribosomes could account for the observed cross resistances to chemically unrelated antibiotics, and that ribosomes and membranes are intimately related or even that mitochondrial ribosomes are part of the inner membrane.

Another mikamycin-resistant mutant is cross resistant to the oxidative phosphorylation inhibitor, oligomycin, as well as to mitochondrial protein synthesis inhibitors (chloramphenicol, lincomycin, carbomycin and tetracycline; Mitchell *et al.*, 1971, 1973). Mitochondria isolated from this mutant are sensitive *in vitro* to these protein synthesis inhibitors and *in vivo* after anaerobic growth. A mitochondrial defect is also indicated by the low aerobic yield of the mutant compared with that of the parent strain (Griffiths *et al.*, 1972). No defect in the stalked particles of the inner membrane was evident in electron micrographs (Watson and Linnane, 1972). The observation that mikamycin can act as a respiratory inhibitor when present at higher concentrations that those necessary to give inhibition of protein synthesis led Dixon *et al.* (1971) to suggest that the action of this antibiotic is on an integrated inner membrane-ribosome system.

Five mutants with *in vivo* resistance to the macrolide antibiotic, spiramycin have been distinguished on the basis of their cross resistance to other antibiotics, their biochemical properties and genetic behaviour (Trembath *et al.*, 1973). Four out of the five are cytoplasmic mutants and recombination analysis performed on crosses between different cytoplasmic markers demonstrated the presence of at least two and possibly three cytoplasmic genetic loci responsible for spiramycin resistance. Attempted correlation of antibiotic resistances of mitochondrial protein synthesis *in vivo* and *in vitro*, again suggest that complex mitochondrial membrane ribosome interactions are involved, furthermore three different mutations, apparently at the same locus, each result in different biochemical behaviour towards spiramycin.

Mutants sensitive to spiramycin which are not cross resistant to erythromycin have been obtained (Thomas and Wilkie, 1968a; Bolotin *et al.*, 1971).

Chloramphenicol-resistant mutants have been described by Coen *et al.* (1970) and by Kleese *et al.* (1972b).

Paromomycin-resistant mutants have been described by Thomas and Wilkie (1968a) and studied genetically (Wilkie and Thomas, 1973; Kleese *et al.*, 1972a; Wolf *et al.*, 1973). This mutation to paromomycin resistance does not confer cross resistance to either chloramphenicol, erythromycin or spiramycin: however interactions at the level of the mitochondrial ribosomes increase the sensitivity to chloramphenicol by about two fold (Wolf *et al.*, 1973). The phenotypic expression of the P^R gene is suppressed by modifying nuclear genes. Cross resistance to neomycin has been shown for certain paromomycin-resistant mutants (Kutzleb *et al.*, 1973). Amino acid incorporation by mitochondria isolated from these strains is also paromomycin resistant.

Chlorimipramine binds to isolated mitochondria and to whole cells of yeast, and leads to a cessation of growth on a glucose-containing medium (Linstead *et al.*, 1974). Strains with a low drug tolerance sometimes show an increase in tolerance up to 5-fold on induction of cytoplasmic *petites*, even in those with no detectable mt DNA. Respiratory deficiency induced by introduction of the nuclear *petite* marker p_7 has no effect on drug tolerance. Anaerobic growth decreases cellular tolerance, whereas severe glucose repression brings about increased resistance in cells. Evidently interaction between nuclear and cytoplasmic factors contribute to the cellular characteristics exhibited towards this drug.

I. Mitochondrial Gene Recombination and Transmission

Rapid advances in the understanding of extrachromosomal genetic systems of eukaryotes have been made possible by the study of mutations to antibiotic resistance. The first successful application of this approach was to the study of chloroplast mutations in *Chlamydomonas reinhardii* (see Sager, 1972 for review). Several types of mutations were discovered in this system, including those conferring resistance to streptomycin, and the use of these mutants enabled recognition of extrachromosomal reassortment and mapping of chloroplast genes (Sager and Ramanis, 1964, 1965, 1970; Gillham, 1965; Gillham and Fifer, 1968). Reassortment of mitochondrial genes in *S. cerevisiae* has also been shown to occur (Thomas and Wilkie, 1968a, b; Coen *et al.*, 1970; Bolotin *et al.*, 1971; Rank and Bech-Hansen, 1972a; Kleese *et al.*, 1972a, b) and recombination experiments have enabled the construction of a linkage map of the mitochondrial genome (Fig. 103).

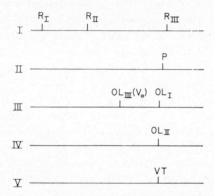

Fig. 103. Recombinational linkage groups on the mitochondrial genome of *Saccharomyces cerevisiae* as determined by recombination experiments. R_I, locus conferring chloramphenicol resistance; R_{II}, R_{III} loci conferring erythromycin resistance; P, locus conferring paromomycin resistance; $OL_{I,II,III}$, loci conferring oligomycin resistance; V_o, V, loci conferring venturicidin resistance; T, locus conferring triethyl tin resistance. Groups I—IV are located on the circular mt DNA, but as yet no definite order can be assigned to them. Linkage Group V is on a separate cytoplasmic DNA molecule which may or may not be mitochondrial. (Reproduced with permission of W. E. Lancashire, R. L. Houghton and D. E. Griffiths.)

In crosses made between mutants of two different phenotypes (i.e. $C^R E^S \times C^S E^R$) both recombinant types appeared, and tetrad analysis strongly suggested that C^R and E^R mitochondrial genes undergo a true genetic recombination at the level of mt DNA (Coen *et al.*, 1970).

The classical model for genetic recombination is based on the formation of new molecules of DNA by crossing over. If however organelles contained a fixed and stable ploidy of DNA molecules, reassortment without molecular breakage and reunion could mimic some aspects of molecular recombination. The final demonstration of true molecular recombination as the explanation for the genetic recombination of organelles has been achieved by physical evidence for differences between the parent mt DNA molecules and the recombined ones (Michaelis *et al.*, 1973). In these experiments, a cytoplasmic *petite* mutant carrying the gene for chloramphenicol resistance was crossed with a cytoplasmic *petite* having the gene for erythromycin resistance. The two *petites* differed in the buoyant densities of their mt DNAs. Diploid *petite* recombinants were obtained, carrying both genes and containing not a mixture of parental DNAs but a new species of mt DNA of intermediate buoyant density. Thus recombinations must have involved breakage and reunion of mt DNA molecules. New suppressiveness, distinct from that of either of the two parent ones, can also result from recombination of mt DNA. Experiments giving evidence for the formation of new species of mt DNA on crossing $\rho^+ \times \rho^-$ strains (Carnevali *et al.*, 1969; Shannon *et al.*, 1972) in the absence of genetic markers are open to other interpretations e.g. instability or degradation of mt DNA in zygotes.

The major difficulty in analysing the mode of segregation and transmission of mitochondrial genes, comes about from the growth of clones by budding rather than by fission, Thus, the zygote may produce twenty buds sequentially, and these buds may themselves produce progeny while the zygote is still producing buds.

The consequences of this complex process of vegetative growth are that unequal partitioning of the cytoplasm, and overlapping vegetative generations occurs. Results of analysis of intra- and inter-clonal distribution of large populations of various cell types in the vegetative diploid progeny obtained after about 20 generations from individual zygotes of a mitochondrial cross (Coen *et al.*, 1970), indicated that the proportions of recombined mitochondrial genotypes revealed a fundamental dichotomy which could not be explained by statistically random processes involving numerous mitochondrial genomes. A single zygote can give rise to at least three and possibly four stable mitochondrial cell types which occur in very different proportions from zygote to zygote in the same cross. Recombinant frequencies both intra-and inter-clonally are non-reciprocal. Two different mechanisms of transmission of a mitochondrial gene were distinguished at the clonal level; one led to a very low degree of transmission and the other to a high degree of transmission. Mitochondrial mutants of the same antibiotic resistance but of different origin differ in the extent to which these two mechanisms operate in their transmission, and this hereditary difference is probably non-chromosomal. Proportions of the different recombinants in individual clones could be fitted to the series: $\frac{1}{2}$, $\frac{1}{4}$, $\frac{1}{8}$ etc.

The proportions of various cell types in the populations change considerably with the numbers of vegetative divisions that have elapsed since the cell copulation event; this variation is very striking during the first few cell divisions. Control experiments showed that these changes could not result from selective pressures due to the differential rate of multiplication of various cell types, and it is in fact due to the kinetics of recombination and segregation of mitochondrial genes. In contrast to those of Thomas and Wilkie (1968a, b) and Wilkie (1970a), the experiments of Coen *et al.* (1970) indicate that anaerobiosis does not modify the recombination and segregation of mitochondrial genes, and the latter investigators attribute the discrepancy to Wilkie's less stringent analytical technique.

From these experiments Coen *et al.* (1970) introduced the concept of polarity as "the deviation from the $\frac{1}{2}/\frac{1}{2}$ proportion of the two allelic forms in the random vegetative descendants of a mitochondrial cross. The greater the deviation, the more polar is the cross". They also defined polarity as "the deviation from the 1:1 ratio of the two types of reciprocal recombinants $C^S E^S : C^R E^R$, and the greater the deviation, the more polar is the two factor cross". In order to explain the observed frequencies of recombination a "polarity point" (O) was proposed to occur on the mitochondrial genome. It was suggested that the location of this polarity point relative to the localization of any ρ^- mutation might determine the degree of suppressiveness of that mutant.

All the antibiotic-resistant mutants examined from strains of "α" mating types were localized very close to the polarity point O, while mutants isolated from "a" strains were further from it. That the polarity characteristics do not segregate during meiosis indicates that mitochondria possess autonomous properties that govern their own hereditary transmission and recombination. This situation can be described as mitochondrial sex, autonomous and distinct from cellular sex, (which is determined by mating-type genes a and α).

These studies indicate a great number of analogies with bacterial genetics and particularly with respect to the mechanisms of sexuality and conjugation. The recombination process of mt DNA is especially relevant in this context, as the parental DNA molecules are separated from each other by the membranes of the organelles. Thus a pairing of homologous mt DNA molecules leading to a genetic cross-over must involve a special mechanism for bringing the two molecules together. Such a mechanism which would resemble bacterial conjugation (with ω^+ analogous to the Hfr factor) has been proposed by Bolotin *et al.* (1971). Markers closest to ω^+ will be transferred preferentially resulting in polarity of transmission.

Homosexual and heterosexual crosses involving three and four mutations located at the loci $R_I R_{III} OL_I$ and P_I and conferring resistance respectively to chloramphenicol, erythromycin, oligomycin and paromomycin were performed in all possible *cis- trans*-configurations (Wolf *et al.*, 1973). P^R and OL^R mutations presented similar features of recombination which were distinct from these shown by C^R and E^R mutations. P^R and OL^R are not included in the $\omega-R_I-R_{II}-R_{III}$ segment and are not linked or only

very loosely linked to it (Fig. 103). P^R and OL^R are not linked. This system
has been used to analyse the features of mitochondrial multifactorial crosses
in terms of the frequency and the distribution of different classes of
recombinants. In homosexual crosses a positive coincidence was observed
for all combinations of three or four markers. In heterosexual crosses the
polarity of recombination depends upon the segment considered. An excess
of the allele of a third marker brought by the ω^- parent was observed among
polar recombinants for two other.

Analysis of recombinants produced in a three-factor cross of cytoplasmically-inherited mitochondrial markers for antibiotic-resistance in *S.
cerevisiae* gave results which could not be used to construct a map of
linkage of gene order (Rank, 1973).

Micromanipulation of early zygote buds shows that recombinant
mitochondrial genomes are formed very rapidly on zygote formation;
initial zygote buds are frequently composed of varying proportions of
recombinant and parental types (Lukins *et al.*, 1973). The possibility that
the formation of recombinant genomes is restricted to the zygote was not
eliminated; the process may continue in the early zygote progeny. The
results obtained were in contrast to those of Coen *et al.* (1970) who found
the distribution of recombinants could be fitted into a series $(\frac{1}{2})^n$, indicating that the recombination occurred in the zygote and that a single unique
event gives rise to a recombinant type which breeds true. Lukins *et al.*
(1973) prefer the interpretation that recombination events occur rapidly
upon zygote formation to give multiple copies of recombinant genomes
which then reaggregate and purify during bud formation. This interpretation offers a different concept of polarity: that this phenomenon is not
related to the segregation of mitochondrial genomes, but rather to the
characteristics of the mitochondrial recombination events which give rise
to the different recombinant genomes in varying proportions. It was
tentatively concluded that segregation of mitochondrial types, both parental
and recombinant, is a random process associated with the transmission of
mitochondria to buds. Further analyses of the polarity of mitochondrial
gene formation and recombination (Howell *et al.*, 1973; Linnane *et al.*,
1973, 1974; Linstead *et al.*, 1974) suggest that polarity is not determined by
a simple mitochondrial sex factor, and that several different interactions
(including possible nuclear effects) are responsible. That the presence of
nuclear gene(s) in some strains also influences the recombinational events
in both homosexual and heterosexual crosses has also been proposed by
Avner *et al.* (1973). This behaviour is superimposed on, and separable
from, the effects due to the mitochondrial gene ω.

Pedigree analysis of crosses between yeast strains carrying multiple
mitochondrial markers (conferring resistance to erythromycin, chloramphenicol, paromomycin and oligomycin) by micromanipulation of buds from

zygotes also indicated complexities in the mechanisms of recombination and and transmission (Wilkie and Thomas, 1973).

The zygote begins to transmit a single mitochondrial type after the first few buds have been formed. This was the parental type in one cross studied (Cross I) but in a second cross (Cross II) it was random, whether the parental or recombinant type became the predominant one. Thus the mixture of mitochondrial types of the early zygote was either short lived, or the contribution of the two parents was grossly unequal. The same was true for the early buds which gave mixed clones, the individual cells of which were pure with regard to mitochondrial type. Zygotes of Cross I tended to stabilize their mitochondrial type sooner than those of Cross II. Using the terminology of Coen *et al.* (1970) and Bolotin *et al.* (1971) Cross I would be termed a heterosexual cross ($\omega^+ \times \omega^-$), whereas Cross II would be a homosexual cross between two ω^- strains. Wilkie and Thomas (1973) suggest that the main difference between these crosses is that in Cross II there is a mechanism for maintaining the integrity of the parental genome in the face of recombinational events. In electron micrographs (Smith *et al.*, 1972) there are indications that mitochondria undergo dediffe-rentiation on zygote formation. This process could result in the liberation of mt DNA molecules which might then recombine. Resynthesis of the mitochondria in the budding zygotes and reincorporation of the new genomes could determine how quickly the mitochondria became stabilized. The continuity and integrity of the parental type in Cross I zygotes may depend on an attachment site for mt DNA on the membrane of these mito-chondria. Wilkie (1972) and Wilkie and Thomas (1973) conclude that until more is known of the molecular mechanisms underlying the behaviour of mitochondrial genotypes in zygotes it is premature to assign any formal mechanisms such as mitochondrial sex by analogy with bacterial conjugation.

Further evidence for control imposed by a nuclear gene on the distri-bution of antibiotic resistance factors comes from the observation that randomization of distribution can be induced by treatment of the newly formed zygotes with cycloheximide for 1 h or with an inhibitor of nuclear RNA synthesis (thiolutin) (Linstead *et al.*, 1974). Blocking of protein synthesis before or during mating does not affect the subsequent distribu-tion of mitochondrial markers.

J. Correlation of Loss of Markers for Drug Resistance with the Induction of the *Petite* Mutation

The survival of the ρ^+ genetic marker is always smaller than the survival of any marker for drug resistance after exposure to ethidium bromide (Deutsch *et al.*, 1974; Uchida and Suda, 1973). The survival of C^R and E^R are similar to each other, whereas that of OL^R is greater. All possible com-binations of Drug^R markers have been found among the ρ^- *petite* cells induced, whilst the only type found among the ρ^+ colonies was the pre-existing one. The loss of the C^R and E^R genetic markers was found to be the most frequently concomitant; correlation between loss of the O^R marker and the other two $Drug^R$ markers was less strong. Similar results were also obtained after UV irradiation.

The conclusions from these experiments are: (1). The ρ^+ factor is the entire mt DNA; and (2). The $Drug^R$ markers are located within it.

The question arising from these conclusions is: how can a mutation, located anywhere along the mt DNA sequence always confer the *petite* phenotype, i.e. lack of *a*- and *b*-type cytochromes leading to respiratory deficiency?

Two types of hypotheses may explain the extreme pleiotropy observed in the cytoplasmic *petites*:

1. The expression of mitochondrial genes is achieved through such a highly coordinated regulatory system, that genetic inactivation of any single gene leads to inhibition of expression of the entire mitochondrial genome. It is known for instance that the whole mt DNA of Hela calls seems to be a single transcription unit (Aloni and Attardi, 1971a).

2. The genes coding for the different components of the mitochondrial protein synthesis system (tRNAs, rRNAs and proteins) are distributed all along the mt DNA sequence. The mutagenic lesions are so large, that the possibility that the whole synthetic machinery will be preserved is virtually zero.

The relationship between loss of the ρ^+ factor and the $Drug^R$ markers suggests that the mutagenic event leading the *petite* mutation is very large (at least 50% of the wild-type mt DNA sequence is lost, Faye *et al.*, 1973), overlapping a significant part of the genome, and rules out the proposals that the *petite* mutation can occur by way of a multiplicity of small events (cf. Borst and Kroon, 1969; Hollenberg *et al.*, 1972a). That the main features of ethidium bromide and UV induction of the *petite* mutation are similar suggest that perhaps *all* mutagenic agents simply trigger the same mechanism. The events leading to the formation of a *petite* may not greatly differ from those produced by the action of mutagens in other organisms, but one is able to recover in *petites* all the mutations which appear lethal in other organisms. Macrolesions occurring during the *petite* mutation may provide an independent method for assessing the topographical relations between mitochondrial genes. The differential loss of different $Drug^R$ markers during *petite* mutation may be explained either by (a), the different nature of base sequences in different regions of the mt DNA, or (b) its dependence on the direction of replication and its distance from the replication site.

The cytoplasmically-inherited "killer" character in yeast is retained under conditions of ethidium bromide treatment in which more than 94%

of detectable mt DNA is lost (Al-Aidroos *et al.*, 1973). Thus it is very unlikely that the "killer" genetic determinants are part of the mito-chondrial genome.

K. Cytoplasmic Mutants Resistant to Agents Inhibitory to the Reactions of Energy Conservation

1. Mitochondrial ATPase of yeast; modifications during chloramphenicol-inhibited growth, glucose repression and in cytoplasmic *petite* mutants

Intact mitochondria from *S. cerevisiae* contain an active Mg^{2+} dependent ATPase which shows two pH optima (at pH 6·2 and 9·5) (Kováč and Weissová, 1968; Schatz, 1968; Somlo, 1968). Similar pH optima are shown by the enzyme from *Candida utilis* (Vignais *et al.*, 1973). The enzymes from these yeasts are less sensitive to inhibition by oligomycin than the mammalian enzyme and sensitivity to this inhibitor is greatest at the alkaline pH optimum. On the other hand, *C. lipolytica* ATPase is much more oligomycin-sensitive at pH 7·0 (Houghton *et al.*, 1973). The ATPase in intact mitochondria is cold stable and oligomycin sensitive, but when detached from the inner membrane it becomes cold-labile and oligomycin insensitive (Schatz *et al.*, 1967). Under these latter conditions however it retains sensitivity to Dio-9 (Guillory, 1964).

Modifications of mitochondrial ATPase have been observed in respiratory deficient cytoplasmic *petite* mutants of *S. cerevisiae* (Kováč and Weissová, 1968; Schatz, 1968; Perlman and Mahler, 1970b). Whereas in the wild-type strains this enzyme is cold-stable and strongly inhibited by oligomycin, the ρ^- mutants contain a mitochondrial ATPase which is cold-labile and oligomycin insensitive. Loss of the projecting inner-membrane subunits have been shown in electron micrographs of negatively-stained prepara-tions (Bachop *et al.*, 1972). Similar changes in the proportion of the enzyme were induced by growth of the wild-type with chloramphenicol, but were not seen on glucose repression (Schatz 1968). Purification of the ATPases of the *petite* and wild-type strains gave enzyme preparations which were indistinguishable on the basis of enzyme properties, inhibition by ADP, Dio-9 and F_1–ATPase inhibitor, stimulation by DNP, sedimentation characteristics, and immunological specificity. When the purified ATPase from the ρ^- mutant or from wild-type yeast (Schatz *et al.*, 1967) was bound to ATPase-deficient SMPs from bovine heart mitochondria the yeast enzyme became cold-stable and sensitive to oligomycin. Therefore it was not the F_1–ATPase which had been altered in the *petite*, but rather that an alteration in the mitochondrial membrane had impaired the binding of the enzyme.

Growth of *C. utilis* in the presence of chloramphenicol gives a 3- to 4-fold increase in the activity of neutral ATPase, and the enzyme is still sensitive to oligomycin (Vignais *et al.*, 1973). A similar change was observed in organisms grown at high glucose concentrations or under conditions of semi-anaerobiosis. Thus the response of this enzyme to physiological manipulation is quite different in the obligately aerobic *C. utilis* from the facultative *S. cerevisiae* (Schatz, 1968). Vignais and co-workers (1973) have also shown that loss of oligomycin sensitivity of ATPase for several different strains of *S. cerevisiae* occurs on glucose repression, and this finding conflicts with those of Schatz (1968).

2. Oligomycin-resistant mutants

One of the oligomycin-resistant mutants examined by Stuart (1970) showed cytoplasmic inheritance. Resistance to oligomycin did not obligately confer resistance to rutamycin or venturicidin. Both nuclear and cytoplasmic genes may govern venturicidin resistance.

A mutant (102P7) resistant to oligomycin was isolated using 2, 6-diaminopurine (Wakabayashi and Gunge, 1970), this resistance was inherited cytoplasmically, but was not linked to ρ as it was not removed by acriflavine nor by the introduction of a nuclear gene which changed ρ^+ to ρ^-. Mitochondrial ATPase in the mutant and parent strains were equally sensitive to oligomycin suggesting that the basis of resistance may be a permeability change or an induced detoxification mechanism. However, in another mutant, oligomycin resistance was correlated with a modified mitochondrial ATPase (Wakabayashi, 1972), although the increased resistance of the mitochondrial ATPase was not very marked. When a mutant with the cytoplasmic oligomycin-resistance determinant (μ) was crossed with a mutant with the erythromycin-resistance determinant (ϵ), the resulting double resistant diploids, when subjected to tetrad analysis, showed a 4:0 (R:S) segregation of both resistances (Wakabayashi and Kamei, 1973a, b). No loss of ϵ or μ was observed in 416 colonies of the double resistant haploid R8. The cross between R8 and the double sensitive N2B gave diploids which segregated in 42% of $\epsilon^R \mu^R$ diploids and 35% $\epsilon^S \mu^S$ diploids.

This showed that the genes ϵ and μ were associated and resided in two incompatible plasmids, probably mitochondria; they were lost on ethidium bromide treatment. However, some of the spontaneous ρ^- mutants of R8 retained resistance to one or to both antibiotics. As the recombinants were found in the cross $\rho^+ \epsilon^R \mu^R \times \rho^+ \epsilon^S \mu^S$, the cross $\rho^- \epsilon^R \mu^R \times \rho^+ \epsilon^S \mu^S$ was analysed. The rate of the transfer of μ^R to ρ^+ was higher that that of ϵ^R. The close association of ϵ^R to ρ^- was shown in the case of ρ^- with high suppressiveness. The transfer of the mitochondrial genes was found to be sequential.

A group of oligomycin-resistant mutants of *S. cerevisiae* have also been selected by Shannon *et al.* (1973). Most are cytoplasmic mutants and the ATPases of their isolated mitochondria have altered sensitivities to oligomycin. Two mutants resistant to growth with oligomycin do not show oligomycin-resistant mitochondrial ATPase. One of these mutants is cytoplasmic and the other nuclear. Studies on the isolated soluble ATPase from wild-type cells shows that the solubilized enzyme is not inhibited by oligomycin whereas the reconstituted enzyme complex is sensitive. ATPases from oligomycin-resistant mutants can also be separated into their components and reconstituted to give oligomycin-resistant complexes. Preparation of hybrid reconstituted ATPase complexes from mixed components of the mutant and wild-type, indicates that the oligomycin resistance is contributed by the membrane fraction of the ATPase. A component of this membrane fraction must therefore be under control of mt DNA.

A series of 62 mutants of *S. cerevisiae* showing high levels of resistance to oligomycin and rutamycin were subdivided into two classes on the basis of their cross resistance to a number of inhibitors and uncouplers of mitochondrial energy-conservation reactions (Avner and Griffiths, 1970, 1973a; Griffiths *et al.*, 1972). Class I mutants show cross resistance to aurovertin, triethyl tin, "1799" (a derivative of hexafluorotriacetone), uncoupling agents and inhibitors of mitochondrial protein synthesis such as chloramphenicol, mikamycin and spiramycin. However, Class II mutants are *specifically* resistant to oligomycin and rutamycin and show no cross resistance to aurovertin, triethyltin or any of the uncoupling agents tested. This result supports biochemical evidence that these agents act at different sites in the energy conserving machinery. Most of the Class I mutants and one of the Class II mutants lose resistance at the non-permissive temperature of 20°C, and could be subclassified on this basis (Tables XLI, XLII). The growth yield of Class I mutants was 15–20% less than that of the wild-type whereas the growth yield of Class II mutants was unaltered; growth rates of most of the mutant strains was normal. No gross disturbances of mitochondrial morphology was evident on examination of any of these mutants by electron microscopy. For instance, stalked particles (inner-membrane subunits) could still be demonstrated (Watson and Linnane, 1972) and the numbers of mitochondria per cell were apparently normal.

Investigation of the two classes of oligomycin-resistant mutants (Avner and Griffiths, 1973b) revealed that all the Class II mutants examined showed typical cytoplasmic inheritance (as judged by the criteria given in Table XLII), and the determinants appeared to be located on mt DNA. It is not known whether the mutated loci code for a structural or regulatory function. Resistance is conferred by at least two groups of distinct non-allelic cytoplasmic determinants. Analysis of Class I mutants revealed a complex involvement of both a nuclear gene and a

Table XLI. Classification of some mutants of *S. cerevisiae* resistant to agents interacting with the energy-conservation mechanism. (Reproduced with permission from Griffiths *et al.* 1972.)

A. Oligomycin-resistant mutants

Class	Levels of resistance × wild-type					
	†Oligomycin	Rutamycin	Aurovertin	TET	"1799"	CAP, Ant-A Mika, Spira, CCCP, TTFB
1A	20	100	R	10–20	>20	2–4
1B*	20	100	R	10–20	>20	2–4
2A	20	100	1	1	1	1
2B*	20	100	1	1	1	1

B. Triethyl tin-resistant mutants

Class	Levels of resistance × wild-type			
	Oligomycin	TET	"1799"	TTFB
1	>20	20	8–12	4
2	1	20	<2	1
3	1	20	8	1

C. TTFB-resistant mutants

Class	Levels of resistance × wild-type				
	TTFB	TET	Oligomycin	"1799"	CCCP
1	6–8	>20	>20	>10	6
2	6–8	1	1	1	6
3	6–8	1	1	8	10

D. Valinomycin-resistant mutants

	Valinomycin	X-206	Nigericin	TET	Oligomycin
Wild-type	2·5	5	5	1·25	0·5
Class 1	>40	>15	>15	10	5
Class 2	>40	5	<5	1·25	0·5

*Class 1B and 2B mutants are temperature sensitive and lose resistance and grow normally at 20°.
†These represent minimal levels of resistance as high oligomycin concentrations were not tested.

Table XLII. Criteria for cytoplasmic inheritance and genetic characteristics of oligomycin-resistant mutants of *S. cerevisiae*. (Reproduced with permission from Griffiths *et al.*, 1972.)

Criteria	Class I OL^R	Class II OL^R
1. Crosses of $OL^R \rho+ \times OL^S \rho+$ should show mitotic recombination—i.e. diploid zygotes give rise to resistant, sensitive and mixed colonies	Anomalous behaviour—all colonies mixed. Resistance allele transmitted at very low rates	Yes
2. Crosses of $OL^R \rho+ \times OL^S \rho-$ give only resistant zygotic progeny	Colonies mixed—low transmission rates of resistance unaltered	Yes
3. $\rho-$ strains derived from OL^R haploid may no longer carry the resistance allele	No loss of resistance seen in any of the strains tested	Yes (in some strains)
4. Demonstration of linkage to other mitochondrial loci	No	Yes
5. Meiotic products of an $OL^R \rho+ \times OL^S \rho+$ cross show 4:0 segregation	No, 2:2 segregation observed	Yes

cytoplasmic element. Resistance to oligomycin could conceivably be caused by one or more of the following factors: (a) changes in the permeability barriers of the cell or the mitochondria so as to limit the entry of the antibiotic; (b) the presence of a cytoplasmic detoxifying mechanism; (c) mitochondrial detoxification including non-specific binding, or (d) modification of a defined drug binding site in the mitochondrial ATP synthetase complex.

Both the ATPase and Pi-ATP exchange reactions of Class II mutants are markedly more resistant to oligomycin than these enzyme activities in the mitochondria of the wild-type strain, and these mutants represent an alteration at the inhibitory or binding site of oligomycin on the ATP synthetase complex. The pH profiles of ATPase in some of the mutants are altered (Houghton *et al.*, 1974; Griffiths *et al.*, 1974). This alteration may involve either a mutant membrane protein and/or changes in lipid interactions in membrane lipoprotein complexes comprising part of the energy conservation system (Griffiths *et al.*, 1972). Proteins which if altered could give changes in oligomycin resistance include F_1–ATPase, the oligomycin sensitivity conferring protein (OSCP) (both synthesized extra-mitochondrially) and a membrane factor (Tzagoloff *et al.*, 1973, see p. 430) which is synthesized in the mitochondria. A lower content of associated ergosterol is characteristic of ATPase purified from two of the Class II mutants (Swanljung *et al.*, 1972).

The mitochondrial ATPase activity of Class I mutants is more oligomycin resistant than that of the wild-type but less markedly so than those of Class II mutants. The Pi-ATP exchange reaction of Class I mutants shows no significant difference in sensitivity to oligomycin from the parental strain. The lowered growth yield of mutants of this Class suggests that the mutation is at the mitochondrial level. It is suggested that Class I mutants may result from lesions in one component co-operatively affecting to a greater or lesser extent a whole range of membrane associated activities (Changeux and Thiery, 1968; Bunn *et al.*, 1970). Further genetic analysis of Class II OL^R mutants by recombination analysis and mapping studies (Coen *et al.*, 1970) have established two specific loci on the mitochondrial genome for oligomycin resistance (Avner *et al.*, 1973). These two loci show a high frequency of recombination with each other. In the majority of oligomycin-resistant strains, the mutation was found to map at the (OL_I) locus. The (OL_I) and (OL_{II}) loci are either unlinked or very weakly linked to each other and both are essentially unlinked from the ω—R_I—R_{II}—R_{III} segment of the mitochondrial genome specifying mitochondrial ribosomal functions (see p. 305). Analysis of crosses involving mutants at either the (OL_I) or (OL_{II}) loci and a series of ρ^- *petites* variously deleted in known mitochondrial genes has shown that the two oligomycin resistance loci are separable, and assuming that the non-deleted segment of mt DNA is continuous, the results suggest the gene order ω–R_I–R_{II}–R_{III}–OL_I–OL_{II}. Oligomycin resistance alleles at both the (OL_I) and (OL_{II}) loci are present in both ω^+ and ω^- strains.

3. Triethyl tin-resistant mutants

Similar studies on triethyl tin-resistant mutants (Lancashire and Griffiths, 1971; Griffiths, 1974) show that again these fall into two general classes (Table XLI). Class I mutants are cross resistant to oligomycin, uncoupling agents, and mitochondrial protein synthesis inhibitors. Class II mutants are specifically resistant to triethyl tin, and show no cross resistance to oligomycin or other agents. Another class (class III) has been defined on the basis of cross resistance to the uncoupling agent "1799" only. Mutants of all three classes are cytoplasmic mutants; *petite* induction in Class III mutants leads to either retention or loss of both markers (TET^R and 1799^R) suggesting that TET resistance and "1799" resistance are closely linked. In contrast with the situation with the oligomycin-resistant mutants it has not been possible to find an alteration in the ATPase of the TET mutants at the mitochondrial level (Griffiths *et al.*, 1974). Studies on Class II and III TET^R mutants indicate a locus for triethyl tin resistance. Recombination analysis of OL_I^R, OL_{II}^R and TET^R mutants suggest that two and possibly three mitochondrially-coded polypeptides are involved in the oligomycin- and triethyl tin-resistance phenomenon. Purification and peptide mapping of the polypeptide subunits of ATP synthetase responsible for interaction with these two inhibitors should be possible. It is of interest in this connection that Tzagoloff and Meagher (1972) have described the isolation of four subunits of the "membrane factor" of ATP synthetase which are mitochondrially synthesized (see p. 430).

4. Venturicidin-resistant mutants

Many of the triethyl tin-resistant mutants are cross resistant to venturicidin, another potent inhibitor of oxidative phosphorylation (Langcake *et al.*, 1974), and this indicates a close association of loci determining resistance for the two inhibitors. Recently however, another class of venturicidin-resistant mutants have been isolated which show no cross resistance to triethyl tin (Lancashire *et al.*, 1974; Griffiths *et al.*, 1974). Recombination analysis shows that this determinant is specifically associated with the OL_I group. Another venturicidin-resistant mutant shows no cross resistance to oligomycin yet maps with OL_I and shows no cross resistance to triethyl tin.

5. Guanidine-resistant mutants

Mutants of *Klyveromyces lactis* resistant to decamethylene diguanidine and octylguanidine have been obtained from cultures grown in the presence of ethidium bromide (Brunner *et al.*, 1973). Both nuclear and cytoplasmic inheritance of resistance has been demonstrated for decamethylene diguanidine.

6. Uncoupler-resistant mutants

Similar studies to those described for the OL^R and TET^R mutants are underway with uncouplers (Griffiths *et al.*, 1972). Again two general Classes of mutants have been obtained (Table XLI). TTFB-resistant mutants are often cross resistant to other uncoupling agents such as CCCP and DNP. The results with "1799" were again anomalous and indicated a difference between this and other uncouplers. Class II and III $TTFB^R$ mutants appear to be cytoplasmic, but the resistance marker is not lost on *petite* induction.

7. Valinomycin-resistant mutants

Investigations on valinomycin resistance exhibits a similar pattern of cross resistance to that seen in the other cases discussed above (Griffiths, 1974; Table XLI). The similarity of behaviour of $TTFB^R$ and VAL^R mutants to that shown by OL^R and TET^R mutants suggests that a similar resistance mechanism is operating in Class II $TTFB^R$ and VAL^R mutants. Thus specific binding sites for both uncouplers and ionophores must exist in mitochondrial membranes, and further studies may cast light on their mechanisms of action. It also suggests that reagents affecting mitochondrial energy conservation reactions react with specific protein binding sites in membranes and act as allosteric effectors leading to a conformational change in the oxidative phosphorylation complex (Weinbach and Garbus, 1969).

8. Mutants resistant to bongkrekic acid

Mutants have also been found which are resistant to bongkrekic acid, a specific inhibitor of the ATP transporter (Cain *et al.*, 1974; Griffiths *et al.*, 1974). TET^R mutants show a two- to three-fold increased resistance to bongkrekic acid. Resistance to both compounds is lost on *petite* induction. Evidence for partial mitochondrial control of the ATP transporter is also provided by the experiments of Perkins *et al.* (1972) and Haslam *et al.* (1973a) see p. 431.

9. Summary of inhibitor-, uncoupler-, and ionophore-resistant mutants of yeast

A summary of inhibitors and uncouplers acting on the energy conservation mechanism and the availability of mutants is presented in Table XLIII.

Table XLIII. Availability of mutants of *S. cerevisiae* resistant to agents interacting with the Energy Conservation Mechanism.

Agent	Site of action	Availability of resistant mutant	Site of allele for resistance
Mercurials	Phosphate carrier	No	?
Aurovertin	F₁–ATPase	Yes	Nuclear
Dio-9	F₁–ATPase	Yes	Nuclear
Oligomycin		Yes	Mitochondrial
Rutamycin		Yes	Mitochondrial
Peliomycin	Membrane subuits	Yes	Mitochondrial
Ossamycin	of ATP synthetase	Yes	Mitochondrial
Venturicidin		Yes	Mitochondrial
DCCD		Yes	?
Triethyl tin			
Triethyl tin	ATP translocase	Yes	Mitochondrial
Bongkrekic acid	,,	Yes	Mitochondrial?
Uncouplers	Proton translocase	Yes	?
"1799"	,,	Yes	Mitochondrial
Valinomycin	K⁺ translocase	Yes	Mitochondrial

A tentative genetic map of loci determining resistance to some of these agents is given in Fig. 103.

L. A Temperature-Sensitive Mitochondrial Mutant

Handwerker *et al.* (1973) have described a temperature-sensitive mutant of *S. cerevisiae* which is converted irreversibly to ρ^- during four to six generations at the non-permissive temperature (35°C). Genetic analyses implicated a mutated mitochondrial gene in the maintenance of the ρ factor in this strain.

II. Extrachromosomal Mutants of Fungi

A. Cytoplasmic Inheritance in *Neurospora crassa*

Sexual reproduction in *N. crassa* is effected by fertilization of proto-perithecia by conidia of opposite mating type (Fig. 104). The conidium contributes only a nucleus at fertilization so the cytoplasmic factors are maternally inherited. Maternal inheritance at the level of mt DNA has been demonstrated in the experiments of Reich and Luck (1966). After fertilization the meiotic products (ascospores) are formed. Interaction between mitochondria can be investigated by making heterokaryons which result from fusion of hyphae between different genetic strains of the same mating type. The resulting mixture of respective cytoplasms and nuclei can be propagated indefinitely, although sectors of cells having only

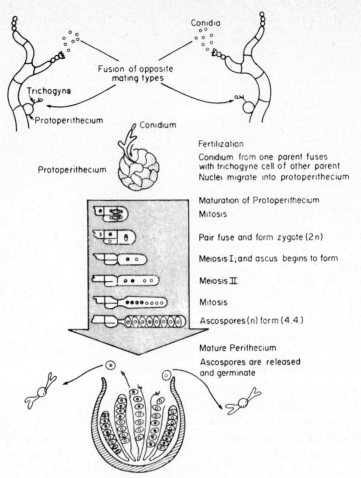

FIG. 104. Life-cycle of *Neurospora crassa*. Single ascospores of both mating types, a and A, produce haploid mycelia which differentiate to form protoperithecia (female) and conidia (male). Fertilization takes place by entrance of a nucleus from conidia or mycelium of one mating type into a protoperithecium of the opposite mating type. The resulting dikaryotic cell divides mitotically to form many dikaryotic cells, each of which becomes a zygote when the two haploid nuclei fuse. Each fused nucleus undergoes meiosis within the ascus producing four haploid nuclei, each of which divides once mitotically. Cell walls are then formed and eight haploid ascospores are relased from the ascus. Nuclear alleles, like A and a, segregate 4:4 in the ascus. (Reproduced with permission from Sager, 1972.)

one type of nucleus (homokaryons) sometimes arise. The terms hetero-cytosome, heteroplasmon, or heteromitochondriote are used for the condition of mixed cytoplasms. Homokaryons can be heterocytosomic. Complementation in heterokaryons is used as a criterion for distinguishing non-allelic genes.

It has never been easy to obtain cytoplasmic mutants of *N. crassa*, partly as a consequence of lack of specific tests; it is highly desirable to do so in this obligate aerobe which should yield mutants with more restricted, and so possibly more readily defined, defects than in the facultatively anaerobic yeasts (Bertrand and Pittenger, 1972).

A summary of some of the cytoplasmic mutants of this organism are presented in Table XLIV (Gillie, 1970). At least 11 of these mutants have abnormal cytochromes. Methods recently devised which facilitate the isolation of respiratory-deficient mutants include: (a) addition of dyes (pontamine sky blue and eosin yellowish), tellurite, or tetrazolium salts to a complete medium containing ethionine (Gillie, 1970). The latter prevents conidiation, and enables the colours of the colonies to be seen clearly. These methods enable nuclear and cytoplasmic cytochrome mutants to be distinguished from wild-type organisms on the basis of colour. (b) Tetra-zolium violet (2:5-diphenyl-3 (1-naphthyl)-tetrazolium chloride) included in modified Nagai medium with ethionine is also recommended for distinguishing cytoplasmic mutants from wild-type (Gillie, 1970). (c) The technique of "inositol-less death" (Flavell and Fincham, 1968a) followed by overlaying with TTC (Edwards *et al.*, 1973).

Several maternally-inherited (*mi*) respiratory deficiencies have been analyzed (Mitchell and Mitchell, 1952), a diagram of inheritance patterns is presented (Fig. 105).

Few antibiotic-resistant mutants of fungi have been described to date: e.g. there have been no reports of oligomycin-resistant mutants of *N. crassa*. This is partly due to the limited permeability and sensitivity of many fungi: for instance a strain of *Aspergillus nidulans* has been tested and found to be insensitive to a wide range of antibiotics which inhibit protein synthesis (G. Turner and R. T. Rowlands, personal communication).

Recombination of mitochondrial marker genes has been shown in *A. nidulans* (Rowlands and Turner, 1974a), but this process has not been described in *N. crassa*.

B. *Poky* Mutants (*mi-1*); *mi-3 mi-4* Mutants

The isolation of *poky*-strains of *N. crassa* and the inheritance of the *poky* character only by way of the maternal (protoperithecial) parent was first described by Mitchell and Mitchell (1952). Another maternally inherited defect was described in *mi-3* (Mitchell *et al.*, 1953).

In the presence of the nuclear gene, *f*, the growth rate of a *poky* strain becomes nearly normal, but the defective cytochrome system remains unchanged, as if the defects were being compensated for by increased activity of some other enzyme system (Mitchell and Mitchell, 1956). When the *f* allele of this gene is replaced by its normal counterpart, the

Table XLIV. Known examples of cytoplasmic inheritance in *Neurospora crassa*. (Reproduced with permission from Gillie, 1970.)

Name	Mutagenic origin*	Criteria of cytoplasmic inheritance†	Abnormal cytochromes	Phenotype	Reference
abn-1	SP	2, 3, 4, 5?	Yes	Low frequency and viability of conidia, female sterile, slow and irregular growth, cultures often die. Growth best on potato dextrose agar	Diacumakos *et al.* (1965); Garnjobst *et al.* (1965)
abn-2	SP	2, 3, 4, 5?	Yes	Slow and irregular growth	Srb (1963)
AC-1	ACR	1	Yes		
Breakdown	SP	2, 5	Not known	Low frequency and viability of conidia, female sterile, slow growth, cultures often produce brown pigment and die	Fitzgerald (1963)
Degenerate	UV	2	Not known	Slow linear growth rate, cultures often die	Lindegren (1956)
mi-1 (poky)	SP	1, 3, 4	Yes	Slow linear growth rate	Mitchell and Mitchell (1952); Mitchell *et al.* (1953)
mi-3	SP	1, 3	Yes	Slow linear growth rate	Gowdridge (1956); Mitchell and Mitchell (1952)

mi-4	SP	1, 3, 4	Yes	Low viability of conidia, female sterile, slow and irregular growth rates, cultures often die	Pittenger (1956)
SG-3 (mi-9)	—	1	Yes	Slow growth	A. M. Srb, personal communication
SG	ACR	1, 3	Yes	Slow germination of spores	Srb (1963)
Stopper (stp)	UV	2, 3, 5	Yes	Female sterile, irregular growth, brown exudate on aerial hyphae	McDougall and Pittenger (1966)
stp-1	SP	2, 3	Yes	Female sterile, irregular growth	Bertrand and Pittenger (1964); Bertrand *et al.* (1968)
stp-2	SP	2	Yes	Female sterile, irregular growth	
UVS-2	SP	2	—	Ultraviolet-sensitive, female sterile	Chang and Tuveson (1967)

* Origin of cytoplasmic variants: ACR, acriflavin-induced mutant; SP, spontaneous mutant; UV, ultraviolet-induced mutant.
† Criteria of cytoplasmic inheritance: 1, the variant shows maternal inheritance; 2, the variant shows non-Mendelian segregations; 3, the variant shows transmission in heterokaryon and related tests; 4, the variant shows "infective" spreading through a culture; 5, the variant shows variation in phenotype on subculturing.

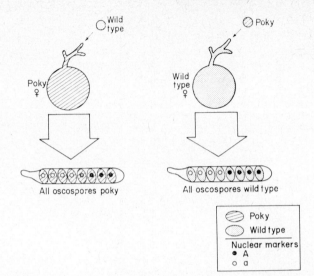

Fig. 105. Maternal inheritance of *poky* (*mi-1*). In reciprocal crosses between wild-type and *poky* strains, all ascospores produce *poky* progeny when the protoperithecial parent is *poky*; and wild-type progeny when the protoperithecial parent is wild-type. Nuclear alleles used as markers segregate 4:4. (Reproduced with permission from Sager, 1972.)

poky character is again fully expressed. No effect of *f* on the other cytoplasmic character, *mi-3* was detected.

Young cultures of *poky* are deficient in cytochromes *a* and *b*, in succinoxidase and cytochrome *c* oxidase activities, and have a large excess (up to 16 fold) of cytochrome *c* (Haskins *et al.*, 1953). The exact composition of the respiratory chain varies with the age of the cultures. Succinoxidase activity is independent of the content of cytochrome *b*. Extracts contained a system, not found in wild-type, which degrades added reduced cytochrome *c* to a green pigment resembling biliverdin. This reaction requires O_2 and succinate, and is inhibited by 10 mM–malonate or 1 mM–CN^-, but not by 1 mM–arsenate or 1 mM–azide. When particulate wild-type preparations were mixed with those from *poky*, the bands of all three cytochromes disappeared within 30 min; the extract from *poky* did not act on haemoglobin, catalase, or haem.

Respiratory deficiency in *N. crassa* does not prevent growth, as alternative respiratory mechanisms are available. Experiments on the respiration of intact mycelium and cell-free extracts (Tissières *et al.*, 1953) confirmed that, whereas the wild-type organism is dependent on the cytochrome system, respiration in *poky* is mediated by a CN^- and azide-insensitive, alternative system (see p. 145). The mutant organism contained twice as much flavin as the parent organism, and the respiration rate of *poky* (but not the wild-type) was doubled by substituting pure O_2 for air.

Since the claim of Woodward and Munkres (1966, 1967) that *mi-1* has altered mitochondrial "structural protein" has been retracted (Zollinger and Woodward, 1972), this mutant has been the subject of intensive investigations. It has a deficiency of small ribosomal subunits (Rifkin and Luck, 1971), and four of its mitochondrial amino-acyl tRNAs are chromatographically distinct from those of wild-type (Bramble and Woodward, 1972). The complement of mitochondrially-synthesized proteins is also abnormal in *mi-1* strains (see p. 414).

The isolation and properties of an altered cytochrome *c* oxidase present in *poky* (*mi-1*) mutants has been described by Edwards and Woodward (1969). The Soret band appears at 410 nm in the oxidized form and shifts to 422 nm on reduction with dithionite. No α-absorption band is seen in the reduced state. It is suggested that a protein component of the cytochrome *c* oxidase complex has an altered configuration due to a genetic change which may result in the altered spectral properties observed. These properties are similar to those of "mitochrome" which has an aldimine linkage between the formyl groups of haem *a* and the ε-amino groups of a Lys residue (Lemberg, 1969). The abnormal cytochrome *c* oxidase is maternally inherited (Woodward *et al.*, 1970). It is suggested that the pleiotropic-producing component which affects more than one mitochondrial enzyme in *mi-1* (malate dehydrogenase, cytochrome *c* oxidase and NADH oxidase are all affected) could be a membrane protein common to several different enzymes.

In mutant *mi-3* cytochromes *c* and *b* are present, but cytochromes $a + a_3$ are replaced by cytochrome "a_1" (α-band at 590 nm, Tissières and Mitchell, 1954) which is probably haemoglobin (Keilin, 1953). The specific activities of cytochrome *c* oxidase and succinoxidase in crude extracts were comparable to those of the wild-type organism.

The mitochondria of mutant *mi-1* are dominant over the mitochondria of wild-type in mixed cytoplasm. (Table XLV; Gowdridge, 1956; Pittenger, 1956; Grindle, M., cited by Woodward *et al.*, 1970). If two strains having similar nuclei, but different cytoplasms, one carrying the *mi-1* mutation and the other wild-type, are mixed and grown as a heterokaryon, the mycelium eventually assumes the characteristics of the *mi-1* strain, and all the mitochondria become respiratory deficient. It was suggested that the rate of replication of the mitochondrial genome determines the rate of mitochondrial division, and that since the *mi-1* mitochondria are deficient in a product(s) of the energy-conservation reactions, it may be this product(s) which is the regulator in the system. The less respiratory deficient mutant *mi-3* is not as dominant as *mi-1* in heterokaryons.

The isolation of mutant *mi-4* was described by Pittenger (1956), who found that the *poky* + *mi-4* heterokaryon has normal growth due to

M

Table XLV. Properties of heterokaryons of *Neurospora crassa* with mixed cytoplasms. (Reproduced with permission from Fincham and Day, 1963.)

Heterocaryon	Growth	Spectrum	Reference
poky +wild-type	normal	normal	Gowdridge (1956); Pittenger (1956)
mi-3 +wild-type	{ sometimes *mi-3* sometimes normal	mutant normal	} Gowdridge (1956)
mi-4 +wild-type	normal	normal	Pittenger (1956)
mi-3 +*poky*	{ usually *mi-3* sometimes *poky*	} mutant	Gowdridge (1956)
mi-4 +*poky*	normal	mutant	Pittenger (1956)

physiological complementation between differently deficient mitochondria. Spectral analysis revealed that this mycelium was still mutant. Homocytosomes recovered from the heterokaryon were either *poky* or *mi-4* but not wild-type. When the heterokaryon was used as the protoperithecial parent and crossed to *poky*, only *poky* or *mi-4* progeny were obtained. These results illustrate the independent continuity of the two types of mitochondria in vegetative cells. In *poky* + *mi-3* heterokaryons there is no physiological complementation but *poky* and *mi-3* homocytosomes can be recovered. In the heterokaryon between the *mi-4* strain carrying the nuclear marker *lys* (Lys-requiring) and wild-type, the mutant phenotype segregated out. Some of the *mi-4* isolates were homokaryotic and Lys-independent. Thus reassortment of the *mi-4* determinant and nuclear genes had occurred without fusion of the heterokaryotic nuclei. The finding that no mutant phenotypes are produced as progeny of crosses between *poky*, *mi-3* or *mi-4* to the wild-type with the latter as the protoperithecial parent, suggests that nuclear genes are not directly involved in producing these mutants.

C. Mutants *abn-1* and *abn-2*

The isolation of two morphologically distinctive, slow growing strains of *N. crassa* (*abn-1* and *abn-2*) was described by Garnjobst *et al.* (1965). These strains have greatly reduced growth rates and often show irregular growth; aerial hyphae are absent, conidia are extremely rare and no protoperithecia are produced. Growth was not improved by adding nutrients; respiration was similar to the wild-type and cytochrome *c* appeared abnormally high, whereas cytochrome *b* was low or absent. The abnormal characteristics are carried by cytoplasmic factors as both *abn* strains gave rise only to normal progeny in crosses with normal strains. The *abn* characteristics appear in heterokaryons, and are transmitted to other genetic strains by heterokaryosis followed by plating of conidia.

Microsurgical techniques (using pipettes with an outer diameter of 3 μm (Wilson, 1961, 1963) were used to inject cytoplasm from normal cells (without detectable transfer of nuclei) into *abn* strains (Garnjobst *et al.*, 1965). This was the first successful application of such techniques to the demonstration of transmission of a cytoplasmically-inherited characteristic.

Nuclear fractions isolated from mutants produced no effect when micro-injected into mutants with complementary requirements; nuclear DNA similarly injected also had no effect (Diacumakos *et al.*, 1965). Mito-chondrial fractions isolated from *abn-1* produced drastic changes in the rate of growth, morphology, reproduction characteristics and cytochrome spectra of normal strains when single hyphal compartments were micro-injected and isolated, whereas the mitochondrial fractions from wild-type produced no such effect. These results again provided evidence for the transmission of biochemical and biological characteristics when mito-chondria are transferred into a new nucleo-cytoplasmic environment.

A particulate fraction, not detectable in wild-type cells has been isolated from the cytoplasm and from lysed mitochondria of *abn-1* (K¨ntzel *et al.*, 1973, 1974) which has altered mt DNA (lowered buoyant density); this fraction carries out mitochondrial protein synthesis. The particles have a buoyant density of around $1\cdot18$ g/cm³, and appear in thin sections as polymorphic vesicles, containing an electron dense nucleoid of 120–170 nm diameter. The particles contain a single-stranded 33S RNA, the synthesis of which is inhibited by ethidium bromide and which is converted to 7–9 S RNS by heat treatment. This RNA differs in base composition from mitochondrial and cytoplasmic rRNAs. The particles also contain phos-pholipids rich in phosphatidyl ethanolamine and cardiolipin and two major protein components, a lipoprotein (mol. wt 15,000, probably identical with a major product of mitochondrial protein synthesis in the wild-type), and a glycoprotein (mol. wt 15,000). Morphologically and chemically these particles resemble certain RNA viruses and have been termed "virus-like particles" although infectivity has not been shown. The 33S RNA hybridizes with at least $4\cdot5\%$ of mt DNA from both wild-type and *abn-1* and is competed away by total mt RNA but not by purified mt rRNA. It is concluded that the mutational alteration of mt DNA in *abn-1* leads to the expression of a hitherto unknown mitochondrial gene for 33S RNA and to the intra-mitochondrial formation of virus-like particles containing mitochondrial gene products.

D. Stopper Mutants

A period of growth between successive UV treatments allowed the selection of an extra-nuclear mutant, stopper (*stp*), characterized by a stop–start growth phenotype, somatic segregation of conidia into a lethal and non-lethal class, and an abnormal cytochrome system (McDougall and Pittenger, 1966). The phenotype can be transmitted from one strain to another by heterokaryosis, but does not pass through a sexual cross. Cytochromes *a* and *b* were not detectable in a hand spectroscope, but a very intense α-band of cytochrome *c* was observed at 550 nm. Five other cytoplasmic mutants [*A18*, (*stp-A*), (*stp-B1*), (*stp-B2*) and (*stp-B*)] with stop–start growth characteristics were isolated from two continuously-growing parent strains (Bertrand and Pittenger, 1969). Mitochondria from these mutants were also deficient for cytochromes *b* and $a + a_3$ and cytochrome *c* oxidase activity, but showed an excess of cytochrome *c* and succinate-cytochrome *c* oxidoreductase activity. Four other cytoplasmic mutants of similar phenotype have been obtained by treatment of conidia and mycelia with *N*-methyl-*N'*-nitro-*N*-nitrosoguanidine (Bertrand and Pittenger, 1972). On the basis of growth and fertility, nuclear suppressors and complementation in heteroplasmons, these workers were able to classify 16 of the available extranuclear mutants of *N. crassa* into three groups: Group I consists of 8 female-fertile variants with both *poky*-like growth and cytochrome defects. Their slow growth is suppressed by the nuclear factor *f*, but not by a second nuclear suppressor, *su-1* (*mi-3*). They complement with Group III mutants in mixed cytoplasms. Group II is represented by a single variant (*mi-3*). It is phenotypically-modified by the *su-1* (*mi-3*) factor but not by *f*. Its cytochrome spectrum is unique, in that it shows deficiency of cytochrome *a*, but cytochromes *c* and *b* are present. It complements in heteroplasmons with Group I and III mutants. Group III includes 7 female-sterile variants with stopper growth pheno-types and the same cytochrome defects as Group I. Group III mutants complement both with Group I and II isolates, but they are unaffected by either *f* or *su-1*.

The cytochrome contents, cytochrome *c* oxidase and succinate cyto-chrome *c* oxidoreductase activities of these mutants are given in Table XLVI.

E. Mutants of *Aspergillus* sp. and *Coprinus*

1. Mutants of Aspergillus *sp.*

A number of spontaneously-occuring, stable oligomycin-resistant mutants of *Aspergillus nidulans* have been isolated (Rowlands and Turner, 1973). Most of the mutations were nuclear, but one (*oli A1*) was cytoplasmic; whereas the nuclear mutants showed no detectable abnormalities on drug-free medium, the growth of the cytoplasmic mutants was impaired. This character appeared to be a secondary effect of the oligomycin-resistance as segregation of the two characters was never observed in any

Table XLVI. Relative cytochrome contents (absorbancy/10 mg mitochondrial protein), activities of cytochrome c oxidase (min^{-1}/mg mitochondrial protein), and succinate: cytochrome c oxidoreductase (μmoles cytochrome c reduced/min/mg mitochondrial protein) in mitochondria from normal and mutant strains. (Reproduced with permission from Bertrand and Pittenger, 1972.)

Mutant	Culture age (hours)	Cytochrome content			Cytochrome c oxidase	Succinate-cytochrome c oxidoreductase
		$a + a_3$	b	c		
[+]	23	0·0351	0·0861	0·1086	56·0	0·532
	36	0·0331	0·0790	0·0877	48·2	0·507
[exn-1]	36	0·0018	0·0126	0·2346	0·7	1·004
	72	0·0103	0·0524	0·1611	13·3	0·912
[exn-2]	36	0·0027	0·0168	0·2374	0·5	1·490
	72	0·0083	0·0444	0·1571	11·2	1·031
[exn-3]	36	0·0016	0·0142	0·2339	0·3	1·138
[exn-4]	36	0·0041	0·0183	0·2493	4·1	1·100
[stp-C]	64	0·0091	0·0505	0·1861	14·6	1·224
	96	0·0082	0·0557	0·1905	13·3	1·048
[poky]	36	0·0038	0·0229	0·2192	4·1	1·006
	72	0·0097	0·0514	0·1657	9·7	0·811
[mi-3]	36	0·0009	0·1384	0·2485	0·9	0·653

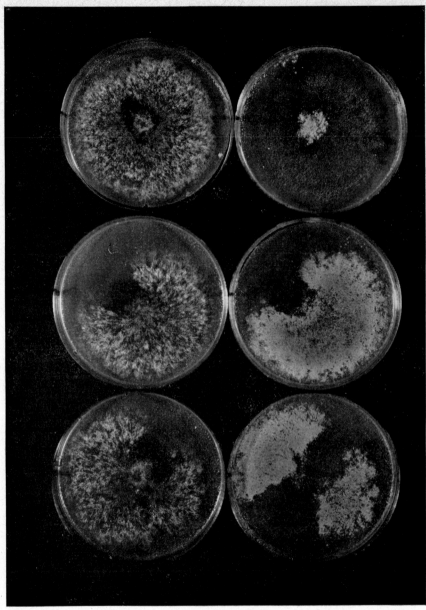

FIG. 106. Sectoring of the extra-nuclear marker (rutamycin resistance) in 8 day-old hetero-karyons of *Aspergillus nidulans*. Each horizontal pair of plates consists of the original heterokaryon (left) and its replica on rutamycin-containing medium (right) obtained by replicating with damp velveteen. Three fates could befall a heterokaryon. In about 40% of the cases the heterokaryotic sectors were all sensitive (top) or all resistant (middle). Otherwise both resistant and sensitive sectors were produced (bottom). These sectors were completely independent of the nuclear markers, as yellow and white conidiating areas could be seen to be well mixed all over the heterokaryons in both resistant and sensitive sections (left-hand column). Random samples of 50–100 conidia from the resistant areas were found to be all resistant and those from the sensitive areas to be all sensitive. (Reproduced with permission from Rowlands and Turner, 1973.)

of the genetic manipulations carried out. The extra-nuclear genetic element specifying the oligomycin-resistance character was not able to coexist within the same mycelia as the wild-type element, and they tended to segregate into sectors consisting almost entirely of one type or another (Fig. 106). The nuclear mutants all mapped on linkage Group VII and showed incomplete dominance in heterozygous diploids, segregating fully resistant homozygous areas.

Of the oligomycin-resistant mutants, all twenty examined (including the extra-nuclear mutant O^R_6 were cross resistant to rutamycin, whereas some were cross resistant to venturicidin and some were not (Rowlands and Turner, 1974b). Strain O^R_1 (nuclear) was unique in that it was initially venturicidin sensitive, but became progressively more resistant with increasing incubation times. None were cross resistant to DCCD, chloramphenicol or cycloheximide.

All of the mutants grew more slowly in the presence of oligomycin than on drug-free medium under the same conditions, but at 37°C and 45°C the growth with oligomycin of strain O^R_{31} (nuclear) was faster than that of the other mutants: at 20°C the growth rate of this strain with oligomycin was similar to that of the other mutants. This observation indicates that the additional growth rate is cold-sensitive. Whereas in the intact organisms at least a fifty-fold increase in resistance to oligomycin was evident, the resistance of ATPase in isolated mitochondria was only increased up to seven fold. This disparity may be due to a decreased permeability to the drug of the plasmamembrane or of the mitochondrial membrane. The effect on ATPase may be a secondary consequence of an altered mitochondrial membrane.

In the cytoplasmic mutant O^R_6, increased oligomycin resistance of the ATPase was only very slight, but in this case the cytochrome content of the mitochondria was altered: the cytochrome c content was increased, without any decrease in cytochromes b and a.

Another extra-nuclear cold-sensitive mutant of this organism (cs67) has been isolated (Waldron, 1973; Waldron and Roberts, 1973); this strain has an altered cytochrome spectrum after growth at restrictive temperatures: it possesses excess cytochrome c and lowered content of b- and a-type cytochromes (G. Turner, personal communication). The genetic behaviour of this locus and that of (oliA1) are identical in sexual crosses and heterokaryons with the wild-type (+). Recombination between the extra-nuclear loci (oli A1) and (cs67) has been demonstrated (Rowlands and Turner, 1974a). The stability of the recombinant strains indicates that physical recombination of genetic material has occurred. The extremely slow growth rate of the double recombinant confirms this; complementation would have resulted in a growth rate faster than that of (cs67). Mixed populations of extra-nuclear genetic elements rapidly segregate into homoplasmic areas in both heterokaryons and homokaryons, and hence rapid expression of new (mutant) or recombinant genotypes will occur.

Cytoplasmic inheritance has been shown for *vegetative death* in *Aspergillus amstelodanii* (Handley and Caten, 1973); this mutant also has an altered cytochrome spectrum.

For other cases of cytoplasmically-inherited mutations of *Aspergillus sp.* (most of which have not been examined for mitochondrial lesions) see Srb (1963), Jinks (1963, 1964), Fincham and Day (1963) and Mahony and Wilkie (1962).

2. A mutant of Coprinus lagopus

A mutant of *Coprinus lagopus* (*acu-10*) is unable to use acetate as the sole source of carbon for growth (Casselton and Condit, 1972). Somatic segregation and non-Mendelian segregation at meiosis indicate that the determinant for this lesion is cytoplasmically inherited. Cytochromes $(a + a_3)$ were not detected in the mutant, but a cytochrome "a_1" band was observed. This may be due to the presence of haemoglobin. Using *acu-10* as a marker it was shown that the migration of nuclei which occurs during dikaryotization is independent of the movement of other organelles.

III. Extrachromosomal Mutants of Protozoa

A. Mutants of Paramecium aurelia

Paramecium aurelia is a large ciliated protozoon which lends itself to the easy detection and analysis of cytoplasmically-inherited characters (Beale, 1954). The large cell size (100 μm) allows the manipulation of single cells, and conjugation occurs between opposite mating types (Sonneborn, 1950; 1970; Beisson *et al.*, 1973).

Cytoplasmically-inherited erythromycin-resistant mutants have been obtained by Beale (1969) in syngen 1 of *P. aurelia*. The effects of five antibiotics on the division and survival of this organism were tested by Adoutte and Beisson (1970). Whereas lincomycin was without effect, and oleandomycin and spiramycin were effective inhibitors of growth only at high levels (>800 μg/ml), chloramphenicol and erythromycin were inhibitory at low concentrations (100–200 μg/ml). In the presence of erythromycin the cells survived for between 10–20 days, although cell division was completely blocked. Under these conditions spontaneously resistant mutants could be individually isolated and used to start erythromycin-resistant clones. UV-irradiated cultures also provided a good source of resistant organisms. The 24 mutants isolated were grouped into 3 classes on the basis of their level of resistance to erythromycin and their thermosensitivity (at 36°C). Most mutants showing a low level of resistance to erythromycin (200 μg/ml) were also thermosensitive. Genetic analysis of three mutants showed that the resistance character is cytoplasmically inherited, as evidenced by its clonal inheritance, its transfer through cytoplasmic bridges between conjugants (Fig. 107), and its non-segregation

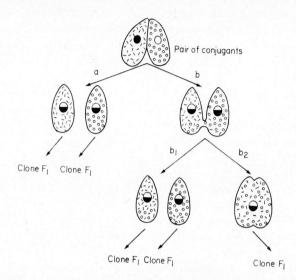

Fig. 107. Genetic exchanges between conjugating paramecia. The different fates of a pair of conjugants are represented. *a*, After completion of nuclear exchanges the two conjugants usually separate without cytoplasmic exchange. *b*, In some cases, a cytoplasmic bridge develops by the end of the conjugation process, through which cytoplasmic material is exchanged. *b₁*, After a variable time (min to h), the bridge is broken, the two conjugants separate and the relative amount of exchanged cytoplasm depends on the duration and width of the bridge. *b₂*, In rarer cases the bridge persists and eventually yields complete fusion of the two mates that reproduce true to type as a doublet. (Reproduced with permission from Adoutte and Beisson, 1972.)

at meiosis (autogamy). The hypothesis that the genetic basis of these mutations is their localization in mt DNA allows a satisfactory interpretation of all the results. Thus mitochondria from a resistant parent can enter the sensitive mate in detectable amounts only when a cytoplasmic bridge occurs. The change from sensitivity to resistance for an organism would then correspond to the differential proliferation of "resistant mitochondria" under the pressure of the selective medium. When sensitive cells receive resistant mitochondria and are allowed to divide in a non-selective medium, those mitochondria would become randomly distributed throughout the fissions.

Whenever an antibiotic-sensitive cell that has just undergone a short cytoplasmic exchange with a resistant partner is placed in the selective medium, it first has a sensitive phenotype. Then, after a lag, it becomes progressively more resistant and reaches full resistance after four fissions. The lag is inversely proportional to the duration and size of the bridge. This observation was made on 200 independant "heteromitochondriotes" from fifteen different crosses (Adoutte and Beisson, 1972). This phenotypic evolution is associated with a rapid genetic purification for the selected mitochondrial character. Thus, under selective pressure, the acquisition of resistance must involve the differential multiplication of resistant mitochondrial genomes along with an elimination of functional sensitive mitochondria.

Various pairwise combinations between resistant and sensitive mitochondria or between two types of resistant mitochondria when followed under non-selective conditions (26°C, no antibiotic) give two distinct kinds of results: (a) the two types of mitochondria show identical "stability" in the mixed population and the cells remain "mixed" for a great many generations; or (b) one type of mitochondria (always the same type for a given combination) is progressively eliminated and "pure" cells of the favoured mitochondrial type are recovered between the twentieth and the fortieth generations. In case (b) the wild-type mitochondria are always the most stable. It is always the most thermosensitive allele which is eliminated from a mixed population.

From these results and those obtained from pedigree analysis it was concluded: (1) that the number of units in each cell is high (as expected for an organism with perhaps thousands of mitochondria); and (2) that each mutant displays a specific rate of transmission in mixed mitochondrial populations. This property is analogous to the phenomenon of "polarity of transmission" of mitochondrial markers in yeast (Coen et al., 1970; Bolotin et al., 1971). In yeast this polarity reflects recombinational events between mitochondrial genomes. Preliminary evidence has been obtained for recombination between C^RE^S and C^SE^R mitochondria to give double-sensitive mitochondria (Adoutte, 1974). No complementation has been observed.

The erythromycin-resistant mutants are deficient in mitochondrial protein synthesis and have poorly-developed mitochondrial inner membranes with lowered amounts of cytochrome c oxidase. The morphological changes progressively produced in the wild-type cells incubated with chloramphenicol or erythromycin include elongation of mitochondria, decreased numbers of cristae and the appearance of lamellar cristae and plates of periodic structure (Adoutte et al., 1972). Three resistant mutants studied, on the contrary, retain normal or nearly normal mitochondrial structure in the presence of the antibiotic to which they are resistant. This observation is in accord with those of Knowles (1971) who studied the effects of erythromycin on sensitive Paramecium after microinjection of mitochondria isolated from a resistant mutant. The most likely explanation is that it is the mitochondrial protein-synthesizing system itself which is rendered resistant by the mutation.

The resistance to erythromycin (250 μg/ml) of mutant 513^R_1 (EDIN) of *P. aurelia* is due to an alteration of mitochondrial ribosomes (Tait, 1972). Isolated mitochondria were incubated with (^{14}C)-erythromycin and then ribosomes were isolated as 80S (monomer) and 55S (single subunit) peaks after lysis with Triton X-100. Erythromycin binds to the mitochondrial ribosomes of the sensitive cells but not to those of the resistant mutant. A control experiment ruled out the possibility that the drug was being bound by the membranes. Cytoplasmic ribosomes did not bind (^{14}C)-erythromycin. Preliminary results with this mutant (Beale *et al.*, 1972) have suggested that the electrophoretic properties of ribosomal proteins may be altered, but this could be a consequence of a primary alteration in rRNA, or could result from a selective loss of proteins during isolation of ribosomes.

Erythromycin-resistant mitochondria from syngen (subspecies) 1 were injected into erythromycin sensitive syngen 7 cells, and the recipient cells were placed in erythromycin to inhibit replication of the sensitive mitochondria (Knowles and Tait, 1972). Clones selected in this way contained a syngen 7 nucleus and a mitochondrial genome which is derived from syngen 1 erythomycin-resistant mitochondria. It has been shown that fumarase from syngen 1 has a different electrophoretic mobility from that of syngen 7 (Tait, 1970a). The fumarase of the selected clones was identical to that of syngen 7, showing that fumarase is not controlled by the mitochondrial genome, and by inference must be controlled by the nucleus. Other mitochondrial enzymes shown to be controlled by nuclear genes in *Paramecium* include β-hydroxybutyrate dehydrogenase (Tait, 1968), and NADP-linked isocitrate dehydrogenase (Tait, 1970b).

Although it is the mitochondrial genome that determines the mitochondrial growth rate it has been demonstrated that a nuclear gene interacts specifically with mitochondria (Adoutte *et al.*, 1973). The wild-type and mutant alleles of this gene are respectively capable of favouring the selective multiplication and proper morphogenesis of the mitochondria previously associated with it, regardless of the mitochondrial genome. Thus the systems of control of drug resistance in *Paramecium* provide a useful system for the evaluation of nucleo-mitochondrial interactions.

The ability to distinguish between fully resistant (E^R) and sensitive (E^S) mitochondria in electron micrographs, has enabled parallel genetic and cytological analyses of the mechanism of acquistion of erythromycin resistance by sensitive cells (Perasso and Adoutte, 1974). Although it is not possible to analyze the same cell by both techniques, several observations suggest that it is valid to assume that all the cells of a clone have a great degree of similarity of their cytoplasms: intense cytoplasmic mixing of mitochondria occurs and the cells divide by binary fission. Sensitive cells, into which a minority of E^R_{102} mitochondria is introduced ($<10\%$), became purely erythromycin resistant, genetically and cytologically, after three or less generations in erythromycin-containing media. Thus an active multiplication of E^R mitochondrial genomes must have occurred, together with an active synthesis of gene products conferring resistance to mitochondria. Elimination of sensitive mitochondria must also have taken place. Good correlation between the extent of purification of the E^R determinant in the cells (as assessed both genetically and cytologically) indicates that there does not exist in detectable amounts genetically E^S but phenotypically E^R mitochondria. Active elimination of sensitive mitochondria does not occur when sensitive cells remain in erythromycin-containing media for up to seven days; these organisms are perfectly capable of resuming growth when returned to normal media. Cells of sensitive origin resume growth in the presence of erythromycin only when they have reconstituted the whole of their mitochondrial population.

Knowles (1971) and Perasso and Adoutte (1974) agree that at least two mechanisms can be proposed to account for the acquisition of erythromycin-resistance: (1) Multiplication of the resistant mitochondria that have initially penetrated the sensitive cell, along with the degeneration of the sensitive mitochondria (with the possible reutilization of some components recovered from the sensitive organelles for assembly of the resistant ones). In this latter connection cytolysosomes may play a part. (2) Multiplication of the initially resistant mitochondria along with the genetic transformation of E^S mitochondria by E^R mt DNA. The cytological situation is not so simple as to enable a clear decision to be made between these two possibilities. The second hypothesis (that of genetical transformation) cannot be excluded, but is not necessary to explain the rapid rate of acquisition of erthryomycin resistance. Hypothesis (1) is regarded as sufficient to explain the observations.

The most important conclusion is that there must be an active multiplication of resistant genomes while the cells are dividing slowly or not at all. The rupture of the normal mechanism ensuring the coordination between mitochondrial multiplication and cell division (Williamson, 1969; Turner and Lloyd, 1971; Lloyd et al., 1971b) enables a great capacity of these organelles to respond to a selective pressure, and emphasizes the important extent of their autonomy with regard to their own multiplication.

B. Mutants of *Tetrahymena pyriformis*

Chloramphenicol-resistant mutants of *T. pyriformis* (CA-101, CA-102 and CA-103) have been obtained (after mutagenesis of the wild-type strain

D 1968-S of syngen 1 with nitrosoguanidine) at a frequency of 10^{-3} mutants per mutated cell (Roberts and Orias, 1973). The mutants are still partially sensitive to chloramphenicol and have a lowered growth rate compared to that of the wild-type in rich medium without chloramphenicol. The determinant is a cytoplasmically-inherited one which is not exchanged during conjugation and provides the first simple cytoplasmic genetic determinant in this species.

C. Cytoplasmic Determinants in Amoebae

Evidence for cytoplasmic DNA-containing organelles in amoebae was obtained in the labelling experiments described by Prescott et al. (1962) and by Rabinowitz and Plaut (1962).

Microinjection of cytoplasmic extracts suggest that neomycin resistance in amoebae has a cytoplasmic determinant (Hawkins and Willis, 1969).

IV. Localization of Extrachromosomal Genes in Chlamydomonas

The localization of extra-nuclear genes in green algae is complicated by the presence of two different classes of DNA-containing organelles. Whereas the linkage group assumed to represent genes of chloroplast DNA has been the subject of intensive investigation, there is as yet little information available on the functions of mitochondrial DNA in *Chlamydomonas reinhardii*. Schimmer and Arnold (1970a, b, c) and Behn and Arnold (1972; 1973) suggest that the gene for streptomycin dependence (*sd*) is mitochondrial, although this interpretation has been questioned by Sager (1972). Gillham and Boynton (1972) suggest that one gene for spectinomycin resistance is mitochondrial.

7

Chromosomal Mutations
Affecting Mitochondrial Functions

I. Introduction

The nuclear genome is the site of almost all the information for the control and specification of components required for the biogenesis of mitochondria. This fact is clearly evident from a consideration of those *petite* mutants of yeast which have grossly altered mt DNA, but still have mitochondria which differ from those of the wild-type yeast only by their loss of some respiratory chain components (cytochromes $a + a_3$, b and c_1), a few inner-membrane proteins, and mitochondrial rRNAs, and tRNAs. The fatty acid composition and the morphology of mitochondrial membranes is not detectably altered (Packer *et al.*, 1973; Fig. 118, Ch. 11). Some 63 Mendelian genes recognized to control functions associated with mitochondria in yeasts were listed by Beck *et al.* (1971). An extended version of this classification is presented in Table XLVII. Assuming that each gene codes for a polypeptide with an average mol. wt of 40,000, then 63 genes would require 44×10^7 daltons of DNA corresponding to an accumulated contour length of 23 μm (Beck *et al.*, 1971). These must represent only a fraction of the total number of nuclear genes required to specify mitochondrial functions. We may note that a haploid strain of *S. cerevisiae* has at least 16 chromosomes (Mortimer and Hawthorne, 1966) containing altogether $1 \cdot 2 – 1 \cdot 3 \times 10^{10}$ daltons of DNA or about 12,000 cistrons (Hartwell, 1970).

II. Chromosomal Mutations Affecting Mitochondrial Functions in Yeast

A. Segregational *Petite* Mutants

Segregational *petite* mutants of *S. cerevisiae* were first discovered and analysed by Chen *et al.* (1950). The results of crosses indicated that although the growth characteristics of the segregational and cytoplasmic *petites* are indistinguishable, unlike the latter the segretational *petites* differ from normal cells by a single nuclear recessive gene, in the presence of which the cytoplasmic "factor" is inactive. This was the first indication

338

Table XLVII. Mendelian genes affecting mitochondrial structure, function and biogenesis in *Saccharomyces cerevisiae*. (Reproduced with permission and modification from Beck *et al.*, 1971.)

Symbols	Class	Cytochromes present	Other properties	Reference
Group I. Genes causing general loss of ability to assimilate non-fermentable substrates.				
p_2p_3	1	c only	all ρ^-, may have porphyrins	Sherman, 1963; Sherman and Slonimski, 1964
p_{12}	1(?)	(c)	all ρ^-, contains pigments like b and c_1 (porphyrins?)	Mackler *et al.*, 1965
ly_6ly_8	1	c	all ρ^-, Lys required	Sherman, 1963; Sherman and Slonimski, 1964
glt_2	1	c	all ρ^-, Glu required	Ogur *et al.*, 1965
$p_1p_6p_7$	2	c	all ρ^+,	Sherman, 1963; Sherman and Slonimski, 1964; Chen *et al.*, 1950
p_8p_{13}	2	(c)	all ρ^+, contains porphyrins or traces of b and c_1	Mackler *et al.*, 1965
$p_{11}p_{14}$–p_{17}	3		no data, or genetic data only	Hawthorne and Mortimer, 1968
gi	4	normal or c only	ρ^- when repressed or grown anaerobically	Negrotti and Wilkie, 1968
p_{ts} (1121)		low cytochrome content at 18°C	ρ^- at 18°C or when grown with chloramphenicol	Weisogel and Butow, 1970
p_{ts} (24, 49, 518-4)		none at 36°C, normal at 25°C	ρ^+ at 18°C	Butow *et al.*, 1973
p_{ts} (1511)		only b_2 present	UQ mutant	Puglisi and Cremona, 1970
58 NW-46 936		c, low $a + a_3$	ATPase and $^{32}P_i$-ATP exchange lost also	Morita and Mifuchi, 1970
881		c, low $a + a_3$	$^{32}P_i$-ATP exchange lost	Ebner *et al.*, 1973b
Group II. Genes causing selective alteration in assimilation of non-fermentable substrates				
pl_1–pl_7	5	$a + a_3 bcc_1$	selective promotion of lactate assimilation	Galzy and Bizeau, 1965
$glg_1 glg_2$	6	$a + a_3 bcc_1$	selective reduction in glycerol assimilation	Beck *et al.*, 1971

Table XLVII. (continued)

Symbols	Class	Cytochromes present	Other properties	Reference
Group III. Genes altering content of one or two cytochromes				
p_4	7	$c\,c_1$ low $a + a_3$ and b	high % ρ^-, $a + a_3$ readily repressed	Sherman, 1963; Sherman and Slonimski, 1964
p_{10}	8	$a + a_3\,c$	$a + a_3$ easily repressed	Reilly and Sherman, 1965
p_5	9	$b\,c\,c_1$	variable % ρ^-	Sherman and Slonimski, 1964
cy_1–cy_6	10	$a + a_3\,b\,c_1$ low c	$a + a_3$ easily repressed in cy_3	Reilly and Sherman, 1965; Sherman, 1964
(1030) (E11) (494)		$b\,c\,c_1$	selective loss of $a + a_3$, cytochrome c oxidase	Ebner et al., 1973b
(495), (396)		$b\,c\,c_1$	Diminished succ-cyt c reductase ATPase, ^{32}Pi-ATP exchange	
Group IV. Genes altering non-fermentable substrate assimilation without deletion of cytochromes				
$p_9\,(op_1)$	11	normal	defective adenine nucleotide translocase, mitochondria show low P/O ratios except with excess ADP, ρ lethal	Douglas and Hawthorne, 1964; Beck et al., 1968; Kováč et al., 1967a
aem_1–aem_8	12	normal	all respire, won't assimilate glycerol	Parker and Mattoon, 1969
aem_9–aem_{11}	13	all low	low respiration, won't assimilate glycerol	Parker and Mattoon, 1969
glt_1	13	all low	requires Glu, lacks aconitase	Ogur et al., 1964
	13	normal	Ubiquinone mutant, won't assimilate glycerol	de Kok, 1973
Group V. Genes altering the TCA cycle				
glt_1	13	all low	lacks aconitase see Group IV	Ogur et al., 1964
glt_2	1	c	lacks aconitase see Group I	Ogur et al., 1965
glt_x	14	normal	assimilates glycerol, Glu auxotrophe linked to glt_1	Beck et al., 1971

Group Va. Genes altering riboflavine synthesis
riboflavine auxotrophes

Gene	No.			Reference
rib_5 rib_7		no data		Oltmanns and Lingens, 1967; Lingens et al., 1967; Bacher and Lingens, 1970, 1971; Oltmanns, 1971

Group VI. Genes altering porphyrin metabolism

Gene	No.			Reference
pop_1 pop_2	15	normal	accumulates free and Zn porphyrins	Pretlow and Sherman, 1967
W–I	16	none	cytochrome synthesis induced by Gly or porphyrin, unstable	Raut, 1953; Yčas and Starr, 1953
cyt		none at 35°C, normal at 20°C	requires protoporphyrin IX at 35°C	Sugimura et al., 1966a; Gunge et al., 1967; Miyake et al., 1972
cyd		none in $cyd\text{-}1\rho^-$	cytochrome synthesis restored by δ-aminolaevulinate	Sanders et al., 1973

Group VII. Genes altering unsaturated fatty acid formation

Gene	No.			Reference
ol_1	17	$a + a_3\, b\, c\, c_1$	requires oleate, respiratory competent	Resnick and Mortimer, 1966; Keith et al., 1969; Proudlock et al., 1969; Haslam et al., 1971, 1973a;
$ol_2\text{-}ol_4$	18	no data	requires oleate, no glycerol assimilation, pleiotropic	Resnick and Mortimer, 1966; Keith et al., 1969; Wisnieski et al., 1970; Wisnieski and Kiyomoto, 1972

Group VIIa. Genes altering sterol formation

Gene	No.			Reference
nys 1–3		no data	nystatin-resistant, accumulate intermediates of sterol synthesis	Ahmed and Woods, 1967; Woods and Hogg, 1969
pol 1–5		no data	polyene-antibiotic resistant	Molzahn and Woods, 1972

Table XLVII. (*continued*)

Symbols	Class	Cytochromes present	Other properties	Reference
Group VIII. Genes altering resistance to antibiotics and other compounds affecting mitochondrial function and/or biogenesis				
CAPR (1–3)	19	normal	chloramphenicol resistant, at least 3 types, 1 dominant	Wilkie et al., 1967; Roodyn and Wilkie, 1968
ERR	20	normal	erythromycin resistant, at least 1 type	Wilkie et al., 1967; Roodyn and Wilkie, 1968
or or₁	21	normal	resistant to oligomycin and venturicidin	Parker et al., 1968; Stuart 1970; Avner et al., 1973
DNPR	22	variable	resistant to DNP, at least 1 type	Parker and Mattoon, 1968
AAR	23	respiration high	resistant to Antimycin A, at least 1 type	Butow and Zeydel, 1968
ROTS		no data	sensitive to Rotenone, at least 2 types	Mounolou, 1973
TETR		no data	resistant to tetracycline, several types	Hughes and Wilkie, 1972
CAN R		no data	resistant to canavanine	Wilkie, 1970a, b, c
b kal		no data	resistant to bongkrekic acid	Lauquin et al., 1973

that "the synthesis of certain respiratory systems in yeast requires the simultaneous presence of a self-reproducing cytoplasmic factor and a nuclear gene".

Segregational (chromosomal) *petites* were further classified as p or cy (Sherman, 1963; 1965). The symbol p is used to denote single-gene mutants which fail to grow on non-fermentable carbon sources such as lactate, ethanol, glycerol, or acetate, but which may or may not have altered cytochrome spectra. A genetic investigation of nine non-allelic genes controlling the ability of yeast to grow on non-fermentable substrates ($p_1 - p_7$, ly_6, ly_8, Hawthorn and Mortimer, 1960) showed that these genes are closely linked (Sherman, 1963). Strains p_8, ly_6 and ly_8 were considered to have genes controlling the retention or synthesis of the cytoplasmic factor; the latter two mutants had a simultaneous requirement for Lys.

The symbol cy denotes single-gene mutants which have altered absorption spectra, but which can utilize non-fermentable carbon sources for growth (Sherman, 1967). Chromosomal respiratory deficient mutants are found at low frequencies, and are therefore much more difficult to isolate than ρ^- mutants. Procedures for their isolation and characterization are described by Sherman (1967).

1. The cy *mutants*

The mutant cy_{1-1} has a decreased content of cytochrome c; this characteristic is controlled by a single gene. The cytochrome c from this mutant differs from normal cytochrome c in primary structure. It was then discovered that normal baker's yeast actually contains two types of cytochrome c: a major component (iso-l-cytochrome c), and a minor one (iso-2-cytochrome c) which accounts for about 5% of the total (Slonimski *et al.*, 1965). The mutant cy_{1-1} contains only iso-2-cytochrome c and it appears that the mutation results in the complete loss of ability to synthesize iso-1-cytochrome c. Low temperature examination of 15,000 clones led to the isolation of 12 cytochrome c-deficient mutants, and genetic studies indicated that these mutations occurred at six different loci $CY_1 \ldots . CY_6$ (Sherman, 1964). The cy_1 mutants contained only iso-2-cytochrome c at concentrations similar to those in the normal strain (Sherman *et al.*, 1965). Mutants $cy_2 \ldots cy_6$ contained reduced amounts of both iso-cytochromes c in approximately the same proportion as was found in the normal strain. Therefore it was suggested that CY_1 is the structural gene for iso-1-cytochrome c, (Slonimski *et al.*, 1965; Sherman *et al.*, 1966), CY_2 and CY_3 are directly involved with cytochrome c synthesis and CY_4, CY_5, and CY_6 cause cytochrome deficiencies by a secondary means. In mutants very deficient in cytochrome c it was not possible to detect an a-type cytochrome.

The synthesis of cytochromes $a + a_3$ was very sensitive to catabolite repression in mutants (cy_{3-5} and cy_{3-6}) and in a double-gene mutant (cy_{1-1} cy_{3-3}), all of which were partially deficient in cytochrome c, and in mutant p_{10} which was deficient in cytochromes b and c_1 (Reilly and Sherman, 1965). Cytochrome $a + a_3$ synthesis behaved normally in other cytochrome deficient mutants (cy_{1-1}, cy_{3-1} and cy_{3-3}). It was suggested that the synthesis of a-type cytochromes is normally regulated by other cytochromes, which are in turn themselves regulated by catabolite repression.

The cytochromes c of seven revertants of the cy_{1-1} mutants were extracted and characterized; none of them contained iso-1-cytochrome c, but all contained large amounts of iso-2-cytochrome c sufficient to compensate for the deficiency (Clavilier $et\ al.$, 1969). Thus none of the revertants resulted from a back mutation of cy_{1-1} and the cy_{1-1} mutation must be a deletion or some other irreversible alteration. Five unlinked gene loci $CY2A$, $CY2B$, $CY2C$, $CY2D$ and $CY2E$, all not linked to the $CY1$ locus controlled the reversions; these five loci are implicated in iso-2-cytochrome c synthesis. Mutations at these loci act as suppressors of cytochrome c deficiency caused by a deletion in the $CY1$ locus, acting not by restoring synthesis of the deficient protein, but by increasing the synthesis of a similar protein. The term "compensator gene" was applied to this type of mutation.

A number of altered forms of iso-1-cytochrome c were obtained from intragenic revertants of cy_1 mutants (Sherman $et\ al.$, 1968). These alterations, except in one case, are all single amino acid substitutions located at six different residue positions. X-ray mitotic recombination frequencies of heteroallelic diploids were used to construct fine-structure maps of CY_1, CY_2 and CY_3. These are the smallest genes reported in yeast. Most of the mutational sites fell within a very small region of the CY_1 gene, and a large proportion of the cy_1 mutants showed no recombination with each other. The groups of cy_1 mutants mapping at the same site have identical phenotypes with respect to the degree of cytochrome c deficiency (Parker and Sherman, 1969).

A summary of the isolation and fine-structure mapping of 16 cytochrome c deficient mutants, cy_{1-1} . . . . cy_{1-16}, and the altered primary structure of iso-1-cytochrome c from intragenic revertants of some of these mutants has been reported (Sherman $et\ al.$ 1970).

Using the cy_1 mutants, Stewart $et\ al.$ (1971) showed that AUG, the codon for Met, immediately precedes the codon for the NH_2-terminal. The residue of iso-1-cytochrome c in messenger RNA of baker's yeast. Mutational alteration in any of the three bases of this AUG triplet prevents the production of the protein. Only by the creation of in-phase Met codons at or near the original location is reversion induced in these mutants of the initiator codon. These results establish the uniqueness of the AUG triplet as initiator codon for *in vivo* cytoplasmic translation in *S. cerevisiae*. Suppression of amber mutants of iso-1-cytochrome *c* results in Tyr substitution (Sherman *et al.*, 1973): amino acid replacements at residue 71 have also been investigated (Stewart and Sherman, 1973). Comprehensive reviews of the genetic studies on cytochrome *c* have been presented by Sherman and Stewart (1971, 1973).

Electron transport and oxidative phosphorylation in mitochondria isolated from strain D-247 (homozygous for the CY_{1-1} gene and containing exclusively iso-2-cytochrome *c* at 5–10% of the total cytochrome *c* found in normal strains) may be efficiently reconstituted by the addition of cytochrome *c* which is tightly bound to the mutant mitochondria (Mattoon and Sherman, 1966). Whereas respiration in mutant mitochondria can be restored equally well by either iso-1- or iso-2-cytochromes *c* from yeast or by horse heart cytochrome *c*, the latter is less effective at restoring phosphorylation.

2. The ρ *mutants*

The respiratory capacity, enzymic activities and content of cytochromes were determined for a number of *p* mutants of *S. cerevisiae* (Sherman and Slonimski, 1964). $p_4\rho^+$ strains respired rather ineffectually and had low concentrations of cytochromes $a + a_3$ and *b*. $p_5\rho^+$ strains were deficient in respiration and cytochromes $a + a_3$. $p_1\rho^+$, $p_6\rho^+$, $p_7\rho^+$, $P\rho^-$ and all $p\rho^-$ strains were deficient in respiration and cytochromes $a + a_3$ and *b*. By using various combinations of the ρ/P genes, cy_1/CY genes and the ρ^+/ρ^- cytoplasmic factor, numerous strains could be obtained which had various alterations in cytochrome content. The *p* mutants are quite distinct from other segregational mutants, (e.g. W–1) which have blocks in the synthesis of protoporphyrin (which can be overcome by the inclusion of Gly in the medium) between glycine and protoporphyrin, and are almost completely devoid of cytochromes and catalase when grown on synthetic medium (Raut, 1953; Yčas and Starr, 1953). The *p* genes are not simply mutant genes controlling cytochrome induction. Alteration of most *p* genes leads to multienzyme deficiencies. These studies emphasized the strong interaction in the control of cytochrome synthesis between the different nuclear and cytoplasmic determinants. Thus the p_3 gene appears to play a part in the control of the retention of the cytoplasmic factor, and all homozygous p_3 strains are ρ^-: ly_6 and ly_8 are also always ρ^-. Haploid cells containing mutant genes p_1, p_2, p_4, p_5, p_6 and p_7 show various degrees of stability of the cytoplasmic factor; the percentage of ρ^- cells in newly isolated segregants of these strains is 55, 99, 40, 1, 1, and 5% respectively.

The cytochrome contents of mutants p_8, p_{12} and p_{13} were compared to that of the parent strain (Mackler *et al.*, 1965). All three contained only traces of cytochromes $a + a_3$; p_8 and p_{13} contained low amounts of cytochromes b, c and c_1, whereas p_{12} contained normal amounts of b- and c-type cytochromes. Respiratory particles from p_{12} contained appreciable amounts of cytochromes c and c_1, but no cytochrome b. The Cu contents of SMPs were normal, they had an elevated content of non-haem Fe and low flavin and UQ contents. The SMPs from the mutants showed very low activities of NADH and succinate oxidases or cytochrome c oxidoreductases, but were able to carry out indophenol reductase reactions linked to succinate and NADH.

A mutant (Z8) with a mutant gene allelic with p_3 and two other mutants non-allelic with this mutant all lacked cytochrome $a + a_3$. However, all showed oxidative phosphorylation with ferricyanide as electron acceptor with the same efficiency as the parent strain (Šubík *et al.*, 1972a, b). A cytochrome b-deficient mutant showed limited respiration because of this deficiency. Respiration was enhanced using a TMPD shunt, and by this means it was shown that phosphorylation efficiency was the same as in wild-type yeast. Cells deficient in both cytochromes a and b were isolated as segregants from a cross between cytochrome a- and cytochrome c-deficient mutants. These cells differed from cytochrome $a + b$-deficient mutants (resulting from either a cytoplasmic or a nuclear single-gene mutation (Šubík *et al.*, 1970)), in that their mitochondria were able to carry out ATP-Pi exchange, showed oligomycin-sensitive ATPase, and were able to synthesize proteins. However, eighteen different nuclear mutants, many of which had multiple respiratory lesions, were all found to retain their ability for mitochondrial protein synthesis although they lack several species of proteins made by wild-type mitochondria (Ebner *et al.*, 1973b). The absence of cytochromes a and b from the respiratory chain does not apparently affect the phosphorylation ability of the mitochondria although these cytochromes are required as redox carriers (Šubík *et al.*, 1972b).

A cross between two cytoplasmic *petite* mutants always results in the formation of zygotes whose vegetative progeny is entirely composed of organisms lacking respiratory enzymes. Crosses between a cytoplasmic mutant and a chromosomal mutant, or between two non-allelic chromosomal mutants yield zygotes, the majority of which yield vegetative progeny with normal respiration (Jakob, 1965). This complementation showed different lag periods in zygotes (as measured by oxygen uptake), for example when either mutants p_5, p_7 or p_1 were crossed with a cytoplasmic mutant, the lag before the onset of respiration was 0.5, 4.8 and 9.7 h respectively. The duration of the lag in crosses between two non-allelic chromosomal mutants was determined by the previous association of the gene and cytoplasmic factor in the haploid parent; thus reciprocal crosses showed very different results. It was suggested that either (a), the primary effect of the mutations is to modify the structure of the mitochondria and thus affect the electron transport system indirectly, or (b) the primary effect is to block the biosynthesis or function of one of the components of the electron transport system. Upon functional complementation in zygotes, an ordered integration occurs within the mitochondria. The initiation of respiration is characterized by a high rate of oxygen-uptake which progressively declines; this overshoot suggests that the respiratory system functions initially in a derepressed state and is then regulated. The phase of rapid oxygen-uptake is accompanied by an extremely high rate of *de novo* synthesis of cytochrome *c* oxidase which is cycloheximide-sensitive (Sels and Jakob, 1967).

Mutant *pet* 936 lacked mitochondrial ATPase, as assayed by enzymic, antibody-binding and radioimmunochemical tests for F_1 subunits (Ebner and Schatz, 1973). It also lacked mitochondrial ^{32}Pi–ATP exchange activity, had reduced content of cytochromes $a + a_3$, b and c_1, and its mitochondrial protein synthesis system did not produce some of the polypeptides normally produced in the wild-type. The three mitochondrially-produced subunits of cytochrome *c* oxidase (see page 422) were produced at only one-tenth of the normal level. This is the first strain of *S. cerevisiae* to be isolated which will not grow anaerobically, even in the presence of fermentable substrates and unsaturated fatty acids. This finding suggests that the mitochondrial ATPase has an essential function even in non-respiring cells. The mutant has a rather unstable mitochondrial genome; the conversion to the double mutant state decreased the growth rate by a factor of 20–30.

3. Mutant op₁ (p₉)

The unsolved problem of the molecular mechanisms underlying oxidative phosphorylation, even in the face of biochemical methods of increasing sophistication has provided the stimulus for the search for mutants with a defective energy-conserving apparatus (Kováč *et al.*, 1967a). In yeast a total absence of respiration leads only to conditional lethality, as fermentative energy supply is able to palliate the deficiency; the same should hold true for a block in oxidative phosphorylation. It was proposed that it would be possible to distinguish between p or ρ^- (*petite*) mutants and oxidative phosphorylation mutants by the criterion that the former do not respire, whereas the latter should have a normal or even higher respiration rate than the wild-type organisms. Nitrous acid mutagenesis of a haploid strain gave a mutant (op_1) which was selected by its ability to grow only very slowly (generation time 30–40 h) aerobically on DL-lactate, but which was capable (unlike ρ^- or p mutants) of reducing tetrazolium. Strain op_1 resulted from a single recessive chromosomal gene mutation, functionally non-allelic to chromosomal respiratory deficient mutants p_1, p_2, p_3, p_4, p_5, p_6 or p_7, and had a defective energy-producing mechanism. Neither this mutant, nor the diploid strain DH 1 (prepared by a cross of two haploid mutant cells of opposite mating types and homozygous for the mutant gene op_1) had marked alterations in cytochromes. Both haploid and diploid organisms showed oxidative metabolism of normal intensity, but no energy was produced which could be used for cellular synthesis. This meant that the growth yields aerobically or anaerobically on glucose were similar (Kormančíková et al., 1969).

Studies with mitochondria isolated from the op_1 strain suggested that the phosphorylation efficiency with all substrates tested was greatly diminished (Ková č and Hrušovská, 1968). Addition of ADP gave no stimulation of oxygen-uptake, but DNP (at concentrations which uncoupled the oxidative phosphorylation in wild-type mitochondria) did increase the rate of respiration of mutant mitochondria: this result suggests that energy-coupling at the level of the respiratory chain could still be operative. The pH 6·2 optimum of the mitochondrial ATPase in the mutant was not modified with respect to oligomycin, azide or fluoride sensitivity; the pH 9·2 component, although reduced in activity, was still inhibitor-sensitive. This difference in activity may be accounted for by the different extent of glucose repression in mutant and wild-type (Somlo, 1970). No uncoupling compound was specifically produced in the mutant yeast. The major conclusions from this work was that a single recessive nuclear gene mutation can lead to a deficiency common to all the phosphorylation steps in the respiratory chain: the possibility that the lesion was in the last step in the mitochondrial phosphorylation system (that is in the adenine nucleotide transporter) was also considered.

Acriflavine treatment of the op_1 mutant leads to loss of multiplication ability due to the production of cytoplasmic ρ^- mutants; cells carrying both ρ^- and op_1 mutations although equipped with fermenting and, at least in part, also with synthetic abilities are incapable of growth and division (Kováčová et al., 1968). As both mutations affect mitochondrial functions, this may suggest the presence of a mitochondrial function related to cell growth and division. This function would not simply be connected with energy supply. The respiration-deficient "petite negative" species of yeast resemble the respiration-deficient mutant of op_1 strain of S. cerevisiae in their inability to multiply continuously (Bulder, 1963).

Doubts as to the justification for classification of mutant op_1 as an "oxidative phosphorylation mutant" arose when Somlo (1970) demonstrated that this strain is capable of building up a considerable and stable steady-state concentration of ATP at the expense of respiratory energy. The level of ATP was only about 20% lower than in the normal strain and the respiration of the mutant cannot therefore be considered to be uncoupled. The steady-state concentration of AMP in the respiring mutant is five times higher than in wild-type cells and therefore the "energy charge" of respiring cells (Atkinson, 1968) is lowered, even when ATP levels are similar to normal. The time course of increasing steady-state ATP levels during respiratory adaptation of wild-type and op_1 cells were very similar in most respects (Somlo, 1971); a marked lag did occur however in the mutant in the appearance of DNP-sensitive and azide-sensitive ATP pools, but several interpretations of these results were possible.

The problem of locating the lesion in the op_1 mutant was finally resolved when Kolarov et al. (1972b) demonstrated unequivocally that this mutant has a modified system for the translocation of adenine nucleotides across the inner mitochondrial membrane. Although low P/O ratios were obtained under standard conditions used for the polarographic measurement of oxidative phosphorylation by the method of respiratory control, these could be raised to normal values (e.g. values approaching two with citrate or succinate) in the presence of unusually high concentrations of ADP. The rate of adenine nucleotide translocation was dependent on the energy level in the mitochondria; when the pool consisted mostly of ADP and AMP, the rate of translocation was 30 times lower than in the mitochondria from the wild-type yeast and the Michaelis constants for the ADP for this process were similar in both types of mitochondria (<10 μM). When the nucleotide pool was enriched in ATP, the translocation rate in mutant mitochondria was as high as in wild-type mitochondria. But under these conditions the Michaelis constant for external adenine nucleotide was more than 100 times higher in the mutant than in the wild-type. Variation in the translocation rates and binding affinities were also produced by agents which affect the energization state of the mitochondria (uncouplers, oligomycin, valinomycin and nigericin) and the modification of the adenine nucleotide carrier was confirmed by detailed studies of its properties.

A nuclear (chromosomal II) mutant of S. cerevisiae (p_9), unable to utilize non-fermentable carbon sources for growth, respires and contains the normal cytochrome complement (Beck et al., 1968). The cytoplasmic respiratory mutant (ρ^-) was abortive in this strain. Mitochondria displayed abnormal oxidative phosphorylation, in that the P/O ratios were low, the Mg^{2+} stimulated ATPase was low and apparently latent, the affinity for ADP was low and response to uncouplers was abnormal (neither DNP nor gramicidin stimulated respiration). A very significant observation was that normal P/O ratios were restored by using high ADP concentrations. A number of interpretations of these results are possible: (a) the altered product of the p_9 gene may be a defective protein (with an altered K_m for ADP?) which normally participates in energy-coupling; (b) the altered p_9 gene product may be an abnormal adenine nucleotide translocase; (c) the overproduction of a normal metabolite or an abnormal metabolite (e.g. an unsaturated fatty acid) gives rise to abnormal mitochondrial function; or

(d) the modification of a membrane protein or enzyme other than the ATP synthetase may give an indirect effect. It is evident that mutants op_1 and p_9 have very similar phenotypes and that neither in fact has a defective oxidative phosphorylation system *per se*.

4. Other mutants with properties similar to those of op₁ and p₉

The isolation of a whole series of mutants which, like p_9 are characterized by a decreased ability to utilize non-fermentable carbon sources for growth, but which possess all the cytochromes and are able to reduce tetrazolium salts, has been described by Parker and Mattoon (1968, 1969). Of 35 mutants examined, at least 19 were the results of mutation in single nuclear genes. At least 11 complementation groups were represented among those 19 mutants, indicating that a large number of nuclear genes control oxidative energy metabolism and that the characteristics of mutants in this class are very diverse and require further investigation.

5. Mutants aem₉, glt₁, glt₂, gltₓ, and glg

Mutant glt_2 lacks aconitase, has blocked Lys and cytochrome synthesis, and has an absolute requirement for Glu (Ogur *et al.*, 1965). Two mutant genes (aem_9), and (glt_1), (Ogur *et al.*, 1964) have also been shown to lower the concentration of all the cytochromes in the cell. Neither mutant assimilates glycerol, and mutant (glt_1) lacks aconitase and requires Glu for growth (Beck *et al.*, 1971). A gene linked to glt_1 [(glt_x), which controls Glu biosynthesis], neither reduces the cytochrome content, nor limits glycerol assimilation. Beck *et al.* (1971) have proposed that gene (glt_x) is a "non-interacting" gene, while (glt_1) and (aem_9) are examples of "interacting" genes which alter mitochondrial biogenesis. Interactions of genes in the background genome of normal strains give up to 2-fold variation of cytochrome content, and the rate of glycerol assimilation can vary four fold among the progeny of diploids obtained by crossing two wild-type cells. Two genes (glg) affecting glycerol assimilation have also been identified.

6. The pop mutants

Pretlow and Sherman (1967) have investigated the accumulation of free porphyrins and Zn porphyrins in the chromosomal mutant pop_1. Uroporphyrin III was present at high concentrations together with heptacarboxylic, hexacarboxylic, and pentacarboxylic porphyrins. The total porphyrin content was 60–100 fold higher than in the parent strain. Mutant pop_2 contained more than 10 times the normal amounts of free and Zn protoporphyrin. These compounds have absorption bands at between 575 and 590 nm.

7. A mutant lacking all cytochromes except cytochrome b₂

A respiratory-deficient mutant (58NW-46) containing no cytochromes a or c was isolated by Morita and Mifuchi (1970). This mutant contained only cytochrome b_2 (L-lactate dehydrogenase) with absorption maxima at 529 and 557 nm; L-lactate dehydrogenase activity was demonstrated in extracts. The presence of cytochrome b_2 in various respiration-deficient mutants has been reported by Gregolin and Megaldi (1961), Sherman and Slonimski, (1964) and Sugimura $et\ al.$ (1966a).

8. Cyd mutants

A new class of mutants deficient in biosynthesis of all cytochromes was isolated from cultures grown in ethidium bromide-containing media; supplementation with δ-aminolaevulinic acid leads to restoration of normal cytochrome c synthesis (Sanders $et\ al.$, 1973). Two genetic determinants, one nuclear and the other mitochondrial, are involved. In cells which possess normal mt DNA, all cytochromes are present and growth on glycerol is possible; the full effect of the cyd_1 mutation is only apparent in ρ^- strains. Complementation between a cytochrome c mutant (cy_{4-1}) and cyd_1 was demonstrated, and the two determinants segregated independently: thus δ-aminolaevulinic acid synthesis is controlled by at least two nuclear genes, and also by one or more genes located in mt DNA. Correlation of action of these genes must involve exchange of information between the nucleo-cytosolic and mitochondrial compartments of the cell.

9. The gi mutant

A mutant of $S.\ cerevisiae$ which gives daughter cells which are respiratory deficient, when grown anaerobically or under conditions of glucose repression, but which gives normal respiratory competent cells in the absence of repression, has been isolated by Negrotti and Wilkie (1968). Although chloramphenicol inhibits the synthesis of membrane-bound cytochromes in this strain, it does not induce the $petite$ mutation in cultures growing under non-repressing conditions. The determinant of this phenomenon (gi) is under the control of a recessive nuclear gene, and spontaneous reversion to normal occurs with low frequency. This determinant is one of the nuclear factors in a system which ensures the normal replication-transmission mechanism of mt DNA.

B. Temperature Sensitive Mutants

1. The cyt mutant

A respiratory deficient mutant of $S.\ cerevisiae$ (cyt) resulting from mutation in a recessive single gene on a chromosome (Gunge $et\ al.$, 1967), lacks all the cytochromes and other haemoproteins when grown at 35°C (Sugimura *et al.*, 1966a). It was originally thought that under these conditions the organism lacks the enzyme for the conversion of coproporphyrin into protoporphyrin, but it was then found that the enzyme coproporphyrino-genase showed a ten-fold higher activity in the mutant than in the wild-type (Miyake and Sugimura, 1968). The spectral characteristics of the porphyrin which accumulates intracellularly suggests that this is a metallo-porphyrin rather than coproporphyrin III (Pretlow and Sherman, 1967). Cells grown at 20°C respired normally and had a normal cytochrome content. The respiratory activity of the mutant growing at 35°C could be restored by the addition of protoporphyrin IX or protohaemin IX to the culture medium (Miyake *et al.*, 1972). The activities of NADH-cytochrome *c* and DCPIP oxidoreductases and succinate-cytochrome *c* and DCPIP oxidoreductases paralleled the changes in respiratory activity of the mutant under the three different conditions of growth as did the complexity of mitochondrial organization.

2. Ubiquinone-requiring mutant (p_{ts}, strain 1511)

A thermosensitive mutant arising from a nuclear mutation has also been described by Puglisi and Cremona (1970). In this organism growth at 36°C results in a block in UQ synthesis and under these conditions cytochromes *b*, *a* + *a*₃ and several respiratory-chain linked enzymes are not formed (D- and L-lactate dehydrogenases, succinate dehydrogenase as well as cytochrome *c* oxidase). Addition of UQ to the medium allows the formation of a complete respiratory chain and growth at the non-permissive temperature.

3. Cold-sensitive respiratory-deficient mutants

A cold-sensitive mutant (1121) of a haploid strain of *S. cerevisiae* was isolated by selection for impaired growth at 18°C on non-fermentable carbon sources (Weisogel and Butow, 1970). Growth on glucose or galactose at either 28°C or 18°C was similar to that of the parent. This cold-sensitive strain was converted to ρ^- with high frequency when grown at 18°C or when grown at 28°C in the presence of chloramphenicol (4 mg/ml). It was concluded that mitochondrial protein synthesis was required for the stability and maintenance of the mitochondrial genome, and that perhaps a mitochondrially produced protein is involved in the replication of mt DNA. The fate of mitochondrial membrane proteins and mt DNA was investigated during *petite* induction at 18°C (Weisogel and Butow, 1971). Of the seven or eight proteins (mol. wts 13,000–45,000) synthesized by the organism in the presence of cycloheximide, the species of mol. wt 41,000 was lost; the species with mol. wt 24,000 showed decreased synthesis whilst that of 13,000 was increased. Growth with chloramphenicol at 28°C also resulted in a decreased synthesis of the component with mol. wt 41,000. Growth at 18°C leads to genetic heterogeneity in the population, as subcloning of randomly selected *petites* indicated two distinct cell types, those with detectable mt DNA and those without. These results confirmed that the alteration in strain 1121 reflects improper formation of a membrane component related to the maintenance of the ρ^+ factor.

Another class of cold-sensitive mutants (24,49,518-4) has been described in which the mitochondrial genome is stable during growth at 18°C, but which nevertheless lose cytochromes and respiratory capacity at low temperature and thus appear phenotypically unstable (Butow *et al.*, 1973). The loss of respiration following growth at 18°C is reversible, as incubation of the mutants under conditions of limited cell growth at 28°C leads to a restoration of respiratory capacity; this constitutes a temperature dependent assembly of the respiratory apparatus and does not involve exactly the same process as respiratory adaptation at 18°C of cells grown anaerobically at 28°C. *In vivo* labelling studies with (^{3}H)-Leu in the presence of cycloheximide reveals major changes in the profiles of products of mitochondrial protein synthesis under the different conditions. Cold sensitivity has a recessive nuclear determinant.

4. Mutant P203ts⁻ having temperature-sensitive mitochondrial ribosomes

A mutant, P203*ts*⁻ has a mitochondrial protein-synthesizing system which is defective at 35°C (Thomas and Scragg, 1973). The deficiency has been demonstrated both in whole cells and in isolated mitochondria; it is irreversible and requires 30 min to become fully apparent. The use of hybrid mitochondrial and *E. coli* systems shows that the temperature-sensitive component is associated with the mitochondrial ribosomes. The exact nature of this component has not been determined, but it was suggested that the recessive nuclear mutation which determines it is associated with ribosomal proteins or chain-elongation factors.

C. Mutants of "*Petite*-negative" Yeasts

Yeast species in which respiratory deficient mutants are hard to obtain have been called "*petite*-negative" (Bulder, 1964a). These species are generally characterized by their limited ability to grow anaerobically (Bulder, 1964b) and by their failure to exhibit the phenomenon of the Crabtree effect (glucose repression of respiration, DeDeken, 1966a; McClary and Bowers, 1968). This effect is usually assessed by RQ measurements [ratio of glucose fermented (measured by CO_2 evolved) to glucose oxidized, (measured by O_2-uptake)]; this ratio is high in glucose-repressed "*petite*-positive" species. The suggestion that *petite*-negative yeasts could not form viable *petite* mutants because such species were unable to obtain sufficient energy for growth by fermentation (Bulder, 1964a) is invalidated by DeDeken's (1966b) demonstration that a number of *petite*-negative species could grow in a concentration of acriflavine sufficient to completely inhibit respiratory enzyme synthesis.

The fission yeast *Schizosaccharomyces pombe* 972h⁻ gives no viable respiratory-deficient mutants on treatment with acriflavine or ethidium bromide, but these mutagens do induce 15–70% of microcolonies which after a few days further develop into normal respiratory-competent colonies (Heslot *et al.*, 1970b). Unstable and reversible respiration-deficient forms have also been described in other *petite*-negative yeasts (Nagai, 1969). Although no viable cytoplasmic *petites* were found, cobalt sulphate-resistant segregational *petites* were obtained; some of these are deficient in cytochromes $a + a_3$.

Thus *S. pombe* appears to be intermediate between *petite*-positive and *petite*-negative yeasts in its properties. The organism does show glucose-repression of mitochondrial enzymes (Heslot *et al.*, 1970a; Poole and Lloyd, 1973) and grows for at least fourteen generations anaerobically. The O_2 requirement for growth can be relieved by KNO_3 (Heslot *et al.*, 1970a).

Chromosomal mutants of *S. pombe* have also been isolated after mutagenesis by N-methyl-N'-nitro-N-nitrosoguanidine or UV, and enrichment using TTC (Wolf *et al.*, 1971). These mutants showed a loss of cytochrome c oxidase activity and succinate-cytochrome c oxidoreductase activities.

Mutant UV119 of *S. pombe* 50 h⁻ yielded mitochondria which showed respiration rates of about 7% of those of the wild-type mitochondria with NADH or succinate as substrates, and of about 45% for ascorbate + TMPD (Bandlow *et al.*, 1974). The major break in electron transport in this mutant appears to be in complex III of the respiratory chain. Oxidation of NADH and succinate was antimycin A- and CN^--insensitive, whilst that of ascorbate + TMPD was much less sensitive to CN^- than the wild-type. The amounts of cytochromes c_1 and $a + a_3$ were about the same in the mutant as in the wild-type with a normal cytochrome $a + a_3:c_1$ ratio. Cytochrome b_{566} was not detectable in low temperature spectra after reduction with dithionite or with various substrates, but cytochrome b_{558} was still present. Even after the addition of UQ, NADH gave very little reduction of b-type cytochromes: UQ_3-cytochrome c reductase activity was very low and succinoxidase was highly stimulated by PMS. Cytochrome b was not reduced during the steady-state oxidation of NADH or succinate, and was reduced only very slowly after anaerobiosis was attained unless PMS was added. Antimycin A showed no binding to SMPs, cytochrome b_{558} of the mutant particles reacted with CO, and

EPR signals of dithionite-reducible iron-sulphur proteins were low compared to the wild-type. ATPase of the mutant mitochondria had a lowered specific activity but was still oligomycin sensitive.

Two mutants of $S.$ *pombe* 972h$^-$, modified in their sensitivities to glucose repression, have been obtained by Foury and Goffeau (1972). Whereas the respiration rate of the wild-type organism grown in the presence of 10% glucose is only one-third of that grown in the presence of glycerol, the respiration rate of strain COB5 after growth with glucose is only 4% that of cells grown with glycerol. This mutant is referred to as a "superrepressed" mutant. Mutant strain COB6, on the other hand, is totally insensitive to glucose repression. The effects of glucose on respiration are paralleled by the effects of glucose on cellular cytochrome content. Full restoration of the respiration and respiratory pigments of the super-repressed strain was achieved in a derepressing medium (containing ethanol) which cannot support cell division. This derepression is inhibited by chloramphenicol and by cycloheximide, and thus involves *de novo* protein synthesis. These mutants provide valuable systems for the study of nuclear and mitochondrial participation in glucose repression and derepression.

Six stable mutants with modified mitochondrial ATPase activity have also been obtained (Goffeau *et al.*, 1972a; Landry and Goffeau, 1972). These were isolated after X-ray or nitrosoguanidine mutagenesis and selected for their ability to grow on glucose but not on glycerol medium.

All the mutants have lost oligomycin-sensitive ATPase. Compared to the wild-type, cells of strains RD14, RD15, M126 and Res 4 had lost about 75% of the particulate Dio-9-sensitive ATPase activity, whereas two other strains (M53 and RD32) had lost more than 95% of the total Dio-9-sensitive ATPase activity per cell. Four of the strains (M126, Res 4, M53 and RD15) were shown to be chromosomal mutants. All mutant strains examined were pleiotropic and showed various cytochrome deficiencies; cytochromes $a + a_3$ were not detectable in any of the mutants, and in some mutants cytochromes c and b showed changes. On tetrad analysis cytochrome deficiencies and ATPase deficiencies segregate together, indicating that the mutant phenotypes are not the result of multiple unlinked mutations.

Other classes of hereditary modifications of mitochondrial ATP synthetase include: (a) loss of oligomycin sensitivity in respiring mutants; (b) deficiency of binding of the Dio-9 sensitive ATPase to mitochondrial membranes; (c) loss of repressivity by glucose of mitochondrial ATPase; and (d) superrepressivity by glucose of mitochondrial ATPase (Goffeau *et al.*, 1972b).

Two environmental modifications are known which reversibly modify the mitochondrial ATPase activity. These are (a) growth at high glucose concentrations which decreases the oligomycin and Dio-9-sensitive ATPase, and (b) induction of oligomycin insensitivity of ATPase which occurs when growth is carried out in the presence of non-mutagenic concentrations of ethidium bromide. This latter effect suggests control of factors conferring oligomycin sensitivity to the Dio-9-sensitive ATPase by mt DNA. On the other hand the chromosomal nature of the mutants with decreased oligomycin sensitivity of mitochondrial ATPase indicates the participation of nuclear DNA in the same control.

In the single-gene nuclear mutant M126 the absorption bands of cytochromes $a + a_3$ and b are not detectable and the mitochondrial ATPase has lost its oligomycin sensitivity (Goffeau et al., 1973a). The Dio-9-sensitive ATPase activity is markedly decreased and the Dio-9-sensitive ATPase in the postribosomal supernatant is similar to that in the wild-type. The ATPase in SMPs from wild-type and mutant are similar in their reactions to anions, cations, inhibitors and low temperature. Both enzymes contain all 5 subunit polypeptides (Goffeau et al., 1973b). Cytochrome c and succinate-PMS reductase are present in the mutant. Growth of the wild-type in the presence of 10 µg/ml ethidium bromide does not produce the ρ^- mutation but leads to the phenotypic production of pleiotropic deficiencies similar to those seen in strain M126. Chloramphenicol has a similar effect. Mutant M126 still possesses mt DNA but has an impaired ability to carry out mitochondrial protein synthesis. It appears therefore that a factor controlled by a nuclear gene participates in the assembly of the products of mitochondrial protein synthesis into the membranes.

The ATP synthetase complex can be purified from S. pombe after extraction with Triton X-100; by this procedure oligomycin sensitivity is not lost (Landry and Goffeau, 1972). F_1-ATPase has been purified from extracts prepared by sonication of the organism.

An oligomycin-resistant nuclear mutant O 77 specifically lacks cytochrome b_{566}, still has cytochrome b_{558}, and is capable of growing on non-fermentable carbon sources (glycerol) with about a half the growth rate of the wild-type (W. Bandlow, personal communication). The ATPase of isolated mitochondria was insensitive to oligomycin (20 µg/mg protein). SMPs prepared from the mutant mitochondria still bind antimycin A despite the lack of cytochrome b_{556}. Thus this cytochrome does not react specifically with antimycin and is not required for electron transport in the mutant. It was suggested that the non-integration of this cytochrome is a consequence of a modified membrane protein which is also responsible for oligomycin sensitivity of ATPase.

Another "petite-negative" yeast Saccharomyces lactis also gives segregational petite mutants (Herman and Griffin, 1968). Of six segregational respiratory-deficient mutants produced after UV irradiation, two had normal cytochromes, and a third had increased levels of all cytochromes. The other three mutants showed a partial or complete loss of cytochromes $a + a_3$, b, c and c_1. Three of the mutants were not genetically identical and occupied loci in the same linkage group, two of these three were cytochrome deficient. Several of the mutants showed pleiotropic defects which affected zygote formation and sporulation. Growth at high or low temperatures, under conditions of increased osmotic pressure, or in media supplemented with fatty acids or sterols, did not relieve these physiological defects.

After extensive ethidium bromide treatment or UV irradiation, no *petite* mutants could be isolated from the obligate aerobe *Hansenula wingei* (Crandall, 1973a); ethidium bromide reversibly inhibits growth on fermentable or non-fermentable carbon sources after one or two generations. Ethidium bromide-resistant mutants were isolated. A glycerol-negative respiratory deficient mutant of this organism which has normal cytochromes has been described by Crandall (1973b). This *glp* mutant is specifically blocked in glucose respiration, but is also pleiotropic, affecting growth rates on other substrates, sporulation, and diploid mitotic segregational frequencies of other genes. It is not an oxidative phosphorylation mutant because the growth yield on glucose is the same as that of the parent strain.

D. Various Auxotrophic Mutants

1. Unsaturated fatty acid-requiring mutants

The effects of depletion of unsaturated fatty acids and ergosterol may be studied in anaerobic cultures of yeasts, as O_2 is required for the synthesis of these components (see Chapter 4). The use of mutants auxotrophic for lipid components also enables manipulation of the lipid composition of mitochondrial membranes under aerobic conditions of growth. Studies on the roles of lipids in respiratory chain and energy-linked functions, and in the molecular structure of membranes, are facilitated by the use of these mutants.

S. cerevisiae strain KD 115 (*ole 1-1*) which cannot synthesize unsaturated fatty acids was isolated by Resnick and Mortimer (1966). When depleted to 20% of their normal content of unsaturated fatty acids, this organism contains cytochromes and respires at normal rates, but under these conditions, it cannot grow on non-fermentable substrates (Proudlock *et al.*, 1969). Isolated mitochondria lacked respiratory control, and showed negligible oxidative phorphorylation. They cannot carry out energy-dependent reduction of NAD by succinate, or the reduction of NADP by NADH. It was concluded that mitochondrial oxidative phosphorylation and energy-linked reactions were lost both *in vitro* and *in vivo* as a con-

N

sequence of unsaturated fatty acid depletion (Proudlock *et al.*, 1971; Linnane *et al.*, 1972a). The loss of the ability for coupled oxidative phosphorylation is reversed *in vivo* by the incorporation of unsaturated fatty acids into mitochondrial membranes. This reversal is not inhibited by inhibitors of protein synthesis, and so loss of oxidative phosphorylation appears to arise solely from a lipid lesion (Haslam *et al.*, 1971). The activity of ATPase in lipid-depleted cells is normal and the enzyme is sensitive to oligomycin and F_1–ATPase inhibitor. The depleted mitochondria have a normal concentration of total phospholipid, but the percentage of phosphatidyl-inositides is decreased from 13% in fully supplemented cells to 4% in depleted cells. A specific role for the phosphoinositides in energy-linked reactions of mitochondria has been suggested by Vignais *et al.* (1963) and Hill *et al.* (1968). Since cation transport is intimately related to oxidative phosphorylation, a further investigation was carried out relating changes induced by unsaturated fatty acid depletion in both systems (Haslam *et al.*, 1973b). Loss of ability for coupled oxidative phosphorylation was accompanied by simultaneous loss of ability for active cation transport and an increased passive permeability of the mitochondria to protons. The uncoupling of oxidative phosphorylation was not due to any accumulation of free fatty acids in mitochondria depleted of unsaturated fatty acids. It was suggested that a change in the physical properties of the lipid phase of the inner mitochondrial membrane is responsible under these conditions for the increased passive permeability to protons, and that it is this which leads to loss of energy-linked reactions.

Another fatty acid Δ^9-desaturase mutant of *S. cerevisiae* (KD20, *ole 1-2*), like *ole 1-1*, is respiratory-sufficient and heteroallelic. A third mutant (KD46, *ole 2*) is a nuclear *petite* and segregates independently of *ole-1* (Resnick and Mortimer, 1966; Keith *et al.*, 1969; Wisnieski *et al.*, 1970). The requirement can be satisfied by several fatty acids differing in double-bond position, steric configuration, chain length, and degree of unsaturation (Wisnieski and Kiyomoto, 1972). Fatty acid specificity is apparently unrelated to respiratory competence; it can be shown that the *cis*-configuration is not an essential structural requirement. The effects of carbon source and temperature were investigated, and electron paramagnetic resonance was used to compare membranes of mutants enriched for different fatty acids. The lipid distribution pattern of the most commonly employed EPR spin-label, 12-nitroxide stearate was compared with that of 18:1 Δ^9cis.

Comparison of EPR spectra of aerobically- and anaerobically-grown spin-labelled yeast, indicates that the signal intensity of aerobic cells decreased on removing O_2 from the culture; CN^- caused a lesser decay of the signal (Nakamura and Ohnishi, 1972). The changes in signal of the spin-labels buried in the membrane were partly related to the respiratory activity of the culture. Results with spin-labelled *Neurospora* mitochondria (Keith *et al.*, 1970) suggests that the hydrocarbon portions of the membranes are relatively fluid and are not extensively restricted in motion by association with proteins.

Mutants of yeast specifically deficient in saturated fatty acid biosynthesis have also been obtained (Schweitzer and Bolling, 1970; Schweitzer and Castorph, 1971; Henry and Keith, 1972; Keith *et al.*, 1973).

2. Sterol-requiring mutants

Mutant strains of *S. cerevisiae* which require ergosterol for growth have been isolated (Resnick and Mortimer, 1966; Bard, 1972; Karst and Lacroute, 1973). These mutants are *petites* and require a fatty acid; some of them also require Met. Six complementation groups have been shown; the mutants do not show a stringent requirement for ergosterol, as stosterol, stigmasterol or cholesterol also support growth. These mutants may be of some use for studies of the role of sterols in the outer mitochondrial membrane.

Ergosterol, the major membrane sterol in yeast has been identified as the binding site for the polyene antibiotic nystatin (Lampen *et al.*, 1972). A genetic analysis of nystatin-resistant mutants of yeast has led to the identification of three genes for nystatin resistance, *nys 1*, *nys 2*, and *nys 3* (Ahmed and Woods, 1967; Woods and Hogg, 1969). Mutants *nys 1* and *nys 3* differed from the sensitive strain (*nys⁺*) in their sterol content: Ultraviolet absorption spectra of the nonsaponifiable material from the sensitive strain revealed the presence of ergosterol and 24 (28)-dehydro-ergosterol (Woods, 1971). In *nys 1* mutants a new sterol was found; *nys 3* contained no detectable ergosterol or 24 (28)-dehydroergosterol but contained another new sterol (tentatively identified as a 28-carbon, zymosterol-like compound, Thompson *et al.* (1971). Conversion of (*nys⁺*) and *nys 3* to *petite* results in loss of 24 (28)-dehydroergosterol in the former and the new sterol in the latter, whereas the new sterol in *nys 1* is only reduced, indicating the participation of functional mitochondria in sterol synthesis. The sterols in ethanol-grown cells of all three genotypes are essentially the same as in glucose-grown cells.

Except in strain *nys 3*, growth of the mutants on ethanol did not appear to place them at a disadvantage compared to wild-type; these observations also indicate that sterol synthesis is in part dependent on mitochondrial function. The formation of ergosterol during respiratory adaptation has also been shown to be sensitive to inhibition by chloramphenicol or acriflavine (Adams and Parks, 1969). Shimizu *et al.* (1971) have shown that one of the enzymes of sterol synthesis is mitochondrial in yeast.

Selection of mutants resistant to other polyene antibiotics (etruscomycin, filipin, pimaricin and rimocidin) enabled allocation to four unlinked genes *pol 1, 2, 3* and *5* (Molzahn and Woods, 1972). All mutants were resistant to nystatin. A fifth gene, *pol 4* was recovered as a double mutant with a strain carrying the *pol 1* mutation: the *pol 4* gene does not confer resistance. *pol 1* and *pol 4* are allelic to *nys 1* and *nys 3*. There are correlations between the polyene used for mutant screening and (a) the extent of cross resistance, and (b) the selection of mutants at particular *pol* genes. 24 (28) Dehydroergosterol was not found in any of the mutants, whilst ergosterol is lacking in *pol 2* and only present at very low levels in *pol 3*. Mutants of *pol 1* excrete sterols into the medium. Metabolic relationship in the sequence *pol 2* → *pol 3* + *pol 5* → pol *1* → *pol 4* → *pol+* is suggested by the epistatic relationships between the mutants. Other reports of nystatin resistance in yeast includes that of Karst and Lacroute (1973).

3. Flavin-requiring mutants

Riboflavin auxotrophes of *S. cerevisiae* were isolated by Oltmanns and Lingens (1967). These mutants have been characterized biochemically (Lingens *et al.*, 1967; Bacher and Lingens, 1970, 1971) and genetically (Oltmanns, 1971), but their potential for the study of flavoprotein participation in mitochondrial electron transport has not yet been investigated.

4. Ubiquinone-deficient mutant

A mutant unable to grow on glycerol was found to be deficient in UQ synthesis (De Kok, 1973). Cytochrome and EPR spectra of SMPs from the mutant were similar to those of wild-type particles. the UQ content was <0.06 nmoles/mg protein in SMPs from the mutant as compared with 7 nmoles/mg protein in the wild-type particles. The antimycin A-sensitive respiration rates (ng atoms/min/mg protein) in mutant and wild-type particles were: TMPD + ascorbate 2010 and 2060, NADH 60 and 1630, succinate 73 and 346, glycerol-1-phosphate 12 and 310 respectively. The respiration rates in mutant particles were stimulated by UQ_1 to values comparable with those in wild-type particles. The UQ-stimulated respiration of intact mutant mitochondria is accompanied by oxidative phosphorylation.

5. Other growth factors

The nutritional requirements of many strains of yeast growing on minimal media provide the possibilities of investigation of the roles of vitamins in mitochondrial development. For instance some strains of *S. carlsbergensis* require inositol, and pantothenic acid. *T. pintolopesii* and *C. bovina* require choline (Cruz and Travassos, 1970). *S. cerevisiae* A.T.C.C. 9371 requires β-alanine. Nicotinic acid, biotin, pantothenic acid and inositol are growth factors necessary for some strains of *Shizosaccharomyces pombe* (e.g. 972h⁻). Using chemostat cultures it is possible to produce cells under conditions of growth limitation for any of these compounds and to study their role in mitochondrial function or development.

Lists of the nutritional requirements of various yeast species and mutants of *S. cerevisiae* are to be found in Sober (1970).

E. Mutants Resistant to Inhibitors of Electron Transport

An acriflavine-induced mutant of *Candida utilis* resistant to inhibition by antimycin A was obtained by Butow and Zeydel (1968). This mutant has a lesion in its respiratory chain (lowered binding constant) rather than a permeability barrier to the antibiotic. It is also resistant to another Site II inhibitor, hydroxyquinoline *N*-oxide, indicating that the sites of action of this compound and antimycin A are very close. From measurements of fluorescence quenching, Berden and Slater (1972) have concluded that antimycin A is bound to Complex III at a distance of 1·9 nm and 2·4 nm from the *b* haem group in the oxidized or reduced complexes respectively.

Mutants of *C. utilis* able to grow in the presence of non-fermentable substrate in the presence of 5 μg/ml/antimycin A have been characterized by Grimmelikhuijzen and Slater (1973). The NADH oxidase of SMPs of one of these mutants is 20 times less sensitive to inhibition than that of the wild-type, as judged by the titre required for 50% inhibition. That the mitochondrial Complex III has been altered in the mutant is indicated by the loss of the characteristic sigmoidicity of the titration curve for antimycin-inhibition. Lowered sensitivity to 4-heptyl-2-hydroxyquinoline-N-oxide was also observed, but sensitivity to treatment with 2,3-dimercaptopropanol and O_2 was similar to that of the wild-type. NADH and TMPD are oxidized at similar rates by particles from the wild-type and mutant, but the latter have only 50% and 20% of the wild-type succinate and α-glycerophosphate oxidase activities respectively. The two types of particle have identical reduced-oxidized spectra at room temperature and the concentration of cytochrome *b* is identical in both strains. There is no impairment of Site II or Site III phosphorylation in the mutant. Binding of the inhibitor to oxidized particles of wild-type and mutant is non-cooperative (K_D's $5\cdot4 \times 10^{-11}$ M, $6\cdot5 \times 10^{-10}$ M, respectively): concentration of binding sites is 0·13 μmole/g protein and 0·22 μmole/g protein respectively. The binding in succinate-reduced wild-type is positively cooperative with a K_D of 1.7×10^{-10} M at zero-bound antimycin and approaching the value for the oxidized particles at increasing inhibitor concentrations. However the lower sensitivity of particles from the mutant cannot be explained entirely by the altered binding characteristics. Attempts to isolate similar mutants in *S. cerevisiae* have not been successful, as those antimycin A-resistant mutants which were selected yielded SMPs insensitive to this inhibitor.

Mutants of *S. cerevisiae* sensitive to rotenone have been obtained by UV irradiation of two genetically well-defined strains 194-5C and iL 126-1B (Mounolou, 1973). The frequency of appearance of sensitive mutants in a haploid population is rather low (about 1 in 1500 colonies), and the isolation procedure was not successful when applied to diploids. These observations are consistent with a control of resistance and sensitivity by nuclear genes and further genetic analysis indicated that two nuclear genes determine sensitivity. In one type of mutant (Class 1), growth on glycerol is only slowed down by rotenone, and NADH oxidase of mitochondria is insensitive to rotenone. A mutant of Class 2 (194-5C/R11/53) grows for 3 generations on glycerol in the presence of rotenone, and then growth stops completely. Growth on fermentable carbon sources is not inhibited by rotenone, indicating that the inhibitor exerts its action specifically on the mitochondrial system. The NADH oxidase activity of isolated mitochondria showed only 27% inhibition at 1 μM–inhibitor concentration.

F. Mutants Resistant to Other Antibiotics and Inhibitors

Two dominant nuclear genes and one recessive gene have been identified conferring resistance to chloramphenicol; for erythromycin one recessive gene for resistance has been recognized (Wilkie *et al.*, 1967; Roodyn and Wilkie, 1968).

The isolation of mutants resistant to oligomycin was first described by Parker *et al.* (1968). After UV irradiation, colonies growing in the presence of oligomycin on glycerol-containing solid medium were used to start resistant clones. Two of these mutants, and a naturally resistant strain, were characterized genetically, and it was found that at least two Mendelian genes control oligomycin resistance which displays partial dominance and additivity in heterozygous diploids. Oligomycin resistance is accompanied by resistance to venturicidin. Oligomycin also partially inhibited cytochrome synthesis and increased the frequency of mutation to cytoplasmic respiratory deficiency (ρ^-). Another report of Mendelian inheritance of oligomycin (and venturicidin) resistance (which resulted in a twenty-fold increased tolerance to the drug) was that of Stuart (1970) (see also p. 312).

Spontaneous mutants resistant to canavanine (a structural analogue of Arg, which inhibits growth on non-fermentative media, decreases respiration and inhibits synthesis of cytochrome $a + a_3$ and b during growth) is controlled by nuclear genes (Wilkie, 1970a, b, c).

Several tetracycline-resistant mutants have been described by Hughes and Wilkie (1972): most of these are chromosomal mutants, although some showed mixedness in $Tet^R \times Tet^S$ crosses. Tetrad analysis seemed to indicate a nuclear origin, and no evidence was obtained to suggest association of the determinant with the ρ factor.

The isolation of mutants resistant to bongkrekic acid, an inhibitor of adenine nucleotide translocation has been described by Lauquin et al. (1973). Resistance is determined by a single nuclear gene bkal (see p. 318).

III. Chromosomal Mutations Affecting Mitochondrial Functions in *Neurospora Crassa*

A. Respiratory-deficient Mutants

The isolation of two chromosomal respiratory-deficient mutants of N. crassa was described by Mitchell et al. (1953).

In mutant C115, there are hardly any detectable a-type cytochromes; experiments with the organism are made difficult by a frequent mutation to (and selection for) a genetically-repressed phenotype (Tissières and Mitchell, 1954). The cytochrome components in mutant C117, cytochromes b and "e" (c_1, α-band at 552–553 nm), when reduced with succinate cannot be reoxidized by air (Tissières and Mitchell, 1954). Respiration is lower in this strain than in any other strains invesitgated, and it is completely azide- and CN^--insensitive.

These nuclear gene mutants C115 and C117, similar in phenotype to poky mutants, do not respond to the gene f, which confers normal growth rates in poky mutants (Mitchell and Mitchell, 1956). However a nuclear gene suppressor (s) restores mutant C115 to normal, not only in growth rate but also in cytochrome content. Suppressor s has no detectable effect on mutants poky, mi-3 or C117. A nuclear suppressor gene of mi-3 (su^-) has been described by Gillie (1970):

Three mutants (cni-1, cni-2 and resp-1) obtained from strain inos-89601 have been characterized. They have rates of CN^--sensitive respiration that are significantly lower than that of the parent strain, and they grow more slowly than the parent strain on either sucrose or acetate media. The succinoxidase rates of mitochondrial fractions from the mutants are lower than in those from the wild-type. Mutants cni-1 and cni-2 yield mitochondria which exhibit cyanide- and antimycin A-insensitivity. Mutant cni-1 is deficient in cytochromes $a + a_3$ and has an excess of cytochrome c (Edwards et al., 1973; Edwards and Kwiecinski, 1973).

B. Malate Mutants

A number of malate mutants of *N. crassa* were examined by Munkres and Woodward (1966). It was shown that the physiological phenotype of these mutants is: (1) a consequence of the genetic alteration of the structure of malate dehydrogenase; and (2) an expression of functional and conformational alterations of mutant enzymes associated with mitochondrial membrane and "structural protein". Thus genetically-determined structural specificity of both enzyme and "structural protein" is important in determining the conformational and functional properties of enzyme-"structural protein" complexes and this leads to the concept of "locational specificity". In these C mutants, mitochondrial malate dehydrogenase is malfunctional (low affinity for malate) because of an abnormal conformation induced by association with the mitochondria. In a second class of mutants (K-mutants) the enzyme is malfunctional because of an altered subcellular location (Munkres *et al.*, 1970): in this case the normally mitochondrially-located enzyme is distributed throughout the cytoplasm. Studies of a repressible glyoxysomal isoenzyme and a constitutive mitochondrial malate dehydrogenase of the prototrophic and mutant strains, indicated that both isoenzymes are encoded by the same nuclear structural gene, and have polypeptide subunits in common (Benveniste and Munkres, 1970).

C. Isoleucine-valine Mutants

The synthesis of Val and Isoleu from pyruvate and α-ketobutyrate in *N. crassa* requires four enzymes which are found to be principally located in the mitochondria of the wild-type (Wagner and Bergquist, 1963; Wagner *et al.*, 1967; Bergquist *et al.*, 1969). These are, acetohydroxy acid synthetase, acetohydroxy acid reductoisomerase, dihydroxyacid dehydratase, and aminotransferase. Structural nuclear genes have been identified for three of these enzymes: *iv-1* for the dehydratase, *iv-2* for the reductoisomerase, and *iv-3* for the acetohydroxyacid synthetase (Wagner *et al.*, 1960; Caroline *et al.*, 1969; Altmiller and Wagner, 1970a, b). Mutations at these loci cause metabolic blocks with the accumulation of intermediates. Complementation between *iv* mutants occurs readily in certain combinations *in vivo* or *in vitro* (Wagner *et al.*, 1967). When the dehydratase purified from either the soluble or mitochondrial fractions of wild-type *N. crassa* is incubated *in vitro* with mitochondria isolated from an *iv-1* mutant strain, Val synthesis is restored to wild-type levels (Leiter *et al.*, 1971). The bulk of this restored synthesis occurs extramitochondrially, as dihydroxyvaline is secreted into the medium, and the subsequent conversion to ketovaline occurs outside the mitochondria. Ketovaline then diffuses back into the mitochondria where it is transaminated to give Val. The shunting of Val precursors in and out of mitochondria has been demonstrated by incubating mitochondria from *iv-1* and *iv-2* mutants in separate dialysis bags. This observation provides a basis for the understanding of growth complementation between organisms when these two *iv*-requiring mutants are cultured together in minimal medium.

Dihydroxyacid dehydratase from an *iv-1* mutant is nearly all found in the non-sedimentable fraction of the extracts (Altmiller and Wagner, 1970a), whereas 60–80% of the enzyme is mitochondrial in the wild-type. Both mitochondrial and cytoplasmic enzymes are the same molecular species (Altmiller and Wagner 1970b; Altmiller, 1972a). It is therefore proposed that an altered "locational specificity site" (Munkres and Woodward, 1966) of the mutant enzyme hinders its incorporation into mitochondria (Altmiller, 1972b).

D. Acetate non-utilizing Mutants

Flavell and Fincham (1968a) have described the isolation of sixty mutants of *N. crassa* which are unable to grow on acetate as the sole source of carbon, but which are able to utilize sucrose. Complementation tests enabled these mutants to be allocated to seven groups, each group representing a different gene (*acu-1* to *acu-7*). Mapping of six of the genes indicated that no two genes are closely linked. Mutations at four of these loci result in poor germination of ascospores. The levels of TCA cycle, glyoxylate cycle and some other enzymes were measured in a wild-type strain and in the seven groups of *acu* mutants both after growth in the presence of sucrose and after a subsequent 6 h incubation in a similar medium containing acetate as the sole carbon source (Flavell and Fincham, 1968b). Incubation of the wild-type strain with acetate medium resulted in a rise in the levels of activity of isocitrate lyase, malate synthase, phosphoenol pyruvate carboxykinase, acetyl-CoA synthetase, NADP-linked isocitrate dehydrogenase, citrate synthase and fumarase. The kinetics of derepression indicated an overall coordination between the synthesis of the two enzymes of the glyoxylate cycle and between various TCA cycle enzymes (Flavell and Woodward, 1970a). This coordination results from the metabolic repressors of each enzyme being in equilibrium or in constant ratio to one another, as the structural genes for each enzyme do not belong to the same operon.

Isocitrate lyase activity was absent in *acu-3* mutants; *acu-5* mutants lacked acetyl-CoA synthetase: hardly any α-ketoglutarate dehydrogenase could be detected in *acu-2* and *acu-7* mutants (Flavell and Fincham, 1968b). In *acu-6* mutants, phosphoenol-pyruvate carboxykinase was either very low or absent. No specific enzyme deficiencies could be attributed to the *acu-1* or *acu-4* mutations.

The metabolic activity of mitochondria may also in part regulate mitochondrial structure by regulating the expression of the nuclear genome or products of the nuclear genome. In a study of the regulation of TCA cycle enzymes (Woodward *et al.*, 1970; Flavell and Woodward, 1970b) have found that both nuclear gene (*cyt-1*) and cytoplasmic gene (*mi-1*) mutations that inhibit the activity of the mitochondrial electron-transport chain cause some derepression of TCA cycle enzymes during growth on a medium containing 1% sucrose. These results are consistent with the hypothesis that ATP, or the cellular energy level is involved in the regulation of TCA cycle enzymes. These respiratory mutants also have an excess of cytochrome c (a nuclear-specified enzyme) over the wild-type organisms. This is also likely to be an effect of the inhibition of electron transport, as growth of the wild-type with antimycin A leads to a similar elevated cellular content of cytochrome c.

E. Auxotrophic Mutants

A list of mutants of *N. crassa* is to be found in Sober (1970). Sterol mutants of *N. crassa* have been isolated by Grindle (1973). Unidentified sterols accumulate, and the mutants have reduced growth rates, abnormal morphology, poor fertility and resistance to a variety of polyene antibiotics. A mutant requiring inositol has been studied by Williams (1971). A choline-less strain was used by Luck (1963a, b) in his classical studies on mitochondrial growth (see p. 454). Several riboflavin-requiring mutants are available (Sober, 1970) but have not been used in mitochondrial investigations.

IV. *Spg* Mutants of *Aspergillus nidulans*

Mutants of *A. nidulans* (*sgp 1–5*) growing more slowly than the parent strain on glucose as the sole source of energy and carbon, have been isolated by Houghton (1970). It was suggested that the abnormal growth and impaired permeability of the mutants to TCA cycle intermediates are due to a limited availability of high energy compounds caused by a lesion in oxidative phosphorylation, but no direct evidence favouring this hypothesis was presented (Houghton, 1971). A list of mutants of *Aspergillus* which includes several acetate non-utilizing and riboflavin-requiring strains is also available (Sober, 1970).

V. Mutants of *Mucor bacilliformis*

Stable mutants of *M. bacilliformis*, which have lost the ability to grow filamentously and to sporulate, occur spontaneously with a frequency of about 1 in 3000 colonies (Storck and Morrill, 1971). These mutants have a yeast-like morphology on solid and liquid media, and reproduce by budding. Inability to form filaments and spores is accompanied by loss of cytochrome c oxidase activity. The mutant has some respiration but obtains most (if not all) of its energy by alcoholic fermentation. Acriflavin or phenylethanol treatment does not result in the production of respiratory deficient mutants of this species (and this is reminiscent of the situation with "*petite*-negative" yeasts, Bulder, 1964a, b), although these agents do prevent filamentous growth (Terenzi and Storck, 1969a, b).

VI. Altered Energy Metabolism in *Schizophyllum commune*

A malfunction of energy metabolism that differs from all others reported occurs in the tetrapolar basidiomycete *Schizophyllum commune* (Hoffman and Raper, 1971). This metabolic defect reduces assimilation of substrate to a half or less of its normal level, whilst leaving oxidative processes apparently unchanged. In this respect it resembles the lesion in the op_1 (p_9) mutants of yeast. The defect is related to the function of the "B incompatibility factor" and occurs in a specific phenotype that results from either of two genotypes; (a) in the common A heterokaryon ($A = B \neq$), in which two compatible wild B factors interact; and (b) in mutant B homokaryons. The nature of the lesion has been investigated with mutant B homokaryons. Unlike the wild-type, both respiration and growth of this mutant is insensitive to concentrations of DNP which are known to uncouple respiration of the wild-type organism. It was suggested that partial uncoupling of a normally functioning oxidative system from the process of energy conservation may occur in mutant B homokaryons. This suggestion was confirmed in a study of isolated mitochondria (Hoffman and Raper, 1972). It was found that mitochondria from the mutant showed a reduced stimulation of succinoxidase on the addition of ADP in the presence of Pi as compared to mitochondria from the wild-type. Further work is necessary to identify the exact nature of this lesion.

Transfer of mitochondria occurs after hyphal fusion in common-A, common-AB, and in fully compatible matings, but not in common-B matings. This process has been followed visually by means of phase-contrast microscopy after vital staining of one of the partners using cobalt chloride (Watrud and Ellingboe, 1973a). Mitochondria which have taken up cobalt *in vivo* may also be distinguished by their increased buoyant densities (Watrud and Ellingboe, 1973b).

8

Mitochondrial Ribosomes and RNAs in Eukaryotic Microooganisms

I. Mitochondrial Ribosomes and Ribosomal RNAs in Eukaryotic Microorganisms

Animal mitochondria contain ribosomes that sediment at about 55S and have a low buoyant density (1·40–1·42). The term "mini-ribosome" which has been applied to these is something of a misnomer, as these ribosomes, despite their slower sedimentation, have a similar mol. wt to those of *E. coli* (2·8 × 10⁶) (O'Brian *et al.*, 1974). They are dissociated to give 39S and 28S subunits and contain two RNA species that sediment at about 16S and 12S. Electrophoresis on acrylamide gels gives S_E values of 17S + 13S or 21S + 15S depending on ionic strength of buffer and temperature (Borst and Grivell, 1971). Separation of ribosomal proteins from *Xenopus* eggs has recently been achieved by two-dimensional acrylamide gel electrophoresis (D. Leister and I. Dawid, personal communication). A summary of these results for both cytoplasmic and mitochondrial ribosomes is presented (Table XLVIII). Whereas ribosomes from bacteria and eukaryotic cytoplasm contain a 5S RNA component in addition to the two high mol. wt species, such a component has not been found in animal or fungal mitochondria, but its presence has not been rigorously excluded

Table XLVIII. Values for the total mol. wts of the ribosome subunits in *Xenopus* oocytes. (Reproduced by permission of D. Leister and I. Dawid.)

Ribosome	Subunit	*No. of proteins	Mean mol. wt	RNA (mol. wt × 10⁻⁶)	†Particle (total mol. wt × 10⁻⁶)
Cytoplasmic	Large	37	22000	1·5	2·6
	Small	34	21000	0·7	1·4
Mitochondrial	Large	41	27000	0·58	1·8
	Small	43	33000	0·32	1·6

* Analysed by two dimensional acrylamide gel electrophoresis.
† Values calculated assume that each protein is only represented once.

as it may cosediment and coelectrophorese with mitochondrial transfer RNAs. 5S RNA has been reported in *Tetrahymena* (Suyama, 1969) and its mitochondrial origin is indicated by hybridization with mt DNA. In the discussion of the mitochondrial ribosomes of eukaryotic microorganisms it will become clear that these differ in many respects from the mitochondrial ribosomes of both animal tissues and bacteria.

A. Mitochondrial Ribosomes and Ribosmal RNAs in Yeasts

Mitochondrial ribosomes have been isolated from several species of yeasts after detergent lysis of purified mitochondrial fractions. The degree of contamination of isolated ribosomes by membranes can be assessed by their E_{260nm}/E_{280nm} ratios: values as high as 1·88 have been reported (Morimoto and Halvorson, 1971). They have been identified as particles which sediment through sucrose gradients at a rate which usually lies between that of their extramitochondrial counterparts and that of ribosomes from *E. coli* (Table XLIX). In contrast to cytoplasmic ribosomes, the mitochondrial ribosomes are split into subunits when the Mg^{2+} concentration is lowered to 0·1 mM. Differences in structure between cytoplasmic and mitochondrial ribosomes in both *S. cerevisiae* (Morimoto and Halvorson, 1971) and *C. utilis* (Yu *et al.*, 1972a) lead to differences in thermal denaturation profiles. After glutaraldehyde fixation, buoyant densities of cytoplasmic and mitochondrial ribosomes of *C. utilis* are 1·53 and 1·48 g/cm³ respectively. Electron microscopy of negatively-stained mitochondrial ribosomes from *C. utilis* (Vignais *et al.*, 1972) show bipartite profiles about 26·5 × 21 × 20 nm (Fig. 108). The 50S subunits display rounded profiles bearing a conspicuous knob-like projection, reminiscent of the large bacterial subunit, whereas the 36S subunits show angular profiles.

Although chloroplast and *E. coli* ribosome subunits can interact to form hybrid ribosomes which are active in polyPhe synthesis directed by poly-(U), yeast mitochondrial ribosomal subunits do not form hybrid ribosomes with either *E. coli* or chloroplast subunits (Grivell and Walg, 1972).

Incubation of isolated mitochondria from *C. utilis* with (^{14}C)-Leu results in the labelling of the 72S mitochondrial ribosomes (Vignais *et al.*, 1972); incorporation of the label is inhibited by chloramphenicol and resistant to cyloheximide. Striping of the incorporated radioactivity from the 72S mitochondrial ribosomes by puromycin, indicates that the (^{14}C)-Leu is incorporated into nascent polypeptide chains. Mitochondrial and cytoplasmic ribosomes can also be distinguished by pulse-labelling of intact cells of *S. cerevisiae* with labelled amino acids in the presence of cycloheximide (Yu *et al.*, 1972b), as under these conditions, only the

Table XLIX.   Mitochondrial ribosomes and ribosomal RNAs in yeasts.

Organism	Ribosome sedimentation coefficients (S)			Ribosomal RNA sedimentation coefficients (S)			Reference
	Ribosome	Large subunit	Small subunit	Large	Small	Mole percent (G + C)	
S. cerevisiae, wild-type W, D273–10B				22·6	16		Wintersberger, 1966a, b, 1967
S. cerevisiae				22·4 (24·6)	17·8 (16·2)		Rogers et al., 1967
S. cerevisiae Fleishmann				23 (24·7)	16 (17·4)		Leon and Mahler, 1968
S. cerevisiae, D261	72 (80)	50 (60)	38 (38)				Schmitt, 1969, 1970
S. cerevisiae. A2111D				25 (25)	15–16 (15–16)		Steinschneider, 1969
S. cerevisiae.				22 (26)	15 (17)	25–27 (45–48)	Fauman et al., 1969

							Reference
S. cerevisiae. iso-N	75 (80)			25 (26)	17 (19)		Stegeman et al., 1970
S. cerevisiae Y55	80 (80)	60				30·2–33·3 (45·2–52·6)	Morimoto and Halvorson, 1971; Morimoto et al., 1971
S. cerevisiae Hansen	80 (80)					38·6 (47·4)	Yu et al., 1972a
S. carlsbergensis. NCYC 74				22	15		De Kloet et al., 1971
S. carlsbergensis NCYC 74	74	50	37	22	15	23 (47·6)	Grivell et al., 1971b; Reijnders et al., 1972
C. krusei	76 (80)	53 (60)	35 (40)				Kaempfer, 1969
C. utilis CBS 1516	72 (78)	50 (61)	36 (37)	21 (25)	16 (17)	33 (50)	Vignais and Huet, 1970; Vignais et al., 1969, 1972
C. parapsilosis	70 (80)	50 (60)	30 (38)			34 (46)	Yu et al., 1972b

Values in parentheses refer to corresponding cytoplasmic ribosomes and RNAs

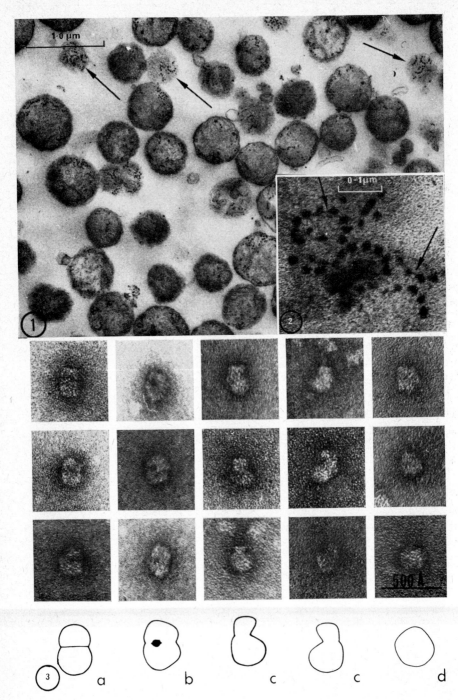

mitochondrial ribosomes become labelled; isotope incorporation is chloramphenicol sensitive. Isolated yeast ribosomes as active in poly-(U)-directed polyPhe synthesis as bacterial ribosomes have been obtained by Grivell *et al.* (1971b). Synthesis is completely dependent on the presence of both subunits, and thus the possibility that the 50S particle represented a "mini-ribosome" in yeast mitochondria was excluded. Scragg *et al.* (1971) have shown that yeast mitochondrial ribosomes can read a bacterial mRNA.

The RNA species derived from mitochondrial ribosomes can be distinguished from those of cytoplasmic ribosomes on the basis of their sedimentation coefficients and molar $(G + C)$ contents (Table XLIX). The large and small RNA species are derived from the large and small ribosome subunits respectively. The low $(G + C)$ contents of mitochondrial RNAs broadly reflect the mole percent $(G + C)$ of the mt DNAs. The mol. wt of *C. utilis* RNAs derived from sedimentation data in EDTA—supplemented gradients were $1·21 \times 10^6$ and $0·71 \times 10^6$ for the 21S and 16S mitochondrial RNAs as opposed to $1·67 \times 10^6$ and $0·80 \times 10^6$ for the 25S and 17S cytoplasmic RNAs (Vignais *et al.*, 1972). Differences in electrophoretic mobilities of RNAs from mitochondria and cytoplasm of *S. cerevisiae* were observed by Stegeman *et al.* (1970), who calculated mol. wts of the mitochondrial species at $1·25 \times 10^6$ and $0·64 \times 10^6$, compared with $1·4 \times 10^6$ and $0·77 \times 10^6$ for the cytoplasmic RNAs. Differences in electrophoretic mobility between the extra- and intra-mitochondrial rRNAs were not observed by DeKloet *et al.* (1971), in extracts of *S. carlsbergensis*, but were seen in both *C. utilis* and *S. cerevisiae* by Yu *et al.* (1972a, b). The behaviour of yeast mitochondrial rRNA on electrophoresis in acrylamide gels is not consistent with its sedimentation properties (the electrophoretic mobility is lower than would be expected, and the change of mobility in different buffers is far greater than the changes in sedimentation coefficients, Forrester *et al.*, 1970).

The electrophoretic mobility relative to cytoplasmic or *E. coli* marker RNA is also strongly dependent on ionic strength and temperature, and exploitation of these properties may be used to separate RNA mixtures that cannot be resolved under standard conditions (Grivell *et al.*, 1971a, b, c). These anomalous properties can be explained on the assumption that these mitochondrial RNAs have a more loosely folded structure,

---

Fɪɢ. 108. (1) Thin section of isolated mitochondria from *Candida utilis* showing numerous mitochondrial ribosomes arranged along the cristae and inner-membrane surfaces. Many mitochondrial ribosomes appear in whorls characteristic of polysomes. Arrows point to grazing sections of mitochondria; polysomes are particularly evident here. Contamination by nonmitochondrial material is negligible. (2) Mitochondrial polysomes in which a dense line can be observed between neighbouring mitoribosomes (arrows). (3) Selected images of mitochondrial ribosomes. Examples of forms a, b, d, and of the two enantiomorphic images of form c are shown. A schematic drawing accompanies each form. (Reproduced with permission from Vignais *et al.*, 1972.)

than for instance *E. coli* rRNA, and they unfold more easily under conditions of low ionic strength or elevated temperature. Caution is necessary to interpret the relative mobility of mitochondrial RNAs on gels in terms of mol. wts: values obtained at various temperatures for those from *S. carlsbergensis* together with values obtained by methods not affected by the secondary structure of the RNA are given in Table L

Table L. Molecular weights of the mitochondrial rRNAs of *S. carlsbergensis* determined by methods not affected by RNA secondary structure. (Reproduced with permission from Reijnders *et al.*, 1973a)
Results are given ±S.D.

Method	10^{-6} × Molecular weight	
	Larger rRNA	Small rRNA
Sedimentation equilibrium	1·26 ± 0·13	0·68 ± 0·08
Gel electrophoresis in urea at 60°C	1·30 ± 0·04	0·70 ± 0·03
Gel electrophoresis in 98 % formamide at 30°C	1·34 ± 0·04	0·71 ± 0·03
Electron microscopy of dimethyl sulphoxide-denatured RNA	1·22 ± 0·10	0·60 ± 0·06

(Reijnders *et al.*, 1973a). No 5S mt RNA has been found in yeast (Reijnders and Borst, 1972) and no methylation of yeast mitochondrial rRNAs has been detected (Grivell, 1974).

Hybridization studies with yeast mitochondrial RNAs and mt DNA have yielded a range of values from 0% to 14% (DeKloet *et al.*, 1971; Wintersberger, 1967; Wintersberger and Vienhauser, 1968; Casey *et al.*, 1972) for the proportion of mt DNA complementary to mitochondrial rRNA. Possible pitfalls in hybridization experiments have been outlined (Reijnders *et al.*, 1972), and in *S. carlsbergensis* it has been shown that maximally only 2·4% of mt DNA hybridizes with RNA from highly purified mitochondrial ribosomes. This result suggests the presence of only one gene for each of the rRNAs on mt DNA. It does not exclude the rather unlikely possibility that the population of mt DNAs is heterogeneous. Separate plateaus observed for the rRNAs from the purified ribosome subunits show that substantial base sequence homologies between these two rRNAs are absent.

The mitochondrial content of RNA is not altered by growth in the presence of chloramphenicol but is drastically reduced in some cytoplasmic *petites* (Wintersberger, 1966a, b). Mitochondrial rRNA is not detectable in unsaturated fatty acid and ergosterol-depleted anaerobically-grown yeast (Forrester *et al.*, 1971). The mitochondrial RNAs of wild-type (*grande*) and a spontaneously mutated cytoplasmic *petite* strain of *S. cerevisiae* have been compared by hybridization to mt DNA (Fauman *et al.*, 1973).

Increasing amounts of (^{3}H)-labelled mitochondrial RNAs were hybridized to filter-bound *grande* and *petite* mt DNAs until saturation levels were obtained. Values were as follows: *grande* (^{3}H) mt RNA—*grande* mt DNA, 15%; —*petite* mt DNA, 7–8%; *petite* (^{3}H) mt RNA—*grande* mt DNA, 13%; —*petite* mt DNA, 26%. Hybridization competition studies showed that *grande* and *petite* mt RNAs each contained sequences not present in the other. The melting temperatures of the four hybrids were also different, and it was concluded that *grande* and *petite* mitochondria each contain transcripts of mt DNA not present in the other, and this probably reflects changes in the mt DNA. An alternative explanation was also considered; that the *petite* is a simple deletion of the wild-type but that a wider range of transcripts of mt DNA were present in the preparations from the *petite*.

In a gradient containing 2 M–LiCl, both *E. coli* ribosomes and mitochondrial ribosomes from *S. cerevisiae* give 40–41S and 24–25S "core" particles arising respectively from large and small subunits (Schmitt, 1971). This finding suggests that the mode of biosynthesis of ribosomes in mitochondria may resemble that in *E. coli*.

B. Mitochondrial Ribosomes and Ribosomal RNAs in *Neurospora crassa*

The mitochondrial ribosomes of *N. crassa* have often been reported to have a sedimentation coefficient of 73·2S, whereas cytoplasmic ribosomes sediment at 76·9S (Küntzel and Noll, 1967; Küntzel, 1969c; Noll, 1970). However, isolation of the monomer from polysome preparations has recently been described (Agsteribbe *et al.*, 1974) and this monomer sedimented at 80S even after run off of nascent peptide chains. Whereas fractionation in EDTA-containing media gave ribosomes with a sedimentation value of 73S, Mg^{2+}-containing media gave a mixture of 73S and 80S ribosomes. Addition of the RNAase inhibitor heparin, to the medium in which the mitochondria were lysed, gave preparations of exclusively 80S monosomes. Only under conditions where polysomes were observed were the 80S ribosomes observed, suggesting that the 80S particles are the native mitochondrial ribosomes and that the 73S particles arise through the action of RNAases during the course of isolation.

As is the case with yeasts, mitochondrial ribosomes can also be distinguished from cytoplasmic ribosomes on the basis of their sensitivity to dissociation to subunits by Mg^{2+} (Rifkin *et al.*, 1967). A protein factor can be isolated from *E. coli* ribosomes which promotes dissociation of bacterial ribosomes into subunits; incubation of mitochondrial ribosomes from *N. crassa* with this dissociation factor also leads to the formation of subunits (Agsteribbe and Kroon, 1973). Under similar conditions no dissociation of cytoplasmic ribosomes from *N. crassa* was observed.

Discrimination between the cytoplasmic and mitochondrial ribosomes is also possible by assay of the peptidyl transferase activities; with mitochondrial ribosomes this reaction is inhibited by chloramphenicol but is unaffected by anisomycin (DeVries *et al.*, 1971). The large and small subunits derived from mitochondrial ribosomes sediment at 50–52S and 37S respectively, whereas their cytoplasmic counterparts sediment at 60S and 37S. Mitochondrial polysomes with sedimentation coefficients of 103, 134, 160 and 186S correspond to dimers, trimers, tetramers and pentamers respectively (Küntzel and Noll, 1967). Incorporation of (^{14}C)-Leu by intact mitochondria leads to the association of acid-insoluble radioactive product with mitochondrial polysomes (Küntzel and Noll, 1967). However, when the mitochondrial suspension was treated with puromycin after amino acid incorporation, no radioactivity could be detected in either monosomes or polysomes (Neupert *et al.*, 1969a). Thus isolated mitochondria incorporate (^{14}C)-Leu into the nascent chains of growing polypeptides and not into ribosomal proteins. The *in vivo* incorporation of (^{14}C)-amino acids into the protein of mitochondrial ribosomes of *N. crassa* is sensitive to cycloheximide and insensitive to chloramphenicol (Küntzel, 1969a; Neupert *et al.*, 1969b). Mitochondrial ribosomes from *N. crassa* are active in poly-(U)-directed polyPhe synthesis; components in the submitochondrial system can be freely interchanged with components from *E. coli*, but not with components from *N. crassa* cytoplasm (Küntzel, 1969; Table LI).

The rRNAs of *N. crassa* mitochondria sediment at 23 and 16S, whereas those of the cytoplasmic ribosomes sediment at 26 and 17S. (Dure *et al.*, 1967; Küntzel and Noll, 1967, but see also Rifkin *et al.*, 1967). The base composition of mitochondrial rRNA corresponds to 35–38% (G+C), whereas that of cytoplasmic RNA is 49–50% (G+C) (Rifkin *et al.*, 1967; Küntzel and Noll, 1967). The observations of anomalous behaviour of yeast rRNAs on electrophoresis are paralleled with *N. crassa* rRNAs (Edelman *et al.*, 1971): the relative mobilities of mitochondrial rRNAs are severely retarded at low salt concentration causing mitochondrial RNA to migrate more slowly than cytoplasmic RNA (Fig. 109). Thus accurate extrapolations from electrophoretic mobilities to values for mol. wts cannot be made. Thermal denaturation patterns of the mitochondrial rRNA are markedly different from those of the cytoplasmic rRNA; this indicates distinct conformations for the two RNAs in solution.

It was concluded from early hybridization studies that *N. crassa* mt DNA has cistrons coding for both 25S and 19S mitochondrial RNAs and that there are four of these per molecule of mt DNA, (Wood and Luck, 1969). However, using improved hybridization techniques, Schäfer and Küntzel (1972) have shown that rRNA extracted from purified mitochondrial ribosomes is complementary to only 2·5% of mt DNA. High mol. wt RNA isolated from whole mitochondria contains minor RNA

Table LI. Incorporation of (^{14}C)-Phe from *E. coli* Phe-tRNA into polyphenyl-alanine in various cell free systems. (Reproduced with permission from Küntzel, 1969b.)

Supernatant enzymes	Ribosomes	Incorporation of p moles/mg RNA	(^{14}C)-Phe %
E. coli	*E. coli*	9·8	100
	N.c. mitochondria	9·2	94
	N.c. cytoplasm	0·2	2
	rat liver cytoplasm	<0·1	< 1
N.c. mitochondria	*E. coli*	4·4	108
	N.c. mitochondria	4·1	100
	N.c. cytoplasm	0·9	23
	rat liver cytoplasm	0·8	19
N.c. cytoplasm	*E. coli*	0·1	1
	N.c. mitochondria	2·8	38
	N.c. cytoplasm	7·4	100
	rat liver cytoplasm	6·8	92
Rat liver cytoplasm	*E. coli*	< 0·1	< 1
	N.c. mitochondria	0·8	7
	N.c. cytoplasm	9·8	90
	rat liver cytoplasm	11·0	100

The mixtures contained 12 O.D.$_{260}$ ribosomes per ml and the following amounts of supernatant enzymes (mg protein/ml): *E. coli* (1·2), *Neurospora* mitochondria (0·9), *Neurospora* cytoplasm (1·1) and rat liver cytoplasm (1·1). The mixtures (0·5 ml) which were incubated at 37°C for 10 min also contained (μmoles/ml) tris pH 7·5 (100); KCl (50), Mg acetate (20), spermidine (2), β-mercaptoethanol (16), glutathione (2), GTP (0·5), ATP (5), phosphoenol-pyruvate (5), pyruvate kinase (20 μg) nineteen amino acids minus Phe (0·04 each), Poly-(U) (120 μg), cytoplasmic tRNA from *Neurospora* (0·6 mg) L-(^{14}C)-Phe (0·5 μCi, sp. act. 325 mCi/mM).

species which saturate an additional 10% of the mitochondrial genome, and it is now agreed that rRNA cistrons are not redundant in *Neurospora* mitochondria.

Non-ribosomal RNA has been isolated from mitochondrial polysomes; this messenger RNA can be used to programme *E. coli* ribosomes for protein synthesis (Agsteribbe *et al.*, 1974). Mitochondrial 5S RNA has not been detected in *N. crassa* (Lizardi and Luck, 1971)) but a 4S RNA possessed 5S RNA activity, in that it promoted reconstitution of 50S bacterial subunits (H. Küntzel and V. A. Erdmann, unpublished).

The basis of the abnormal mitochondrial phenotype of the cytoplasmic mutant *poky* (*mi-1*) is an interference in the synthesis or assembly of the small subunits of mitochondrial ribosomes (Rifkin and Luck, 1971). Thus, during exponential growth, mitochondrial fractions were deficient in small subunits. and the ratio of 25S/19S mitochondrial rRNA was higher than in wild-type. In the stationary phase these small subunits are more abundant and are present in ribosomal monomers. This change was correlated with the return of mitochondrial cytochrome content to amounts approaching those of the wild-type mitochondria. Other slow growing mutants with abnormal cytochromes (*abn-1*, *mi-3*, C117 and STL-7 (*lys-5*)) all showed normal ratios of large/small ribosomal subunits. A specific type of mitochondrial RNA (33S) has been isolated from the *abn-1*

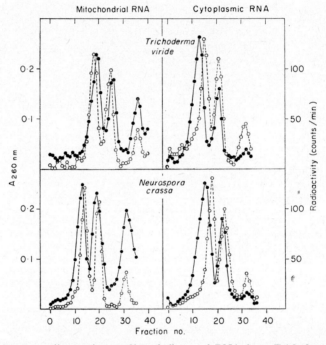

Fig. 109. Sucrose sedimentation profiles of ribosomal RNA from *Trichoderma, Neurospora*, and *E. coli*. RNA samples were layered over 5 to 20% linear sucrose gradients made up in 50 mM–NaCl, 50 mM–acetate buffer, pH 5·7. Centrifugation was carried out for 5·5 h at 38000 rev/min in the SW 39 rotor of a Spinco preparative ultracentrifuge at a chamber temperature of 4°C. Fractions were collected through a hole punctured in the bottom of the tube. (^{3}H)-labelled *E. coli* rRNA, (O– – – –O), was included in each gradient; counts were for 10 min. Absorbance, ●——●. (Reproduced with permission from Edelman *et al.*, 1971.)

mutant by Küntzel *et al.* (1973; 1974, see p. 327). Alteration of mt DNA in this mutant leads to the expression of a mitochondrial gene for 33S RNA and the formation of virus-like particles containing mitochondrial gene products.

C. Mitochondrial Ribosomes and Ribosomal RNAs in *Aspergillus nidulans*

It has been reported that the sedimentation coefficients of mitochondrial ribosomes and their subunits in *A. nidulans* are 67, 50 and 32 S (Edelman *et al.*, 1970), but it is suggested that correction of these values would lead to identical values to those for *N. crassa* and yeasts (Borst and Grivell, 1971). The rRNAs from mitochondrial ribosomes sediment at 23·5 and 15·5S, whereas those from cytoplasmic ribosomes sediment at 26·5 and 17·0S. Again, anomalous mobility on electrophoresis was observed (Edelman *et al.*, 1971). Thermal denaturation measurements enabled calculation of the contributions of (A + U) and (G + C) residues to the ordered structure of rRNAs; these were evaluated as 27 and 32% for the heavy and light mitochondrial components and 55·5 and 51% for the corresponding cytoplasmic components (Edelman *et al.*, 1970). The nucleotide fragments produced on digestion of *A. nidulans* mitochondrial and cytoplasmic rRNAs have been compared by two-dimensional iono-phoresis (Verma *et al.*, 1971a). Major differences were found between the two RNA species and between the heavy and light components within each rRNA sample. Verma *et al.* (1970) have determined the chain lengths of rRNAs by electron microscopy. The mean chain lengths of the two mitochondrial rRNA components were 0·47 and 0·91 µm, whereas those of cytoplasmic ribosomes were 0·40 and 0·72 µm. Assuming an inter-nucleotide spread of 2·45 Å, these values correspond to mol. wts of $1·27 \times 10^6$ and $0·66 \times 10^6$ for the two mitochondrial RNAs. The precision of these estimates depends upon the actual internucleotide spacing in the urea-spread RNAs, and this value is not constant for all RNAs (Granboulan and Scherrer, 1969). However it is clear that mitochondrial rRNA is larger in size than bacterial-type rRNA but smaller than cytoplasmic rRNA.

D. Mitochondrial Ribosomal RNAs in *Trichoderma viride*

The anomalous electrophoretic mobility properties of mitochondrial rRNAs from *T. viride* have been described (Edelman *et al.*, 1971). The (G + C) content of the heavy and light components of the mitochondrial rRNAs is 31·5 and 35·5% respectively, whereas that of the cytoplasmic counterparts is 50 and 49%. CD measurements indicate that mitochondrial rRNA has a less ordered structure than cytoplasmic rRNA (Verma *et al.*, 1971b).

E. Mitochondrial Ribosomes and Ribosomal RNAs in *Tetrahymena pyriformis*

An 80S mitochondrial ribosome has been identified in extracts of the mitochondria of *T. pyriformis* (Chi and Suyama, 1968, 1970; Suyama, 1969). Lowering the Mg^{2+} concentration to 0·1 mM led to dissociation of the 80S cytoplasmic ribosomes into 60 and 40S subunits.

Mitochondrial ribosomes did not dissociate under these conditions, but addition of EDTA or further lowering the Mg^{2+} concentration gave rise to a single 55S subunit peak. Buoyant densities of mitochondrial and cytoplasmic 80S ribosomes after glutaraldehyde fixation were unusually low (1·46 and 1·56 g/cm³ respectively). The 55S subunits yielded two density species (1·52 and 1·46 g/cm³) in a 1 : 2 ratio. The 60 and 40S cytoplasmic ribosome subunits had densities 1·57 and 1·53 g/cm³ respectively. Although the mitochondrial and cytoplasmic ribosomes are both 80S, they can be distinguished by gel electrophoresis, the mitochondrial ribosomes migrating more slowly (Stevens et al., 1974). The two classes of ribosome also have different fine structure as revealed by negative staining. The cytoplasmic ribosomes measure $27·5 \times 23·0$ nm, with an electron opaque spot which appears on the left-hand side when the small subunit is on top. The mitochondrial ribosomes measure $37·0 \times 24·0$ nm, an electron dense line delineates the two almost equal subunits. An earlier report that the mitochondrial ribosomes appear smaller than the cytoplasmic ribosomes in T. pyriformis (Swift et al., 1968) should be discounted. The 55S subunits have a similar morphology and cannot be separated into two classes by electrophoresis.

The numbers of mitochondrial ribosomes observed in thin sections of T. pyriformis vary from 160/μm³ in exponentially growing organisms, to 50/μm³ in cells from the stationary phase of growth.

Plots of S values against electrophoretic mobility distinguishes eukaryotic and prokaryotic ribosomes; on this basis the ribosomes of T. pyriformis do not clearly fall into either class; further work on ribosomes of other ciliates should prove very interesting (Stevens et al., 1974).

Mitochondrial and cytoplasmic rRNAs isolated from their respective 80S ribosomes were 21 and 14S (mol. wts 0·9 and $0·47 \times 10^6$, Reijnders et al., 1973b), and 26 and 17S respectively. 5 and 4S RNAs were also present (Chi and Suyama, 1968, 1970; Suyama, 1969). The (G + C) contents of the large and small mitochondrial rRNAs were 27·9% and 30·6% respectively, compared with 43·2% and 49·2% for the corresponding cytoplasmic species. 21 and 14S mitochondrial rRNAs were complementary to 3·8% and 1·9% of the mt DNA respectively, whereas no significant complementarity exists between mt DNA and cytoplasmic rRNAs.

The hybridization value of 6·3% (RNA/DNA $\times$ 100) for mitochondria rRNA is equivalent to a mol. wt of approximately 2×10^6 which clearly exceeds the total mol. wt of rRNA ($=1·35 \times 10^6$); thus there is a possibility of the existence of redundant rRNA cistrons in mt DNA (but see p. 394).

F. Mitochondrial Ribosomes in Crithidia luciliae

Two classes of ribosomes differing in sedimentation coefficients and sensitivities to Mg^{2+} have been isolated from the trypanosomatid proto-zoan, *Crithidia lucilhae* (Thirion and Laub, 1972a, b). The mitochondrial ribosomes (50S) are chloramphenicol and erythromycin sensitive, but cycloheximide-insensitive as shown by incorporation of (^{3}H)-Leu into the proteins of isolated mitochondria. Growth for 48 h with ethidium bromide (20 µg/ml) gave dyskinetoplastic organisms: mitochondria from these organisms gave a five-fold lower incorporation rate of labelled Leu.

G. Mitochondrial Ribosomes and Ribosomal RNAs in *Euglena gracilis*

The sedimentation coefficients of ribosomes from the mitochondria, cytoplasm, and chloroplasts of *E. gracilis* (var. bacillaris and strain z) are 71–72S, 86–87S and 69S respectively (Avadhani and Buetow, 1972a). Nine or ten peaks of mitochondrial polyribosomes were seen in gradients, and were disaggregated by EDTA or RNAase (Fig. 110). Subunits of the mitochondrial ribosome sediment at 50 and 32S and the rRNA of these subunits is 21·4 and 16S (Table LII). These values are higher than the values 14S and 11S published by Krawiec and Eisenstadt (1970b). The (G + C) content of mitochondrial rRNA is 29·8% as opposed to a value of 50·0 for chloroplast rRNA and 55·7 for the cytoplasmic species (Avadhani and Buetow, 1972a). The mitochondrial polyribosomes can participate directly in protein synthesis without supplementary messenger RNA being added; this system is chloramphenicol sensitive and cycloheximide resistant.

II. Mitochondrial Transfer RNAs in Eukaryotic Microorganisms

Mitochondria contain tRNA species that are distinct from those in the cell sap: hybridization studies indicate that these differences are not due to secondary alterations of cytoplasmic tRNAs imported into the mito-chondria. The size of tRNAs has not measurably decreased during evolution (Attardi and Attardi, 1971; Dawid and Chase, 1972). In Hela cells, 12 genes appear to be the upper limit of the coding capacity of mt DNA for tRNA species; a minimum of 33 tRNAs are necessary to read the 61 amino acid-specifying codons. It is not yet completely certain that the mitochondria of animal cells utilize only this set of endogenously-produced tRNAs, although the possibility that they import the remaining species from the cytosol seems unlikely from results obtained on the incorporation of amino acids into proteins (Attardi *et al.*, 1974). At least one mitochondrial-specific methylase for tRNAs has been shown in Hela cells (Klagsbrun, 1973).

A. tRNAs of *Neurospora crassa* mitochondria

Counter current distribution and reversed-phase column chromatography

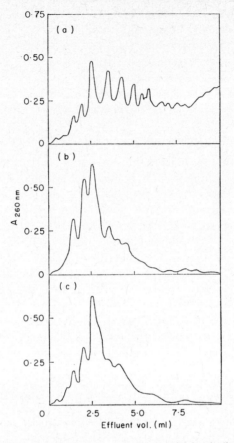

FIG. 110. Sucrose-density-gradient pattern of mitochondrial polyribosomes from *Euglena gracilis* (SM-L1). (a) Untreated polyribosomes; (b) EDTA-treated polyribosomes; (c) RNAase-treated polyribosomes. (Reproduced with permission from Avadhani and Buetow, 1972a.)

has shown the presence in *N. crassa* of distinct and separable species of mitochondrial and cytoplasmic tRNAs for fifteen amino acids (Barnett and Epler, 1966; Barnett and Brown, 1967; Brown and Novelli, 1968; Epler, 1969). The cytoplasmic Leu-tRNAs correspond to UC-, UG-, and ambiguously to U-containing polymers in ribosomal binding, whereas mitochondrial Leu-tRNA responds to only UC-containing polymers, and hence the coding properties are quite different (Epler and Barnett, 1967). *Neurospora* mitochondria were also found to contain amino acyl synthetases for all fifteen amino acids tested (Barnett *et al.*, 1967). Comparison of mitochondrial and cytoplasmic Asp-, Phe-, and Leu-tRNA synthetases by column chromatography, and for utilization of tRNAs isolated from both

Table LII. Sedimentation coefficients of ribosomes and rRNA.
(Reproduced with permission from Avadhani and Buetow, 1972a.)

Cytoplasmic, chloroplast and mitochondrial ribosomes and rRNA were isolated from *Euglena* (variety *bacillaris*). Sedimentation coefficients ±s.D. were calculated on the basis of four independent determinations. The values given in parentheses are the corresponding values for various ribosomes and rRNA from *Euglena* (strain z).

Type of particle or molecule	Sedimentation coefficients (S)		
	Cytoplasm	Chloroplast	Mitochondria
Ribosome			
Monomeric	86·9 ± 0·5 (86·3)	69·4 ± 0·9 (69·2)	71·3 ± 0·9 (71·7)
Large subunit	67·2 ± 0·9 (67·4)	49·8 ± 0·7 (50·1)	50·1 ± 0·6 (50·4)
Small subunit	45·9 ± 0·6 (50·2)	29·3 ± 0·5 (29·4)	32·4 ± 0·7 (32.2)
Ribosomal RNA			
Large subunit	24·4 ± 0·5 (24·6)	22·1 ± 0·5 (22·4)	21·4 ± 0·6 (21·3)
Small subunit	20·1 ± 0·4 (19·9)	16·5 ± 0·6 (16·8)	15·9 ± 0·5 (16·3)

mitochondria and cytoplasm, confirmed that mitochondria contain unique enzymes and tRNAs for these three amino acids. The absence of detectable mitochondrial Leu-tRNA synthetase activity from mycelial and mitochondrial extracts of mutant 45208 t, a temperature-sensitive auxotrophe, together with the altered enzymic characteristics of the cytoplasmic enzyme, suggest that both the mitochondrial and the cytoplasmic enzymes are functionally controlled by the same nuclear gene (Gross *et al.*, 1968). It was suggested that the differing physical and enzymic properties of the two enzymes results from modification of the cytoplasmic enzyme after import into the mitochondria. Further evidence for this proposal was obtained by a study of "reversion" of the 45298 t mutation which resulted not only in the reappearance of mt Leu-tRNA synthetase activity but also in the production of a cytoplasmic synthetase with an affinity for Leu intermediate between mutant and wild-type (Weeks and Gross, 1971).

As the mechanism for the initiation of protein synthesis in bacteria involves formyl-Met-tRNA, both mitochondrial and cytoplasmic extracts were analysed for this amino acyl tRNA (Epler *et al.*, 1970). It was shown that a Met-tRNA from mitochondria can be formylated by an extract of *E. coli*, but that the corresponding cytoplasmic tRNA cannot be. A second mitochondrial Met-tRNA was also present which cannot be formylated. Mitochondrial Met-tRNA synthetase is specific, and aminoacylates only the mitochondrial tRNAs and not the cytoplasmic one. The mitochondrial extracts contain a formylase which reacts only with the mitochondrial formyl Met-tRNA. A further similarity between the tRNAs of organelles and of prokaryotes is that both do not contain a fluorescent base which is present in the cytoplasmic tRNAs of eukaryotes adjacent to the 3'adenosine of the anticodon and necessary for normal recognition (Fairfield and Barnett, 1971).

Poky (mi-1) mutants of N. crassa contain different mitochondrial amino-acyl tRNA species from wild-type strains (Brambl and Woodward, 1972).

B. tRNAs of Yeast Mitochondria

Wintersberger (1965) showed that yeast mitochondria contain amino acyl tRNA synthetases and tRNAs. Mitochondrial tRNAs were compared with their cytoplasmic counterparts (Accoceberry and Stahl, 1971). Mito-chondrial Arg-, Lys-, and Phe-tRNAs are chromatographically distinct from those found in the cytoplasm, whereas mitochondrial Isoleu-tRNA contains only one species identical with the corresponding cytoplasmic tRNA. A method for the preparation of total mitochondrial aminoacyl tRNA synthetases has been published (Accoceberry et al., 1973). Mito-chondrial (^{3}H)-formyl Met-tRNA was prepared using a formyl THFA: Met-tRNA transformylase from the mitochondria of S. cerevisiae (Halbreich and Rabinowitz, 1971). This formyl-Met-tRNA hybridized with mt DNA but not with yeast nuclear of E. coli DNAs. Unlabelled mitochondrial tRNA (but not cytoplasmic tRNA) competed in this reaction. Yeast cytoplasmic formyl Met-tRNA (formylated with the E. coli enzyme) and E. coli formyl-Met-tRNA do not hybridize with mt DNA. It was concluded that yeast tRNA^{fMet} is a gene product of the mitochondrial genome. The hybridization of mitochondrial Leu-tRNA and Val-tRNA with mt DNA and nuclear DNA of a grande strain of yeast was studied by Casey et al. (1972). The tRNAs were charged in vitro with amino acids of high specific radioactivities, and hybridization was achieved in formamide at pH 5·0 and 33°C to minimize deacylation of the tRNA. Both mitochondrial (^{3}H)-Leu- and (^{3}H)-Val-tRNAs hybridized with mt DNA, but not with yeast nuclear DNA or E. coli DNA. Unlabelled yeast mitochondrial tRNAs competed in the reaction, whereas the cytoplasmic tRNAs of yeast did not. Apparent hybridization saturation curves were observed, and RNAase and alkaline digestion of the (^{3}H)-Val-tRNA-mt DNA hybrid yielded (^{3}H)-valyl-adenosine and (^{3}H)-Val respectively, indicating the functional validity of the hybridization system. Subclones of spontaneously-mutated petite mutant (RI-6) were heterogeneous with respect to hybridization of their mt DNAs with grande mitochondrial Val- and Leu-tRNAs (Cohen et al., 1972). Hybridization of mt DNA of all petite subclones with grande mitochondrial Leu-tRNAs, resulted in apparent saturation levels that were equal to, or usually 2–6 times greater than that obtained with grande mt DNA.

These data suggest reiteration of segments of petite mt DNA containing the Leu-tRNA section, or alternatively a stable microheterogeneity of the mt DNA of each subclone (Rabinowitz *et al.*, 1974). On the other hand the ability to hybridize with *grande* mitochondrial Val-tRNA was lost by mt DNA in some subclones and partially retained in others. Mitochondrial Leu- and Val-tRNAs isolated from some *petite* strains retained their capacity to be amino acylated *in vitro*, and were able to hybridize with mt DNA of *grande* and some *petite* strains. Labelled mitochondrial tRNAs (but not cytoplasmic tRNAs) competed with the hybridization of both *grande* and *petite* tRNAs with *grande* and *petite* mt DNAs.

The number of 4S-RNA genes on mt DNA of *S. carlsbergensis* has been determined in hybridization experiments by Reijnders and Borst (1972). A plateau of 0·9 µg RNA hybridized with 100 µg mt DNA indicated that yeast mt DNA contains at least 20 genes for tRNA, i.e. in principle a full set in contrast to the smaller number reported for animal mt DNAs. Rabinowitz *et al.* (1974) have reached a similar conclusion for *S. cerevisiae* as all 14 of the mt tRNAs studied hybridized with *grande* mt DNA. These results are also compatible with the idea that yeast mitochondria may not contain several tRNAs for one amino acid, but do not exclude the possibility that additional tRNAs specified by nuclear genes are imported into the mitochondria. Hybridization studies have enabled mapping of 11 of the tRNA cistrons with reference to genetically defined mitochondrial antibiotic resistance markers of yeast mt DNA (Rabinowitz *et al.*, 1974).

C. tRNAs of *Tetrahymena pyriformis* Mitochondria

In *T. pyriformis*, mitochondrial Leu-tRNA and its synthetase both differ from the extramitochondrial Leu-tRNA and synthetase (Suyama and Eyer, 1967). Antisera directed towards the Leu-tRNA synthetase purified from the mitochondria did not exhibit a positive immunological reaction when tested against the cytoplasmic enzyme by the following criteria: (a) inhibition of charging activity; (b) precipitation by immunodiffusion; or (c) binding by affinity chromatography (Chiu and Suyama, 1973). The *E. coli* enzyme did not bind to the anti-mitochondrial enzyme affinity column, and therefore these enzymes do not share common subunits or possess close structural similarities. Only 0·7% hybridization of mitochondrial tRNA with mt DNA was observed (Suyama, 1967; 1969) and it was suggested that a large portion of mitochondrial tRNA is not transcribed from mt DNA. These results contrast with those obtained with *N. crassa* and yeast.

Multiplicity of tRNA species plays important roles in protein synthesis and its regulation. Several isoaccepting tRNA species have been detected in mitochondria of *T. pyriformis* (Chiu *et al.*, 1974). Of the seven isoaccepting Leu-tRNA species identified in whole cell-tRNA by reverse phase chromatography, three species were exclusively found in the mitochondria and preferentially charged by the mitochondrial Leu-tRNA synthetase. These three mitochondrial Leu-tRNAs correspond to the codons CUG, CUA and CUU respectively, with Leu-tRNA (CUA) as the major species. Neither the mitochondrial nor the cytoplasmic tRNA responds to UUA or UUG. Hybridization experiments indicated that all three Leu-tRNAs differ in primary structure, and are coded from different segments of mt DNA. Isoaccepting species of tRNAs were demonstrated for other amino acids (number of species in parenthesis: Phe (2), Val (2), and Lys (2). In spite of six possible codons, only one Arg-tRNA was resolved. These results indicate that a rather extensive codon degeneracy is operative in mitochondria.

D. tRNAs of *Euglena gracilis* Mitochondria

A specific Isoleu-tRNA associated with the mitochondrial fraction of *E. gracilis* can be aminoacylated by either the mitochondrial or the cytoplasmic amino-acyl synthetases (Kislev *et al.*, 1972). Mitochondrial tRNA synthetases specific for His, Isoleu, Leu, Phe and Met were different from those obtained from the chloroplasts of green *E. gracilis* (Kislev and Eisenstadt, 1972).

III. Synthesis of Mitochondrial RNA in Eukaryotic Microorganisms

The detailed studies of mitochondrial RNA synthesis carried out with Hela cells (Attardi and Attardi, 1971; Aloni and Attardi, 1971a, b, c) give a more complete picture of the transcription of mt DNA than those yet available for microbial systems. The H and L strands of HeLa mt DNA were separated in alkaline CsCl gradients. Pulse-labelled and long-labelled RNAs were isolated, and hybridization experiments indicated that both H and L strands are equally transcribed in the case of pulse-labelled RNA, whereas the transcript of the complete H strand is only found in long-labelled RNA. Thus strand selection in transcription is achieved by transcribing both strands and rapidly degrading (or exporting) 98% of the L-strand transcripts. The transcription products have been used to map mt DNA; the H strand bears genes for 9 tRNAs, whereas the L strand has the two ribosomal RNA genes and also those for 3 tRNAs. Electron microscopy of ferritin-labelled tRNA-mt DNA hybrids confirms the location of these genes (Attardi *et al.*, 1974). In rat liver (Borst and Aaij, 1969) and in *Xenopus* eggs (Dawid, 1972), long-labelled mt RNA also hybridized with the H strand of mt DNA, whereas 4S RNA or Tyr- and Ser-tRNAs hybridize with the L strand (Nass and Buck, 1970).

Recent experiments with the Hela cells suggest that messenger RNA is also transcribed; mitochondrial RNA containing poly-(A) hybridizes with mt DNA (Attardi *et al.*, 1974). In ovarian mitochondria of *Xenopus* the *in vitro* transcription product is bound to the DNA template through a DNA/RNA hybrid region: released product sediments at between 4 and 12S (Dawid *et al.*, 1974). In this system mitochondrial RNA polymerase initiates at several different sites of the template predominantly with ATP, and 2-fold less frequently with GTP. Initiation with pyrimidine nucleotides is rare.

Studies on mitochondrial RNA synthesis in microbial systems have largely concentrated on characterization of the DNA-dependent RNA polymerases. These enzymes can be demonstrated in isolated intact mitochondria and are assayed by the incorporation of nucleotide triphosphates into high mol. wt RNA. In mitochondria from mammalian tissues (Wintersberger, 1964a, b; Neubert and Helge, 1965) from *Neurospora crassa* (Luck and Reich, 1964) and from *Saccharomyces cerevisiae* (Wintersberger, 1966b) this reaction is inhibited by actinomycin D, which binds to the DNA template.

A. RNA Synthesis in *Neurospora crassa* Mitochondria

Mitochondrial RNA polymerase has been isolated as an electrophoretically-pure protein from *N. crassa* by Küntzel and Schäfer (1971). It is a single polypeptide chain (mol. wt 64,000) which aggregates to a form which sediments at 24S in glycerol gradients. Although RNA synthesis is rifampicin-resistant in intact or sonicated mitochondria (Herzfeld, 1970), the purified enzyme is inhibited by this antibiotic. The polymerase is not inhibited by α-amanitine, and thus resembles bacterial RNA polymerases in its sensitivity to inhibitors. The primitive structure of the mitochondrial enzyme resembles that of the bacteriophage T7 specific enzyme and this simplicity may reflect the fact that both enzymes have to transcribe a relatively small genome with only a few cistrons or operons. Enzyme-bound DNA is used as a template, and a preference for native mt DNA from *N. crassa* over denatured DNA is indicated.

Intact mt DNA (25–26 μm contour length) can also be transcribed *in vitro* with RNA polymerase from *E. coli* (Schäfer *et al.*, 1971), and the transcription product has the low $(G + C)$ content characteristic of mt RNA. After the formation of an initial complex between the enzyme and mt DNA, only a limited number of bound enzymes can start RNA synthesis in the presence of rifampicin, and the sigma factor is required to confer resistance to this inhibitor. By comparing the initial rates of rifampicin-resistant transcription of phage and mt DNAs under saturating enzyme conditions, the number of specific binding sites for the enzyme per mitochondrial genome (mol. wt $50–60 \times 10^6$) was estimated at 8 or 9. The transcription product synthesized in the presence of rifampicin was characterized by competitive hybridization and RNAase digestion. About 50% of the sequences hybridizing with mt DNA are homologous to RNA isolated from intact mitochondria; 35% of the product consisted of rapidly renaturing double-stranded regions characteristic of rRNA, and 5% slowly renaturing double strands. Thus bacterial RNA polymerase recognizes the mt DNA sequences resembling phage promotor sites, and enzymes bound to these specific sites confer rifampicin-resistance and transcribe asymmetrically cistrons coding for mt RNA.

A rifampicin-resistant RNA polymerase from mitochondria of *N. crassa* has been purified by Wintersberger (1972), and this study included purification of enzymes with similar properties from yeast and liver mitochondria. The relationship between this enzyme and that of Küntzel and Schäfer (1971) is not clear but it is distinctly different from the nuclear RNA polymerases (Table LIII).

The mechanism of mitochondrial rRNA synthesis has been investigated in *N. crassa* by continuous labelling with (5-^{3}H) uracil and pulse chase experiments (Kuriyama and Luck, 1973). A short-lived 32S mito-chondrial RNA ($2\cdot4 \times 10^6$ daltons) was detected along with two other short-lived components; one slightly larger than large subunit rRNA (25S; $1\cdot28 \times 10^6$ daltons) and the other slightly larger than small subunit rRNA (19S; $0\cdot72 \times 10^6$ daltons). It was suggested that 32S RNA is a precursor of both large and small rRNAs, as both mature rRNAs compete with 32S RNA in hybridization with mt DNA. Intermediates in the maturation of large and small subunit rRNAs may involve components with mol. wts about $1\cdot6 \times 10^6$ and $0\cdot9 \times 10^6$ respectively. The precursor like the mature species, has a strikingly low (G + C) content, and 22% of the molecule is lost during the maturation process which is inhibited in growing cells by 50 μM–ethidium bromide.

Mature mitochondrial RNAs are methylated to about one half the extent of their cytoplasmic counterparts and by comparison are richer in methylated bases in 2-O-methyl groups; the 32S precursor is also exten-sively methylated (Kuriyama and Luck, 1974), In "*poky*" mitochondria there is a deficiency of small mitochondrial ribosomal subunits, the pro-cessing of the precursor RNA appears to be abnormal and the mature mitochondrial RNAs are strikingly under-methylated.

B. RNA Synthesis in Yeast Mitochondria

Yeast mitochondria contains a rifampicin-resistant RNA polymerase (Wintersberger, 1970; Wintersberger and Wintersberger, 1970a; Tsai *et al.*, 1971; Benson 1972; Eccleshall and Criddle, 1972; Table LIII). The enzyme is membrane-bound and can be solubilized by mechanical dis-ruption or treatment of the mitochondrial membranes with detergents. Tsai *et al.* (1971) solubilized the enzyme from yeast mitochondrial membranes using 0·5 M–KCl and obtained two forms of DNA-dependent

Table LIII. Mitochondrial DNA-dependent RNA polymerases in *Neurospora* and yeast

	Mol. wt		Inhibitor sensitivity		Reference
	intact enzyme	subunits	rifampicin	α-amanitine	
Neurospora crassa					
	500,000	64,000	+	−	Küntzel and Schäfer, 1971.
			−	−	Wintersberger, 1972
					D. O. Woodward, unpublished results
Yeast					
I	500,000	2 large	−	+	Eccleshall and Criddle, 1974
II	500,000	3 large	−	+	
III	500,000	2 large	+	−	
	< 200,000	65–70,000	+	−	Scragg, 1971
			−	−	Scragg, 1974
					Wintersberger and Wintersberger, 1970a
I Ia			−	−	Benson, 1972
II			−	+	

RNA polymerase which differ in template specificity and metal-ion dependency. The mitochondrial enzyme differed from the nuclear enzyme with respect to antibiotic specificity and template specificity. Only the nuclear enzyme was α-amanitine-sensitive and the relative activities of the purified nuclear and mitochondrial polymerases towards their homologous DNAs were consistent with their *in vivo* functions. Possible contamination of mitochondrial enzyme by enzymes of bacterial or nuclear origin was ruled out in these experiments. A *petite* mutant with no detectable mt DNA had greatly diminished levels of mitochondrial DNA-dependent RNA polymerase activity, whereas a *petite* mutant with only a small change in mt DNA base composition had normal amounts of the enzyme.

The absence of protein synthesis in mitochondria isolated from nuclear (p), cytoplasmic (ρ^-) and double ($p\rho^-$) mutants, results from a block at the level of translation rather than transcription, as incorporation of (^{3}H)-UTP proceeds in all cases (Kuzela and Fečiková, 1970).

As in *N. crassa* the question of rifampicin-sensitivity of the mitochondrial RNA polymerase has not been finally settled, as other reports (Scragg, 1971; 1974) describe the isolation of a polymerase sensitive to the inhibitor; this enzyme prefers denatured DNA as a template, but also functions with yeast or rat liver mitochondrial DNAs. Of the total high mol. wt RNA transcribed, 70% was single stranded and this transcription product coded for proteins in an *E. coli* cell-free system which were similar to *in vivo* products, as determined electrophoretically and antigenically. The RNA polymerase was not detectable in a *petite* mutant which has no detectable mt DNA.

Eccleshall and Criddle (1974) have reported the isolation of three distinct high mol. wt mitochondrial DNA-dependent RNA polymerases from yeast mitochondria. All three enzymes are rifampicin insensitive; the proportions and yields of the enzymes was greatly influenced by the presence of antioxidant (2-6-di-*t*-butyl-4-hydroxy-methyl phenol) and protease inhibitors during isolation. Mitochondrial enzyme I was identical to nuclear enzyme I, and may therefore be a nuclear contaminant. Mitochondrial enzyme II was similar to nuclear enzyme II but has a large extra polypeptide. The yield of mitochondrial enzyme III is greatly reduced under isolation conditions giving rise to increase in enzyme I, and is sensitive to attack by proteases during purification.

It is generally agreed that the yeast nuclear RNA polymerase activity is α-amanitine sensitive and rifampicin resistant (Ponta *et al.*, 1971). Comparison of the properties of the various enzymes isolated from *Neurospora* and yeast are presented in Table LIII. The effects of ethidium bromide on the purified RNA polymerases has not been tested.

The *in vitro* transcription of mt DNA in yeast by the DNA-dependent

RNA polymerases of yeast and *E. coli* has been studied by Michaelis *et al.* (1972). Both the *E. coli* enzyme and the yeast mitochondrial enzyme transcribe the ribosomal genes of mt DNA as shown by competition with unlabelled mitochondrial rRNA. A cytoplasmic *petite* lacked 50–60% of the wild-type genome as shown by hybridization of the RNA product of transcription of the *petite* with mt DNA of the wild-type.

The RNA polymerase of yeast mitochondria can recognize some transcriptional signals which are normally recognized by the *E. coli* enzyme (Richter *et al.*, 1972). When a coupled system of transcription and translation was programmed with DNA of phage T_3, two specific phage enzymes, lysozyme and an *S*-adenosyl Met cleaving enzyme (SAMase) were synthesized. Synthesis of these two enzymes is inhibited by fusidic acid and actinomycin D, but not by cycloheximide or the bacterial protein synthesis inhibitor, thiostrepton. When T_7 phage DNA was used, no SAMase was made, as this phage does not carry the SAMase structural gene.

The *in vivo* transcription of the mitochondrial genomes of respiratory deficient cytoplasmic and nuclear *petites* has been compared by hybridization of total cellular RNAs with nuclear and mt DNAs (Fukuhara *et al.*, 1969). When ρ^+ RNA was hybridized to ρ^+ and ρ^- mt DNA, the homologous cross ρ^+RNA $\times$ ρ^+ mt DNA formed roughly twice as much hybrids as the heterologous cross (ρ^+ RNA $\times$ ρ^- mt DNA). The nuclear mutation p_7 did not introduce a change in mt DNA base sequence detectable in these experiments. 4S RNA was found to contain RNA species which hybridize with mt DNA, and the degree of hybridization was very different for ρ^+ and ρ^- 4S RNA when they were hybridized with either ρ^+ or ρ^- mt DNA.

Both aerobically-grown and anaerobically-grown lipid-supplemented wild-type *S. cerevisiae* contain RNA transcribed from mt DNA (Fukuhara 1967, 1968). Mitochondria isolated from anaerobically-grown cells depleted of both unsaturated fatty acids and ergosterol appear to lack mitochondrial rRNA and this suggests that the control of its synthesis is related to the composition and structure of the membranes (Forrester *et al.*, 1971). It is suggested that in this case receptor sites for the mitochondrial DNA-dependent RNA polymerase may be missing as the activity of the enzyme is low under these conditions (Linnane *et al.*, 1972a; Ward *et al.*, 1973).

The amount per cell of transcripts of mt DNA increase during the respiratory adaptation of anaerobically-grown cells (Fukuhara, 1967, 1968; Schuit and Balzer, 1971) even in cytoplasmic *petite* strains (Fukuhara *et al.*, 1969).

Glucose repression also influences the activity of mitochondrial DNA-dependent RNA polymerase (South and Mahler, 1968). Highly purified yeast mitochondria were treated with digitonin to give a preparation capable of incorporation of any of the four ribonucleoside triphosphates in the presence of the other three. Glucose-repressed cells yielded mitochondria with low activities, whereas those from glucose-derepressed cells showed increased isotope incorporation rates. Incorporation of labelled uracil by whole cells into a 11S mitochondrial RNA component occurs most rapidly under conditions of glucose derepression (Leon and Mahler, 1968).

Mitochondrial polysomes active in protein synthesis have been isolated from yeast sphaeroplasts undergoing glucose-derepression (Cooper and Avers, 1974). An eight minute pulse with (^{3}H)-uracil in isolated mitochondria gives (^{3}H)-labelled RNA associated with mitochondrial polysomes. When extracted from the polysomes this RNA binds to poly-(U)-sepharose (unlike rRNA or tRNA), suggesting that the rapidly-labelled species is yeast mitochondrial messenger RNA containing tracts of poly-(A). Poly-(A) has been found in the RNA from mammalian mitochondria (Avadhani et al., 1973) but not generally in the RNA of prokaryotes (Sheldon et al., 1972).

The specificity of mitochondrial peptide chain initiation for formyl-Met has enabled the incorporation of labelled formyl groups to be used as an operational test for functional mitochondrial messenger RNA (Mahler et al., 1974).

C. RNA Synthesis in *Tetrahymena pyriformis* Mitochondria

RNA molecules in the process of transcription from mt DNA have been revealed in electron micrographs (Fig. 111); the strand of messenger RNA in mitochondrial polysomal aggregates has also been demonstrated (Fig. 112, Charret and Charlier, 1973).

The effects of a number of antibiotics on RNA synthesis by phosphate-swollen mitochondria from *T. pyriformis* have been tested (Coolsma, 1971). Actinomycin was completely inhibitory, whereas α-amanitine, rifamycin, and streptolydigin were all inactive. Treatment by a range of mild extraction procedures led to inactivation of the mitochondrial DNA-dependent RNA polymerase and it was suggested that the enzyme is firmly attached to a membrane.

Suyama and Eyer (1968) were unable to confirm the *in vitro* synthesis of tRNA or rRNA in mitochondria isolated from *T. pyriformis*. A relatively large mol. wt RNA (14–18S) was synthesized which hybridized specifically with mt DNA. Although net synthesis of tRNA was not shown, the mitochondria contained a tRNA fraction which can be labelled with (^{3}H)-ATP but not with (^{3}H)-UTP as a precursor, suggesting the presence of an enzyme system which accomplishes the terminal addition of adenylate to mitochondrial tRNA molecules.

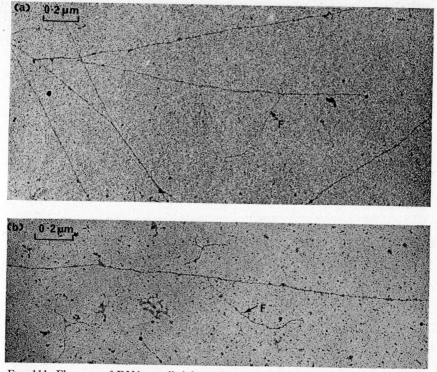

Fig. 111. Flaments of DNA expelled from the mitochondria of *Tetrahymena pyriformis* by osmotic shock. Long straight filaments of mt DNA bear frail short (0·5 μm) filaments (F) which are thought to be molecules of mt RNA in the process of transcription. (Reproduced with permission from Charret and Charlier, 1973.)

The synthesis of mitochondrial RNAs has also been studied in whole cells (Millis and Suyama, 1972). After incorporation of (^{3}H-)-uridine, analysis of RNA isolated from a mitochondrial fraction indicated that both mitochondrial 21S and 14S RNAs were radioactive; the 2:1 mass ratio expected for the large and small mol. wt rRNA was indicated by the absorbance profiles. The specific radioactivity of mitochondrial 14S was much higher than that of 21S RNA and the specific activity of the two mitochondrial rRNA species was 5–10 times greater than the corresponding cytoplasmic RNA species. In the presence of cycloheximide, inhibition of the synthesis of both cytoplasmic and mitochondrial rRNAs (21S + 14S) was 10% and (78% + 60%) respectively. It was not certain what proportion of these labelled products represent messenger RNAs. These results were interpreted as an indirect effect of the interdependence of the cytoplasmic and mitochondrial protein synthesis systems on the synthesis of RNAs known to be transcriptive products of the mitochondrial genome. The origins of mitochondrial RNAs synthesized *in vivo* were also examined

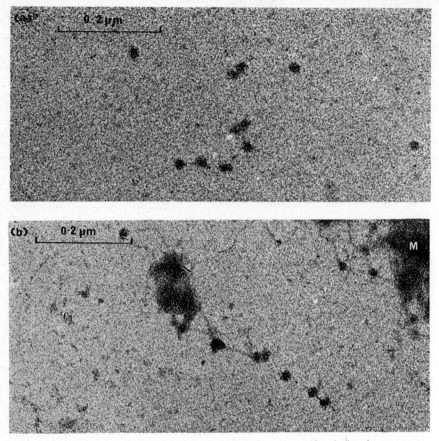

FIG. 112. Polysomes from the mitochondria of *Tetrahymena pyriformis* showing messenger RNA. In (b) the mitochondrial membranes are also seen closely associated with polysomal aggregates. (Reproduced with permission from Charret and Charlier, 1973.)

by DNA-RNA hybridization studies (Suyama 1967; Chi and Suyama, 1970). Whereas both 21S and 14S RNAs from mitochondrial ribosomes hybridized with mt DNA, mitochondrial tRNA did not to any significant extent, and it was concluded that a large proportion of the tRNA found in mitochondria isolated from *T. pyriformis* is not transcribed from mt DNA The mitochondrial rRNAs hybridize efficiently only to the H-strand of the mt DNA; the hybridization plateau value suggests that 1·0 ribosomal cistron per DNA molecule is present (Schutgens *et al.*, 1973).

D. RNA Synthesis in *Blastocladiella emersonii* Mitochondria

The mitochondrial RNA polymerase from *B. emersonii* is rifampicin-sensitive and α-amanitine-resistant (Horgen and Griffin, 1971 a, b).

E. RNA Synthesis in *Physarum polycephalum* Mitochondria

Mitochondria isolated from *P. polycephalum* incorporate (^{3}H)-UTP into an acid insoluble product which is sensitive to RNAase (Grant and Poulter, 1973). Isotope incorporation was sensitive to actinomycin D and rifampicin, and in the presence of rifampicin the major part of the acid insoluble radioactivity disappeared with a half-life of 2–3 mins. A parallel decrease in (^{14}C)-Leu incorporating ability suggested a messenger function for the product of (^{3}H)-UTP incorporation. This product was shown by acrylamide gel chromatography to be distinct from mitochondrial rRNA.

F. RNA Synthesis in *Leishmania tarentolae* Kinetoplasts

Purified kinetoplasts from *L. tarentolae* contain two unique species of RNA which sediment at 9 and 12S in sucrose gradients. The synthesis of these species is inhibited by ethidium bromide. Labelled 9 and 12S RNAs specifically hybridize to minicircles of kinetoplast DNA. Whether these RNA species represent messenger RNA or rRNA is not yet clear (Simpson, 1973).

G. RNA Synthesis in an Amoeba

The incorporation of (^{3}H)-uracil into the mitochondria of enucleated amoebae proceeds more rapidly, and to a greater extent, than in control organisms (Perasso, 1973). This observation suggests that mitochondrial transcription may be restrained in normal cells by repressors of nuclear origin.

H. Inhibitors of RNA Synthesis in Mitochondria

It is evident from this survey that conflicting results have been obtained with regard to the rifampicin-inhibition of mitochondrial DNA-dependent RNA synthesis. Alternative explanations have been offered to explain these discrepancies. There may be two different enzymes present, although the small genome size makes this seem unlikely. The low mol. wt of purified enzyme suggests that this may be part of a larger rifampicin-sensitive complex. Sigma-like cofactors may be lost during some purification procedures. Sensitivity to rifampicin also depends upon whether the nucleotide incorporation reflects initiation or chain elongation.

Other inhibitors of mitochondrial RNA synthesis include daunomycine (Evans *et al.*, 1973), cordycepin (which may be 3'deoxyadenosine, Zylber *et al.*, 1971), and lomofungin. These agents may also interfere with nuclear RNA synthesis (Fraser *et al.*, 1973; Cannon *et al.*, 1973). Acridines and ethidium bromide also inhibit mitochondrial RNA synthesis (South and Mahler, 1968; Zylber and Penman, 1969; Zylber *et al.*, 1969;

Fukuhara and Kujawa, 1970; Yu *et al.*, 1972a), but interpretation of results obtained with these inhibitors is not straightforward due to their mutagenic properties and possible direct inhibitory action on protein synthesis (Sinohara, 1973). Multiple effects of rifamycin derivatives on macromolecular synthesis occur in animal cells (Busiello *et al.*, 1973).

In ts-136, a temperature-sensitive mutant of *S. cerevisiae*, it is possible to inhibit selectively nuclear and mitochondrial RNA synthesis by exposure to 36°C and ethidium bromide respectively (Mahler and Dawidowicz, 1973). The ability of mitochondrial polyribosomes to form nascent polypeptide chains was used in an assay for functional messenger RNA. Ethidium bromide but not temperature-shifts had a measurable effect on the mitochondrial system, whereas the reverse was the case for the cell sap system. These investigators concluded that transcription of mt DNA is sufficient, and that import of messenger RNA transcribed from nuclear chromosomes made no measurable contribution to mitochondrial protein synthesis. Mitochondrial polysomes decayed in 25 min in both the wild-type and the mutant.

The half-life of mitochondrial messenger RNA has been studied in yeast by following the decay of mitochondrial (cycloheximide-resistant) protein synthesis after blocking RNA synthesis with ethidium bromide (Weislogel and Butow, 1971) or with acriflavin (Schatz *et al.*, 1972): both methods indicate a half-life of less than 10 min. Over 50% of the ethidium-bromide resistant RNA found in animal mitochondria contains poly-(A) sequences and has a half-life of 45 min. In this system the mitochondrially-synthesized RNA (EB sensitive) had a half-life of 15 min (Avadhani *et al.*, 1973).

Chloramphenicol has little or no inhibitory effect on synthesis of yeast mitochondrial rRNA *in vivo* (Wintersberger, 1967). The observation that yeast grown with chloramphenicol yields mitochondria still capable of carrying out protein synthesis suggest that the synthesis of mitochondrial tRNAs is not affected by this antibiotic either (Davey *et al.*, 1969).

9

Mitochondrial Protein Synthesis in Eukaryotic Microorganisms

I. Introduction

That mitochondria isolated from mammalian tissues can incorporate amino acids into proteins was first reported by McLean *et al.* (1958). After about ten years of controversy it became evident that although bacterial contaminants can contribute to this activity, *in vitro* protein synthesis is a genuine property of mitochondria isolated from a wide variety of sources (Beattie, 1971a). Early attempts to identify the products of *in vitro* incorporation of labelled amino acids suggested that enzymes such as malate dehydrogenase or cytochrome *c* are not produced (Roodyn *et al.*, 1961), and this observation has been extended to include all the readily solubilized mitochondrial proteins (Wheeldon and Lehninger, 1966; Beattie *et al.*, 1967; Kadenbach, 1967a, b). Labelled proteins were located in a membrane fraction obtained by extraction of intact mitochondria with the non-ionic detergent Triton X-100 (Roodyn, 1962). A "structural" protein fraction contained over 22% of the incorporated radioactivity. Fractionation of *in vitro* labelled mitochondria to give inner- and outer-membranes revealed that the predominant products were "structural' proteins of the inner-membrane; outer-membrane or soluble proteins of the matrix acquired very little radioactivity (Beattie *et al.*, 1967; Neupert *et al.*, 1967).

The analysis of the products of protein synthesis by isolated mitochondria is only one of the approaches used to identify the proteins translated on mitochondrial ribosomes and encoded by mt DNA. Several drawbacks to the usefulness of this approach are apparent: (a) Even under optimal conditions the rates are considerably slower than the estimated capacity of *in vivo* synthesis; the extent of synthesis is also limited. (b) Control normally imposed by the interaction with the cytoplasm is absent. (c) Integration of products into their correct sites may not occur in the absence of cytoplasmically-synthesized proteins and lipids. Thus *in vitro* incorporation may give an incomplete picture of the *in vivo* capacity of the intrinsic mitochondrial system of protein synthesis. However Coote and

Work (1971) have shown that the products formed *in vitro* and *in vivo* are similar, and so perhaps the major objection to the use of isolated fractions containing mitochondria may be in the limited rate and extents to which proteins are produced.

The second approach to the identification of the biosynthetic origin(s) of mitochondrial proteins is the study of effects of antibiotic inhibitors specific for either the mitochondrial or cytoplasmic protein synthesis systems *in vivo* on the development of enzymes and proteins. Pulse-labelling with amino acids in the presence of chloramphenicol or cycloheximide has provided a great deal of information on the contributions of the two systems, especially now the emphasis has shifted to the study of specific, well-characterized mitochondrial enzymes. A major criticism of this approach is the existence of control mechanisms responsible for the close coupling of the intrinsic and extrinsic systems for the synthesis of mitochondrial proteins (Mahler *et al.*, 1968). In this case primary inhibition of the microsomal system by cycloheximide would be expected to lead to secondary inhibition of the mitochondrial protein synthesis system (Ashwell and Work, 1968). Another possible criticism to the use of protein synthesis inhibitors (specially at high concentrations) is their side-effects on energy production (Freeman and Haldar, 1967). However, cycloheximide even at a concentration of 100 μg/ml has little effect on electron transport (e.g. on beef heart mitochondrial NADH oxidase, Ashwell and Work, 1968) or on the energy metabolism of intact cells (e.g. that of *Euglena gracilis*, Kirk, 1970).

Some organisms (e.g. *Saccharomyces fragilis*, Rao and Grollman 1967; Battamere and Varquez 1971, and several species of amoebae, Hawkes and Holberton, 1973) are resistant to cycloheximide. Inhibition of cytoplasmic protein synthesis in ρ^- yeast mutants and in anaerobically-grown yeasts also requires high levels of this inhibitor (Biliński and Jachymczyk, 1973).

Two other methods of identifying the sites of synthesis of mitochondrial proteins are (1) identification of modified gene products in mutants showing cytoplasmic inheritance, and (2) identification of those mitochondrial proteins still produced in mutants totally lacking mt DNA, which therefore must be coded by nuclear genes and synthesized on cytoplasmic ribosomes. These approaches have been dealt with in Chapter 6.

II. Protein Synthesis in Isolated Mitochondria

A. Properties of *in vitro* Systems

A mitochondrial system of protein synthesis was reported in the large subcellular fraction of *Tetrahymena pyriformis* by Mager (1960). Incorporation of amino acids into mitochondrial proteins was ATP-dependent, and was significantly enhanced by addition of a mixture of TTP, CTP and

UTP; its specific activity was three-to four-fold lower than that of the microsomal preparation from this organism. The saturating level of (^{14}C)-Leu required for optimal incorporation was 0·5 mM, nearly 5 times higher than the corresponding value for the microsomal system, and the presence of a supernatant fraction was not necessary for the activity of the mitochondrial system. Two distinctive characteristics which sharply differentiated between the mitochondrial system and its cytoplasmic counterpart were, (a) resistance to treatment with RNAase, and (b) 30–60% inhibition by chloramphenicol (25–150 µg/ml). Experiments with chlortetracycline produced analogous results.

Amino acid incorporation into protein by mitochondria isolated from yeast was first described by Wintersberger (1965). One of the major problems in studies of protein synthesis with isolated mitochondria has been the possibility of a contribution to the observed activity by contaminating bacteria (Roodyn, 1966; Beattie, 1971a). Completely germ-free preparations of mitochondria isolated from *S. carlsbergensis* were found to incorporate amino acids, and contamination by up to 100 organisms/mg of mitochondria protein made no significant contribution to the activity (Grivell, 1967). In this system, a linear uptake of (^{14}C)-Leu occurred for 30 min in the presence of an oxidizable substrate, together with either ATP or ADP, and Mg^{2+}. Rates of incorporation of up to 55 pmoles/mg mitochondrial protein/30 min were observed, and 90% inhibition by choramphenicol occurred at 50µg/ml. Insensitivity to RNAase at concentrations far above those required to abolish microsomal incorporation, also indicated that the cell sap system played no significant part in the observed amino acid incorporation. The *in vitro* effects of a number of antibacterial antibiotics on the mitochondrial and cytoplasmic protein synthesis systems of *S. cerevisiae* were tested by Lamb *et al.* (1968). The mitochondria were found to show maximal rates of amino acid incorporation in an ATP-supported system of 80 pmoles (^{14}C)-Leu/mg protein/20 min (Table LIV). Amino acid incorporation by bacteria contaminating the preparations was insignificant. Chloramphenicol, lincomycin and the macrolide antibiotics, erythromycin, spiramycin, carbomycin and oleandomycin, all known inhibitors of bacterial protein synthesis at the level of the 50S ribosome subunit, were shown to inhibit specifically amino acid incorporation by mitochondria, but to be without effect on the cytoplasmic ribosomal system. Conversely, cycloheximide (a glutarimide antibiotic from *Streptomyces griseus*, which does not inhibit bacterial protein synthesis), does specifically inhibit the cytoplasmic ribosomal system by acting at a transfer step subsequent to the formation of the ternary complex between mRNA, amino acyl-tRNA and the ribosome (Siegel and Sisler, 1965). It had no effect on mitochondrial amino acid incorporation (Table LV). The phenotypic effects of growth of organisms

Table LIV. Energy requirements for optimal *in vitro* incorporation of (^{14}C)-Leu into mitochondrial protein. (Reproduced with permission from Lamb *et al.*, 1968.)

Additions (mM)	Leu incorporation counts/min/mg protein
None	0
ATP (1)	110
ADP (1) + Succinate (2–10)	190
ADP (1) + Pyruvate (5–10) + Malate (0·5–2)	570
ATP (1) + PEP (5) + 30 μg Pyruvate kinase	1135
ATP/PEP/Pyruvate kinase + GTP (0·2)	1125
ATP/PEP/Pyruvate kinase + Amino acid mixture	1115

The standard incubation medium contained 40 mM tris HCl (pH 7·4), 5 mM KH_2PO_4, 100 mM KCl, 8 mM $MgCl_2$ and 100 mM sorbitol. Each 1 ml system contained 0·2 μCi of (^{14}C)-leucine (40 μCi/μmole) and approximately 1 mg of mitochondrial protein. Incubations were for 15 min at 30°C, with additions as indicated; in addition 0·25 mM NAD, 0·25 mM NADP and diphosphothiamine (10 μg/ml) were added to the succinate and pyruvate-malate supported experiments.

with these antibiotics has already been discussed in Chapter 4 (p. 222). Inhibition of *in vitro* protein synthesis by mitochondria isolated from *Candida parapsilosis* by chloramphenicol, lincomycin, erythromycin and also by ethidium bromide and euflavine was observed by Kellerman *et al.* (1969).

Mitochondria-like structures from *S. cerevisiae* grown anaerobically in the presence of galactose, Tween 80 and ergosterol contain an active protein synthesizing system (Davey *et al.*, 1969). In contrast, the amino acid incorporating system of mitochondrial equivalents from cells grown anaerobically with glucose in the absence of lipid supplements was so small as to be accounted for by slight contamination with bacteria and cytoplasmic ribosomes (Watson *et al.*, 1970). Aeration of these cells for 2 h gave rise to organelles which on isolation showed the choramphenicol-sensitive and ATP-dependent incorporation characteristic of aerobic mitochondria.

Anaerobically-grown cytoplasmic *petite* mutants yield mitochondria which are unable to incorporate amino acids into proteins (Kužela and Grečna, 1969). Mitochondria isolated from a variety of aerobically-grown cytoplasmic *petite* mutants have also been examined for evidence of their ability to synthesize proteins *in vitro* (Kellerman *et al.*, 1971). In several different types of *petites*, with widely differing degrees of suppressiveness (even when grown under conditions of glucose derepression in chemostat cultures) there was a complete inability to synthesize protein. It was therefore concluded that there is a general loss of mitochondrial protein synthesizing ability associated with the cytoplasmic *petite* mutation.

Intact mitochondria from *Neurospora crassa* incorporate (^{14}C)-Leu, and lysis with Triton X-100 (a detergent which does not lyse possible

Table LV. Differential effect of antibiotics on amino acid incorporation into protein by yeast cytoplasmic ribosomes and mitochondria. (Reproduced with permission from Lamb *et al.*, 1968.)

Antibiotic concentration (mM)	Mitochondria	Cytoplasmic ribosomes
	p moles Leu incoporated/mg protein/20 min	
None	47	110
	Percentage inhibition by antibiotics	
Chloramphenicol		
0·03	73	0
0·10	87	0
0·50	95	0
Erythromycin		
0·001–1·5	74	0
Spiramycin		
0·03	66	2
0·50	75	4
Oleandomycin		
0·03–0·5	53	5
Carbomycin		
0·03–0·1	77	3
Lincomycin		
0·03	60	0
0·10	69	2
0·50	88	5
Cycloheximide		
0·03–0·5	0	99

bacterial contaminants) yields labelled mitochondrial polysomes (Küntzel and Noll, 1967). The proportion of acid-insoluble radioactive product in the soluble fraction increases with increasing incubation times due to the completion and release of nascent polypeptide chains. Puromycin treatment after amino acid incorporation releases incorporated radioactivity completely from monosomes and polysomes (Neupert *et al.*, 1969a).

Mitochondria from a streptomycin–bleached mutant SB3 of *Euglena gracilis* strain z, incorporated labelled amino acids into proteins (Kislev and Eisenstadt, 1972). Chloramphenicol and puromycin inhibited this incorporation, whereas cycloheximide produced little effect. Protein synthesis was resistant to added RNAase and DNAase. Labelling of mitochondrial polysomes of *Euglena* has also been demonstrated (Avadhani *et al.*, 1971).

B. Antibiotic Inhibitors

The sites of inhibition by antibiotics of bacterial ribosomal function are shown in Fig. 113, and the specificity of their action in various *in vitro*

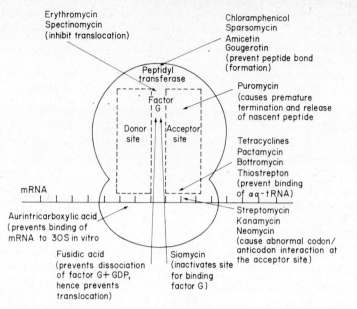

Fig. 113. Probable sites of action of some antibiotics which interfere with synthesis of proteins. (Reproduced with permission from Mandelstam and McQuillen, 1973.)

Table LVI. Effects of inhibitors on *in vitro* amino acid incorporation by yeast mitochondria, yeast cytoplasmic ribosomes, and *Escherichia coli* ribosomes. (Reproduced with permission from Linnane and Haslam, 1971.)

Drug	Yeast mitochondria	Yeast cytoplasmic ribosomes	*E. coli* ribosomes
Chloramphenicol	+	−	+
Lincomycin	+	−	+
Carbomycin	+	−	+
Spiramycin	+	−	+
Oleandomycin	+	−	+
Mikamycin	+	−	+
Cycloheximide	−	+	−
Neomycin	+	±	+
Paromomycin	+	±	+
Streptomycin	±	±	+
Kanamycin	±	±	+
Viomycin	±	±	+
Cryptopleurine	+	+	±
Tylocrebrine	+	+	±
Euflavine	+	±	±
Proflavine	+	±	±
Ethidium bromide	+	±	±
Prothidium bromide	+	±	±

Symbols: + extensive inhibition by low concentrations of the drug; ± inhibition by high concentrations of the drug; − no effect.

systems of amino acid incorporation are presented in Table LVI. The use of chloramphenicol as a specific inhibitor of mitochondrial protein synthesis has been questioned, as at the high concentrations sometimes used in experiments *in vivo*, it also may inhibit respiration in the vicinity of the first site of energy conservation (Freeman and Haldar 1967, 1968; Firkin and Linnane 1968; Freeman, 1970). However the oxidative metabolism of mitochondria isolated from *S. cerevisiae* is not affected by 2 mg/ml chloramphenicol (Ball and Tustanoff, 1970); that some of the effects reported for organisms which have all three sites of phosphorylation are secondary effects of disturbed energy metabolism remains a possibility. Only the D(−)isomer is active in blocking mitochondrial protein synthesis. Complete inhibition of mitochondrial protein synthesis *in vitro* is obtained at concentrations at which there is no effect on oxidation or phosphorylation (Kroon, 1969). The site of action is at the peptidyl-transferase stage of mitochondrial protein synthesis (DeVries *et al.*, 1971). A number of analogues of chloramphenicol are available (Ringrose and Lambert, 1973). Strains of yeast sensitive to low concentrations of antibiotics may be used to minimize the possibility of effects other than those directly upon mitochondrial protein synthesis (Wilkie *et al.*, 1967).

Anisomycin, like cycloheximide, does not inhibit bacterial or mitochondrial protein synthesis but does interfere with protein synthesis on cytoplasmic ribosomes (Lietman, 1970). Emetine (an ipecac alkaloid) acts as an inhibitor of cytoplasmic protein synthesis, it does not inhibit mitochondrial protein synthesis *in vivo* (Perlman and Penman, 1970), but does inhibit *in vitro* (Lietman 1970; Chakrabarti *et al.*, 1972).

The effects of three phenanthrene alkaloids, tylophorine, tylocrebrine and cryptopleurine on the protein synthesis systems of yeast mitochondria, yeast cytoplasmic ribosomes, and the ribosomes from *E. coli* were compared by Haslam *et al.* (1968). All three alkaloids inhibit the incorporation of (^{14}C)-Leu into protein on yeast cytoplasmic ribosomes: high concentrations of tylocrebrine and cryptopleurine inhibit the mitochondrial system. Very high concentrations of cryptopleurine were required to produce any inhibition of the *E. coli* ribosomal system and the other two alkaloids were not inhibitory to this system.

The protein synthesizing system of rat liver mitochondria resembles that of yeast mitochondria in being sensitive to chloramphenicol and resistant to cycloheximide, but differs from the yeast system in being insensitive to erythromycin and linomycin (Firkin and Linnane, 1969). These mitochondria are however sensitive to carbomycin and to a lesser degree to spiramycin and tylosin (Kroon *et al.*, 1974). It appears that they are impermeable to the macrolides with the smaller lactone ring, but permeable to those with the larger lactone ring. Thus carbomycin is concentrated within the mitochondria and competes with chloramphenicol for the sites of inhibition of mitochondrial ribosomes, whereas erythromycin does not.

Aminoglycoside antibiotics (which affect bacterial protein synthesis by reaction with the 30S ribosomal subunit) can be divided into two groups on the basis of their action on *in vitro* protein synthesis by cytoplasmic and mitochondrial systems of *S. cerevisiae* (Davey *et al.*, 1970). The first group, neamine, kanamycin, streptomycin and viomycin have only minor effects (Table LVI). The second group neomycin B, neomycin C, and paromomycin inhibited *in vitro* incorporation of amino acids into proteins by both mitochondrial and cytoplasmic ribosomes, although the mitochondrial system was considerably more sensitive to these antibiotics. The mitochondrial system of rat liver was insensitive to the neomycins and paromomycin. It is thus evident that phylogenetic differences in mitochondrial protein synthesis systems from different sources exist with regard to ribosomal binding of antibiotics; the possibility that these differences are a consequence of permeability barriers to the antibiotics seemed unlikely (Towers *et al.*, 1972). It was proposed that the evolution of the ribosome through the yeast mitochondrial ribosome to the mammalian mitochondrial ribosome may be expressed in the loss or modification of certain antibiotic binding proteins of the mitochondrial protein synthesis system and the reduction of its functional complexity to a corresponding minimum.

Binding of a non-inhibitory antibiotic (e.g. erythromycin in rat liver mitochondria) can prevent inhibition of protein synthesis by the exclusion of an inhibitory antibiotic (e.g. carbomycin, Towers *et al.*, 1973a). This shows that lack of inhibition is not due to exclusion by permeability barriers and suggests that modification rather than loss of antibiotic binding proteins has occurred during evolution. However, from other data, Kroon *et al.* (1974) suggests that the evidence against an explanation of many of these results on the basis of permeability barriers to the non-inhibitory antibiotics is not unequivocal.

A complete independence of action of three macrolides has been established in genetic studies on yeast mitochondria (Trembath *et al.*, 1973); thus mutants have been isolated sensitive to erythromycin alone, spiramycin alone, to both spiramycin and carbomycin, or to all three compounds. The genetics of antibiotic resistance has been discussed in Chapter 6.

Some antibiotics that affect mitochondrial protein synthesis also inhibit mitochondrial respiration. Thus mikamycin, which blocks mitochondrial protein synthesis at concentrations of the order of 1 µg/ml or less, also inhibits mitochondrial respiration at a site between cytochromes *b* and *c*, when present at 25–50 µg/ml (Dixon *et al.*, 1971). Carbomycin, oleandomycin, paromomycin, and choramphenicol, also interfere with respiration at high concentrations (Linnane and Haslam, 1971) and it is suggested that mitochondrial ribosomes are closely associated with, or an integral part of the inner mitochondrial membrane, Further support for this suggestion (which could account for the difficulty of separation of mitochondrial ribosomes from the membrane and the necessity for the use of detergents) comes from genetic studies on antibiotic resistance of yeast (see Chapter 6), and from the observation that temperature-induced phase changes in mitochondrial membranes are accompanied by a transition in the Arrhenius plots for the rate of mitochondrial protein synthesis (Towers *et al.*, 1973b).

C. Poly-(U)-directed Polyphenylalanine Synthesis on Isolated Mitochondrial Ribosomes

1. *In* Neurospora crassa

A submitochondrial poly-(U)-directed poly-Phe-synthesizing system containing *Neurospora* mitochondrial ribosomes was investigated by Küntzel (1969b). Table L1 summarizes the results obtained with homologous and heterologous systems from *Neurospora* mitochondria, *Neurospora* cytoplasm, *E. coli* and rat liver cytoplasm. It is evident that the mitochondrial and cytoplasmic systems of *Neurospora* show incompatibility, and this observation reflects a differing specificity of mitochondrial and cytoplasmic ribosomes in their interaction with chain elongation factors. That the mitochondrial system can be exchanged with the bacterial system, but not with the cytoplasmic system from the same cell or from rat liver, suggests functional similarities between mitochondrial and bacterial ribosomes and chain elongation factors.

2. *In yeasts*

An active submitochondrial protein synthesizing system was also obtained from the mitochondria of *S. cerevisiae* (Scragg *et al.*, 1971; Morimoto *et al.*, 1971). Amino acid incorporation is dependent upon ribosomes and an ATP-generating system (Table LVII). It is stimulated by poly-(U) or by phage R17 RNA, and is sensitive to RNAase, chloramphenicol, streptomycin and erythromycin, but is not inhibited by cycloheximide (Table LVIII). Higher rates of poly-(U)-directed poly-Phe synthesis have been obtained with mitochondrial ribosomes from yeast by Grivell *et al.* (1971b). When combined with supernatant factors from *E. coli* these ribosomes synthesize poly-Phe at rates up to 3·7 nmoles/mg RNA/30 min, a value approaching that obtained with ribosomes from *E. coli*. Poly-(UG) stimulated incorporation of Phe, Leu, and Gly into acid-insoluble material, and Leu incorporation was as sensitive to chloramphenicol inhibition as it was in the case of *E. coli*. ribosomes. Yeast mitochondrial ribosomes translate phage MS2 RNA, but amino acid analysis of the product suggests mis-

Table LVII. Poly-(U)-directed poly-Phe synthesis by a submitochondrial system from yeast. (Reproduced with permission from Morimoto *et al.*, 1971.)

Additions	pmoles (^{14}C)-Phe incorporated at 30°C for 30 min/mg ribosomal protein
Complete system	4·0
− poly-(U)	2·7
+ 100 μg poly-(U)	14·4
− ribosomes	0
− ATP, UTP, CTP, GTP	1·02
− nucleotide and PEP, PEP kinase	0
+ 5 μg ribonuclease	1·18

The reaction mixture (0·20 ml) contained: 28·5 μmoles tris HCl pH 7·6, 1·0 μmoles magnesium acetate, 2·0 μmoles KCl, 2·0 μmoles dithiothreitol, 0·2 μmoles spermidine, 0·5 μmoles ATP, 0·15 μmoles each of UTP, GTP and CTP, 0·6 μmoles PEP, 5 μg PEP kinase, 25 μg of poly-(U), 20 μg of mitochondrial tRNA, 0·5 μg of (^{14}C)-L-Phe, 50–100 μg of supernatant enzyme and 100–300 μg of ribosomal protein.

Table LVIII. The effect of various inhibitors upon poly-(U)-directed cell free synthesis from the cytoplasm and mitochondria of yeast. (Reproduced with permission from Morimoto *et al.*, 1971.)

Inhibitor		% inhibition	
		Cytoplasmic	Mitochondrial
Cycloheximide	0·1 mM	85	0
Chloramphenicol	5·0 mM	17	47
Streptomycin	0·2 mM	25	47
Erythromycin	0·1 mM	25	52

The reaction mixture for the mitochondrial system is described in the legend to Table LVII except that the magnesium concentration was 7 mM.

reading; thus inability for recognition of the initiation sites leads to the production of proteins different from those normally translated (Grivell and Reijnders, 1973).

Although hybrid ribosomes containing chloroplast and *E. coli* subunits are active in poly-Phe synthesis, yeast mitochondrial subunits do not form hybrid ribosomes with either chloroplast or *E. coli* subunits (Grivell and Walg, 1972).

3. *In* Tetrahymena pyriformis

The 80S mitochondrial ribosomes from *T. pyriformis* serve as the functional monosome in (^{14}C)-Phe incorporation (Allen and Suyama, 1972). The rate of incorporation in the presence of poly-(U) was 126 pmoles/h/per mg ribosomal protein: incorporation was chloramphenicol-sensitive and cycloheximide-resistant. In the absence of poly-(U) some incorporation into high mol. wt proteins was still evident, presumably under the direction of endogenous messenger. Incompatibility between the ribosome and supernatant fractions of mitochondrial and cytoplasmic systems was demonstrated.

4. In Euglena gracilis

Intact mitochondrial polysomes have been isolated from *E. gracilis* (Avadhani and Buetow, 1972a, b); these are directly active in an *in vitro* protein synthesis system in the presence of ATP, GTP and supernatant enzymes from isolated mitochondria, and in the absence of added messenger RNA (Table LIX). Mitochondrial ribosomes exhibited little activity

Table LIX. Requirements for protein synthesis by polyribosomes. (Reproduced with permission from Avadhani and Buetow 1972a.)

The complete system included cytoplasmic, chloroplast, or mitochondrial polyribosomes (equivalent to 40 μg of RNA), the S-150 fraction from *Euglena*, and other components as described. The cytoplasmic, chloroplast and mitochondrial polyribosomes incorporates 12000, 8900 and 9400 c.p.m., respectively. Bovine pancreatic RNAase (deoxyribonuclease-free) was added to a concentration of 20 μg/tube; EDTA was added to a 10 mM concentration. Chloramphenicol and cycloheximide were added at a concentration of 25 μg/tube. All the additions were made at zero time of incubation. The percentage activity was calculated on the basis of control values, which were considered to be 100% active.

Addition or omission	Activity (% of control)		
	Cytoplasm	Chloroplast	Mitochondria
None	100	100	100
− Five unlabelled L-amino acids	45	37	41
− ATP	9	7	8
− Phosphoenol-pyruvate	16	21	21
+ RNAase	29	30	28
+ EDTA	20	24	25
+ Chloramphenicol	91	48	47
+ Cycloheximide	22	93	90

in the absence of added RNA, but phenol-extracted mitochondrial RNA serves as an efficient message. The mitochondrial ribosomal system requires supernatant enzymes (either from mitochondria or *E. coli*) for complete activity. A similar "pH 5·4 fraction" from *Euglena* cytoplasm was 50% efficient, whilst that from rat liver cytoplasm was only 12% efficient. Incorporation of labelled amino acids by the mitochondrial protein synthesis system was sensitive to inhibition by chloramphenicol and was cycloheximide-resistant.

D. Mitochondrial Peptide Chain Initiation

Bacteria and mitochondria have similar mechanisms for the initiation of protein synthesis. Ribosomes from the mitochondria of *Neurospora crassa* are able to recognize, bind and translocate formyl-Met-tRNA in response to the codon AUG (Sala and Küntzel, 1970). Chloramphenicol and sparsomycin inhibit the translocation stage of initiation. A procedure which is known to remove initiation factors from the ribosomes of *E. coli* (i.e. washing the mitochondrial ribosomes with 1 M–NH$_4$Cl) renders them incapable of binding formyl-Met-tRNA. This capacity is restored by the addition of initiation factors from *E. coli*. Cytoplasmic ribosomes did not bind formyl-Met-tRNA, and bacterial initiation factors had no effect. Thus the similarities between mitochondrial and bacterial protein synthesis extend to the mechanism of polypeptide chain initiation, in that the specific reaction of mitochondrial ribosomes and the bacterial initiation factors F$_1$ and F$_2$ can occur.

Isolated yeast mitochondria can synthesize formyl-Met peptidyl-puromycin in the presence of a system for the incorporation of amino acids, a formyl group donor and puromycin (Bianchetti *et al.*, 1971). This indicates that endogenous messenger(s) programme polypeptide chains starting with formyl-Met, and confirms the physiological importance of formyl-Met-tRNA in the initiation of protein synthesis.

The incorporation of labelled formyl groups in the presence of puromycin into labelled formyl-Met puromycin has been used as a continuous regenerative assay of chain initiation (Mahler *et al.*, 1974). Chain initiation does not take place in any of the ρ° mutants tested. Although mitochondrial protein synthesis is inhibited by ethidium bromide, peptide chain initiation is not affected (Mahler *et al.*, 1974).

E. Mitochondrial Peptide Chain Elongation

Early work on cross-complementation between bacterial ribosomes and supernatant fractions prepared from mammalian tissues or eukaryotic microbes in poly-(U)-directed Phe-incorporating systems, gave confusing results because of the failure to distinguish between cytoplasmic and mitochondrial elongation factors (Rendi and Ochoa, 1962; Nathans and Lipmann, 1961; So and Davie, 1963; van Etten *et al.*, 1966; Richter *et al.*, 1968; Ciferri *et al.*, 1968; Albrecht *et al.*, 1970; Tiboni *et al.*, 1970). It was evident in these experiments that transfer factors active on both 70S and 80S ribosomes were present in supernatant fractions from yeast and from *Prototheca zopfii*. Separation of mitochondrial and cytoplasmic peptide chain elongation factors in *S. fragilis* and in *S. carlsbergensis* was achieved by Richter and Lipmann (1970), who showed the presence of two factors (G$_{mit}$ and T$_{mit}$) analogous to peptidyl translocase and amino acyl-tRNA binding factors in mitochondrial extracts. The mitochondrial and cytoplasmic factors differed in their mol. wts and chromatographic mobility, and an antiserum raised to the cytoplasmic factors did not react with the factors from mitochondria. Mitochondrial factors and ribosomes were interchangeable with bacterial ribosomes and factors, but cytoplasmic ribosomes responsed only to the binding factor, T_{mit}. Cytoplasmic ribosomes also responded to bacterial T. It was concluded that mitochondrial chain elongation factors are of a prokaryotic rather than a eukaryotic type. Absolute ribosome specificity for the two sets of transfer factors was shown in the systems from *S. fragilis* (Pirani *et al.*, 1971), and from *S. cerevisiae* (Scragg, 1971), and it was suggested that the reports that T_{mit} can function with G_{cyto} and cytoplasmic ribosomes, (Krikso *et al.*, 1969; Richter and Lipmann, 1970) may have been the result of cross contamination or incomplete removal of elongation factors from ribosome preparations. Although mitochondrial supernatants crossreacted with *E. coli* ribosomes, *E. coli* supernatants showed far more activity with either *E. coli* or mitochondrial ribosomes (Morimoto *et al.*, 1971; Scragg, 1971): it was shown that T_{mit} is twice as unstable as *E. coli* T. T_{mit} was separated into Tu and Ts. Different functions for Ts in the stimulation of (^{3}H)-GDP exchange with Tu or Tu-GDP were evident between the yeast mitochondrial and *E. coli* Ts factors.

Mutants of *S. cerevisiae* containing no detectable mt DNA (obtained by ethidium bromide treatment) still contained the complete set of mitochondrial elongation factors T and G (Richter, 1971; Scragg, 1971); these enzymes were identical in function, chromatographic behaviour and immunological properties with their counterparts in the parent strain. The 70S-specific factors were also still present after growth of cells with chloramphenicol (Parisi and Cella, 1971), and therefore the mitochondrial peptide chain elongation factors are encoded on nuclear DNA and translated on cytoplasmic ribosomes. The synthesis of mitochondrial chain elongation factors T and G in *S. fragilis* is repressed in the presence of high concentrations of glucose or under anaerobic conditions, but stimulated during growth with chloramphenicol (Richter, 1973). Two complementary peptide chain elongation factors (T and G) have also been isolated from the mitochondria of *Neurospora crassa* (Grandi and Küntzel, 1970). Both factors are specific for 70S ribosomes and can be crossed with T and G factors from *E. coli*. The T activity of mitochondrial polymerizing enzymes varies with different preparations and is completely lost after freezing intact mitochondria prior to extraction. The instability of T_{mit} accounts for the relative inactivity of the combination [T (Neurospora) + G(*E. coli*)]. Although G_{mit} is interchangeable with the bacterial G factor, the two factors differ in their response to fusidic acid (Grandi *et al.*, 1971). This antibiotic inhibits protein synthesis in bacteria and in the cytoplasm of eukaryotes by interaction with G factors; it does not however affect the function of mitochondrial G factor in *Neurospora*.

Two G factors which function with *E. coli* ribosomes have been isolated from light-grown *Chlorella vulgaris* (Ciferri *et al.*, 1974). One of these two factors is probably located in the mitochondria, as one of them is missing in dark-grown cells and in a mutant lacking photosynthesis. Fusidic acid was used to titrate the amount of elongation factor in crude extracts of this mutant. Nalidixic acid, rifampicin, erythromycin, or choramphenicol all inhibited growth of the mutant but gave no inhibition of the production of elongation factor, again suggesting that this factor is coded by nuclear DNA and translated on cytoplasmic ribosomes.

F. Products of *in vitro* Mitochondrial Protein Synthesis

1. *In* Neurospora crassa

Early studies on *N. crassa* suggested that the intrinsic system of protein synthesis within the mitochondrion synthesizes a "structural' protein (Woodward and Munkres, 1966), but this concept was evidently an oversimplification. Protein fractions from many sources which had been identified as homogeneous structural protein of the inner mitochondrial membrane were found to contain many different proteins including denatured F_1–ATPase (Lenaz *et al.*, 1968a, b; Schatz and Saltzgaber, 1969b; Senior and McLennan, 1970; Kaplan and Woodward, 1973).

Mitochondria isolated from *N. crassa* under optimal conditions incorporate 300 pmoles of labelled Leu/h/mg protein (Sebald *et al.*, 1968; 1971). Nearly all the radioactivity incorporated is associated with the mitochondrial membrane proteins, and analysis by gel electrophoresis indicated that of the 25 discrete bands identified, only five were highly labelled. In mitochondria from the cytoplasmic *mi-1* mutant (*poky*) the rate of incorporation was only 20% that of the wild-type mitochondria. Analogous protein bands were obtained from the mutant mitochondria; although quantitative differences were evident in several bands the remarkable feature was that no labelling of a specific protein ("band 4 protein") occurred (but see p. 414).

An outer mitochondrial membrane fraction prepared from *N. crassa* mitochondria, labelled *in vitro* with radioactive amino acids, contained only negligible incorporated radioactivity (Neupert and Ludwig, 1971).

The *in vitro* transcription products of mt DNA from *N. crassa* contain messenger RNAs that are translated by a submitochondrial system into at least four distinct protein species with mol. wts ranging from 11,000 to 180,000 (Blossey and Küntzel, 1972). Gel electrophoresis of proteins synthesized by mitochondrial ribosomes *in vitro*, in the presence or in the absence of mitochondrial transcription products, indicated that, whereas the latter (probably representing the run-off of mitochondrial polysomes) closely resembled the pattern of *in vivo* products obtained in the presence of cycloheximide, the pattern of DNA-dependent translation products was completely different. Several alternative explanations were proposed for this discrepancy, that the DNA-dependent translation products (a) are artefacts, (b) represent precursors of the *in vivo* products, (c) are identical with the *in vivo* products but have different aggregational properties, or (d) that they represent true mitochondrial gene products not found *in vivo* in the mitochondria.

2. In yeast

The products of *in vitro* incorporation of (^{14}C)-amino acids into proteins of mitochondria isolated from *S. carlsbergensis* are associated with a membrane fraction (Yang and Criddle 1969, 1970) which was further subfractionated. The proteins were solubilized by reduction and carboxymethylation, or by sulphonation in the presence of SDS and separated by disc gel electrophoresis. A predominant component ("CS-1") had the highest specific radioactivity, and accounted for 50–60% of the total incorporated radioactivity; incorporation into this component was strongly inhibited by chloramphenicol. Actinomycin D, rifamycin, and ethidium bromide had no effect on *in vitro* incorporation; incorporation into membrane components other than CS-1 was not very sensitive to chloramphenicol. Mitochondria-like structures from glucose-repressed anaerobically-grown cells showed only one-tenth of the incorporating activity of fully functional mitochondria, and limited sensitivity to chloramphenicol. Another approach to the problem of the isolation of specific membrane protein fractions is to extract hydrophobic proteins found in association with membrane lipids directly into chloroform: methanol mixtures (Soto *et al.*, 1969; Cattell *et al.*, 1970; Kadenbach, 1971b; Kužela *et al.*, 1973). Kadenbach (1971b) has reported the presence of a single hydrophobic peptide of mol. wt 2000 as the sole product of liver mitochondrial protein synthesis *in vivo*. Proteolipid accounts for up to 3% of total yeast mitochondrial protein (Murray and Linnane, 1972). The incorporation of (^{14}C)-Leu into proteolipid is inhibited by erythromycin and mikamycin, but not by cycloheximide. It is stable to a puromycin chase, which solubilizes 40% of the total incorporated radioactivity. The proportion of total incorporated radioactivity ranges between 20 and 40% of that in completed polypeptides, depending on the growth phase of the culture, and therefore it is evident that proteolipid is not the only product of mitochondrial protein synthesis in yeast.

III. Mitochondrial Protein Synthesis in vivo

A. Contribution of Mitochondrial Protein Synthesis to the Total Synthesis of Mitochondrial Proteins

The selective inhibition of the intra-mitochondrial protein synthesis system by chloramphenicol, and of cytoplasmic protein synthesis by cycloheximide, has been used to study the contribution of the mitochondrial system *in vivo*. Hawley and Greenawalt (1970) described a procedure which was designed to inhibit maximally the synthesis *in vivo* of all proteins of *Neurospora crassa* except those synthesized in the mitochondria. The cells were preincubated for 15 min with 100 μg/ml of cycloheximide prior to the introduction of (^{14}C)-Leu. Of the protein synthesized by whole cells, only a small fraction (2·15%) was completely resistant to inhibition by cycloheximide and sensitive to inhibition by chloramphenicol. Subcellular fractionation by sucrose-gradient centrifugation indicated that the cycloheximide-resistant, chloramphenicol-sensitive protein synthesizing system cosedimented with cytochrome *c* oxidase activity; this observation confirmed that it was the mitochondria which were the *in vivo* sites of cycloheximide-resistant protein synthesis. It was estimated that the protein synthesized *in vivo* in cycloheximide-inhibited cells was about 14·5% of the total mitochondrial protein. Mitochondrial protein synthesis in this system was over 100 times more rapid than the fastest rates reported *in vitro*, and was linear for at least 30 min. A figure of between 8 and 11% for the proportion of total mitochondrial protein synthesized by the intrinsic mitochondrial system was obtained in other similar experiments with *Neurospora* (Sebald *et al.*, 1971; Swank *et al.*, 1971).

Only 11% of the total mitochondrial protein is synthesized at a chloramphenicol-sensitive site in intact suspensions of *S. cerevisiae* growing in the presence of 1% glucose (Henson *et al.*, 1968a). Using pulse-labelling and pulse-chase experiments with Phe as a tracer, derepressing yeast cells incorporate this precursor into their protein with a lag of less than 1 min in a linear fashion for periods ranging up to 60 min. The incorporation can be diluted by the addition of unlabelled amino acid, and the specific activity of acid-soluble material becomes constant within 90 s. Cycloheximide inhibits within 1 min; maximal inhibition is 98% at 2 μg/ml, and the inhibition is reversible.

Chloramphenicol does not inhibit incorporation by repressed cells, but gives a maximum inhibition of 22% at 4 mg/ml in derepressing cells, a concentration that blocks derepression completely. The non-additivity of the effects of the two inhibitors is interpreted in terms of some form of coupling between the two systems of protein synthesis (Mahler *et al.*, 1971a).

In galactose-grown yeast 8–9% of the total mitochondrial protein (accounting for 15% of the total mitochondrial membrane protein) is synthesized *in vivo* by the cycloheximide-resistant system of protein synthesis (Schweyen and Kaudewitz, 1970): cycloheximide-resistant protein synthesis was not detectable in a cytoplasmic *petite* mutant. Kellerman *et al.* (1971) have shown that the contribution *in vivo* of mito-chondrial protein synthesis to the synthesis of mitochondrial total protein is decreased from 15% in catabolite-derepressed yeast to as little as 5% in catabolite-repressed cells.

The promitochondria of *S. cerevisiae* grown anaerobically in the presence of lipid supplements may be selectively labelled *in vivo* by incubating the cells with radioactive Leu in the presence of cycloheximide under nitrogen (Schatz and Saltzgaber, 1969a). The incorporation of labelled Leu under these conditions was inhibited by chloramphenicol, and was not detectable in cytoplasmic *petite* mutants. Promitochondria from lipid-depleted anaerobically-grown *S. cerevisiae* contain only trace amounts of isotope after pulse-labelling of cells in the presence of cycloheximide (Gordon and Stewart, 1972) and erythromycin does not inhibit this in corporation. Oxygenation of a non-growing cell suspension of lipid-depleted anaero-bically-grown yeast leads to a rapid reactivation of mitochondrial protein synthesis accompanied by the rapid synthesis of unsaturated fatty acids and ergosterol.

The contribution of mitochondrial protein synthesis to total synthesis of cell proteins may be higher in the trypanosomatid protozoan *Crithidia fasciculata* than in *Neurospora* or yeast (Laub and Thirion, 1972). In this case, concentrations of more than 0·8 mg/ml chloramphenicol gave 50% reduction in the capacity of whole cells to synthesize protein.

Another approach to the specific labelling of mitochondrial translation products, based on the observation that the mitochondrial protein syn-thesis system utilizes formyl-Met in chain initiation, is to use radioactive formate to specifically label nascent polypeptide chains in sphaeroplasts of *S. cerevisiae* (Mahler *et al.*, 1972; Dawidowicz and Mahler, 1972). In the presence of puromycin, quantitive conversion into formyl-Met peptidyl-puromycin deviatives occurs. It was shown by this method that mitochondrial protein synthesis is decreased under conditions of glucose represssion and was absent from a strain lacking mt DNA. As mito-chondria are deficient in deformylase and other enzymes involved in post-translational processing, N-terminal formyl-Met residues are retained subsequent to the release of completed polypeptide chains and their inte-gration into the inner membrane (Mahler *et al.*, 1974). These residues can be identified, either directly after complete enzymatic proteolysis, or as formate after mild acid hydrolysis, and therefore provide a direct test for mitochondrial translation products not dependent on the use of specific inhibitors.

Selective labelling of mitochondrially-synthesized proteins without use of inhibitors is also possible in a temperature-sensitive mutant (171 ts⁻) of *S. cerevisiae* in which only the cytoplasmic system is inactive at 36·5°C (Schweyen and Kaudewitz, 1971).

B. Membrane Proteins Synthesized *in vivo* by Mitochondrial Protein Synthesis

1. *In* Neurospora crassa

In *N. crassa* the mitochondrial system of protein synthesis is active for at least 1 h after inhibition of the extra-mitochondrial system in intact mycelia by cycloheximide (Sebald *et al.*, 1969). This indicates that the two systems are not closely coupled in the presence of the antibiotic, and permits specific labelling of mitochondrial products *in vivo*. Amino acid incorporation into extractable mitochondrial proteins is inhibited to the same extent (97%) as incorporation into proteins of the cytosol, and cycloheximide-resistant incorporation is found almost exclusively in mitochondrial membrane proteins. Sonication was used to remove soluble proteins, and after lipid extraction, insoluble membrane protein was dissolved in phenol-formic acid-water and analysed by acrylamide gel electrophoresis. Twenty-five protein bands were revealed on staining with amido black; only six of these were radioactively labelled (accounting for 5–8% of the total mitochondrial proteins). These labelling patterns were essentially the same as those obtained on analysis of mitochondrial membranes labelled *in vitro* (Sebald *et al.*, 1968, 1971), although the comparative electrophoretic mobilities and the individual proportions of proteins synthesized under the two conditions were not reported. SDS-acrylamide gel electrophoresis confirmed that only six or seven protein components are synthesized in cycloheximide-inhibited *N. crassa* (Swank *et al.*, 1971a). Of the total mitochondrial protein, 10% was accounted for by three proteins labelled under these conditions; these were of mol. wts 33,500, 27,700 and 17,500 (relative proportions 64:20:16). About 1% of the total protein was found in three other labelled components with mol. wts 11,000, 21,000 and 25,000. In similar experiments with the *mi-1* mutant, three components were also detected which accounted for the major part of the cycloheximide-resistant incorporation. However the relative proportions and electrophoretic mobilities of these differed from those in the wild-type. These experiments thus failed to substantiate the claim of Sebald *et al.* (1968) of a missing cycloheximide-resistant protein from mutant *mi-1*, or the conclusion of Neupert *et al.* (1971) that the mutant is defective in mitochondrial protein synthesis, although as pointed out by Swank *et al.* (1971a) differences in the strains of *Neurospora* used, growth conditions and analytical methods may account for the contradictory results obtained. Similarly the detection of only one major labelled product of mitochondrial protein synthesis may be a consequence of different experimental techniques (Birkmayer, 1971a, b). Swank *et al.* (1971a, b) have isolated mitochondrial proteins of mol. wts between 2500 and 10,000, comprising 12% by weight of total mitochondrial protein. These were synthesized on cytoribosomes but were not detectable in the cytosol; they were apparently not products of turnover or incomplete synthesis. The possible functions of these "miniproteins" have been discussed by Munkres *et al.* (1971).

The sites of synthesis of proteins of the inner- and outer-membranes of mitochondria of *N. crassa* have been investigated by Neupert and Ludwig (1971). Polyacrylamide gel electrophoresis enabled the resolution of about 20 protein bands from the inner-membrane, whereas the outer-membrane showed essentially only one band. When mycelia were incubated in the presence of radioactive amino acids and cycloheximide, hardly any incorporation of label into mitochondrial outer-membranes was observed. It was concluded that the vast majority, and possibly all, of the outer-membrane protein is synthesized by the cytoplasmic system, and that polypeptide chains formed by mitochondrial ribosomes are integrated into the inner mitochondrial membrane.

Selective pulse-labelling of the nascent translation products on mitochondrial ribosomes and in the inner mitochondrial membrane by radioactive Leu in cycloheximide-inhibited *N. crassa* showed that up to 85% of the radioactivity associated with mitochondrial ribosomes is found at polymeric ribosomes which cannot be degraded by RNAase and tend to form aggregates (Michel and Neupert, 1973, 1974). Aggregation appeared to be due to the highly hydrophobic nature of the nascent polypeptides, and these products showed a broad range of apparent mol. wts. Material associated with monomeric ribosomes had an apparent mol. wt of 27,000 and probably represents peptidyl-tRNA. Products in the membrane after short chase periods were of mol. wts in the range 8,000–12,000, whereas longer chases gave products of mol. wts 30,000 and 36,000. The conversion of the low mol. wt products to higher mol. wt products during the chase was temperature-dependent. It was suggested on the basis of these results that: (1) the polypeptides at monomeric mitochondrial ribosomes represent completed chains which are transported by the ribosomes to the membrane; (2) the original translation products are converted at the membrane into products of higher mol. wt which are incorporated into complexes (such as cytochrome *c* oxidase, cytochrome *b*, ATPase); and (3) the reason for the existence of a mitochondrial translation system is that the products of this system are so hydrophobic that they have to be delivered to the inner membrane from the matrix side as they cannot be transported through cytoplasm and outer membranes.

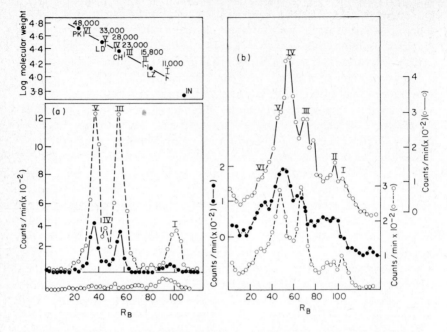

FIG. 114. (a) Electrophoretic profiles of labelled mitochondrial membranes prepared from *Saccharomyces cerevisiae* (diploid strain Z65) showing the effects of different concentrations of erythromycin in the presence of 100 µg/ml cycloheximide. ○---○, Control + cycloheximide; ●——● erythromycin (200 µg/ml) + cycloheximide; ○——○, erythromycin (3 mg/ml) + cycloheximide. Cells were grown to late logarithmic or early stationary phase in 1% yeast extract (Difco), 2% Bactopeptone, 2% glucose, at 25°C. In some cases, inhibitors and labelled precursors were added directly to growing cultures at cell densities of about 5×10^7 cells/ml. More usually, however, the cells were washed once in phosphate buffer (0·05 M, pH7·4) and resuspended at between 10^8 and 10^{11} cells/ml in incubation medium (0·05 M–phosphate buffer, pH7·4, ethanol (2–4%), 0·5% glucose and 100 µg/ml cycloheximide). After incubation for 15 min at 25°C, labelled amino acid was added (^{35}S-methionine, specific activity of 19–24 Ci/mmol, Amersham) at 3–5 µC/ml, and incorporation continued for 1–2 h. The cells were then washed twice by centrifugation in ice-cold TME (tris HCl buffer, 0·05 M, pH 7·4, mannitol 0·25 M, EDTA 2 mM). The pellet was suspended in 2 vol TME, 1 vol of acid-washed glass beads (0·8 mm diameter) was added and the cells broken in a CO_2-cooled Braun shaker for 40–60 s. Breakage, estimated by microscopic examination, was around 20–40%. The broken cell suspension was diluted three fold with ice-cold TME and centrifuged at 5,000 g for 5 min. The supernatant was layered on a 52% sucrose cushion and centrifuged at 20,000 g for 10 min in a Spinco SW27 rotor at 4°C. The pellet was resuspended in Na_3PO_4 (about 20 mM, pH between 11·0 and 11·5) at a protein concentration of about 5 mg/ml, and again centrifuged at 1,500 g for 15 min in the cold. This treatment reduced the total label in the pellet by only about 10% but removed a good deal of the total protein content, thus enabling the loading of more radioactivity on the gels without overloading with unlabelled protein. Controls showed the loss of label at this stage had no significant effect on the observed final gel pattern. The translucent yellow pellet obtained at this stage was resuspended in sodium phosphate buffer (0·01 M, pH8·0) containing 0·1% SDS and 0·01 M–mercaptoethanol, and heated at 90°C for 2 min to solubilize the membranes. The solution was dialysed against the same buffer containing casamino acids (0·5 mg/ml Difco) at room temperature overnight. SDS polyacrylamide (10%) disc gels

2. In Aspergillus nidulans

Cycloheximide-resistant incorporation of radioactive Leu in mycelia of
A. nidulans was into four mitochondrial membrane components (mol. wts
13,000, 18,000, 27,000 and 40,000) (Turner, 1973).

3. In yeast

Electrophoretic analysis of membrane proteins from mitochondria isolated
after incubation of *S. cerevisiae* with a radioactive amino acid in the
presence of cycloheximide, revealed label in only a few well defined bands
(Thomas and Williamson, 1971; Fig. 114). These bands (I-VI) corresponded to proteins of mol. wts ranging from 11,000 to 48,000. Radioactivity was also incorporated from (2-^{3}H)-glycerol into two of the bands
(I and VI), and a portion of the incorporated label was soluble in chloro-
form/methanol suggesting that these products may have a lipid content
and may consist of lipoproteins. The synthesis of much of the material in
band I was more sensitive to erythromycin inhibition than that in the
other bands. Daunorubicin, a DNA-binding drug, blocked all incorporation except into band I, and in certain strains ethidium bromide had a
similar effect. Mutant P203, which has temperature-sensitive mitochondrial
ribosomes, incorporates label only into band I at 35°C, and has a normal
pattern of incorporation at permissive temperatures. Thomas and Williamson (1971) have pointed out that the pattern of incorporation is remarkably
simple and similar to that found in similar experiments with mammalian
cells (Coote and Work, 1971), and that this is surprising in view of the
size difference between the mitochondrial genomes of mammalian and
yeast mitochondria. If each of the bands represents a single polypeptide
species, then the total amount of information needed to specify them
amounts to about 1 μm of DNA. Rather more information was required to
specify all the protein products detected by Mohar-Betancourt and
García-Hernández (1973). Tzagoloff and Akai (1972) have observed
five major radioactive bands (mol. wt range 7,800–45,000) which account
for over 60% of the radioactivity incorporated from L(4,5-^{3}H)-Leu after
acrylamide gel electrophoresis of mitochondrial membrane proteins isolated
from cycloheximide-inhibited *S. cerevisiae*. The mitochondrial products
can be extracted from the membrane with acidic chloroform-methanol. A

were run in 6 mm × 150 mm glass tubes at 5 mA/gel for 20 h. The gels were usually
sliced with a multiple razor blade device after freezing, and the gel fragments (about
50/gel) extracted with 0·5 ml hydrogen peroxide before adding 5 ml scintillation
fluid. (b) Electrophoretic profiles of labelled mitochondrial membranes prepared from
diploid strain Z65, showing the effects of different concentrations of ethidium bromide in
the presence of 100 μg/ml cycloheximide. O– – –O, Control + cycloheximide; ●———●,
ethidium bromide (2 μg/ml) + cycloheximide; O———O, ethidium bromide (10 μg/ml)
+ cycloheximide. (Reproduced with permission from Thomas and Williamson, 1971.)

polymeric protein (mol. wt 45,000) can be depolymerized to give protein of mol. wt 7,800 and this protein appears to be the major product of mitochondrial protein synthesis. It has been suggested that the polymer accumulates in the presence of cycloheximide, but that normal yeast mitochondria contain the low molecular weight form. Tzagoloff and Akai (1972) suggest that this hydrophobic protein may play a structural role in the assumbly of multienzyme complexes.

Proteins missing in mitochondrial membranes of *petite* mutants correspond to the mitochondrial products observed in membranes labelled in the presence of cycloheximide (Groot *et al.*, 1972). Electrophoretic comparison of polypeptides labelled *in vivo* in the promitochondria of cycloheximide-inhibited cells in the presence or absence of O_2 indicates that O_2 controls the synthesis of a specific class of protein (mol. wts 42,000 and 34,000 (Groot *et al.*, 1972). This effect of O_2 is inhibited by glucose. These results suggest that the promitochondrial system can respond to O_2 and to glucose during respiratory adaptation by a mechanism independent of the cytoplasmic system of protein synthesis.

Ibrahim *et al.* (1973) have shown that 80–90% inhibition of *in vitro* mitochondrial protein synthesis occurs in the presence of chloramphenicol, erythromycin or ethidium bromide, whereas actinomycin D and cordycepin gave only 20% inhibition, and rifamycin or cycloheximide were without effect. The rate of amino acid incorporation *in vitro* was not subject to glucose repression, but the cycloheximide-insensitive incorporation into mitochondrial membrane proteins *in vivo* was lowered more than 90% in repressed cells and the products were different from those in derepressed cells. A two-fold increase in labelling of the two components of highest mol. wts as well as the peak of lowest mol. wt was observed in mitochondrial membranes from repressed cells. Stimulation of protein synthesis *in vivo* and *in vitro* was observed in partially derepressed cells which had been grown with chloramphenicol for 3 h, washed and allowed to grow for 1 h in fresh medium without the antibiotic. Results suggested that the accumulation of mitochondrial proteins in the cytoplasm stimulated the synthesis of certain other specific mitochondrial proteins under these conditions. Depletion of the cytoplasmically-synthesized protein pool by growth with cycloheximide followed by growth in the absence of this inhibitor resulted in a 50% decrease in the incorporation rate *in vitro*.

The synthesis of glycoproteins has been demonstrated in isolated mammalian mitochondria. Rat liver mitochondria incorporate labelled monosaccharides (fucose, mannose, glucose and galactose, but not xylose) from their UDP derivatives into at least four glycoproteins which have been identified by SDS polyacrylamide gel electrophoresis (Bosmann and Martin, 1969; Bosmann, 1971a). Altered synthesis of glycoproteins has been shown in hepatoma mitochondria (Bosmann and Myers, 1974).

Cyclic AMP and dibutyryl cyclic AMP at high concentrations increased protein and RNA synthesis in isolated mammalian mitochondria: only the former compound stimulated glycoprotein synthesis (Bosmann, 1971b). Similar studies with microbial mitochondria have not been performed.

10

Synthesis of Mitochondrial Components by Cooperation Between the Mitochondrial and Cytoplasmic Systems

I. Synthesis of Enzyme Complexes

A. Synthesis of Cytochrome c Oxidase

1. In Neurospora crassa

Hyphae of *N. crassa* were labelled with radioactive Leu in the presence of cycloheximide and the membrane protein of their mitochondria separated on oleyl polymethacrylic acid resin (Weiss *et al.*, 1971). Double labelling was used, so that the profile of (^{14}C) radioactivity represented total membrane protein, whereas that of (^{3}H) radioactivity represented those proteins synthesized by the cycloheximide-insensitive system. The highest (^{3}H/^{14}C) ratio was found in a fraction containing cytochromes $a + a_3$, which proved to be a pure and enzymically-active cytochrome c oxidase. By means of SDS gel electrophoresis the enzyme was separated into five polypeptides with mol. wts 30,000, 20,000, 13,000, 10,000 and 8,000, and one of these polypeptides (mol. wt 20,000) had an exceptionally high content of (^{3}H)-Leu.

More recently cytochrome c oxidase (containing 10 nmoles haem *a*/mg protein) isolated from *N. crassa* has been resolved into seven polypeptides (mol. wts 36,000, 28,000, 18,000, 17,000, 13,000, 11,000 and 8,000) on polyacrylamide gel electrophoresis in the presence of SDS (Sebald *et al.*, 1972). *In vivo* incorporation of labelled amino acids into single polypeptides was investigated after pulse-labelling in the presence and absence of chloramphenicol. Marked inhibition (90–95%) of incorporation of amino acids into all seven polypeptides occurred in the presence of 4 mg/ml chloramphenicol, whereas labelling of whole mitochondrial membrane protein was inhibited by only 30%. After washing out the inhibitor, and allowing hyphal growth to continue, the four smaller polypeptides were highly labelled, whereas the other polypeptides showed only small increases in radioactivity. It was concluded that the four small sized polypeptides of cytochrome c oxidase are synthesized, but not integrated, into the functional enzyme complex in the presence of chloramphenicol. After pulse-labelling of growing $N. crassa$ the appearance of label follows a different time course for each of the polypeptides of cytochrome c oxidase (Schwab et al., 1972). At least four independent pools of precursor poly-peptides were distinguished, and the pool sizes ranged from 2% to 25% of the amount of the corresponding polypeptides present in cytochrome c oxidase. The pool sizes and half-lives of the precursors (min in paren-theses) were as follows: for polypeptides of mol. wt 36,000 10% (23); 28,000, 5% (10); 18,000, 2% (3·5); 13,000, 20% (37); 11,000, 25% (47); 8,000, 4% (7). Serial pools of precursors were evident in the case of the largest polypeptide. The smallest pool was assigned to a polypeptide of mitochondrial origin; the direct transfer of a newly-synthesized peptide chain from mitochondrial ribosomes to the mitochondrial membrane would then constitute the rate-limiting step in the assembly of cytochrome c oxidase. The most recent estimates for the mol. wts of the seven sub-units of cytochrome c oxidase of $N. crassa$ are: (1) 41,000; (2) 28,500; (3) 21,000; (4) 16,000; (5) 14,000; (6) 11,500; and (7) 10,000 (Sebald, 1974; Sebald et al., 1973; Weiss et al., 1973; Weiss, 1974). Components 1, 2 and 3 are products of mitochondrial translation and have a high content of apolar amino acids and a low content of basic amino acids. A stoichio-metry of 1:1:1 for subunits 1, 2 and 3 was obtained assuming a mol. wt of 150,000 for the whole cytochrome c oxidase complex. On the basis of haem a content, a mol. wt of about 70,000/haem group was determined. Thus the smallest structural unit of cytochrome c oxidase contains 2 haem groups and has a mol. wt of 150,000. Precursor proteins of cytochrome c oxidase have been shown in Cu-depleted cells (Schwab, 1974).

Only four mitochondrially-synthesized membrane proteins have been detected in the experiments of Lansman et al. (1974): of these the three highest mol. wt (38,000, 30,000 and 20,000) were components of cyto-chrome c oxidase. These major products of mitochondrial protein syn-thesis became associated with the enzyme only after cytoplasmic products were made available. Correspondence between the in vivo and the in vitro products of mitochondrial protein synthesis (both pulse-labelled and pulse-chase labelled) and with the subunits of cytochrome c oxidase in terms of mol. wts and solubility properties in acidic chloroform-methanol led Rowe et al. (1974) to suggest that the mitochondrially-synthesized subunits of cytochrome c oxidase are representative of the total mito-chondrial protein synthesis, and that these products may function generally in the association of enzymes with the membrane.

P

2. In yeast

Cytochrome $a + a_3$ has been purified 35–40 fold from SMPs of baker's yeast by cholate fractionation, ammonium sulphate fractionation, followed by DEAE-cellulose chromatography in the presence of Triton X-100 (Mason and Schatz, 1973a). The purified cytochrome (containing 10 nmoles of haem a/mg protein) was free of other haemoproteins and was separated into six bands by electrophoresis in polyacrylamide gels in the presence of SDS (Fig. 115). These polypeptides had the following mol.

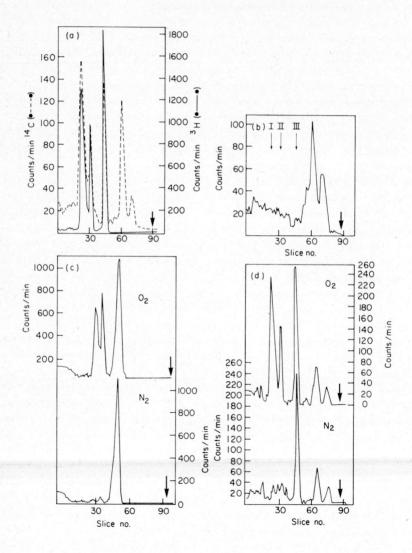

wts: 42,000, 34,800, 23,000, 14,000, 12,500 and 9,500. All six were specifically associated with cytochrome c oxidase, even after solubilization of the mitochondrial inner-membrane, as they behaved as a single species during DEAE-cellulose chromatography and sucrose-gradient centrifugation, and on reaction with rabbit antiserum raised to the holoezyme. Component V (mol. wt 12,500) contained two distinct polypeptide species. Subunit-specific antisera have been prepared for each of the subunits (Ross *et al.*, 1974). That the biosynthesis of cytochrome c oxidase in yeast requires the products of both mitochondrial nad cytoplasmic translation was first demonstrated by Huang *et al.* (1966) and by Clark-Walker and Linnane (1966, 1967). Confirmation of these results was obtained by a detailed analysis of the short- and long-term effects of chloramphenicol and ethidium bromide on the development of respiratory components during glucose derepression (Mahler *et al.*, 1971a), and growth (Mahler and Perlman, 1971), of various species of yeast. Evidence for mitochondrial precursors of this enzyme during the respiratory adaptation of *S. carlsbergensis* was obtained by Chen and Charalampous (1969, 1973). Immunological evidence for the presence of cytoplasmic products as components of cytochrome c oxidase was obtained in a cytoplasmic *petite* mutant lacking the functional enzyme (Kraml and Mahler, 1967). Direct analysis of the contributions of the two systems of protein synthesis to the synthesis of cytochrome c oxidase has now been achieved. Yeast cells were labelled with (^{3}H)-Leu in the presence of specific inhibitors of either the mito-

---

FIG. 115. (a) Cycloheximide-resistant labelling of cytochrome c oxidase. Yeast cells were grown aerobically for 8–10 generations with 0·025 μCi of L-(U^{14}-C)-Leu per ml. The cells were washed and labelled for 60 min with (^{3}H)-Leu in the presence of cycloheximide. The isolated mitochondria (1·54 × 10^5 c.p.m. of ^{14}C and 1·38 × 10^5 cpm of ^{3}H per mg of protein) were fractionated and subjected to immunoprecipation with an antiserum against native yeast cytochrome c oxidase. An aliquot of the immunoprecipitate (3,150 c.p.m. of ^{14}C and 13,620 c.p.m. of ^{3}H) was analyzed by SDS acrylamide gel electrophoresis. (b) Erythromycin-resistant labelling of cytochrome c oxidase. Resting yeast cells were labelled with (^{3}H)-Leu for 60 min as outlined in (a) except that cycloheximide was replaced by 4 mg of erythromycin per ml. In this particular experiment cytochrome c oxidase was isolated by immunoprecipitation from an extract which had been carried out through a DEAE-cellulose chromatography step. An aliquot of the immunoprecipitate (2,136 c.p.m. of ^{3}H) was analyzed by SDS-acrylamide gel electrophoresis. (c) Effect of anaerobiosis on the labelling of cytochrome c oxidase polypeptides in the presence of cycloheximide. Yeast cells were labelled with L-(4,5-^{3}H)-Leu in the presence of cycloheximide, except that one-half of the cells was labelled aerobically and the other half was labelled anaerobically. From each aliquot, cytochrome c oxidase was isolated by immunoprecipitation and analyzed by SDS-acrylamide gel electrophoresis. (d) Effect of anaerobiosis on the synthesis of cytochrome c oxidase polypeptides in the absence of antibiotics. The experiment was identical to that described in (c) except that the cells were labelled in the absence of cycloheximide. The mitochondria were isolated and subjected to immunoprecipitation with an antiserum against the small cytochrome c oxidase components. Aliquots of the immunoprecipitates (aerobic sample, 3,500 c.p.m.; anaerobic sample 1,980 c.p.m.) were analyzed by SDS-acrylamide gel electrophoresis. (Reproduced with permission from Mason and Schatz, 1973b.)

chondrial or cytoplasmic protein synthesis systems. Cytochrome c oxidase
was then isolated from crude mitochondrial extracts by immunoprecipitation and analyzed by SDS acrylamide gel electrophoresis (Schatz et al.,
1972; Mason and Schatz, 1973b). Labelling of the three large components
was found to be insensitive to cycloheximide and sensitive to erythromycin (Fig. 115). Labelling of the components with mol. wts 42,000 and
34,800 was strictly dependent on the presence of O_2. The four smallest
polypeptides were not labelled in the presence of cycloheximide but were
labelled in the presence of erythromycin or acriflavin. Therefore it was
concluded that of the seven polypeptides of cytochrome c oxidase, three
were translated on mitochondrial ribosomes, four on cytoplasmic ribosomes. It was not clear in these experiments whether anaerobiosis inhibited
the synthesis or the integration of two of the polypeptides into cytochrome
c oxidase. Similar results have been reported by Tzagoloff et al. (1973).
Glucose-repressed yeast was incubated sequentially in chloramphenicol and
cycloheximide $+(^{14}C\text{-})$Leu-containing media. SMPs were isolated and
extracted with Triton X-100 under conditions which solubilize cytochrome c oxidase. The Triton extract was treated with antiserum specific
for purified cytochrome c oxidase. The antibody precipitate was depolymerized in SDS and analyzed on polyacrylamide gels. Comparison of the
profile with that obtained on electrophoresis of cytochrome c oxidase
purified from yeast grown with (^{3}H)-Leu, showed that the mitochondrial
protein system contributes the three high mol. wt subunits (Table LX).
Tzagoloff et al. (1973) and Rubin and Tzagoloff (1973b) conclude that
these mitochondrial products are the same as those reported by Schatz

Table LX. Subunit proteins of cytochrome oxidase. (Reproduced with permission
from Tzagoloff et al., 1973)

Subunit	Molecular weight			
	Yeast % acrylamide		Beef % acrylamide	
	7·5	10	7·5	10
1	35,000	40,000	34,500	40,000
2a	24,500	27,300	19,000	22,500
2b	22,000	25,000		
3a	14,500	13,800	13,500	15,000
3b	12,000	13,000		11,200
4	9,500	10,200	9,800	9,800
5		9,500	7,300	7,300

The molecular weights were calculated from the migration values of the proteins relative
to molecular weight standards on 7·5% and 10% polyacrylamide gels in the presence of
sodium dodecyl sulphate. The ratio of acrylamide to bis-acrylamide was 37 for the 7·5%
gels and 54 for the 10% gels.

et al. (1972) and Mason and Schatz (1973b); discrepancies in mol. wts are probably due to the different conditions of gel electrophoresis employed by the two groups of investigators. Further discrepancies between these results and those of Mahler *et al.* (1974) are evident and remain to be resolved; labelling experiments with formate confirmed that three of the smallest subunits are not mitochondrial translation products.

Separation and purification of the subunits has been achieved by a procedure involving acetone treatment, and use of the chaotropic agent, guanidine thiocyanate (Rubin and Tzagoloff, 1973a). All the haem was recovered with the largest mitochondrially-synthesized subunit (subunit 1). This subunit (mol. wt 38,000) contained 62% non-polar amino acid residues and showed the absorption maxima (at 595 nm and 438 nm) characteristic of a denatured cytochrome *c* oxidase. Tzagoloff *et al.* (1974) suggests that the purified cytochrome *c* oxidase with a mol. wt of 150,000 has two haem groups. The ease of removal of the non-covalently bound haem groups during the separation of the subunits makes their definitive location within the cytochrome *c* oxidase complex no easy task.

Another report indicates a method of preparing cytochrome *c* oxidase with a haem *a* content up to 14·6 nmoles/mg protein (min mol. wt of 69,000) which contains only two polypeptide species of mol. wt 14,000 and 11,500 (Komai and Capaldi, 1973).

The relationship between the spectrophotometrically-distinguishable cytochromes *a* and *a*$_3$ in the quaternary structure of the complex is not yet clear; two distinct haemoproteins have never been separated and cytochrome *a*$_3$ may represent a conformational modification of cytochrome *a* (Wainio, 1970). The two haems are very closely positioned and interact very markedly (Leigh and Wilson, 1972; Wilson and Leigh, 1972; Wilson *et al.*, 1972). The observations that the ratio of cytochrome *a*$_3$/cytochrome *a* varies during respiratory adaptation (Cartledge *et al.*, 1972; Chen and Charalampous, 1973) and also during the cell-cycle of glucose-repressed *S. pombe* (Poole *et al.*, 1974) suggest the possibility of a precursor-product relationship between cytochrome *a* and *a*$_3$ regulated by a complex control mechanism (Fig. 116). However, an alternative explanation for these results could be that the independent syntheses of cytochromes *a* and *a*$_3$ occur so as to produce separate and distinct haemoproteins.

As alternative approach to the understanding of the mechanisms of assembly of cytochromes is to block their formation genetically and then to analyze the lesions leading to defective phenotypes.

Three types of mutants have been examined (Ebner and Schatz, 1973; Ross *et al.*, 1974):

(1) Pleiotropic nuclear *petite* mutants (*pet* ρ^+). All 18 examined retain their ability for mitochondrial protein synthesis, but have lost cytochromes *a* + *a*$_3$, *b*, and *c*$_1$, and also several species of polypeptides characterized

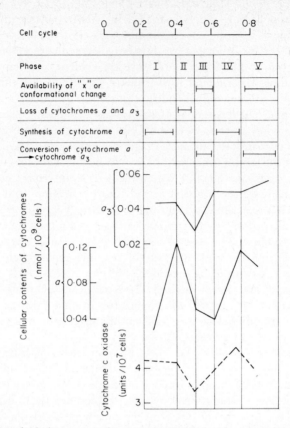

FIG. 116. Hypothetical sequence of events contributing to the expression of cytochromes *a* and *a₃* during the cell cycle of *Schizosaccharomyces pombe*. (Reproduced with permission from Poole *et al.*, 1974.)

by SDS electrophoresis. As several enzymes are affected by a single gene mutation (e.g. in mutant *pet* 7), presumably it is an "organizational" or "assembly" component which is defective. It was suggested from results with another similar mutant which also lacked ATPase (*pet* 936), that the primary lesion was in F_1–ATPase and that "organizer proteins" may be well-recognized enzymes of the inner-membrane necessary for the assembly of other components including those made by the mitochondrial protein synthesis system (Ebner and Schatz, 1973). This mutant does not grow anaerobically even in the presence of unsaturated lipids and fermentable substrates.

(2) Nuclear mutants in which the synthesis or integration of mitochondrially-synthesized subunits of cytochrome *c* oxidase is prevented. Some of these mutations can be suppressed by amber suppressors, and are thus

probably caused by a change in a protein which is nuclear-coded, is translated on cytoplasmic ribosomes, and has some regulatory function in the integration process.

(3) A cytoplasmic *petite* mutant lacked not only the three mitochondrially-synthesized polypeptides, but also the smallest of the four cytoplasmically-synthesized subunits of cytochrome *c* oxidase. This shows the close interaction between the two systems. The remaining three subunits were only loosely attached to the inner-membrane. Extra-chromosomal mutations can thus impair or completely prevent the integration of cytoplasmically-synthesized cytochrome *c* oxidase subunits, and the mitochondrially-synthesized polypeptides are necessary for tight binding of the cytoplasmically-synthesized subunits (Ebner *et al.*, 1973a, b).

B. Synthesis of Cytochrome *b*

A membrane protein containing 16 nmol cytochrome b/mg protein has been prepared from *N. crassa* by chromatography on oleyl-polymethacrylic acid resin (Weiss, 1972). Spectral examination of the purified cytochrome b (α-band in the reduced state at 558 nm at 77°K, 561 nm at room temperature) failed to identify it specifically as either the b_{562} or b_{556} observed *in vivo*. The specific haem content correlates with the relative amount of a 30,000 mol. wt polypeptide in the purified preparations and this suggests that this polypeptide carries the haem group (Weiss *et al.*, 1973; Weiss and Ziganke, 1974a, b; Lorenz *et al.*, 1974). This polypeptide is labelled by incorporation of amino acids in the presence of cycloheximide, but is not labelled in the presence of chloramphenicol, indicating that it is a mitochondrial translation product. The pool size of mitochondrial precursors is about 5% of the polypeptide found in the membrane and the precursor half-life is about 10 min. The delayed appearance of a pulse of (^{3}H)-Leu in the polypeptide, and of a pulse of (^{59}Fe) in the haem group of cytochrome b also indicates the involvement of precursors in the assembly of the haemoprotein.

C. Synthesis of a Component Associated With Cytochrome c_1

Cytochrome c_1 is another mitochondrial component whose formation or integration is dependent on the mitochondrial protein synthesis system (Ross *et al.*, 1974). This cytochrome was purified from yeast mitochondria to a haem content of 27 nmoles/mg protein; it had a mol. wt of about 43,000, and its reduced form did not react with CO or O_2. It was resolved by SDS polyacrylamide gel electrophoresis into two polypeptides (mol. wts 26,000 and 18,000), and the haem was exclusively associated with the larger subunit. The incorporation of (^{3}H)-Leu into cytochrome c_1 of intact yeast cells was cycloheximide sensitive and acriflavine insensitive,

indicating that both polypeptides are synthesized on cytoplasmic ribosomes. It is suggested that the addition of the haem moeity, or integration
of the haemoprotein into the membrane, only occurs when the two cytoplasmically-synthesized polypeptides have been combined with one or
more mitochondrially-synthesized polypeptide.

D. Synthesis of Components of the ATP Synthetase Complex

Rutamycin-sensitive ATPase has been purified from yeast; its activity is
stimulated by the addition of phospholipid, and the purified enzyme is
composed of F_1 and a lipoprotein fraction which binds F_1 and is necessary
for conferral of rutamycin sensitivity (Tzagoloff, 1969a). A second protein
necessary for this similar to F_6 of bovine heart (Fassaden-Raden and
Hack, 1972). The ATPase complex is tightly associated with the inner
mitochondrial membrane and detergents are necessary for its solubilization. The complex extracted in the presence of Triton X-100, and purified
by centrifugation through glycerol gradients, consists of an oval particle
10×15 nm with a mol. wt approximately 460,000 (Tzagoloff and
Meagher, 1971).

The subunit proteins of the purified ATP synthetase complex have
been analysed by polyacrylamide gel electrophoresis in the presence of
SDS. At least nine different mol. wt species are present (Fig. 117). Five

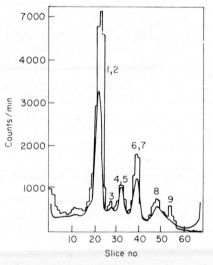

FIG. 117. Gel of the rutamycin-sensitive ATPase complex purified from yeast grown on
(^{3}H)-Leu. Yeast were grown aerobically on 0·8% glucose medium containing (^{3}H)-
Leu. The ATPase complex was purified from the labelled mitochondria and separated
on 7·5% acrylamide gels. The gel was stained with amido black, scanned in a spectro-
photometer at 650 nm (curved trace) sliced into 1 mm sections and counted (blocked trace).
(Reproduced with permission from Tzagoloff and Meagher, 1971.)

of the proteins are subunits of the soluble ATPase (F_1), and the others are components of a lipo-protein factor which alter the catalytic and physical properties of F_1. Subunit 8 consists of two different proteins, only one of which is a subunit of F_1–ATPase. In addition to the eight subunits, ATP synthetase also contains another protein which migrates in the low mol. wt region of the gel. The mol. wt of the ATP synthetase proteins are shown in Table LXI (Tzagoloff and Meagher, 1971).

Table LXI. Molecular weights of the subunit proteins of mitochondrial ATPase. (Reproduced with permission from Tzagoloff and Meagher, 1971.)

| Subunit | Molecular weight | |
	F_1	Rutamycin-sensitive complex
1	58,500	58,500
2	54,000	54,000
3	38,500	38,500
4	31,000	31,000
5	Absent	29,000
6	Absent	22,000
7	Absent	18,500
8	12,000	12,000
9	Absent	7,500

Subunit 9 corresponds to the protein which is not stained in gels of the whole complex.

The ATP synthetase complex of yeast mitochondria has been resolved into three functional parts, each of which is necessary for the reconstitution of rutamycin sensitivity (Tzagoloff, 1970). These are: (a) the F_1–ATPase, which is water soluble, not inhibited by rutamycin and cold-labile (Schatz et al., 1967; Tzagoloff, 1969b); (b) OSCP, a single protein necessary for the reconstitution of oligomycin (rutamycin) sensitivity which has been purified from yeast as a homogeneous protein (Tzagoloff, 1970); and (c) membrane factor; this can only be solubilized with detergents and contains four subunits. Resolution of constituent subunits has also been achieved by Capaldi (1973).

The biosynthesis of the ATP synthetase complex depends on the participation of both the cytoplasmic and mitochondrial protein synthesizing systems, and has been investigated by taking advantage of the rapid synthesis which occurs during glucose derepression in yeast (Tzagoloff 1969b, 1970, 1971; Tzagoloff and Meagher, 1972; Tzagoloff et al., 1973).

(a) Biosynthesis of F_1-ATPase. The ATPase activity of mitochondria increases during derepression, and this increase can be shown to be accompanied by an increase in the F_1–ATPase content to the mitochondrial membranes. The increase of ATPase in the mitochondrial fraction is inhibited by chloramphenicol, and under these conditions soluble ATPase accumulates in the postribosomal supernatant (Tzagoloff, 1969b). This soluble ATPase has been partially purified, and its properties indicate that it is identical with F_1 (Tzagoloff et al., 1972). Cycloheximide also inhibits the increase of ATPase activity in the mitochondrial fraction, but in this case no accumulation of non-mitochondrial ATPase was observed. It was concluded that F_1–ATPase is synthesized by the cytoplasmic system of protein synthesis. All five subunits of F_1–ATPase became labelled when derepression was carried out in the presence of chloramphenicol and (^{14}C)-Leu.

(b) *Biosynthesis of OSCP.* As was the case with F_1–ATPase, derepression of yeast cells in the presence of chloramphenicol leads to the accumulation of soluble OSCP. (Tzagoloff, 1970). The amount of OSCP present in the cytoplasmic fraction was sufficient to bind all of the soluble ATPase also present in that fraction; the two enzymes are not present as a physical complex. Cycloheximide blocked the synthesis of OSCP.

(c) *Biosynthesis of the membrane factor.* SMPs from glucose-repressed cells bind only small amounts of soluble ATPase components (F_1 and OSCP) in an *in vitro* reconstitution assay (Tzagoloff, 1971). The capacity of the particles to bind these components increases during derepression, suggesting that a membrane factor is synthesized during this process. The membrane factor does not increase when derepression is carried out either in the presence of chloramphenicol or cycloheximide. If however sequential incubations are carried out in derepression media containing first chloramphenicol and then cycloheximide, increase of membrane factor does occur. It was suggested that the membrane factor is synthesized by the mitochondrial protein synthesizing system and that this synthesis is stimulated by products of cytoplasmic protein synthesis. The net *in vivo* incorporation of (^{3}H)-Leu in the presence of cycloheximide by mitochondria was three times greater when products of the cytoplasmic system were allowed to accumulate by preincubation in the presence of chloramphenicol (Tzagoloff, 1971). SMPs from these cells were extracted with Triton X-100 under conditions which solubilize the ATP synthetase complex, and the extract was treated with antiserum to the rutamycin-sensitive ATPase. The precipitate formed, when analysed by polyacrylamide gel electrophoresis, showed the presence of at least four radioactive components; the major product was a low mol. wt protein, probably subunit 9 (Tzagoloff and Meagher, 1972); other products were subunits 5, 6 and a protein which has a similar mobility to subunit 8 of F_1–ATPase. Confirmation of the identity of these products of the mitochondrial protein synthesis system was achieved by the demonstration that they comigrate with known subunits of the rutamycin-sensitive ATPase. They comprise the least soluble part of the ATP synthetase complex, and play an important role in the integration of F_1 and OSCP into the membrane during the assembly of this complex. Subunit 9 can be extracted from the antibody precipitate with chloroform-methanol (Tzagoloff et al., 1973; Sierra and Tzagoloff, 1973). It has a minimal mol. wt of 8900, an extremely high percentage (76%) of non-polar amino acids, and a high affinity for phospholipids. It is possibly related to the proteolipid which specifically reacts with DCCD. (Cattell et al., 1970; Stekhoven et al., 1972; Capaldi, 1973).

E. Synthesis of Components of the Adenine Nucleotide Translocase

Indirect evidence for the participation of the mitochondrial transcription and translation of a component of the adenine nucleotide translocase comes from studies on the effects of physiological and genetic manipulation on ATP uptake by isolated yeast mitochondria (Perkins et al., 1972; Haslam et al., 1973a).

Cells grown anaerobically with 4% galactose, Tween 80 and ergosterol or cells grown aerobically with 2% glucose (catabolite repressed) yielded mitochondria with a normal ATP uptake system, but mitochondria from a ρ° mutant grown aerobically with 4% glucose, or from cells grown with erythromycin, showed loss of sensitivity to atractyloside and loss of the high affinity binding site. However, the adenine nucleotide transporter does not appear to be totally eliminated, as some low-affinity ATP binding remains. Alternative interpretations of these results are either, (a) loss of all detectable mt DNA by mutation or elimination of mitochondrial protein synthesis prevent the normal synthesis or expression of the atractyloside binding sites or, (b) extensive reorganization of the membrane components in the ρ° mutant or after growth with erythromycin leads to non-specific loss of inhibitor-sensitivity. All types of mitochondria from the cells grown aerobically were intact as judged by their cytochrome c content and the volumes of their sucrose-impermeable spaces. It is suggested that the adenine nucleotide transporter system is under a dual genetic control in the same way as other complex inner membrane functions, but the finding that the high-affinity binding adenine nucleotide transporter is retained in a respiratory-deficient strain of S. cerevisiae (Kolarov et al., 1972a) contradict this conclusion. A nuclear gene is altered in the mutant op1 to give two- to four-fold decrease in the translocation rate of ADP (Kolarov et al., 1972b). In this case it also remains to be proven that the lesion is actually in the transporter itself and not in the organization of the membrane environment of the enzyme.

II. Mitochondrial Participation in Lipid Synthesis

Many of the enzymes involved in the synthesis of complex mitochondrial lipids are extramitochondrial in location (Wilgram and Kennedy, 1963); most if not all mitochondrial phosphatidyl choline, phosphatidyl ethanolamine and phosphatidyl inositol of rat liver occurs on endoplasmic reticulum (Schneider, 1963; Jungalwala and Dawson, 1970b; Dennis and Kennedy, 1972). Mitochondria have little ability to synthesize nitrogen-containing phosphoglycerides (McMurray and Dawson, 1969; Jungalwala and Dawson, 1970a), and their synthesis is limited to components of the outer-membrane (Bygrave, 1969). Endogenous diglyceride precursors are apparently limiting, as lecithin synthesis *can* occur in isolated rat liver mitochondria when diglycerides are generated by the addition of phospholipase (Van Schijndel *et al.*, 1974). Respiration-dependent incorporation of inorganic phosphate into phospholipids of isolated mitochondria has been demonstrated (Garbus *et al.*, 1963). Exchange of phospholipids between microsomal fractions and isolated mitochondria (Kadenbach, 1968; McMurray and Dawson, 1969) may involve an exchange apoprotein which ferries phospholipid as a lipoprotein complex into the mitochondria (Dawson, 1966; Wirtz and Zilversmit, 1968); transfer of phospholipids from "microsomal" synthetic sites to mitochondria has also been shown in intact cells (Jungalwala and Dawson, 1970b).

Malonyl-CoA, long chain fatty acids, phosphatic acid, acyldihydroxyacetone phosphate, sphingomyelin and cardiolipin can all be synthesized intramitochondrially (Hülsmann, 1962; Kiyasu *et al.*, 1963; Hajra and Agranoff, 1967; Sribney, 1968; Shephard and Hübscher, 1969).

Cardiolipin is synthesized in mammalian mitochondria by a pathway similar to that found in bacteria (Davidson and Stancev, 1971).

1. sn-glycerol-3-P $+$ 2 acyl-CoA $\rightarrow$ phosphatidic acid
2. phosphatidic acid $+$ CTP $\rightarrow$ CDP-diglyceride
3. sn-glycerol-3-P $+$ CDPdiglyceride $\rightarrow$ phosphatidylglycero-phosphate
4. phosphatidyl glycerophosphate $\rightarrow$ phosphatidylglycerol
5. phosphatidylglycerol $+$ CDP diglyceride $\rightarrow$ cardiolipin

The first two steps of this pathway have been found in both mitochondria and mitochondria-free supernatants of *S. carlsbergensis* (Johnson and Paltauf, 1970).

The synthesis of mevalonic acid from (1^{14}C)-acetate was demonstrated in a mitochondrial fraction of aerobically-grown *S. cerevisiae* incubated with ATP, CoA, Mg^{2+} and an NADPH-generating system (Shizimu *et al.*, 1973). 3-Hydroxy-3-methylglutaryl-CoA reductase was uniquely associated with the mitochondria, whereas acetoacetyl-CoA thiolase was also detected in other fractions. The levels of both these enzymes was suppressed by growth in the presence of chloramphenicol. The further metabolism of mevalonate *via* farnescyl pyrophosphate to give ergosterol is catalysed by soluble and microsomal enzymes.

III. Mitochondrial Participation in Haem Synthesis

In rat liver the first and last enzymes of haem biosynthesis (δ-amino-laevulinate synthase and ferrochelatase) are located within the inner mitochondrial membrane (Jones and Jones, 1969, 1970). Both enzymes are inhibited by protohaem. The matrix space is readily penetrated by Gly but not by δ-aminolaevulinate. Increased haem synthesis induced by allylisopropyl acetamide is accompanied by increased mitochondrial protein synthesis (Beattie, 1971b). δ-Aminolaevulinate synthase is the rate-limiting step of haem biosynthesis in mammalian tissues (Granick, 1966) and plays a significant role in the regulation of tetrapyrrole biosynthesis (Lascelles, 1964).

This enzyme has been detected in cell free extracts of anaerobically-grown *Candida utilis*, which contains little or no haem (Porra et al., 1972a). No activity was found after aeration of these cells for 1 h, nor was it found in aerobically-grown cells. It was suggested that an inhibitor of the enzyme may have been present in the extracts, as the enzyme has been demonstrated in mitochondria of aerobically-grown *S. cerevisaie* (de Barreiro, 1967; Porra et al., 1972b). The formation of porphyrins by semi-anaerobically-grown *S. cerevisiae* was first demonstrated by Stich and Eisgruber (1951). Porra et al. (1972b) have shown that the level of δ-aminolaevulinate synthase is about ten-fold higher in semi-anaerobic than in aerobic yeast and is correlated, not with haemoprotein content which is higher in aerobic cells, but with the greater ability of semi-anaerobic cells to accumulate and excrete porphyrins. The enzyme is associated mainly with the mitochondria of aerobic cells, but with the cytosol of semi-anaerobic cells, and its activity in aerobic cells enhanced by the addition of protohaemin IX; in other situations this compound produces feedback inhibition (Lascelles, 1964). Control of biosynthesis of porphyrins occurs partly by the separation of enzymes between cytosol and mitochondria (Sano and Granick, 1969) and by regulation of the passage of porphyrin intermediates across the mitochondrial membrane. The overproduction of porphyrins by semi-anaerobic cells may be associated: (a) with the increased δ-aminolaevulinate synthase activity; and (b) with the presence of this enzyme in the cytosol, thus removing the constraints on the passage of δ-aminolaevulinate across the inner mitochondrial membrane (Porra et al., 1972b).

Ferrochelatase has been detected in the "promitochondrial fractions" obtained from *S. cerevisiae* grown under strictly anaerobic conditions (Schatz, 1965; Criddle and Schatz, 1969). The specific activity of ferrochelatase is similar in mitochondria from wild-type, chloramphenicol-grown and the *mi-1* mutant of *Neurospora crassa*. It is also similar in wild-type and cytoplasmic *petite* strains of *S. cerevisiae* (Birkmayer and Bücher, 1969).

During respiratory adaptation of yeast there is substantial *de novo* biosynthesis of haem *a*, cryptohaem *a* and other haems as shown by incorporation of labelling from Gly and δ-aminolaevulinate (Barrett, 1969). Labelled Gly is equally available for incorporation into the haem and protein moeities of cytochrome *c*. After the development of the cytochromes, porphyrin synthesis is continued, the main porphyrin synthesized is coproporphyrin; coprobiliverdin is also produced and excreted into the medium. All the enzymes necessary for the synthesis of protohaem have been detected in extracts of both aerobically- and anaerobically-grown *S. cerevisiae* (Labbe, 1971). During respiratory adaptation in the presence of glucose there is a decrease in the specific activity of δ-aminolaevulinate synthase during the first h of aeration; the enzyme remained at an undetectable level for 9 h (Labbe *et al.*, 1972). Loss of one or more enzymes of the protohaem synthesis pathway between porphobilinogen and protoporphyrin also occurred and resynthesis *de novo* was observed only after the exhaustion of glucose from the adaptation medium. Loss of these enzymes was also observed when cells were incubated anaerobically with glucose. Respiratory adaptation in the presence of galactose also leads to the disappearance of δ-aminolaevulinate synthase, but the porphobilinogen to protoporphyrin step is not lost even when an addition of glucose was made to the adapting culture. These results suggest that either the rate of glycolysis or the pool size of glycolytic intermediates may have a regulatory effect on haem biosynthesis and hence cytochrome formation during respiratory adaptation.

Aeration of non-proliferating cell suspensions of *S. cerevisiae* grown aerobically or anaerobically in the presence of 2·6 μM–$ZnSO_4$ leads to the accumulation of Zn-protoporphyrin (Ohaniance and Chaix, 1966; Gilardi *et al.*, 1971). Intramitochondrial accumulation of this product results in the inhibition of respiratory activity of both whole cells and isolated mitochondria in the case of the aerobically-grown cells, and an inhibition of cytochrome biosynthesis and respiratory adaptation when the anaerobically-grown cells are aerated.

The regulation of haem biosynthesis has also been studied in *Neurospora crassa* (Muthukrishnan *et al.*, 1969, 1972). Growth of the fungus in iron-deficient medium does not lead to the accumulation of porphyrins. δ-Aminolaevulinate dehydratase is inducible on addition of iron and repressed by protoporphyrin. This enzyme is allosterically regulated by coproporphyrinogen III.

Detailed kinetic studies of the incorporation of (^{59}Fe) and (^{14}C)-Lys into mitochondrial cytochrome c by Kadenbach (1969, 1970, 1971a, b) showed that apocytochrome c is synthesized on extra-mitochondrial ribosomes and then transferred to the mitochondria as a protein-phospholipid complex; the formation on the holoenzyme subsequently occurs on the inner membrane. Experiments on the incorporation of δ-(^{14}C)-aminolaevulinate also ruled out the possibility that protoporphyrinogen might be attached to the apoprotein at the endoplasmic reticulum. In the protohaemin-requiring slime-mould *Physarum polycephalum*, experiments on the incorporation of (^{59}Fe)-haemin indicate that the synthesis of the cytochrome c holoenzyme must proceed by direct attachment of haem to the apoprotein rather than by the intermediate formation of a protoporphyrinogen-apoprotein complex (Colleran and Jones, 1973). Whether there are special terminal synthetic routes for the formation of haemoproteins with non-covalently attached haem groups is not yet determined.

Extramitochondrial Synthesis of Mitochondrial Components; Mitochondrial Assembly, Growth and Division

I. Role of the Cytoplasmic Protein Synthesis System in Mitochondrial Development

A. Role in the Production of Mitochondrial Membrane Structure and Enzymes

As discussed in the previous chapter, all the available evidence suggests that the mitochondrial protein synthesis system is involved in the elaboration of only a few hydrophobic proteins of the inner-membrane. The proteins of the outer-membrane, the vast majority of the proteins of the inner-membrane (including, for instance, those enzymes inducible in pre-existing mitochondria, Lloyd et al., 1968), and the soluble enzymes of the matrix, are all synthesized on cytoplasmic ribosomes. Thus organelles which in many respects are not very different from fully-functional mitochondria develop in yeasts, even in the complete absence of mt DNA or the mitochondrial protein synthesis system (Fig. 118; Linnane and Haslam, 1970; Perlman and Mahler, 1970b; Kellerman et al., 1971). The membrane morphology of mitochondria after chloramphenical-inhibited growth of some species of fungi (Marchant and Smith, 1968a) algae (Lloyd et al., 1970b) or protozoa (Turner and Lloyd, 1971) is not grossly altered (see Chapter 4). These organelles have poorly developed inner-membranes, but they do possess respiratory chains (with modified proportions of respiratory components), enzymes of the tricarboxylic acid cycle, and normal anion transport mechanisms (Perkins et al., 1972); they also have intact outer-membranes.

B. Role of the Cytoplasmic Protein Synthesis System in mt DNA replication

Inhibition of cytoplasmic protein synthesis by cycloheximide in both S. cerevisiae and E. gracilis produces a marked inhibition of nuclear DNA synthesis, whereas mt DNA synthesis proceeds (Grossman *et al.*, 1969; Richards *et al.*, 1971). In these organisms the synthesis of both nuclear and mt DNA are not affected by growth in the presence of chloramphenicol for up to five generations, and it is concluded that products of mitochondrial protein synthesis are not required for mt DNA synthesis. This conclusion is also evident from work on cytoplasmic *petite* mutants which completely lack a functional system of protein synthesis but show continued replication of their defective mt DNAs (Kužela and Grečna, 1969; Coen *et al.*, 1970; Linnane and Haslam, 1971; Scragg, 1971).

Another observation which leads to a similar conclusion is the reversibility of induction of the *petite* phenotype by acridines in *Candida parapsilosis* (Kellerman *et al.*, 1969), and by ethidium bromide in *Kluyveromyces lactis* (Luha *et al.*, 1971). In these experiments, inhibition of synthesis of mitochondrial gene products did not lead to inability to synthesize functional mitochondria once the inhibitors were removed. However, other work has produced contradictory evidence. High concentrations of chloramphenicol or erythromycin will induce *petite* mutant formation in *S. cerevisiae* after growth for extended periods in the presence of these antibiotics (Williamson *et al.*, 1971a). The appearance of *petites* is accompanied by the sudden disappearance of mt DNA. It was postulated that a "replication factor" is necessary for the replication of mt DNA, and that this is a product of mitochondrial protein synthesis. High concentrations of the antibiotics inhibit the formation of a replication factor, but the results of this inhibition are not manifest until pre-existing replication factor is diluted out during growth. *Petite* induction is accompanied by degradation of pre-existing mt DNA. That a factor produced by the mitochondrial protein synthesis system is required for mt DNA replication has also been suggested by the observation that a cold-sensitive mutant is converted into *petite* when grown at 18°C or 28°C in the presence of chloramphenicol (Weislogel and Butow, 1970, 1971). At 18°C the direct phenotypic result of a cold-sensitive assembly step is the improper incorporation of a protein into a membrane; this leads to a defective process of mt DNA replication and a loss of mt DNA. Presumably it is the synthesis of this protein factor (possibly similar to the one postulated by Williamson *et al.*, 1971a) which is inhibited by chloramphenicol at 28°C.

C. Evidence for the Extramitochondrial Synthesis of the Machinery Involved in Mitochondrial Protein Synthesis

1. Enzymes of the mitochondrial protein synthesis system

All the components of the mitochondrial protein synthesis system (with the exception of mitochondrial messenger RNA, ribosomal RNAs and some transfer RNAs) are coded for by nuclear DNA and translated on

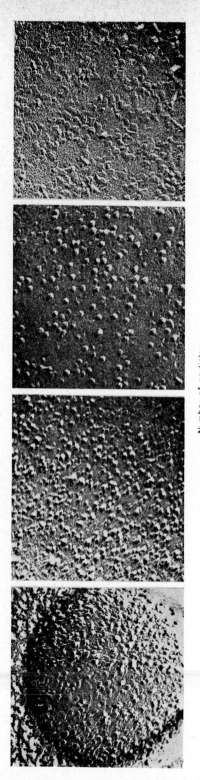

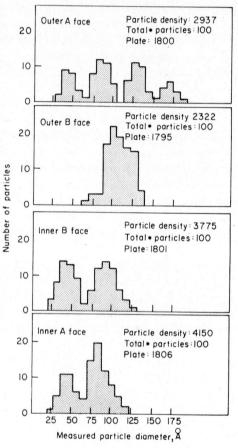

(a) Wild-type mitochondria

Outer A face — Particle density: 2937 — Total particles: 100 — Plate: 1800

Outer B face — Particle density 2322 — Total particles: 100 — Plate: 1795

Inner B face — Particle density: 3775 — Total particles: 100 — Plate: 1801

Inner A face — Particle density: 4150 — Total particles: 100 — Plate: 1806

Number of particles

Measured particle diameter, Å

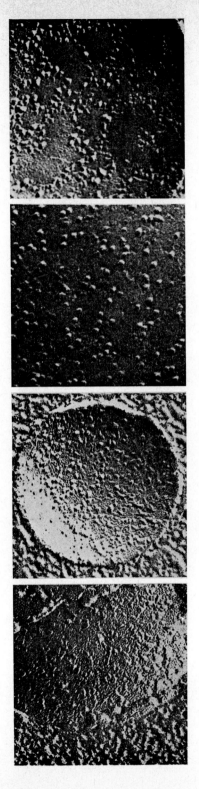

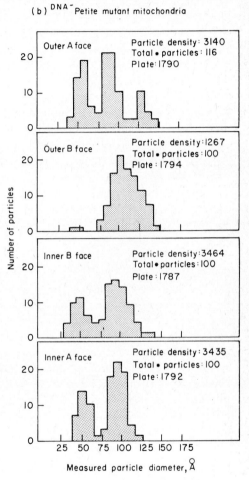

(b) $^{DNA^-}$ Petite mutant mitochondria

Outer A face — Particle density: 3140, Total • particles: 116, Plate: 1790

Outer B face — Particle density: 1267, Total • particles: 100, Plate: 1794

Inner B face — Particle density: 3464, Total • particles: 100, Plate: 1787

Inner A face — Particle density: 3435, Total • particles: 100, Plate: 1792

Number of particles

Measured particle diameter, Å

FIG. 118. Characteristic fracture faces of outer- and inner-membranes from mitochondria isolated from (a) wild-type yeast (b) from mutant yeast without mt DNA together with histograms of particle size (diameter) distribution in fracture faces of mitochondria. (Reproduced with permission from Packer *et al.*, 1973.)

cytoplasmic ribosomes. Mitochondrial peptide chain elongation factors (G and T) are still synthesized in chloramphenicol-inhibited yeast or *Chlorella* (Cifferi *et al.*, 1974) and in cytoplasmic *petite* mutants (Parisi and Cella, 1971), even in strains which have no detectable mt DNA (Scragg, 1971; Richter, 1971). One of the amino-acyl-tRNA synthetases (for Leu, Gross *et al.*, 1968), Met-tRNA transformylase (Barath and Küntzel, 1972a) and mt RNA polymerase (Barath and Küntzel, 1972b) are all synthesized extra-mitochondrially. The experiments of Davey *et al.* (1969) suggest that the total machinery for mitochondrial protein synthesis is synthesized by the cytoplasmic system. Yeast grown for five generations in the presence of 2% glucose and chloramphenicol (4 mg/ml) contained no detectable cytochromes $a + a_3$, b, or c_1; growth with lincomycin (1·5 mg/ml) for the same number of cell divisions gave rise to cells which still retained some cytochrome b. Yet, after washing out the antibiotics, isolated mitochondria still possessed the ability to incorporate labelled amino acids into proteins. As both antibiotics are potent inhibitors of mitochondrial protein synthesis, all the protein remaining in the cells must have been formed by the non-mitochondrial protein synthesis system. Similar results have also been obtained on growing cells for up to 20 generations in the presence of chloramphenicol (Plummer *et al.*, 1974).

2. Site of synthesis of mitochondrial ribosomal proteins

(a) *In* Neurospora crassa. The mitochondrial ribosomal proteins of *N. crassa* were separated using carboxy-methyl cellulose chromatography (Küntzel, 1969, a, b; Gualerzi, 1969). Comparison with the chromatograms obtained for the cytoplasmic ribosomal proteins indicated that the two sets of proteins were very different, although the technique was not sensitive enough to conclude that none of the proteins were shared by both classes of ribosome. The *in vivo* synthesis of mitochondrial ribosomal proteins from (^{3}H)-Lys was 97% inhibited in the presence of cycloheximide. Similar results were reported by Neupert *et al.* (1969b, 1971) who also concluded that the vast majority, if not all the protein components of mitochondrial ribosomes are synthesized extra-mitochondrially; however in these experiments it was not impossible that one or two of the mitochondrial ribosomal proteins (out of more than 50) were products of the mitochondrial protein synthesis system. Three complementary approaches to this problem were used by Lizardi and Luck (1972). (a) Tritium-labelled proteins made by isolated mitochondria were compared to (^{14}C)-labelled mitochondrial ribosomal proteins by cofractionation in a procedure using isoelectric focusing and polyacrylamide gel electrophoresis. None of the peaks of the *in vitro* product corresponded exactly to any of the mitochondrial ribosomal protein peaks. (b) Cells were labelled with (^{3}H)-Leu in the presence of chloramphenicol; mitochondrial ribosomal subunits were subsequently isolated and their proteins fractionated by isoelectric focusing followed by gel electrophoresis. The labelling of every single mitochondrial ribosomal protein was found to be insensitive to chloramphenicol. (c) In the presence of anisomycin, (an effective *in vivo* selective inhibitor of cytoplasmic protein synthesis) the labelling of virtually all mitochondrial ribosomal proteins is inhibited to the same extent as the labelling of cytoplasmic ribosomal proteins. Lizardi and Luck (1972) concluded from these results that most, if not all, of the 53 structural proteins detected in mitochondrial ribosomal subunits in *N. crassa* are synthesized by cytoplasmic ribosomes. They stressed that these conclusions apply only to those proteins tightly associated with ribosomes and hence consistently recovered under high salt isolation conditions. The possibility remains that synthesis of membrane binding proteins (e.g. an anisomycin-resistant protein sometimes found in association with the small subunit) or factors for protein synthesis may require a role for the mitochondrial protein synthesis system.

The cytoplasmic *"poky"* mutation of *N. crassa* results in the abnormal assembly of ribosomal subunits in mitochondria (Rifkin and Luck, 1971). No gross defects in ribosomal RNA cistrons were evident and a mitochondrial ribosomal protein mutation was suspected.

(*b*) *In yeasts.* The seemingly conclusive results of Davey *et al.* (1969) which clearly indicated that all the machinery of the mitochondrial protein synthesis system is synthesized on extra mitochondrial ribosomes in yeast conflicts with the observation that a single cytoplasmically-determined mutation in *S. cerevisiae* results in mitochondrial protein synthesis becoming resistant to erythromycin (Linnane *et al.*, 1968c; Thomas and Wilkie, 1968a, b). By analogy with antibiotic resistance in bacteria which often results from changes in single ribosomal proteins, it was suggested that mitochondrial resistance might similarly involve an alteration of a mitochondrial ribosomal protein. Implicit in this suggestion was that the cytoplasmic determinant for erythromycin resistance is located on mt DNA, and therefore at least one of the mitochondrial ribosomal proteins would have to be encoded by mt DNA. Other possible explanations are (i) Erythromycin resistance could result from a modified large subunit rRNA (methylation of adenine) as in *Staphylococcus aureus* (Lai and Weisblum, 1971). Such mutations of rRNA are uncommon in bacteria, possibly because of the redundancy of their cistrons. It seems that there is no such redundancy of cistrons coding for mitochondrial rRNA. (ii) Structural genes for mitochondrial ribosomal proteins are present in mt DNA but are translated cytoplasmically. (iii) Enzymes coded by mt DNA could bring about specific modification of some mitochondrial ribosomal proteins once they have been imported into the mitochondria. A definitive answer as to whether some mitochondrial ribosomal proteins are chemically modified variants of their cytoplasmic counterparts awaits further structural comparisons e.g. by peptide mapping or antibody cross-reactivity studies.

Mutants resistant to chloramphenicol, erythromycin, spiramycin or paromomycin, as a result of mutation in mt DNA have been studied; the response of mitochondrial ribosomes to antibiotics acting on the large ribosomal subunit was facilitated by the use of the modified fragment reaction (Grivell, 1974). Complete characterization of mutants necessitated study of the response to antibiotics in a cell free system programmed by a natural messenger RNA (e.g. phage MS2 RNA). Several mutants resistant to chloramphenicol or erythromycin were shown to have altered mitochondrial ribosomes, but unlike analogous mutants of bacteria, resistance to antibiotics did not seem to be associated with deleterious effects on ribosomal function, and did not arise as a result of impermeability of the mitochondrial membrane. All attempts to demonstrate (by acrylamide gel electrophoresis) an alteration in a ribosomal protein as the result of mutation to antibiotic resistance have proved negative, and the possibility that such resistance is due to a change in rRNA remains open. Preliminary results indicate an absence of methylated RNA in yeast mitochondrial rRNA, therefore such a change cannot involve an alteration in RNA methylase.

Direct analysis of the site of synthesis of yeast mitochondrial ribosomal proteins *in vivo* in cycloheximide-inhibited cells gave similar results to those found with *N. crassa*. Schmitt (1970) showed that most if not all of the ribosomal proteins of the mitochondria were different from those of the cytoplasm. These results were confirmed using carboxymethyl cellulose chromatography (Schmitt, 1971); inhibition of cytoplasmic protein synthesis by cycloheximide inhibited the incorporation of ($^{35}SO_4$) into ribosomal proteins of the mitochondria. Chloramphenicol had a small effect on the synthesis of both mitochondrial and cytoplasmic ribosomal proteins.

When yeast cells are labelled *in vivo* with radioactive Leu in the presence of cycloheximide, and then allowed to grow for one generation in the absence of the inhibitor to allow integration of mitochondrially-synthesized precursors, a significant incorporation of radioactivity into 73S ribosomes was detected (Groot, 1974); labelling was in a single protein from the 38S ribosomal subunit. Despite this, cells grown in the presence of erythromycin retain cycloheximide-resistant protein synthesis, and mitochondria isolated from these cells show protein synthesis which is chloramphenicol sensitive. It was concluded therefore that all proteins necessary to constitute an active mitochondrial ribosome are synthesized on cytoplasmic ribosomes, and the labelled protein found in the 38S subunit is not essential for ribosomal function.

(c) *In* Tetrahymena pyriformis. The effects of chloramphenicol and cycloheximide on the synthesis of mitochondrial and cytoplasmic ribosomes in *T. pyriformis* were studied by Millis and Suyama (1972). They found that the synthesis of mitochondrial ribosomes was completely abolished by 1 mM–cycloheximide but not by 0·5 mM–chloramphenicol. The formation of some mitochondrial ribosomal proteins (in the mol. wt range 13,000–25,000) were particularly sensitive to chloramphenicol inhibition, whereas the formation of cytoplasmic ribosomal proteins was not differentially inhibited by chloramphenicol. Mitochondria isolated from cycloheximide-poisoned cells still retained a functional mitochondrial protein synthesis system, as evidenced by their ability for *in vitro* incorporation of (^{14}C)-Leu; on the other hand, the ability to incorporate (^{3}H)-uridine into mitochondrial ribosomes was greatly diminished under these conditions. These results suggest that some mitochondrial ribosomal proteins are synthesized in mitochondria and others by cytoplasmic protein synthesis, and are clearly in disagreement with data obtained from experiments with *N. crassa* and yeasts. The synthesis of those proteins produced by the mitochondrial system appeared to be subject to control by a product (membrane factor?) of the cytoplasmic system of protein synthesis. However the assembly of mitochondrial ribosomes was observed in the absence of cytoplasmic protein synthesis.

It is evident that the problem of the biosynthetic origins of *Tetrahymena* mitochondrial ribosomal proteins requires further study by both analytical and genetic approaches before general conclusions can be made.

II. Interaction of The Extrinsic and Intrinsic Systems of Synthesis of Mitochondrial Proteins

Interaction between the mitochondrial and cytoplasmic protein synthesis systems by way of control mechanisms is necessary for the organized production and integration of mitochondrial components. Although mitochondrial protein synthesis can proceed after cytoplasmic protein synthesis has been blocked (see p. 412), experiments with inhibitors have clearly indicated a degree of coupling between the two systems. Thus during the processes of respiratory adaptation of anaerobically-grown yeast (Chen and Charalampous, 1969, 1973; Yu *et al.*, 1968; Rouslin and Schatz, 1969) and aerobic catabolite-derepression (Henson *et al.*, 1968a, b; Tzagaloff, 1969b), interplay between cytoplasmic and mitochondrial translation processes are necessary, and interference with either system ultimately leads to derangement of mitochondrial components. This derangement results from reduced rates or cessation of synthetic processes or from lack of integration of products into the membranes.

A number of schemes have been proposed for control circuits which operate between mitochondria and cytoplasm (Williamson, 1970; Lloyd *et al.*, 1971b; Puglisi and Algeri, 1971; Barath and Küntzel, 1972a).

The observations that ρ^- yeast cells cannot utilize galactose as the carbon source for growth and are no longer inducible, for the enzymes of the galactose pathway (Douglas and Pelroy, 1963; Douglas and Hawthorne, 1964) have been used as starting points for the investigation of possible interaction between an exported mitochondrial translation product (repressor) and the *gal* system (Puglisi and Algeri, 1971). Experimental evidence was obtained which strongly favoured the concept of coordinated nucleo-mitochondrial regulation of these enzymes (Fig. 119). A similar

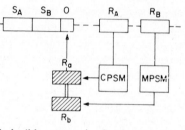

FIG. 119. Model of an inducible system in *Saccharomyces cerevisiae* O, S_B and S_A: operon; R_A regulator gene, whose product (R_a) is synthesized on CPSM (cytoplasmic protein synthesis machinery); R_B regulator gene, whose product R_b, is synthesized on MPSM (mitochondrial protein synthesizing machinery); R_a—R_b the holorepressor. (Reproduced with permission from Puglisi and Algeri, 1971.)

mechanism has also been invoked to explain the dependence of sporulation, conjugation, and morphogenesis in the dimorphic yeast *Endomyces capsulatis* on mitochondrial protein synthesis (Puglisi and Algeri, 1974). A mitochondrially-produced protein may be necessary for cell division in *Polytomella caeca* (M. Cantor and D. Lloyd, unpublished results).

The direct control of expression of the mitochondrial protein synthesis system by cytoplasmic products has been demonstrated in a temperature-sensitive yeast mutant (ts187) by Mahler *et al.* (1974). Protein synthesis (actually chain initiation) in the cytosol is blocked at the non-permissive temperature; under these conditions mitochondrial chain initiation (measured by incorporation of labelled formate) continues, but integration of the products of mitochondrial protein synthesis into the inner-membrane ceases. Therefore a continual availability of cytoplasmically-synthesized polypeptides is required for this integration. These results are reminiscent of those obtained in studies of the assembly and integration of the components of ATPase during glucose derepression in yeast (Tzagoloff, 1969b).

The kinetics of appearance of enzyme activities characteristic of the various segments of the respiratory chain have been investigated during glucose derepression of *S. cerevisiae* (Kim and Beattie 1973; Beattie *et al.*, 1974). Asynchrony of assembly of the various enzyme complexes was observed, for example succinate dehydrogenase and oligomycin-sensitive ATPase increased uniformly, whereas NADH dehydrogenase and NADH-cytochrome *c* oxidoreductase showed a lag of 13 h before starting to increase in activities. Whereas the increase in activity of both cytochrome *c* oxidase and NADH-cytochrome *c* oxidoreductase was inhibited completely by chloramphenicol, it was not affected for several hours after the addition of cycloheximide. These results suggest that the synthesis of mitochondrial proteins on cytoplasmic ribosomes may precede that on mitochondrial ribosomes, and that the cytoplasmically-translated proteins may accumulate in mitochondria. This accumulation leads to control of mitochondrial protein synthesis. Thus the rate of mitochondrial protein synthesis (measured by amino acid incorporation *in vitro* under optimal conditions) was stimulated when yeast cells had grown for several hours in chloramphenicol and were then allowed to grow for one to three hours in fresh medium with or without cycloheximide. SDS-polyacrylamide gel electrophoresis indicated marked differences in the proteins labelled *in vivo* by mitochondria under these different conditions. The time course of development of enzyme activities in the presence of the inhibitors indicated that mitochondrial synthesis of proteins constituting the b and c_1 regions of the respiratory chain may be subject to control by proteins synthesized in the cytoplasm. No indications of a similar control mechanism were observed for the NADH dehydrogenase or cytochrome *c* oxidase.

III. Assembly of Mitochondria

A. Import of Proteins and Phospholipids

Direct demonstration of extra mitochondrial synthesis of a mitochondrial enzyme has been achieved in mammalian systems in the case of cytochrome *c* (González-Cadavid and Campbell, 1967; Freeman *et al.*, 1967; González-Cadavid *et al.*, 1968; Davidian *et al.*, 1969). Some evidence suggests that it is the apoprotein which is synthesized on microsomes and that the haem group is inserted at, or within, the mitochondrion (Kadenbach, 1970).

Transfer *in vitro* of proteins from microsomes to mitochondria has been reported by Kadenbach (1967a, b). Some phospholipids are also synthesized in the microsomes (Wilgram and Kennedy, 1963; McMurray and Dawson, 1939). Therefore phospholipids must also be imported into mitochondria (Dawson, 1973). Identical kinetics of transfer of labelled phospholipids and proteins from microsomes to mitochondria led Kadenbach (1968) to suggest that phospholipid-protein complexes are involved in the import of proteins into mitochondria. Alternative proposals for possible mechanisms for protein import include: (a) diffusion of proteins to high affinity binding sites in the mitochondrial membranes; (b) facilitated penetration of nascent proteins by virtue of hydrophobic amino acid sequences added by a process of post-translational modifications; (c) synthesis in lipoprotein vesicles that fuse with growing mitochondria (Borst et al., 1967). Morphologically distinct vesicles which may have this function have been observed in fractionated extracts of S. carlsbergensis during the rapid phase of respiratory adaptation (Cartledge and Lloyd, 1973). (d) Synthesis of mitochondrial enzymes on rough endoplasmic reticulum or on specialized cytoplasmic ribosomes which are in close contact with the outer membrane of mitochondria, followed by vectorial discharge of translation products through the mitochondrial membranes. Close association of cytoplasmic polyribosomes with mitochondria are especially marked in situations where mitochondrial development is proceeding rapidly e.g. during the respiratory adaptation of S. carlsbergensis (Cartledge and Lloyd, 1973). The unusually high E_{260nm}/protein ratios in mitochondrial fractions during the rapid phase of respiratory adaptation confirms the evidence obtained from electron micrographs for this polysome-mitochondrial association. Of the total complement of cytoplasmic polyribosomes in growing yeast cells, 10–15% are closely associated with mitochondria (Schmitt, 1969; Kellems and Butow, 1972; Kellems et al., 1974). Mitochondria from cells starved for 1 h have about half as many bound ribosomes as those from growing cells, and this observation correlates with the finding that the organelles from starved cells are capable of binding more ribosomes in vitro. Binding of ribosomes leads to a marked increase in the buoyant density of the mitochondria in sucrose gradients. Complete removal of the bound polysomes in vitro necessitated washing with 1·5 M–KCl and 1 mM–puromycin, implicating a role of nascent polypeptide in the polysome-membrane interaction.

Incorporation of (^{3}H)-Leu by mitochondria containing bound polysomes in the presence of chloramphenicol followed by puromycin discharge, showed that 53% of the nascent polypeptides remained associated with the mitochondria.

It is evident that details of the molecular mechanisms whereby high mol. wt polypeptide components penetrate a membrane which is impermeable to NAD^+ and has specialized transport mechanisms for adenine nucleotides, fatty acids and small anions and cations remain to be elucidated.

B. Possible Import of RNAs

Experiments with synthetic high mol. wt polynucleotides have suggested that *Xenopus* egg mitochondria have a mechanism for their uptake; thus poly-(U)-directed poly-Phe synthesis sensitive to chloramphenicol but resistant to cycloheximide and RNAase was observed in intact mitochondria. RNAs with a high degree of secondary structure were taken up only to a limited extent (Swanson, 1971). Template RNA of nuclear origin can also interact with the mitochondrial protein synthesis system (Gaitskhoki *et al.*, 1973). It has even been considered possible that the mitochondrial protein synthesis system may function solely in the translation of messenger RNA imported from the nucleus and that mt DNA may not contain a single structural gene coding for a protein (Dawid, 1970; 1972).

The observation that mitochondrial RNA contains minor species which hybridize with at least 20% of the mitochondrial genome in *N. crassa* in addition to the stable RNA cistons (Schafër and Küntzel, 1972) suggests that mt DNA does in fact code for messenger RNA. Direct evidence for the mitochondrial synthesis of messenger RNA is provided by the *in vitro* transcription system of Blossey and Küntzel (1972).

Evidence against messenger RNA import in yeast comes from the demonstration that specific inhibition of transcription of mt DNA by ethidium bromide or acridines immediately blocks all mitochondrial protein synthesis (Penman *et al.*, 1970); Fukuhara and Kujawa, 1970). Therefore imported messenger RNA cannot contribute to any great extent, although if import was closely coupled to mitochondrial transcription and the half-life of the imported messenger was very short, then its existence might escape detection. It is also possible that ethidium bromide may also have an inhibitory effect on nucleotide uptake or on translation. The absence of substantial amounts of imported nuclear transcripts in extracts of total mitochondrial RNA from yeast was indicated in hybridization experiments (Reijnders *et al.*, 1972), and in intact rat liver, chicken liver, yeast, and *Tetrahymena* mitochondria no stimulation of poly-Phe synthesis by exogenous poly-(U) was detectable (Grivell, 1974; Borst and Grivell, 1973; L. A. Grivell and V. Metz, unpublished results). The limited extent of mitochondrial protein synthesis *in vivo* in terms of the few identified products also suggests that messenger RNA import is not of any great importance, but whether it occurs at all must still be considered an open question.

C. Possible Export of Mitochondrially-Synthesized RNAs and Proteins

The possibilities of export of mitochondrially-synthesized messenger RNAs and proteins must also be considered, although the small size of the mitochondrial genome, and the limited number of products of mitochondrial protein synthesis, lead one to predict that these would represent information and products for the elaboration of mitochondria themselves. There is no evidence for the export of RNA except for the reports of Attardi and Attardi (1967, 1968), and alternative explanations could account for these results (Borst, 1972). Neither is there experimental evidence for the export of mitochondrial proteins, although several hypotheses for the control of nuclear-mitochondrial interaction have been put forward which require the mitochondrial production of regulatory proteins which can interact with the nucleus (Williamson, 1970; Turner and Lloyd, 1971; Lloyd *et al.*, 1971b; Puglisi and Algeri, 1971; Barath and Küntzel, 1972a). There is no experimental support for the suggestion that mt DNA may code for "structural" protein that is also required for other cell membranes (Woodward and Munkres, 1966; Munkres and Woodward, 1966).

A diagram summarizing processes of import and export of components is presented in Fig. 120.

D. *In vitro* Reconstitution of Mitochondrial membranes from Purified Components

The *in vitro* reconstitution of fractionated membrane components to give functional complexes may provide clues to the mechanisms of organelle assembly *in vivo*. The formation of membranes *in vitro* can proceed by spontaneous combination of components; little or no directive influence other than the complementarity of surfaces appears to be necessary for the assembly process. Whether or not the actual assembly of organelle membranes *in vivo* is also an externally-undirected process is not yet clear.

The supramolecular assembly which carries out the function of electron transport and oxidative phosphorylation is localized in the inner-mitochondrial membrane and consists of four electron transport complexes and the oligmycin-sensitive ATPase complex. These complexes account for the bulk of the membrane mass of fully-functional mitochondria (MacLennan, 1970) and consist of high mol. wt enzymes (200,000–500,000) consisting of multiple species of proteins and phospholipids (Green and Tzagoloff, 1966). The electron transport carriers are organized in chains or clusters in a highly hydrophobic environment, and measurements of reaction kinetics indicate that haemoproteins are closely enough associated so that only rotational movement is necessary for their interaction (Chance, 1967). The elucidation of the molecular organization of the mitochondrial membranes in relation to the mechanism of energy-coupling is still a formidable problem. One approach to this problem is to fractionate the membrane using reagents such as cholate, deoxycholate, *tert*-amyl alcohol, butanol, petroleum ether or cyclohexane which cleave protein-lipid and/or lipid-lipid bonds. Fragmentation of the respiratory chain

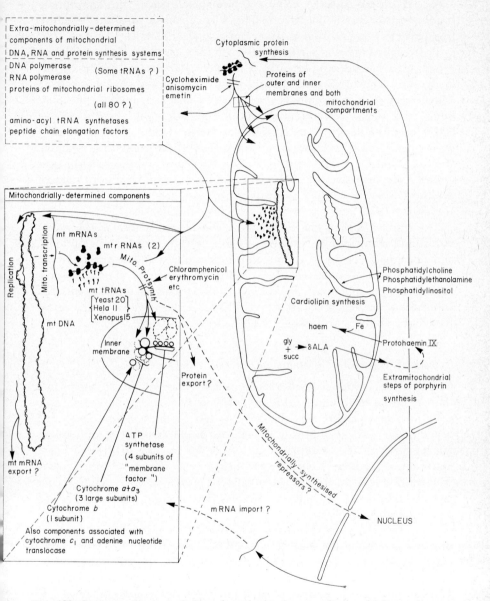

FIG. 120. Summary diagram of sites of synthesis of some mitochondrial components, their import and export.

into complexes has been achieved by these methods (Green and Fleischer, 1962). When the isolated complexes are mixed under the appropriate conditions they reassemble to form a product capable of integrated electron transport (Fowler and Hatefi, 1961; Hatefi *et al.*, 1962).

The reconstitution *in vitro* of the respiratory chain from its individual components has been reported by Yamashita and Racker (1969). In this study the reconstitution of succinoxidase complex on mixing succinate dehydrogenase, cytochromes *b*, *c* and *c*$_1$, cytochrome *c* oxidase, phospholipids, and UQ$_{10}$ took several hours. Reconstitution of phosphorylating SMPs has been described by Kagawa and Racker (1966). Morphologically the reconstituted particles were indistinguishable from the starting material, and consisted of characteristic vesicular structures; vesicularization was essential for oxidative phosphorylation. The successful reconstitution of the third Site of energy conservation has been reported by Racker and Kandrach (1971) and that of Site I by Ragan and Racker (1973).

The fraction CF$_0$ is the hydrophobic residue isolated from the inner-mitochondrial membrane after the removal of cytochromes, flavoproteins, and phospholipids, by bile salts and ammonium sulphate precipitation. This fraction may represent the structural backbone of the membrane; mixed with phospholipids, F$_1$–ATPase and coupling factors, the complex formed had the oligomycin-sensitive ATPase activity typical of the native membrane (Bulos and Racker, 1968).

E. Role of "Structural" Proteins and Proteolipids

Criddle *et al.* (1962) claimed to have isolated "structural protein" from mitochondrial membranes; this protein combined in a one to one ratio with purified cytochromes, self-polymerized, bound phospholipids and was thought to have an organizational role in the assembly of compounds (analogous with the "morphopoetic principles in bacteriophage", Edgar and Lielausis, 1968; Kellenberger, 1966). The validity of this concept depends upon the isolation of a non-catalytic protein which is homogeneous and does not contain denatured enzymes (Lenaz *et al.*, 1968a, b; Schatz and Saltzgaber, 1969b).

Thus the claims of Munkres and Woodward (1966) that structural proteins determine the localization and activity of mitochondrial malate dehydrogenase have been questioned because of the uncertainties involved in the biochemical characterization of structural proteins. Two classes of mutants have more recently been studied and the results obtained add further weight to the original hypothesis. In the "C mutants" malate dehydrogenase is malfunctional because of an abnormal association with a structural protein, and in the "K mutants" the enzyme has an altered subcellular location (Munkres *et al.*, 1970). The structural protein of

Neurospora mitochondria has been resolved into six monomeric oligo-peptides of mol. wts in the range 2600–8800. These comprise 10–13% of mitochondrial protein (Munkres *et al.*, 1971; Swank *et al.*, 1971). These "miniproteins" are not incomplete proteins or degradation products and are not produced in the presence of cycloheximide. It is proposed that the localization of malate dehydrogenase depends on its ability to bind to the aggregates of these "miniproteins". Highly-purified malate dehydrogenase has been isolated which has an oligopeptide of mol. wt 3000 associated with it.

It has been suggested (Capaldi and Green, 1972) that all membranes contain two distinct classes of proteins: (a) highly hydrophobic ("intrinsic") proteins which form the basic structure of the membrane by their strong interaction with phospholipids, and (b) water soluble ("extrinsic") proteins which are also firmly bound to the membrane but which are not essential for the maintenance of membrane structure. Tzagoloff (1972) has suggested on the basis of his work on the ATPase complex, that the intrinsic membrane proteins of the inner-mitochondrial membrane are generated by combination of cytoplasmically-synthesized products with small hydrophobic polypeptide products of the mitochondrial protein synthesis system, whereas all extrinsic proteins are synthesized entirely extra-mitochondrially. Furthermore these intrinsic proteins (proteolipids containing esterified fatty acids) interact strongly with phospholipids. The ability for spontaneous assembly of multienzyme complexes into membrane structures may depend on this type of interaction.

F. Growth of Mitochondrial Membrane at the Molecular Level

The mechanisms involved in the growth of the mitochondrial membrane are unknown. It is not yet clear whether multienzyme complexes are assembled outside the membrane, and subsequently incorporated as preassembled units, or whether assembly occurs within the membrane itself. Flexibility of composition of the membrane with regard to individual respiratory chain components would suggest that assembly occurs as an intra-membrane process. One hypothesis for the mechanism of membrane growth (Tzagoloff, 1972) proposes that the mitochondrial membrane has a number of growing points which consist of the whole protein-synthesizing machinery of the mitochondrion. This machinery synthesizes the hydrophobic "handles' (low mol. wt peptides which combine with phospholipids) to form the basic substructure of the complexes and also of the mosaic of components of which the membrane is composed. The incorporation of the extrinsic proteins into the complexes could then be a self-assembly process of the type demonstrated by *in vivo* reconstruction experiments (Kagawa and Racker, 1966; King and Takemori, 1964).

G. Assembly as a Discontinuous Process Through the Cell-cycle

Although the ratio of rate of synthesis of mitochondrial proteins to that of total cell protein may not vary through the cell-cycle in some cases (Kahn and Blum, 1967), the activities of some individual mitochondrial enzymes and cytochromes do not increase continuously; they may increase in a stepwise pattern or they may show oscillations (see Chapter 4). Furthermore, increases in individual enzymes or cytochromes may occur at different times in the cell-cycle so that the process of assembly of mitochondria consists of a series of asynchronous steps. The composition of mitochondria is far from constant throughout the cell-cycle and the exact time course of its assembly can only be charted in synchronous cultures. This task has only recently been undertaken and our appreciation of the temporal sequences of events involved is very rudimentary.

H. Energy Requirements for Mitochondrial Assembly

Little is known about the origin of the energy required for the maintenance of the structure of mitochondria and for their biogenesis (Kováč, 1972). The energy may be generated within the mitochondria themselves or be derived from extramitochondrial sources. During respiratory adaptation, synthesis and assembly of components is not inhibited by CN^- (Slonimski, 1953a, b), antimycin A (Yčas, 1956) or by oligomycin (Galleoti et al. 1968). When S. cerevisiae DTX 11 is grown aerobically on glucose in the presence of antimycin A or oligomycin, cells with a normal content of cytochrome c but deficient in cytochrome b and containing virtually no cytochromes $a + a_3$ are produced (Kováč et al., 1970a). Diminished synthesis was interpreted as being due to the enhanced catabolic repression that ensued as a consequence of the elimination of the Pasteur effect by antimycin A or oligomycin. This is confirmed by the observation that when galactose replaces glucose in this experiment, neither antimycin A nor oligomycin, nor even a combination of both, prevents the synthesis of any of the cytochromes (Kováč, 1972), although the respiration of both glucose- and galactose-grown cells is equally sensitive to inhibition by these antibiotics. Erythromycin (2 mg/ml) did not significantly inhibit the formation of the respiratory pigments when growth was on galactose; only the combined action of erythromycin and antimycin A was able to eliminate cytochrome $a + a_3$ synthesis. Excess cytochrome c was found under these conditions. The frequency of *petite* induction was not enhanced under these conditions and inhibition was reversible; cells grown in the presence of inhibitors were able to resume normal growth on nonfermentable substrates once the inhibitors were removed. A similar effect of the inhibitors was seen in mutant DH1, which has a modified adenine nucleotide translocation mechanism (rate of translocation $2 \cdot 5 \times$ slower than wild-type, Kolarov et al., 1972b). Complete blockage of adenine nucleotide translocation across the inner-mitochondrial membrane by use of atractyloside was not possible, as yeast cells are impermeable to this inhibitor.

These results suggest that energy generated by the mitochondria themselves is not obligatory for mitochondrial biogenesis and that the energy can be supplied by cytosolic reactions. Neither are processes which derive energy directly from the high energy state (reversed electron transport, ion transport etc.) obligatory for mitochondrial biogenesis. However Luzikov (1973) proposes that the maintenance of mitochondrial organization does require active mitochondrial energy conservation (see p. 191). Diminished transport of cytosolic ATP into mitochondria by a defective adenine nucleotide translocase does not interfere with the synthesis of a complete respiratory chain. Neither the functioning of the respiratory chain nor the maintenance of the high energy state is necessary for the maintenance and normal replication of mt DNA. Thus the presence of an energy-transfer system in "promitochondria" (Groot et al., 1971) must have a function other than assuring the replication of mt DNA. A mutant lacking F_1–ATPase will not grow anaerobically (Ebner and Schatz, 1973).

However, when ATP synthesis in the mitochondria of aerobically-grown yeasts is inhibited, and simultaneously the entry of cystosolic ATP into mitochondria is prevented by bongkrekic acid, mass formation of respiratory-deficient mutants takes place (Šubik et al., 1972a). The multiplication of the respiratory-deficient mutants in complex glucose-containing medium is arrested by bongkrekic acid. Thus the continual presence of ATP inside the mitochondria is an absolute requirement for the normal replication of mt DNA and for a function related to cellular multiplication.

IV. The Formation and Development of Mitochondria

A. Growth and Division During Exponential Cell Growth

There have been many different hypotheses to explain the developmental origins of mitochondria, the early cytological workers attempted to distinguish between three often-stated alternatives: (a) origin from many diverse and unrelated membrane structures; (b) origin de novo, from precursors of molecular size; or (c) origin by the processes of growth and division of pre-existing mitochondria. All the various intracellular membranes have at one time or another been implicated as precursors of mitochondrial membranes, e.g. the cytoplasmic membrane (Robertson, 1964), the nuclear membrane (Bell and Mühlethaler, 1964) or the membrane of peroxisomes (Rouiller, 1960). These ideas, which are discredited in the light of the findings that different membranes have different biochemical constitutions and enzymic complements, have been reviewed by Novikoff (1961). The possibility of a *de novo* origin for mitochondria was for some time apparently supported by the erroneous conclusions that membranous structures were completely absent in yeast after anaerobic growth (see Chapter 1).

The genetic continuity of extrachromosomal inheritance, the semi-autonomy of the mitochondrion with regard to its unique apparatus for biosynthesis, and the independence of replication of mitochondrial and nuclear DNAs, all suggest that mitochondria must be formed by the growth and division of existing mitochondria. This theory is supported by a great deal of morphological evidence, although in only a few studies have quantitative morphometric methods been applied. Thus Bahr and Zeitler (1962) and Claude (1965) both suggested that the dumb-bell-shaped mitochondria observed in rat liver represent division stages in the life cycles of the mitochondria. Some striking electron micrographs of the formation of cross-partitions during the division of hepatic mitochondria which accompanied the recovery of mice from riboflavine deprivation have been presented by Tandler *et al.* (1969); another situation which has provided a clear picture of mitochondrial division is the fat body of insects after emergence from the pupae (Larsen, 1970). Division of the mito-chondria of *Physarum polycephalum* is preceded by the "division" of the DNA-rich "nucleoid" (Guttes *et al.*, 1966). In *Tetrahymena pyriformis* the mitochondria divide across their long axes during exponential growth in pace with division of the organisms, so that the numbers of mito-chondria per organism is kept at a constant minimal value (Elliott and Bak, 1964). Cessation of growth of the culture led to an increase in mitochondrial numbers per cell, and a change in intracellular distribution.

Biochemical evidence to support the theory of growth and division of mitochondria was first provided by the experiments of Luck (1963 a, b) who grew a choline-requiring strain of *Neurospora crassa* (*chol*-1, #34486) with labelled choline to obtain organisms with labelling in the lecithin of mitochondrial membranes. After transfer to nonradioactive medium, purified mitochondrial fractions were obtained during the subsequent logarithmic growth period, and the distribution of label among the individual mitochondria was determined using quantitative radioauto-graphy. The results indicated that choline-containing phospholipid was sufficiently stable to serve as a satisfactory marker through three mass doubling cycles. Analysis indicated that a random distribution of radio-activity occurred in mitochondrial populations isolated from fully-labelled cells and also from the cells undergoing three subsequent doubling cycles in the presence of unlabelled choline. Control experiments indicated that the random distribution did not result from fractionation procedures;

it was also unlikely that it was effected *in vivo* by repeated mitochondrial fissions and fusions. These data, which fitted a dispersive distribution of label, excluded the possibility of *de novo* formation and did not favour the hypothesis of origin from non-mitochondrial membranes. It was evident that mitochondrial mass increases by a continuous addition of newly-synthesized components, including choline-containing phospholipids, to the existing mitochondrial structure, and that the number of individuals in the mitochondrial population increases by division of organelles. This division process would distribute the label at random so that all the pre-existing membrane would be transmitted nearly uniformly to all progeny.

When the mutant was grown on a medium containing 1 μg/ml choline chloride, it yielded mitochondria which contained less lipid than normal and therefore had a higher buoyant density (Luck, 1965a). After transfer of cells from a choline-poor to a choline-rich medium, the lipid content of the mitochondria increased. Over a ninety-minute period, during which time there is an increase of mitochondrial protein mass of about 50% over that initially present, the mitochondria changed density as a single population (Luck, 1965b). If new mitochondria had arisen *de novo*, one would have expected a population of light lipid-replete mitochondria to have appeared alongside the original heavy lipid-depleted population. In all these experiments the possibility exists that exchange or turnover of lipids between different membrane structures is extremely fast and might be sufficient to cause the randomization of label. Although Luck attempted to rule out this possibility it cannot be entirely eliminated, but the conclusion that mitochondria arise from pre-existing mitochondria on the basis of this and other evidence is probably valid.

Evidence for division of mitochondria also comes from work on the physical continuity of mitochondrial DNA through the growth and division cycles of *N. crassa*, yeast, and *T. pyriformis* (see Chapter 5). The exact relationship between the timing of the replication of mitochondrial DNA and the division of mitochondria cannot yet be defined with any precision.

The addition of chloramphenicol (40 μg/ml) to Hela cells blocks the formation of cytochrome *c* oxidase, without having any effect on the pre-existing enzyme; the numbers of mitochondria per cell remain constant for up to four generations in the presence of the antibiotic (Storrie and Attardi, 1973). Staining with 3,3'-diaminobenzidine indicated that over this period all the mitochondrial profiles retained cytochrome *c* oxidase activity, while the intensity of staining along the inner-membranes decreased as the experiment proceeded. These results indicate that mitochondria grow by the random insertion of components into the membranes and subsequently divide.

When chloramphenicol (100–500 μg/ml) is added to an exponentially-growing culture of *T. pyriformis*, there is a sudden fall in growth rate, followed after two cell generations by the complete inhibition of growth (Fig. 121; Turner and Lloyd, 1971). No direct inhibition of respiration of isolated mitochondria occurs at this concentration of the antibiotic (D. Lloyd, unpublished). The organisms so produced have respiratory-deficient mitochondria which have a defective energy-conservation mechanism. A great reduction in the organization of the normally densely-packed cristae, results in a decreased intensity of staining by the osmium fixative; the entire mitochondrial population is affected. The organisms produced after growth for two generations have four times the mito-chondrial numbers of normal cells. The median $S_{20, W}$ for these chlor-amphenicol-inhibited organelles in a sucrose gradient is 10,600, as opposed to a value of 90,000 for normal mitochondria (Poole et al., 1971b), and radii calculated on the basis of this sedimentation data are 0·15 μm and 0·45 μm respectively. These calculated radii correspond well with measure-ments made on electron micrographs of isolated mitochondria (0·125 μm for chloramphenicol-inhibited mitochondria, 0·53 μm for normal organelles). In organisms grown to stationary phase in the absence of chloramphenicol, there is no reduction in size or increase in numbers of organelles. The chloramphenicol-induced abnormality is reversible; after a short lag cells grow normally when transferred to growth medium which does not contain the antibiotic. These observations strongly suggest the existence of a control mechanism in normal growth which ensures that mitochondria divide when they attain a certain size, and that coordination occurs between cell division and mitochondrial division so as to maintain the mitochondrial population of the obligate aerobe within certain limits (Turner and Lloyd, 1971). Chloramphenicol interferes with this control mechanism in T. pyriformis and permits mitochondrial division to outpace cell division (for instance in Fig. 121, three to four mitochondrial genera-tions have been produced during two cell divisions). A hypothesis (based on the ideas of Williamson, 1970) which can account for the possible mechanism for the breakdown of intracellular control produced by chloramphenicol is illustrated in Fig. 122. It assumes that the equal

---

FIG. 121. Growth of *Tetrahymena pyriformis* in the presence and absence of chloram-phenicol; changes in mitochondrial size and numbers. The proteose-peptone-liver extract medium (1 litre) was inoculated at zero time with 200 organisms/ml and the cultures were grown with forced aeration. Chloramphenicol (500 μg/ml) was added at time indicated by arrow to one culture (b) and led to a reduction in growth rate and complete inhi-bition of growth after two cell-generations. Electron micrographs (ultra thin sections of osmium-fixed samples) show the mitochondrial morphology in the control culture during exponential growth (a_1), in the stationary phase (a_2) and after chloramphenicol inhibition (b). Measurements of average mitochondrial radii (r_{av}) were made on isolated mitochondria in electron micrographs and calculated from $S_{20, W}$ values obtained in the BXIV-zonal centrifuge (in parentheses). (Reproduced with permission from Lloyd et al., 1971b.)

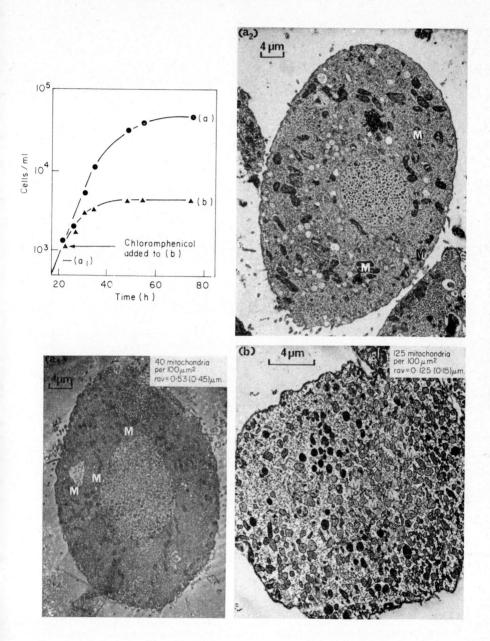

(a₂)

4 µm

M

M

M

(a₁)

Cells/ml

Time (h)

Chloramphenicol added to (b)

(a)

(b)

40 mitochondria per 100 µm²
$r_{av} = 0.53\,(0.45)\,\mu m$.

M

M

M

(b)   4 µm

125 mitochondria per 100 µm²
$r_{av} = 0.125\,(0.15)\,\mu m$.

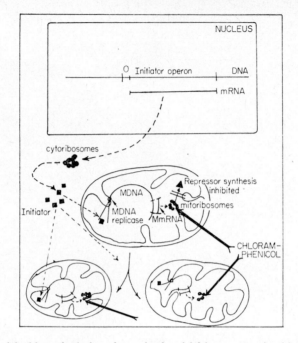

FIG. 122. Model of hypothetical nuclear-mitochondrial interactions in chloramphenicol-inhibited *Tetrahymena pyriformis*. It is proposed that chloramphenicol inhibits mitochondrial protein synthesis. One of the mitochondrial gene products is a repressor which in normal cells can combine with a nuclear operator (O) controlling the extra-mitochondrial synthesis of an initiator (■) of mitochondrial DNA (MDNA) replication. When chloramphenicol is present the block in repressor formation leads to a permanent switching on of the initiator operator and mitochondrial DNA replication is released from nuclear control. This leads to the formation of many more mitochondria per cell than are normally present. These mitochondria exhibit pleiotropic respiratory deficiency, and the resulting reduced capacity for energy conservation in this obligate aerobe leads to a reduced growth rate and then to complete cessation of growth, (MmRNA: mitochondrial messenger RNA). (Reproduced with permission from Lloyd *et al.*, 1971b.)

distribution of mt DNA to daughter mitochondria (Parsons and Rustad, 1968; Charret and André, 1968) necessitates the semiconservative replication of mt DNA before mitochondrial fission can occur. A similar hypothesis has more recently been developed to account for the induction by chloramphenicol or ethidium bromide of enzymes involved in the expression of the mitochondrial genome in *Neurospora crassa* (Barath and Küntzel, 1972a).

B. Growth and Division of Mitochondria in Synchronously Dividing Cell Cultures

An extension of the hypothesis developed from experiments with *Tetra-*

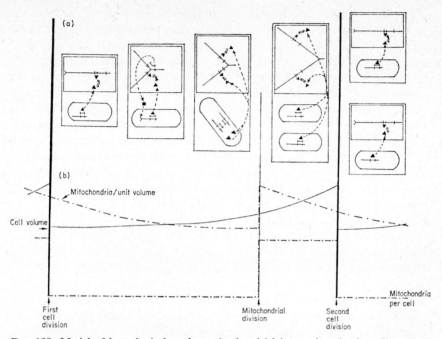

FIG. 123. Model of hypothetical nuclear-mitochondrial interactions in the cell-cycle of *Tetrahymena pyriformis* and theoretical changes in mitochondrial numbers/unit volume. (a) Symbols used are the same as in Fig. 123. The diagram illustrates how initiator of mitochondrial DNA replication could be produced only at one stage in the cell-cycle, due to the effective halving of repressor concentration when a replication fork passes through the nuclear operator. The *status quo* is restored when the capacity for mitochondrial repressor synthesis is doubled by duplication of the gene coding for this protein. (b) Theoretical variations in mitochondrial numbers per cell and per unit cell volume. (Reproduced with permission from Lloyd *et al.*, 1971b.)

hymena would predict a synchronous division of all the mitochondria of a single cell (Fig. 123, Lloyd *et al.*, 1971b). Counting of mitochondrial numbers in a cell by serial sectioning of organisms for examination by electron microscopy is a time-consuming task; a more useful approach in future studies may include high voltage electron microscopy of intact cells, or the use of a Coulter counter for the electronic counting and sizing of isolated organelles (Gear and Bednarek, 1972). Synchrony of mitochondrial division in several discrete zones of mycelium of defined age behind the growing tips of hyphae of *Neurospora* has been reported by Hawley and Wagner (1967) who observed "cupping" of mitochondria followed by division to give closely associated mitochondria. A linear rate of increase in mitochondrial profiles per cell section during the cell-cycle of *S. cerevisiae* led Cottrell and Avers (1971) to suggest that the mitochondria showed "synchronous development but not simultaneous division" but these conclusions require re-evaluation in the light of the recent observation of a single mitochondrion in yeast (Hoffmann and Avers, 1973). The claim for synchronous division of mitochondria in synchronous cultures of *Shizosaccharomyces pombe* (Osumi and Sando, 1969) is unfortunately not supported by a statement of the statistical significance of the measurements of mitochondrial numbers. However, the observations of McCully and Robinow (1971) also suggest that long cells (which are close to nuclear division) possess long mitochondria which may stretch the entire length of the cell. They remain long during mitosis and appear to divide just before the inward growth of a transverse septum results in the formation of two daughter cells. Continued fragmentation of the mitochondria after closure of the septum appears to produce the short mitochondria characteristic of cells in the early growth phase. The most clear instance of mitochondrial division occurring at a set time in the sequence of events which constitutes the cell-cycle is still that observed in the small green flagellate *Micromonas pusilla* by Manton and Parke (1960), where the question is simplified to its limit by the presence of only a single mitochondrion per cell. A similar sequence occurs in the Kinetoplastidae which also have only one mitochondrion per cell (Anderson and Hill, 1969); these organisms also provide the only unequivocal case of the occurrence of a discrete S phase for a mitochondrial (kinetoplast) DNA. The evidence for and against periodic synthesis of mt DNA in the cell-cycles of various organisms has already been discussed (see Chapter 5).

C. Turnover of Mitochondrial Components in Non-proliferating Cell Suspensions

The rate of turnover of components of a microorganism is at a minimum under conditions that favour maximal growth rates (e.g. 0.2% per h in yeast cultures growing aerobically with 3% glucose and with a mean generation time of 90 min (Bartley and Birt, 1970)). Non-proliferating suspensions of some eukaryotes remain viable for extended periods (up to 3 months in the case of *Prototheca zopfii* under suitable conditions), and these conditions give maximal turnover rates (about 0.7% per h). Alterations of mitochondria and their components in the absence of net cell growth have been discussed in Chapter 4; these studies have included few measurements of turnover rates of mitochondrial components. The question which obviously arises is whether the various mitochondrial constituents form a unit with its own distinctive rate of turnover, or whether the individual components are selectively replaced without destruction of the whole organelle. The original measurements of Fletcher and Sanadi (1961) indicated that the protein and lipid components of rat liver mitochondria all turned over at the same rate, giving a half-life of

10 days, but subsequent studies have revealed that this is an over-simplification, and that the turnover of individual protein and phospholipid components is such that each has its own distinctive half-life (Ashwell and Work, 1970; Bartley and Birt, 1970). Apart from the autophagic ingestion and subsequent digestion of whole mitochondria in lysosomes, there are evidently more subtle mechanisms for the controlled and selective release and replacement of individual components of the organelle. Thus selective loss of cytochromes $a + a_3$ occurs during the encystment of *Hartmanella* (A. H. Chagla, A. J. Griffiths and D. Lloyd, unpublished results), selective loss of certain enzymes has been observed during glucose repression of yeast (Chapman and Bartley, 1968), and selective loss of electron-transport components accompanies specific nutrient limitation (see Chapter 4). The loss of cytochrome c oxidase activity in certain stages of the cell-cycle has been shown to be accompanied by loss of spectrophotometrically-detectable haemoprotein (Poole *et al.*, 1973), and must therefore involve a very specific process of degradation.

Neutral proteinases located in yeast mitochondrial fractions have been implicated in mitochondrial turnover (Bartley and Birt, 1970). Endonucleases have also been isolated from mitochondrial fractions from *Neurospora* (Linn and Lehman, 1965a, b; 1966). The activities of these enzymes must be held in check by an inhibitory structure or component. It is proposed that they may be continuously functional, degrading mitochondrial enzymes at a relatively steady state. Different enzymes will have different susceptibilities to hydrolysis, and this might be one factor which results in a range of half-lives for different mitochondrial proteins. An energy-requirement for the maintenance of mitochondrial organization is indicated by the observation that any interruption of oxidative phosphorylation results in *in vivo* degradation of mitochondria (Luzikov, 1973, but see Kováč, 1972, p. 452). The actual mechanism of the degradation process is unknown, and the factors directly involved remain to be investigated. Thus any mitochondrial protein may be considered as being "in temporary occupation of a particular binding site and in kinetic equilibrium with other similar but less stable molecules not so bound" (Ashwell and Work, 1970). The half-life of total mitochondrial protein in *Tetrahymena* suspended in an inorganic salts solution is about 24 h (G. Turner and D. Lloyd, unpublished); in this case extensive lysosomal destruction of whole mitochondria is evident in cytochemical studies (Levy and Elliott, 1968; Nilsson, 1970).

D. Modification of Existing Mitochondria by Insertion of New Components

The insertion of newly-synthesized components into the pre-existing mitochondria of cells in the absence of net cell growth may also take place;

R

the part played by the turnover of mitochondrial constituents in this process has not yet been assessed. Thus the adaptation of *Prototheca zopfii* and *Polytomella caeca* to propionate assimilation involve the integration of extra-mitochondrially-synthesized enzymes into the existing mitochondrial structure (Lloyd and Venables, 1967; Lloyd *et al.*, 1968). The recovery of *Candida utilis* from iron- or sulphate-limited growth (Garland, 1970) or from copper-limited growth (Light, 1972a) also involve synthesis and insertion of newly-synthesized respiratory chain components in the absence of net cell growth, although it would seem that the conditions employed would favour extensive turnover of existing components. Respiratory adaptation of facultatively-anaerobic yeasts provides the most extensively studied system of mitochondrial modification which can occur in the absence of increase in cell mass or numbers. In this case the rapid assembly of mitochondria may involve the insertion of newly-formed components into an existing framework, but this hypothesis has very little experimental support and the molecular mechanism of assembly is still open to debate (see Chapter 4).

PART III

EVOLUTION

12

The Evolution of Mitochondria

I. Introduction

"The danger of orthodoxy is that the conviction of being in possession of an undisputable truth may lead us to an all too easy abandonment of further search for truth" (Lemberg and Barrett, 1973).

The pathway of evolution by which present day eukaryotes acquired their mitochondria has been the subject of speculative discussion for more than eighty years. Two major hypotheses have been developed. The first proposes that mitochondria are derived from ancestral endosymbiotic bacteria which populated large primitive anaerobic host cells. This hypothesis has recently become very popular as a consequence of the advances in the molecular biology of the mitochondrion which have revealed striking (if somewhat superficial) analogies between various features of the biosynthetic machinery of prokaryotes and energy-yielding organelles. The second (plasmid) hypothesis suggests that mitochondria have evolved from the specialized mesosomal membranes of a large, highly advanced, aerobic prokaryote. A third hypothesis proposes that prokaryotes are degenerate eukaryotes; this possibility must be borne in mind but will not be discussed.

II. Similarities and Differences Between Bacteria and Mitochondria

Throughout the previous chapters attention has been drawn to the analogies which exist between the biosynthetic apparatus of bacteria and mitochondria. Table LXII presents a resumé of these similarities, but also emphasizes some striking dissimilarities. The proponents of the endosymbiont hypothesis would have us believe that these differences have arisen by divergence over the long course of evolution; indeed it is surprising that such great similarities are still evident between the mitochondria, so completely integrated into the administration and government of the nucleo-cytoplasmic hierarchy, and the free living and highly

Table LXII. A comparison of the characteristics of mitochondria and bacteria.

	Mitochondria	Bacteria
A. General Properties		
1. Size range	generally 0·5–2 μm	generally 0·5–2 μm
2. Shape	variable, often highly branched or reticulate	restricted by cell wall
3. Cell wall	absent (outer membrane resembles e.r.)	present except in pleomorphic L-forms
4. Growth	by increasing size, then dividing	by increasing size, then dividing
5. Mean generation time	as little as 1 h (in yeast) regulated by MGT of cell	as little as 10 min (*Beneckia*)
6. Autonomy	very limited but can grow and divide after transfer to another cell	complete
B. Characteristics of DNAs		
1. Size range	generally within range 0·3–60 μm	chromosome (*E. coli*) 1100 μm, plasmids 1–60 μm
2. Characteristics	*circular	circular, many physical characteristics resemble mt DNA
3. Nearest neighbour base frequencies	*more like bacterial than nuclear DNA	
4. Histone	*none associated with DNA	none associated with DNA
5. Membrane attachment site for DNA	*yes	yes
6. Requirement of this site for replication	?	yes
7. DNA polymerase	*membrane bound	membrane bound
8. DNA replication	not continuous at least in some species (trypanosomes)	continuous through cell cycle
C. Genetic Properties of DNAs		
1. Genes	* ω, ribosomal RNAs, rRNAs antibiotic resistance	mostly on the chromosome; sex factors, antibiotic resistance factors on plasmids
2. Recombination	yes	yes
3. Conjugation	?	yes, specialized F pili

* Indicates differences between mitochondrial systems and nucleo-cytosolic systems.

	Mitochondria	Bacteria
D. Transcription		
1. Mode	* asymmetrical	symmetrical
2. DNA dependent-RNA polymerase	* rifampicin sensitive, * amanitine insensitive	same
3. t½ messenger RNA	* evidence for short-lived mRNA	short-lived mRNA
4. In vitro stimulation by DNAs	* Bact. DNA stimulates mito. prot. synth. phage DNA stimulates mito. prot. synth. (some misreading)	mt DNA stim. bact. prot. synth.
5. Streptolydigin action	not inhibitory	inhibitory
E. Translation Ribosomes		
1. Ribosome (S values)	* range from 50–80	70
2. Large Subunit (S values)	* 33–58	50
3. Small Subunit (S values)	* 25–40	30
4. Large Subunit RNA (mol. wt)	* 0.65–1.28×10^6	1.10×10^6
5. Small Subunit RNA (mol. wt)	* 0.36–0.79×10^6	0.56×10^6
6. Large Subunit rRNA (G + C)	* 25–46 %	52–53 %
7. Small Subunit rRNA (G + C)	* 27–38 %	52–54 %
8. Specified by	* mt DNA	bacterial DNA
9. Presence of 5S RNA in ribosomes	* not detected (except in Tetrahymena, equivalent 4S RNA in Neurospora which is poorly homogeneous with bacterial 5S RNA)	present
10. Methylation of rRNA	* in Neurospora, not detected in yeast	yes
11. Formation of hybrid ribosomes	* large subunit does not exchange for bacterial large subunit	hybrid ribosomes formed between widely divergent species
12. Chloramphenicol, macrolide antibiotics, lincomycin, mikamycin, erythromycin etc.	* bind to the large subunit and inhibit protein synthesis (phylogenetic differences which may not be permeability differences)	same

Table LXII. (*continued*)

	Mitochondria	Bacteria
13. Aminoglycoside antibiotics		
kanamycin, streptonycin, viomycin }	* not inhibitory	inhibitory
paromomycin, neomycin B C }	* inhibitory (bind to small subunit)	inhibitory
14. Phenanthrene alkaloids		
tylophorine	not inhibitory	not inhibitory
tylocrebrine	*} inhibitory at high	not inhibitory
cryptopleurine	*∫ concentrations	inhibitory at very high concentrations
15. Euflavine, proflavine, ethidium bromide	* inhibit p.s.	little effect
16. Inhibitors of cytoplasmic p.s. cycloheximide emetin anisomycin	* not inhibitory	not inhibitory
17. Amino acid analogues	canavanine or *p*-fluoro–Phe very effective inhibitors in some organisms	inhibitory
18. tRNAs	* at least 16 distinct from cytostolic ones isoaccepting species, codon degeneracy	complete set
19. Originate from	* at least 12 (in some species perhaps all) from mt DNA	bact. DNA
20. Contain fluorescent base	* yes	yes
21. tRNA-amino acyl-synthetases	* distinct from cytosolic enzymes	—
22. Methylase	* 2-methyl-adenine-tRNA methylase present	same
23. Transformylase	* present	present
Peptide Chain Inititation		
24. Initiator codon	* AUG for *N*-formyl methionine	AUG for *N*-formyl methionine

	Mitochondria	Bacteria
25. Binding of N-formyl Met tRNA to ribosomes	* lost on washing with M—NH$_4$Cl to remove initiation factors, can restore with *E. coli* factors (F$_1$F$_2$F$_3$) binding does not respond to phage RNA	same
26. Translocation step	* sensitive to chloramphenicol and sparsomycin	same

Peptide Chain Elongation

	Mitochondria	Bacteria
27. Factors	* T and G	T and G
28. Poly-(U)-directed poly-Phe synthesis	can occur when mitochondrial ribosomes and bacterial supernatant mixed	*vice versa*, but complete factor interchangeability not shown, as T mit labile
29. Effect of 120 mM NH$_4$Cl	* 90 % inhibition	maximum stimulation
30. Fusidic acid	* not inhibitory	inhibitory

Products

	Mitochondria	Bacteria
31. Protein synthesis products	* hydrophobic membrane polypeptides, no enzymically active subunits of enzymes	membrane proteins and enzymes
32. Translation of viral information	* viruses reported, not certain that mitochondrial machinery used in viral multiplication	yes (bacteriophage infections)

F. *Energy Producing Mechanisms*

	Mitochondria	Bacteria
1. Membranes	respiratory chain and energy conservation machinery integral part of inner-membrane	mechanisms integral part of lipoprotein membrane
2. Membrane "subunits"	9 nm subunits on inner surface of inner-membrane	9 nm subunits on inner surface of membrane
3. Permeases and ATPases	in inner-membrane	in membrane
4. Compartmentation	ATP-generating system isolated from cytosol by permeability barrier important in control	ATP-generating system exposed to cytosol, but some evidence for functional compartmentation of cytosol
5. Amount of energy-generating machinery	proportional to cristal development and mitochondrial numbers	proportional to occupancy of membrane by respiratory chains and area of membrane
6. Facility for increased membrane area	by increased complexity of cristal invagination	by complexity of invagination to form mesosomes

Table LXII. (*continued*)

	Mitochondria	Bacteria
7. Variability of composition as response to environmental modification	stoichiometry of respiratory chains very variable	show greater variability than mitochondria
8. Phospholipids	implicated in membrane organization and hence functional integrity	similar
9. Polyunsaturated fatty acids	non-essential to structure, function, or development	not detectable
10. Cardiolipin	important constituent	similar
11. Sterols	not detectable in inner-membrane	not detectable in nearly all prokaryotes
12. Keto acid oxidase complexes	similar size, structure and mechanisms in mitochondria and bacteria	
13. Cytochromes	typically a, a_3, b's, c, c_1	wide variety of a-, b- and c-type oxidases
14. Amino acid sequence of cytochrome c		much greater divergence between species
15. Oxidases	a_3 sole cytochrome oxidase except in *Tetrahymena* (a_{620} + a_3); cytochrome c peroxidase in yeast and *Crithidia*	variety of terminal oxidases b-type (o) as well as a-type
16. Quinones	UQ	UQ in Gram −ve NQ in Gram +ve
17. Electron transport inhibitors	antimycin A, cyanide, azide and CO nearly effective	classical inhibitors not always effective high concentrations required
18. Alternative pathways of electron transport	usually inducible	very common
19. Uncouplers	effective except in resistant mutants	not always effective (permeability?)
20. Inhibitors of energy conservation	Effective except in resistant mutants	effective in some species (strain dependent, sometimes permeability barriers present)
21. Ionophores	effective	often effective
22. Efficiency of oxidative phosphorylation	P/O ratios up to 3	same
23. Respiratory control	demonstrated in all intact mitochondrial preparations	also present in bacteria, difficulties of demonstrations are preparative

versatile present day bacteria. The champions of the plasmid hypothesis must be more concerned about the evolutionary divergence between mitochondria and the nucleo-cytoplasm which they believe to have taken place, and they must contemplate the selective pressures which act on the nuclear and mitochondrial genomes and give rise to evolutionary modification.

III. The Endosymbiont Hypothesis

Altmann (1890) was the first to suggest that mitochondria may represent bacteria living within the cell membranes of higher organisms. The "bioblasts" were thought to be capable of a free-living existence and arose by the growth and division of preexisting "granules". The gross morphological similarities between mitochondria and bacteria, and also between chloroplasts and blue-green algae, have been repeatedly referred to ever since (Famintzin, 1907; Mereschowsky, 1910). The first clear and detailed presentations of the endosymbiont hypothesis were by Portier (1918) and by Wallin (1927), who maintained that symbiotic relationships may give rise to the generation of new species, and that mitochondria represent bacteria which have come to live symbiotically within the cell. Unfortunately these investigators went further than this, claiming to have cultivated mitochondria *in vitro* (a claim repeated as recently as 1965 by Vogel and Kemper), and as a result, their proposals were not well received. However Nass (1969) has cited Wilson's (1928) prophetic statement: "To many, no doubt, such speculations may appear too fantastic for present mention in polite biological society, nevertheless it is within the range of possibility that they may some day call for more serious attention".

The startling differences in ultrastructural features between bacteria (and blue-green algae) and all other living cells became more apparent after the introduction of the electron microscope as a tool in biological research. Dougherty (1957) introduced the terms "prokaryotic" and "eukaryotic" and the differences between the two cell types became defined in precise terms (Stanier, 1961; Murray, 1962; Stanier and van Neil, 1962). Despite some assertions to the contrary (Mazia, 1965), similarities at the molecular level (e.g. the universal uniformity of the genetic code) confirm that all living organisms are branches of a common stem of cellular evolution, but as pointed out by Stanier *et al.* (1963) "the basic divergence in cellular structure which separates the bacteria and blue-green algae from all other cellular organisms probably represents the greatest single evolutionary discontinuity to be found in the present day living world".

The endosymbiont theory attempts to explain this discontinuity by proposing that cells without nuclei or specialized energy-yielding organelles were the first to evolve, and that all eukaryotic cells represent the descendants of a symbiotic union of several cell types (Ris, 1961; Sagan, 1967; Roodyn and Wilkie, 1968; Nass, 1969; Margulis, 1970; Flavell, 1972; Schnepf and Brown, 1972).

A great diversity of examples of symbiotic liaisons which are perpetuated on a stable hereditary basis may be cited, e.g. between a protozoon (*Paramecium*) and green alga (*Chlorella*) (Karakashian *et al.*, 1968), between a protozoon (*Myxotrichia paradoxa*) and three different species of bacteria (Grimstone and Cleveland, 1964), and between *Paramecium* and its cytoplasmic factors (kappa, mu, lambda and sigma particles, Gibson, 1970). Associations such as these often confer selective advantages on the host (as yet imprecisely ascertained) in the face of competition from well-established species.

The retention of a degree of autonomy of present day organelles has been demonstrated: chloroplasts have been found to be capable of surviving ingestion into mouse fibroplasts (L cells) without loss of morphological integrity, photosynthetic activity, or DNA, for up to five days and through five generations of the host cell (M. M. K. Nass, 1969b). Mitochondria have also been shown to survive uptake without digestion in autophagic vacuoles, but their survival has not been studied as carefully as that of chloroplasts; they also survive microinjection from one cell line into another (see Chapter 6). Chloroplasts have also been reported to be capable of undergoing (or completing) one division in culture (Ridley and Leech, 1970; Giles and Sarafis, 1971).

Two versions of the endosymbiont hypothesis have been proposed: (1) that the symbiotic union may have been achieved by way of a colonial association of several distinct cell types (S. Nass, 1969); or (2) that a sequence of implantations led to the development of, (a) cells possessing mitochondria, then (b) to cells with mitochondria and a flagellum, and (c) to cells with mitochondria, flagellum and chloroplasts (Sagan, 1967; Margulis, 1970; Fig. 124). Thus the invasion of the primitive anaerobic host by bacteria which had acquired the advantages of an aerobic metabolism produced the counterparts of present day mitochondria and made possible the evolution of the entire kingdoms of higher life. In turn flagella were produced from ancestral symbiotic spirochaetes, and chloroplasts were eventually derived from blue-green algal invaders.

The evolutionary time scale over which these events may have occurred is as follows (Sagan, 1967; Margulis, 1970). The origin of heterotrophic obligately anaerobic bacteria is believed to have occurred some 3×10^9

Fig. 124. Hypothesis of the origin and evolution of eukaryotic cells. Ranges of DNA base ratios (%G + C) superimposed on overall phylogeny. (Reproduced with permission from Margulis, 1970.)

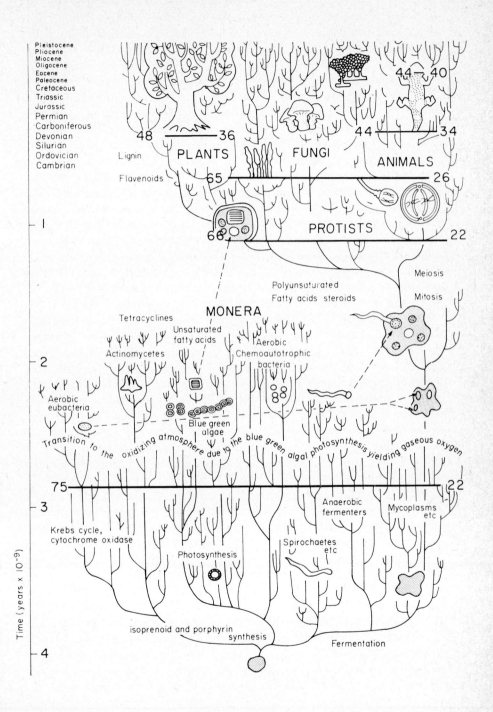

years ago. The evolution of porphyrins was followed by the development of photosynthetic Light System I. The gradual production of O_2 in the atmosphere accompanied by elaboration of a functional photosynthetic Light System II by various species of blue-green algae which are richly represented in the fossil records of the Middle Precambrian Era (some $2 \cdot 7 \times 10^9$ years ago). The increasing partial pressure of O_2 in the atmosphere conferred a differential survival value on the mechanisms of protection against oxygen-toxicity in organisms which had always been obligate anaerobes: protective metabolic mechanisms may have involved carotenoid compounds, specialized electron transport pathways (e.g. high affinity O_2-scavenging systems associated with bioluminescence), metalloproteins (superoxide dismutases) and haemoproteins (catalase). The evolution of the first aerobic bacterial cells must have occurred at the time when O_2 had accumulated to a considerable extent ($1 \cdot 8 \times 10^9$ years ago, Cloud, 1965; Schopf, 1967); these bacteria may have possessed electron transport chains capable of accepting electrons from ammonia or sulphide. Roodyn and Wilkie (1968) have suggested that the most primitive respiratory organelles found in some present day cells, the peroxisomes ("fossil organelles", de Duve and Baudhuin, 1966; de Duve, 1969), which may have originally evolved as oxygen-protective structures designed to compartmentalize reactions involving H_2O_2, were gradually replaced by the energy-conserving, thermodynamically-advantageous mitochondria over the past 10^9 years of eukaryote evolution. The postulated ancestral peroxisome was a particle of general metabolic importance, which shared with the mitochondrion the ability to oxidize fatty acids, L-amino acids, and a number of carbohydrate derivatives. It could even break down purines and D-amino acids that are normally not catabolized by mitochondria. However, peroxisomes could never oxidize α-ketoacids or succinate, and so lacked the key functions which allow mitochondria to degrade many metabolites all the way to CO_2 through the TCA cycle, never possessed any energy-conserving machinery, and had a low affinity for O_2. All versions of the endosymbiont hypothesis possess an inherent paradox (Stanier, 1970) which may be stated as follows. Without chloroplasts or mitochondria a cell of the eukaryotic type is dependent on fermentation for its energy supply, and so this primitive and inefficient mechanism of energy generation was the only one available to the evolving eukaryotic cell-line until the capabilities of respiration and photosynthesis were implanted in it by acquisition of prokaryotic endosymbionts. This implies that "most of the structural complexities of eukaryotic cells arose in a metabolically primitive cell line, while the main thrust of progressive biochemical evolution was being expressed in parallel prokaryotic cell-lines, accompanied by a far smaller degree of structural evolution". Stanier has suggested that the progressive structural evolution of the eukaryotic cell received its initial impetus from the acquisition of a novel cellular property, the capacity to perform endocytosis; this property has not been shown in any contemporary prokaryotic cell-line, even in the largest prokaryotic organisms. The eukaryotic capacity for endocytosis meant that this group of organisms were not obliged to undergo the wide diversification of modes of energy-yielding metabolism which gives bacteria their unrivalled contemporary variety in this respect. The picture then is of a group of predatory eukaryotes which had evolved a considerable amount of structural complexity; over a long evolutionary period a shift from predation to predation plus endosymbiosis took place repeatedly in many different cell-lines.

The mitochondrial electron transport system of eukaryotes is fairly uniform in its constituent components throughout a wide range of organisms; it is within the Phylum Protozoa that the greatest departures from this uniformity are found (see Chapter 3). This situation contrasts with the diversity of electron transport systems found in contemporary prokaryotes. This implies that the origin of mitochondria occurred soon after the evolutionary development of the respiratory chain in prokaryotes. Stanier (1970) also suggests (in contrast to the views of Margulis, 1970) that mitochondria were the last (not the first) symbiotic components established in eukaryotic cells, and that it was this event which brought to a close the purely cellular phase of evolution. The subsequent expression of evolutionary change occurred mainly at the level of the organism.

Early events in the intra-cellular evolution of the prokaryotic precursors of mitochondria would have included loss of the cell wall, which was completely dispensable, and the abstraction of some genetic determinants of endosymbiont function from the prokaryotic genome and deposition of these in the nuclear genome of the host. This step meant an enforced and stable hereditary future partnership; endosymbionts had become organelles.

IV. The Plasmid Hypothesis for The Non-symbiotic Origin of Mitochondria

A. Introduction

Although the symbiotic theory for the origin of eukaryotic cells has become extremely popular in recent years, there are a number of considerations which do not make it unquestionable. The alternative model which envisages the evolution of eukaryotes directly from an advanced type of prokaryote has been proposed from many independent sources (Work, 1968; Lloyd, 1969; Allsopp, 1969; Attardi and Attardi, 1969; Hughes et al., 1970; Perlman and Mahler, 1970b; Bell, 1970; Mahler et al.,

1971a). The hypothesis has been restated and amplified in a fascinating article by Raff and Mahler (1972), and further discussed by Uzzell and Spolsky (1973), Raff and Mahler (1973) and de Duve (1973).

B. Criticisms of the Endosymbiont Hypothesis

Close examination of the endosymbiont hypothesis reveals the following shortcomings.

1. The oldest fossil records of eukaryotic cells are those from the Beck Springs Dolomite of California ($1 \cdot 2$–$1 \cdot 4 \times 10^9$ years old, Cloud *et al.*, 1969); the accumulation of free O_2 in the atmosphere corresponded with the appearance of oxidized red beds ($1 \cdot 8 \times 10^9$ years ago, Cloud, 1968). If we accept the validity of these estimates, then the most primitive eukaryotes may have appeared 4×10^8 years after the appearance of O_2. The endosymbiont hypothesis would have us believe that the original hosts for the prokaryotic invaders were anaerobic cells.

2. The postulated protoeukaryote, although structurally advanced, was primitive and inefficient in its respiratory metabolism. It would thus have been at a great disadvantage in competition with prokaryotes having highly efficient energy-yielding processes.

3. The eukaryotic cell is not an anaerobic cytoplasm containing an aerobic respiratory organelle (Cohen, 1970); characterization of the metabolic pathways of the cytoplasm show that the eukaryotic cell is, and has always been, basically aerobic. Thus, catalase, only partly located within the peroxisomes, is ubiquitous in the lower eukaryotes, and may be involved in protecting aerobic cells against H_2O_2 generated by oxidative enzymes (de Duve and Baudhuin, 1966). Superoxide dismutase may play a similar role in protection against the superoxide radical (Welsiger and Fridowich, 1973a, b). Present day anaerobic eukaryotes are few, and are probably the results of secondary adaptation to specialized environments (e.g. the trichomonads, Müller, 1973), hypermastiginids, holotrich ciliates of the rumen (Hungate, 1967) and the fungus *Aqualinderella* (Emerson and Held, 1969)). Yeast has an absolute requirement for O_2, in that without it the synthesis of oleate and steroids is blocked; strictly anaerobic growth depends on the addition of these components (Andreason and Stier, 1953). Alternative aerobic and anaerobic pathways exist for synthesis of mono-unsaturated fatty acids, tyrosine, nicotinic acid, carotenoids, and porphyrins; steroids and polyunsaturated fatty acids have no anaerobic pathways. Eukaryotes and advanced prokaryotes use the aerobic pathway for synthesis of monounsaturated and hence polyunsaturated fatty acids. On the other hand the majority of both anaerobic and aerobic bacteria use the evolutionary, older anaerobic pathway (Bloch, 1962) for making monounsaturated fatty acids, and cannot convert these into polyunsaturated fatty acids. The universal presence of sterols in eukaryotes

suggests that the aerobic pathway for their production was present in the ancestral protoeukaryote.

4. There is a considerable evolutionary divergence between mitochondrial and bacterial cytochromes (Kamen and Horio, 1970; Gel'man *et al.*, 1967). Bacterial cytochromes are more varied, and a single species often has a multiplicity of oxidases. This is often a characteristic of bacteria which can utilize a variety of terminal electron acceptors. Mitochondrial cytochrome *c* reacts poorly or not at all with many bacterial oxidases, and similarly mitochondrial cytochrome *c* oxidase does not react rapidly with bacterial cytochromes *c*. Although mitochondrial and bacterial cytochromes *c* have a common prosthetic group, they differ widely in their primary sequences, isoelectric points, and redox potentials. Cytochrome *P450* the terminal oxidase in bacterial hydroxylases (mixed function oxidase reactions), is generally associated with the endoplasmic reticulum of eukaryotes rather than with mitochondria. This suggests that the cells ancestral to eukaryotes already possessed their own cytochrome components.

5. The degree of divergence between sequences in eukaryotic and prokaryotic cytochromes *c* is comparable to that between eukaryotic and prokaryotic tRNAs, and is significantly greater than divergences between different eukaryotic kingdoms (McLaughlin and Dayhoff, 1970). This suggests that the gene for cytochrome *c* has resided in the nuclear genome ever since the divergence of eukaryotic from prokaryotic cells and does not support the idea of a late transfer of this gene from symbiont to host nucleus.

6. Although mitochondrial and bacterial systems of protein synthesis have many features in common, they are also quite different one from another in many respects (Table LXII). It may be that the differences between the two systems of protein synthesis in the eukaryotic cell have evolved during their long coexistence as different units of selection; they have thus diverged both from their common ancestral prokaryotic characteristics and from each other.

7. That the mitochondrial genome together with a mechanism for its replication, transcription, and translation has been conserved over 1.2×10^9 years of eukaryotic evolution suggests that it is not a mere evolutionary relic of an originally endosymbiotic prokaryote, but that it has arisen because it confers a selective advantage in cells which were able to synthesize a few crucially-important insoluble hydrophobic proteins at the sites of mitochondrial membrane assembly.

8. The integration of large tracts of endosymbiont genome into the unrelated genome of the host is mechanistically difficult to envisage.

9. There is no known case where the reduction in complexity of the metabolic apparatus which accompanies parasitism has continued to a point where the genome is incapable of coding, either for much of its own biosynthetic machinery (particularly its own DNA polymerase), or for

the majority of its proteins. Mt DNA is known to specify only rRNAs, tRNAs and messenger RNA for a few hydrophobic proteins but not the mitochondrial DNA- or RNA-polymerases, ribosomal proteins, polypeptide chain initiation or elongation factors.

10. The suggestion that the evolutionary development of flagella (and hence, cilia, centrioles, and the mitotic apparatus) owes its origin to an invasive spirochaete does not take into account the lack of homology between bacterial flagellin and the microtubule protein of eukaryotes (Smith and Koffler, 1971).

11. The proposal for a multiple origin for chloroplasts from three separate groups of photosynthetic prokaryotes (Raven, 1970) would seem an improbable one, as two of the postulated prokaryotic types are not in evidence today.

C. Bacterial Mesosomes and Plasmids

In general it is true to say that the respiratory and photosynthetic machinery in prokaryotes reside in the polyfunctional cytoplasmic membrane. Some exceptions to this have been found: e.g. in the blue-green algae and the green photosynthetic bacteria there are no connections between the cytoplasmic membrane and the specialized photosynthetic membranes; "gas vacuoles" which are also found in some species in both groups represent another class of prokaryotic organelle quite distinct from the cytoplasmic membrane. All other membraneous structure in prokaryotes (mesosomes) appear to arise by invagination of the plasmamembrane (Fig. 125). Mesosomes occur in aerobes and in anaerobes, and it is suggested that they may carry out a variety of functions (Gel'man et al., 1967; Hughes et al., 1970; Burdett, 1972; Reusch and Burger, 1973). Inability to detect enzymes distinct from those of the cytoplasmic membrane itself suggests that they may simply represent a development for increased surface area of the polyfunctional membrane rather than be specialized to perform any particular function. It has also been suggested that some mesosomes may bear the replication sites for the bacterial chromosome or plasmids and that they may be involved (by analogy with the nuclear membrane and spindle of eukaryotes) in the segregation of genetic material at cell division.

Plasmid DNA molecules constitute the independent extrachromosomal genetic elements in prokaryotes and have many features in common with mt DNAs. All plasmids which have been isolated and examined can exist as covalently closed circular duplex molecules (with supercoiled helices) ranging in length from 0·4–60 μm (0·8–116 $\times$ 10^6 daltons) with

FIG. 125. Mesomes in methane-oxidizing bacteria seen in a section (a), and in a negatively-stained autolysed cell (b). (Reproduced with permission of R. Whittenbury from Hughes et al., 1970.)

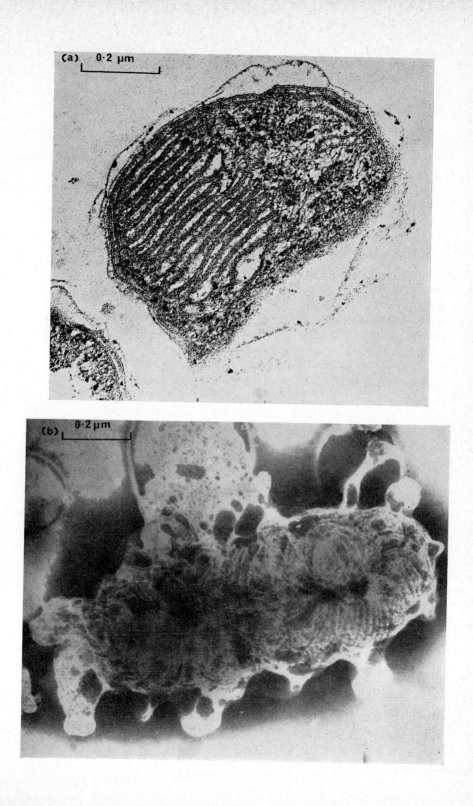

(a)   0·2 μm

(b)   0·2 μm

most of them 25 μm in length (Meyer, 1973). Oligomeric forms and catenates have been described; these are often generated when protein synthesis is inhibited. Supercoiled and oligomeric forms of the bacterial chromosomal DNA have not been demonstrated. Plasmids replicate autonomously (i.e. they may be considered as replicons) and independently of chromosomal DNA replication, although their numbers per organism are regulated. Selective inhibition of synthesis of plasmid DNA can be achieved by the use of agents inducing mitochondrial (but not nuclear) mutations in eukaryotes. Thus plasmids can be eliminated ("cured") by exposure to Co^{2+}, acridine dyes and ethidium bromide; the specificity of action of these agents may reside in membrane-DNA attachment sites or in the supercoiled helix structure of the plasmid DNA. It has not been demonstrated that plasmid replication uses a DNA polymerase distinct from that used by the chromosomal DNA. Most plasmids can be considered as consisting of two parts, one part concerned with the replication of the replicon and its distribution to daughter cells, and the other part carrying genetic determinants that confer properties on the host cell but which are not themselves concerned directly with plasmid survival (Richmond, 1970). Plasmids carry colicin genes and genes controlling resistance to many antibiotics including erythromycin, chloramphenicol, penicillin, streptomycin, kanamycin, neomycin, sulphonamides, tetracycline, etc. (Novick, 1969). Some of these genes affect protein synthesis by altering ribosomal proteins; others affect the induction of inactivating enzymes (penicillinase; acetylase for chloramphenicol etc.). There is no evidence for either rRNA or tRNA genes on plasmids, but there is one example of a tRNA gene being carried on a transducing phage (Smith *et al.*, 1966), and there is no *a priori* reason for assuming that any bacterial genes could not be borne on plasmids (Meyer, 1973). Some plasmids can be involved in recombinational events with other plasmids and with the chromosome (Novick, 1969). Thus great genetic diversity is possible in prokaryotes, and identical genetic markers carried extrachromosomally and on the chromosome are capable of an independent evolution. Plasmids carrying certain markers have a great selective advantage to the organism growing under certain environmental conditions (e.g. a cell having a plasmid which carries a chloramphenicol-resistance factor growing in cultures containing the antibiotic.)

D. The Mechanism of Mitochondrial Evolution According to the Plasmid Hypothesis

According to the plasmid hypothesis the ancestral prokaryote which gave rise to eukaryotes was a large highly developed aerobic bacterium. The necessity for an increased surface of respiratory membrane in this large organism led to the production of invaginations of the plasmamembrane to give the villi-like projections typical of present day bacterial mesosomes. Eventually the formation of closed respiratory vesicles, free in the cytoplasm became complete. The permeability barriers and compartmentation generated in this way was advantageous, as it afforded the basis for a more sophisticated regulation of respiratory metabolism (Hughes *et al.*, 1970). However this arrangement was not without its difficulties, as certain respiratory components (like cytochrome *c* oxidase) required synthesis *in situ*, and the membrane enclosing the respiratory apparatus was impermeable to ribosomes or ribosomal RNAs. It was thus advantageous to place a whole system for protein synthesis within the particle; this was achieved by incorporating a stable plasmid containing the appropriate genes for ribosomal components into the respiratory organelle. This process is analogous to the mechanism for gene sequestration (Perlman and Mahler, 1970b; Mahler *et al.*, 1971a) which occurs during the generation of multiple replicates of rRNA which are packaged in nucleoli during amphibian oogenesis (Brown and Dawid, 1968). This arrangement permits a closer regulation of key gene products at appropriate points in space and time.

The plasmid model suggests that genes at one time present in the nucleus are now borne on mt DNA, and if this is the case, then it is not surprising that genes which are involved in making different subunits of a single protein and are coordinately controlled are now to be found at two separate locations within the cell. The selective advantages of having multiple copies of the mitochondrial genome are three fold: (1) the arrangement provides possibilities for extrachromosomal recombination; (2) it protects against the possible modification of key components of the respiratory apparatus by mutation; and (3) it makes the assembly of the energy-coupling membranes (in which the lay-out of components is crucial to function) more easily accomplished ("vectorial assembly" of the inner mitochondrial membrane may be obligatory). Some recent experiments suggest that copies of mitochondrial genes are not be to found in the nucleus of present day obligate aerobes, e.g. in *Tetrahymena* (see p. 262); if this is so, evolution may have eliminated the "master-copy" of the genes after sequestration and amplification. The original sequestration may have involved a large section of the nuclear genome containing many more genes than are found in present day mitochondria; these could have been reduced to a bare minimum by a process of mutation, deletion, or exchange with the nuclear genome.

V. Some Experimental Approaches

Direct experimental evidence enabling a choice to be made between the two hypotheses of the evolutionary origin of mitochondria will not be

Table LXIII. Base composition of high molecular weight mitochondrial rRNA, cytoplasmic rRNA and mitochondrial DNA. (Reproduced with permission from Freeman et al., 1973.)

Organism	Mitochondrial rRNA												Ref.	Cytoplasmic rRNA GC%	Ref.	Mitochondrial DNA GC%	Ref.
	Smaller component (or total)*						Larger component										
	Moles %						Moles %										
	A	U	G	C	|G−C|+|A−U|	GC%	A	U	G	C	|G−C|+|A−U|	GC%					
Saccharomyces	38·4	34·5	16·1	11·0	9·0	27·1	40·3	34·6	14·0	11·0	8·7	25·2	1	49·0	1,2,3	24·0	20
	33·3	36·4	17·4	12·9	7·6	30·3	32·6	34·0	17·2	16·1	2·3	33·3	2				
	36·1	34·5	17·7	12·5	6·8	30·2							2				
	38·7	38·6	14·0	8·7	5·4	22·7							3				
Neurospora crassa	31·8	31·7	20·8	16·6	4·3	36·4	33·9	31·9	19·1	15·0	6·1	34·1	4	51·0	4,5	41·8	21
	27·2	29·8	22·9	14·8	10·7	37·7							5				
Aspergillus nidulans	34·5	34·0	18·0	13·5	9·0	31·5	36·0	33·5	18·0	12·5	8·0	30·5	6	53·7	6	29·9	6
Trichoderma viride	31·5	33·0	22·5	13·0	11·0	35·5	35·5	33·0	19·5	12·0	10·0	31·5	7	49·7	7		
Tetrahymena pyriformis	33·8	35·7	18·8	11·8	9·9	30·6	36·9	35·2	16·2	11·7	6·2	27·9	8	45·2	8	24·6	22

Euglena gracilis	39·5	33·2	14·9	12·5	8·7	27·4	36					9	54·4	9	31·6	23
Xenopus laevis	34	23	21	22	12	43	24	20				10	62	16	44·8	26
Mouse (L cells)	34·1	26·3	17·4	22·2	12·6	39·6	37·7	28·8	16·0	17·3	10·2	40	11	63·4	17	39·0	24
Rat (liver)	38·3	26·1	15·2	20·3	17·3	35·5					33·3	12			18	
(hepatoma cells)	30·5	22·9	21·6	24·9	10·9	46·5						13				
	31·5	27·5	16·4	24·5	12·1	40·9										
Hamster (BHK-21 cells)	33·2	27·4	18·4	20·8	8·2	39·2	33·0	27·0	17·7	22·3	10·6	40·0	12	65·2	19	39·0	25
	33·3	23·4	18·4	24·9	16·4	43·3	33·9	24·4	18·1	23·6	15·0	41·7	14			47	26
Human (HeLa cells)	32·7	22·6	18·3	26·3	18·1	44·6	32·0	23·4	18·9	25·7	15·4	44·6	15				27

* If no values are given for the larger component then the value in these columns is for total rRNA.

1. Fauman et al. (1969)
2. Morimoto and Halvorson (1971)
3. Reijnders et al. (1972)
4. Rifkin et al. (1967)
5. Küntzel and Noll (1967)
6. Verma et al. (1971b)
7. Edelman et al. (1971)
8. Chi and Suyama (1970)
9. Krawiec and Eisenstadt (1970a)
10. Dawid and Chase (1972)
11. Bartoov et al. (1970)
12. Montenecourt et al. (1970)
13. Bartoov (1971)
14. Vesco and Penman (1969)
15. Attardi and Attardi (1971)
16. Dawid (1966)
17. Lane and Tamaoki (1967)
18. Kirby (1965)
19. Darnell (1968)
20. Borst and Kroon (1969)
21. Luck and Reich (1964)
22. Suyama (1966)
23. Edelman et al. (1966)
24. Nass (1969)
25. M. Davidson and K. B. Freeman, unpublished results
26. Clayton and Vinograd (1967)
27. Corneo et al. (1968)

easily obtained, although some insight into the subsequent pathway of evolutionary change has been gained in recent work. It has been suggested that work on organelle origins may be more easily performed on chloroplasts than on mitochondria (Cohen, 1973).

The evolution of mt DNA has occurred so rapidly that no detectable sequence homologies are evident between the mt DNA of *Xenopus laevis* and that of yeast (Dawid and Wolstenholme, 1968) or even between the mt DNAs of *Tetrahymena* and *Paramecium* (Flavell and Jones, 1971a, b). Great changes are also evident in kinetoplast DNAs of Kinetoplastidae. The primitive free-living forms (e.g. *Bodo*) have a large amount of kDNA with the fibres arranged in visible order. Forms parasitic in one invertebrate host (e.g. *Crithidia*) have less kDNA, whereas the most advanced forms (parasitic in two hosts) have smaller masses of kDNA arranged in precise configurations (Simpson, 1973). A detailed study of the molecular changes and loss of information which have occurred during the evolution of these parasites, might lead to a general explanation of the evolutionary constriction on the amount of genetic information in other mitochondria. A start in this direction has been made by Newton *et al.* (1973a); comparison of base sequence homologies between kDNA from *Trypanosoma mega* and *Crithidia luciliae* (by hybridization with complementary RNAs synthesized on purified kDNAs) suggests that the evolution of base sequence homologies in these two organisms is not highly restricted. These studies suggest that it is hardly worthwhile to look for homologies between mt DNA and bacterial DNAs, although Cummins *et al.* (1967), on the basis of nearest neighbour analyses of RNA copies of the mt DNA of *Physarum polycephalum*, have shown absence of the low CpG frequency which characterizes all eukaryotic nuclear DNAs, and suggest therefore that slime mould mt DNA is more akin to bacterial DNA.

While the size of the mitochondrial genome has decreased (Borst and Flavell, 1972), the combined mol. wts of the rRNAs has decreased from 1.7×10^6 in *E. coli* to 0.9×10^6 in animal mitochondria (Robberson *et al.*, 1971; Dawid and Chase, 1972). It is surprising therefore that the rRNAs of yeast mitochondria are much larger than bacterial rRNAs and even larger than yeast cell-sap rRNAs. Thus the present bacterial ribosome and the yeast mitochondrial ribosome could have evolved from a common and now extinct bacterial ancestor, or yeast and animal mitochondrial ribosomes may represent different lines in divergent evolution from an *E. coli*-like endosymbiont (Reijnders *et al.*, 1973b). No large scale base sequence homologies between the rRNAs of lower eukaryotic mitochondria and those of *E. coli* have been detected, but possible base sequence homologies between mitochondrial rRNAs from animals, protozoa and fungi on the one hand, and from bacteria on the other, may enable us to distinguish between the two alternatives. The proportion of

C increases and of U decreases on going from lower eukaryotes to verte-
brates (Table LXIII, Freeman *et al.*, 1973).

The rate of evolutionary change of cytochrome *c* corresponds to one
preserved codon mutation in 26×10^6 years; this rate of alteration is
much slower than that of many other proteins and makes cytochrome *c*
very suitable for the study of long term evolution. There is now strong
evidence that there has been a common ancestor for all cytochromes *c*,
certainly for all eukaryotic cytochromes *c*, constituting a divergent homologous (rather than a convergent analogous) evolution (Lemberg and
Barrett, 1973; McLaughlin and Dayhoff, 1973).

The differences observed between amino acid sequences of the cytochromes *c* of various bacterial species is often greater than those observed
between organisms belonging to different eukaryotic phyla. Cytochrome
c_2 from *Rhodospirillum rubrum* is convincingly homologous with eukaryotic
cytochrome *c* (Dus *et al.*, 1968). Comparison of this cytochrome c_2 with
the cytochrome c_{551} of *Pseudomonas aeruginosa* and horse heart cytochrome *c* on the basis of amino acid sequences and tertiary structure, as
determined by X-ray analysis, led Dickerson *et al.* (1971) to propose that
most of the observed structural and functional differences can be explained
by postulating the deletion of one 16-residue hair-pin-shaped loop (residues
31–46) and seven other residues in the horse cytochrome *c*. A detailed
study of the cytochromes of some lower eukaryotes might prove very
interesting, as some of these depart markedly from the general uniformity
in their properties. Several of the Protozoa (e.g. *Astasia*, *Euglena*, *Tetrahymena* and *Crithidia*) have a cytochrome oxidase which does not react with
reduced mammalian cytochrome *c*, and their own *c*-type cytochromes
show spectral and structural anomalies (see Chapter 3).

Sequence homologies between mitochondrial and bacterial cytochromes
c can be explained in two ways. Either, (1) the structural gene for cytochrome *c* was originally part of the genome of the bacterial endosymbiont,
i.e. one of those many codons which were transferred to the nucleus as the
autonomy of the symbiont was gradually lost; or (2) the gene was present
in an aerobic prokaryotic ancestor and was not one of those sequestered
into the plasmid associated with the evolving mesosome.

Sequence analyses of mitochondrial gene products would also be
interesting, although even if homologies were to be found, similarly
opposed arguments could be applied. It is quite remarkable that a similar
spectrum of polypeptides are produced by fungal and mammalian mitochondria. There is as yet no evidence that the five-fold larger mitochondrial
genome in fungi is capable of coding for any more proteins. It appears
that the 4·6 μm mt DNA of mammalian mitochondria may represent the
minimal size necessary to specify all the products which must be produced
at the membrane assembly site or within the permeability barrier of the

inner mitochondrial membrane. The continuing survival of this arrangement in the most highly evolved organisms is evidence for its selective
advantage.

References

Aaronson, S. (1973). *Microbios* **7**, 231.

Accoceberry, B. and Stahl, A. (1971). *Biochem. biophys. Res. Commun.* **42**, 1235.

Accoceberry, B. and Stahl, A. (1972). *C. r. hebd. Séanc Acad. Sci., Paris, Ser. D.* **274**, 3135.

Accoceberry, B., Schneller, J. M. and Stahl, A. (1973), *Biochimie* **55**, 291.

Adams, B. S. and Parks, L. W. (1969). *J. Bact.* **100**, 370.

Adoutte, A. (1974). *In* "Biogenesis of Mitochondria", (A. M. Kroon and C. Saccone, eds.), pp. 263–271. Academic Press, London and New York.

Adoutte, A. and Beisson, J. (1970). *Mol. Gen. Genet.* **108**, 70.

Adoutte, A. and Beison, J. (1972). *Nature, Lond.* **235**, 5338.

Adoutte, A., Balmefrézol, M., Beisson, J. and André, J. (1972). *J. Cell Biol.* **54**, 8.

Adoutte, A., Sainsard, A., Rossignol, M. and Beisson, J. (1973). *Biochimie* **55**, 793.

Agar, H. D. and Douglas, H. C. (1957). *J. Bact.* **73**, 365.

Agsteribbe, E. and Kroon, A. M. (1973). *Biochem. biophys. Res. Commun.* **51**, 8.

Agsteribbe, E., Kroon, A. M. and Van Bruggen, E. F. J. (1972). *Biochim. biophys. Acta* **269**, 299.

Agsteribbe, E., Datema, R. and Kroon, A. M. (1974). *In* "Biogenesis of Mitochondria", (A. M. Kroon and C. Saccone, eds.), pp. 305–314. Academic Press, London and New York.

Ahmed, K. A. and Woods, R. A. (1967). *Genet. Res.* **9**, 179.

Ainsworth, P. J. and Tustanoff, E. R. (1972). *Biochem. biophys. Res. Commun.* **47**, 1299.

Aitken, W. B. and Neiderpruem, D. J. (1970). *J. Bact.* **104**, 981.

Al-Aidroos, K., Somers, J. M. and Bussey, H. (1973). *Mol. Gen. Genet.* **122**, 323.

Albertsson, P. A. (1960). "Partition of Cell Particles and Macromolecules". John Wiley and Son Inc., New York.

Albrecht, V., Prenzel, K. and Richter, D. (1970). *Biochemistry N.Y.* **9**, 316.

Alexander, R., Brand, F. C. and Alexander, G. J. (1965). *Biochim. biophys. Acta* **111**, 318.

Alexeieff, A. G. (1916). *C. r. Séanc. Soc. Biol.* **79**, 1072.

Allen, N. E. and MacQuillan, A. M. (1969). *J. Bact.* **97**, 1142.

Allen, N. E. and Suyama, Y. (1972). *Biochim. biophys. Acta* **259**, 369.

Allsopp, A. (1969). *New. Phytol.* **68**, 591.

Aloni, Y. and Attardi, G. (1971a). *J. Molec. Biol.* **55**, 251.

Aloni, Y. and Attardi, G. (1971b). *J. Molec. Biol.* **55**, 271.

Aloni, Y. and Attardi, G. (1971c). *Proc. natn. Acad. Sci. U.S.A.* **68**, 1157.

Altmann, R. (1890). "Die Elementarorganismen und ihre Beziehungen zu den Zellen". Veit Co., Leipzig.

Altmiller, D. H. (1972a). *Biochem. biophys. Res. Commun.* **49**, 1000.

Altmiller, D. H. (1972b). *Genetics, Princeton* **71**, 52.

Altmiller, D. H. and Wagner, R. P. (1970a). *Archs Biochem. Biophys* **138**, 160.

Altmiller, D. H. and Wagner, R. P. (1970b). *Biochem. Genet.* **4**, 243.

Andersen, V., Birch-Andersen, A., Skovbjerg, H. and Hellung-Larsen, P. (1970). *Acta path. Microbiol. Scand.* Section A **78**, 537.

Anderson, E. (1962). *J. Protozool.* **9**, 380.

Anderson, E. (1967). *In* "Research in Protozoology", (T. T. Chen, ed.), Vol. 1, pp. 3–40. Pergamon Press, Oxford.

Anderson, E. and Beams, H. W. (1959). *J. Morph.* **104**, 205.

Anderson, E. and Beams, H. W. (1960). *J. Protozool.* **7**, 190.

Anderson, J. A., Sun, F. K., McDonald, J. K. and Cheldelin, V. H. (1964). *Archs Biochem. Biophys* **107**, 37.

Anderson, J. G. and Smith, J. E. (1971). *J. gen. Microbiol.* **69**, 185.

Anderson, N. G. (1966). *Science, N.Y.* **154**, 103.

Anderson, W. and Hill, G. C. (1969). *J. Cell. Sci.* **4**, 611.

André, J. and Marinozzi, V. (1965). *J. Microscopie* **4** 615.

Andreason, A. A. and Stier, T. J. B. (1953). *J. cell. comp. Physiol.* **41**, 23.

Andresen, N. (1956). *C. r. Trav. Lab. Carlsberg*, **29**, 435.

Appleby, C. A. and Morton, R. K. (1959). *Biochem. J.* **71**, 492.

Arcos, J. C., Argus, M. F., Sardesai, M. and Stacey, P. E. (1964). *Biochemistry, N.Y.* **3**, 2041.

Armstrong, J.McD., Coates, J. H. and Morton, R. K. (1961). *In* "Haematin Enzymes" (J. E. Falk, R. Lemberg and R. K. Morton, eds.), pp. 385–388. Pergamon Press, New York.

Arnberg, A. C., Van Bruggen, E. F. J., Schutgens, R. B. H., Flavell, R. A. and Borst, P. (1972). *Biochim. biophys. Acta* **272**, 487.

Arnberg, A. C., Van Bruggen, E. F. J., Flavell, R. A. and Borst, P. (1973a). *Biochim. biophys. Acta* **308**, 276.

Arnberg, A. C., Flavell, R. A. and Van Bruggen, E. F. J. (1973b). *In* "Proceedings 1st John Innes Symposium", (R. Markham, B. Bancroft, D. R. Davies, D. A. Hopwood and R. W. Horne, eds.), p. 341, North Holland, Amsterdam.

Arnold, C. G., Schimmer, O. Schötz, F. and Bathelt, H. (1972). *Arch. Mikrobiol.* **81**, 50.

Asano, K. (1972). *J. Biochem., Tokyo* **72**, 737.

Ashwell, M. A. and Work, T. S. (1968). *Biochem. biophys. Res. Commun.* **32**, 1006.

Ashwell, M. and Work, T. S. (1970). *A. Rev. Biochem.* **39**, 251.

Ashworth, J. M., Duncan, D. and Rowe, A. J. (1968). *Expl. Cell. Res.* **58**, 73.

Atkinson, A. W. Jr. (1972). Ph.D. Thesis, Queens University, Belfast.

Atkinson, D. E. (1968). *Biochemistry, N.Y.* **11**, 4030.

Attardi, B. and Attardi, G. (1967). *Proc. natn. Acad. Sci. U.S.A.* **58**, 1051.

Attardi, B. and Attardi, G. (1968). *Proc. natn. Acad. Sci. U.S.A.* **61**, 261.

Attardi, B. and Attardi, G. (1971). *J. molec. Biol.* **55**, 231.

Attardi, G., Constantino, P. and Ojala, D. (1974). *In* "Biogenesis of Mitochondria", (A. M. Kroon and C. Saccone, eds.), pp. 9–30. Academic Press, London and New York.

Attardi, G. and Attardi, B. (1969). *In* "Problems in Biology: RNA in Development", (E. W. Hanly, ed.), pp. 102–131. University of Utah Press, Salt Lake City.

Avadhani, N. G. and Buetow, D. E. (1972a). *Biochem. J.* **128**, 353.

Avadhani, N. G. and Buetow, D. E. (1972b). *Biochem. biophys. Res. Commun.* **46**, 773.

Avadhani, N. G., Lynch, M. J. and Buetow, D. E. (1971). *Expl. Cell. Res.* **69**, 226.

Avadhani, N. G., Kuan, M., Vanderlign, P. and Rutman, R. J. (1973). *Biochem. biophys. Res. Commun.* **51**, 1090.

Avers, C. J. (1967). *J. Bact.* **94**, 1225.

Avers, C. J., Rancourt, M. W., Lin, F. H. and Pfeffer, C. R. (1964). *J. Cell. Biol.* **23**, 7A.

Avers, C. J., Pfeffer, C. R. and Rancourt, M. W. (1965). *J. Bact.* **90**, 481.

Avers, C. J., Billheimer, F. E., Hoffmann, H. P. and Pauli, R. M. (1968). *Proc. natn. Acad. Sci. U.S.A.* **61**, 90.

Avner, P. R. and Griffiths, D. E. (1970). *FEBS Lett.* **10**, 202.

Avner, P. R. and Griffiths, D. E. (1973a). *Eur. J. Biochem.* **32**, 301.

Avner, P. R. and Griffiths, D. E. (1973b). *Eur. J. Biochem.* **32**, 312.

Avner, P. R., Coen, D., Dujon, B. and Slonimski, P. P. (1973). *Mol. Gen. Genet.* **125**, 9.

Bacher, A. and Lingens, F. (1970). *J. biol. Chem.* **245**, 4647.

Bacher, A. and Lingens, F. (1971). *J. biol. Chem.* **246**, 7018.

Bachofen, V., Schweyen, R. J., Wolf, K. and Kaudewitz, F. (1972). *Z. Naturf.* **27b**, 252.

Bachop, W., Boatman, E. S., Mackler, B. and Ladda, R. L. (1972). *Trans. Am. microsc. Soc.* **91**, 169.

Baernstein, H. D. (1953). *Ann. N.Y. Acad. Sci.* **56**, 982.

Baernstein, H. D. (1963). *J. Parasitol.* **49**, 12.

Bahr, G. F. and Zeitler, E. (1962). *J. Cell Biol.* **15**, 489.

Bak, A. L., Christiansen, C. and Stenderup, A. (1969). *Nature, Lond.* **224**, 270.

Baker, E. G. S. and Baumberger, J. P. (1941). *J. cell. comp. Physiol.* **17**, 285.

Baker, J. E. (1963). *Archs Biochem. Biophys* **103**, 148.

Balcavage, W. X. and Mattoon, J. R. (1967). *Nature, Lond.* **215**, 166.

Balcavage, W. X. and Mattoon, J. R. (1968). *Biochim. biophys. Acta* **153**, 521.

Balcavage, W. X., Beale, M., Chasen, B. and Mattoon, J. R. (1968). *Biochim. biophys. Acta* **162**, 525.

Balcavage, W. X., Beck, J. C., Beck, D., Greenawalt, J. W., Parker, J. H. and Matton, J. R. (1970). *Cryobiol.* **6**, 385.

Balcavage, W. X., Lloyd, J. L., Mattoon, J. R., Ohnishi, T. and Scarpa, A. (1973). *Biochim. biophys. Acta* **305**, 41.

Baldwin, H. H. and Rusch, H. P. (1965). *A. Rev. Biochem.* **34**, 565.

Ball, A. J. S. and Tustanoff, E. R. (1968). *FEBS Lett.* **1**, 255.

Ball, A. J. S. and Tustanoff, E. R. (1970). *Biochim. biophys. Acta* **199**, 476.

Ball, A. J. S. and Tustanoff, E. R. (1971). *In* "Autonomy and Biogenesis of Mitochondria and Chloroplasts", (N. K. Boardman, A. W. Linnane and R. M. Smillie, eds.), pp. 466–480. North Holland, Amsterdam.

Ball, A. J. S., Janki, R. M. and Tustanoff, E. R. (1971). *Can. J. Microbiol.* **17**, 1125.

Band, R. N. and Mohrlok, S. (1969). *J. gen. Microbiol.* **59**, 351.

Bandlow, W. (1972). *Biochim. biophys. Acta* **282**, 105.

Bandlow, W., Wolf, K., Kaudewitz, F. and Slater, E. C. (1974). *Biochim. biophys. Acta* **333**, 446.

Barath, Z. and Küntzel, H. (1972a). *Proc. natn. Acad. Sci. U.S.A.* **69**, 1371.

Barath, Z. and Küntzel, H. (1972b). *Nature New Biology* **240**, 195.

Bard, M. (1972). *J. Bact.* **111**, 649.

Barnes, R., Colleran, E. M. and Jones, O. T. G. (1973). *Biochem. J.* **134**, 745.

Barnett, W. E. and Brown, D. H. (1967). *Proc. natn. Acad. Sci. U.S.A.* **57**, 452.

Barnett, W. E. and Epler, J. L. (1966). *Cold Spring Harb. Symp. quant. Biol.* **31**, 549.

Barnett, W. E., Brown, D. H. and Epler, J. L. (1967). *Proc. natn. Acad. Sci. U.S.A.* **57**, 1775.

Barrett, J. (1969). *Biochim. biophys. Acta* **177**, 442.

Bartley, W. and Birt, L. M. (1970). *In* "Essays in Cell Metabolism", (W. Bartley, H. L. Kornberg, and J. R. Quayle, eds.), pp. 1–44. Wiley-Interscience, London.

Bartley, W. and Tustanoff, E. R. (1966). *Biochem. J.* **99**, 599.

Bartnicki-Garcia, S. and Nickerson, W. J. (1961). *J. Bact.* **82**, 142.

Bartoov, B. (1971). Ph.D. Thesis, McMaster University, Canada.

Bartoov, B., Mitra, R. S. and Freeman, K. B. (1970). *Biochem. J.* **120**, 455.

Battamere, E. and Varquez, D. (1971). *Biochim. biophys. Acta* **254**, 316.

Baudhuin, P. P., Evrard, P. and Berthet, J. (1967). *J. Cell Biol.* **32**, 181.

Baudras, A. and Spyridakis, A. (1971). *Biochimie* **53**, 943.

Baur, E. (1909). *Z. Vererblehre* **1**, 330.

Bayne, R. A., Muse, K. E. and Roberts, J. F. (1969a). *Comp. Biochem. Physiol.* **30**, 61.

Bayne, R. A., Muse, K. E. and Roberts, J. F. (1969b). *Comp. Biochem. Physiol.* **30**, 1049.

Beale, G. H. (1954). "The Genetics of *Paramecium aurelia*". Cambridge University Press, London.

Beale, G. H. (1969). *Genet. Res.* **14**, 341.

Beale, G. H., Knowles, J. K. C. and Tait, A. (1972). *Nature, Lond.* **235**, 396.

Beattie, D. S. (1971a). *Subcell. Biochem.* **1**, 1.

Beattie, D. S. (1971b). *Archs Biochem. Biophys* **147**, 136.

Beattie, D. S., Basford, R. E. and Koritz, S. B. (1967). *Biochemistry, N.Y.* **6**, 3099.

Beattie, D. S., Lin, L. H. and Stuchell, R. N. (1974). *In* "Biogenesis of Mitochondria", (A. M. Kroon and C. Saccone, eds.), pp. 465–476. Academic Press, London and New York.

Beaufay, H. (1966). "La Centrifugation en Gradient de Densité. Application a l'Etude des Organites Subcellulaires". Ceutrick, Louvain.

Beaufay, H. and Berthet, J. (1963). *Biochem. Soc. Symp.* **23**, 66.

Beaufay, H., Jacques, P., Baudhuin, P., Sellinger, O. Z., Berthet, J. and deDuve, C. (1964). *Biochem. J.* **92**, 184.

Bech-Hansen, N. T. and Rank, G. H. (1973). *Mol. Gen. Genet.* **120**, 115.

Beck, C. and von Meyenburg, H. K. (1968). *J. Bact.* **96**, 479.

Beck, J. C., Mattoon, J. C., Hawthorne, D. C. and Sherman, F. (1968). *Proc. natn. Acad. Sci. U.S.A.* **60**, 186.

Beck, J. C., Parker, J. H., Balcavage, W. X. and Mattoon, J. R. (1971). *In* "Autonomy and Biogenesis of Mitochondria and Chloroplasts", (N. K. Boardman, A. W. Linnane, R. M. Smillie, eds.), pp. 194–204. North Holland, Amsterdam.

Bégin-Heick, N. (1973). *Biochem. J.* **134**, 607.

Bégin-Heick, N. and Blum, J. J. (1967). *Biochem. J.* **105**, 813.

Behn, W. and Arnold, C. G. (1972). *Mol. Gen. Genet.* **114**, 266.

Behn, W. and Arnold, C. G. (1973). *Arch. Mikrobiol.* **92**, 85.

Beisson, J., Beale, C. H., Knowles, J., Sainsard, A., Adoutte, A. and Tait, A. (1973). Proceedings of the 13th International Congress of Genetics, Berkeley.

Bell, P. R. (1970). *Symp. Soc. exp. Biol.* **24**, 109.

Bell, P. R. and Mühlethaler, K. (1964). *J. Cell Biol.* **20**, 235.

Bendall, D. S. and Bonner, W. D. Jr. (1971). *Pl. Physiol., Lancaster* **47**, 236.

Bensley, R. R. and Hoerr, N. (1934). *Anat. Rec.* **60**, 449.

Benson, R. W. (1972). *Fedn. Proc. Fedn. Am. Socs exp. Biol.* **31**, 427.

Bent, H. K. and Moore, R. H. (1966). *Symp. Soc. gen. Microbiol.* **16**, 82.

Benveniste, K. and Munkres, K. D. (1970). *Biochim. biophys. Acta* **220**, 161.

Berden, J. A. and Slater, E. C. (1972). *Biochim. biophys. Acta* **256**, 199.

Bergquist, A., Labrie, D. A. and Wagner, R. P. (1969). *Archs Biochem. Biophys* **134**, 401.

Bernardi, G. and Timasheff, G. N. (1970). *J. molec. Biol.* **48**, 43.

Bernardi, G., Carnevali, F., Nicolaieff, A., Piperno, G. and Tecce, G. (1968). *J. molec. Biol.* **37**, 493.

Bernardi, G., Faures, M., Piperno, G. and Slonimski, P. P. (1970). *J. molec. Biol.* **48**, 23.

Bernardi, G., Piperno, G. and Fonty, G. (1972). *J. molec. Biol.* **65**, 173.

Bertoli, E., Barbaresi, G., Castelli, A. and Lenaz, G. (1971). *Bioenergetics* **2**, 135.

Bertrand, H. and Pittenger, T. H. (1964). *Genetics, Princeton* **50**, 235.

Bertrand, H. and Pittenger, T. H. (1969). *Genetics, Princeton* **61**, 643.

Bertrand, H. and Pittenger, T. H. (1972). *Genetics, Princeton* **71**, 521.

Bertrand, H., McDougall, K. H. and Pittenger, T. H. (1968). *J. gen. Microbiol.* **50**, 337.

Bianchetti, R., Lucchini, G. and Sartirana, M. L. (1971). *Biochem. biophys. Res. Commun.* **42**, 97.

Biggs, D. R. and Linnane, A. W. (1963). *Biochim. biophys. Acta* **78**, 785.

Biggs, D. R., Nakamura, H., Kearney, E. B., Rocca, E. and Singer, T. P. (1970). *Archs Biochem. Biophys.* **137**, 12.

Biliński, T. and Jachymczyk, W. (1973). *Biochem. biophys. Res. Commun.* **52**, 379.

Billheimer, F. E. and Avers, C. J. (1969). *Proc. natn. Acad. Sci. U.S.A.* **64**, 739.

Birkmayer, G. D. (1971a). *Eur. J. Biochem.* **21**, 258.

Birkmayer, G. D. (1971b). *Z. physiol. Chem.* **352**, 761.

Birkmayer, G. D. and Bücher, Th. (1969). *FEBS Lett.* **5**, 28.

Birky, C. W. Jr. (1973). *Genetics* **74**, 421.

Birnboim, H. C. (1971). *J. Bact.* **107**, 659.

Blamire, J., Cryer, D. R., Finkelstein, D. B. and Marmur, J. (1972). *J. molec. Biol.* **67**, 11.

Bleeg, H. S., Lethbak, A., Christiansen, C., Smith, K. E. and Stenderup, A. (1972). *Biochem. biophys. Res. Commun.* **47**, 524.

Bloch, K. (1962). *Fedn. Proc. Fedn. Am. Socs exp. Biol.* **21**, 1058.

Blossey, H. Ch. and Küntzel, H. (1972). *FEBS Lett.* **24**, 335.

Blum, J. J. and Bégin-Heick, N. (1967). *Biochem. J.* **105**, 821.

Boell, E. J. (1945). *Proc. natn. Acad. Sci. U.S.A.* **31**, 396.

Boell, E. J. (1946). *Biol. Bull. mar. biol. Lab., Woods. Hole* **91**, 238.

Bolotin, M., Coen, D., Deutsch, J., Dujon, B., Netter, P., Petrochilo, E. and Slonimski, P. P. (1971). *Bull. Inst. Pasteur, Paris* **69**, 215.

Bonner, B. A. and Machlis, L. (1957). *Pl. Physiol., Lancaster* **32**, 291.

Borque, D. P. and Naylor, A. (1972). *Pl. Physiol., Lancaster* **49**, 826.

Borst, P. (1970). *Symp. Soc. exp. Biol.* **24**, 201.

Borst, P. (1972). *A. Rev. Biochem.* **41**, 333.

Borst, P. (1974a). *Biochem. Soc. Trans.* **2**, 182.

Borst, P. (1974b). *In* "Biogenesis of Mitochondria", (A. M. Kroon and C. Saccone, eds.), pp. 147–156. Academic Press, London and New York.

Borst, P. and Aaij. C. (1969). *Biochem. biophys. Res. Commun.* **34**, 358.

Borst, P. and Flavell, R. A. (1972). *In* "Mitochondria: Biogenesis and Bioenergetics, (S. C. Van den Burgh, P. Borst, L. L. M. Van Deenen, J. C. Riemersma, E. C. Slater and J. M. Tager, eds.), FEBS Symposium Vol. 28, pp. 1–19. North Holland, American Elsevier, Amsterdam.

Borst, P. and Grivell, L. A. (1971). *FEBS Lett.* **13**, 73.

Borst, P. and Grivell, L. A. (1973). *Biochimie* **55**, 801.

Borst, P. and Kroon, A. M. (1969). *Int. Rev. Cytol.* **26**, 108.

Borst, P., Kroon, A. M. and Ruttenberg, G. J. C. M. (1967). *In* "Genetic Elements, Properties and Function", (D. Shugar, ed.), p. 81. Academic Press and Polish Scientific Publishers, London and Warsaw.

Borst, P., Schutgens, R. B. H., Van Bruggen, E. F. J., Weijers, P. J. and Arnberg, A. C. (1973a). 4th International Congress of Protozoology, Clermont-Ferrand.

Borst, P., Flavell, R. A., Sanders, J. P. H., and Mol, J. N. H. (1973b). Proceedings of the 9th International Congress of Biochemistry, Stockholm. A 3a7.

Bosmann, H. B. (1971a). *Cytobios* **4**, 121.

Bosmann, H. B. (1971b). *Nature New Biology* **234**, 54.

Bosmann, H. B. and Martin, S. S. (1969). *Science, N.Y.* **164**, 190.

Bosmann, H. B. and Myers, M. W. (1974). *In* "Biogenesis of Mitochondria", (A. M. Kroon and C. Saccone, eds.), pp. 525–536. Academic Press, London and New York.

Bostock, C. J. (1969). *Biochim. biophys. Acta* **195**, 579.

Boulter, D. and Derbyshire, E. (1957). *J. exp. Bot.* **8**, 313.

Boveris, A. and Chance, B. (1973). *Biochem. J.* **134**, 707.

Bowers, B. and Korn, E. D. (1968). *J. Cell Biol.* **39**, 95.

Bowers, W. D., McClary, D. O. and Ogur, M. (1967). *J. Bact.* **94**, 482.

Bowman, I. B. R., Srivastava, H. K. and Flynn, I. W. (1972). *In* "Comparative Biochemistry of Parasites", (H. van de Bosch, ed.), pp. 329–343. Academic Press, New York and London.

Brack, Ch., Delain, E., Riou, G. and Festy, B. (1972a). *J. Ultrastruct. Res.* **39**, 568.

Brack, Ch., Delain, E. and Riou, G. (1972b). *Proc. natn. Acad. Sci. U.S.A.* **69**, 1642.

Bracker, C. E. and Grove, S. N. (1971a). *Cytobiologie* **3**, 229.

Bracker, C. E. and Grove, S. N. (1971b). *Protoplasma* **73**, 15.

Brambl, R. M. and Woodward, D. O. (1972). *Nature New Biology* **238**, 198.

Brandes, D., Buetow, D. E., Bertin, I. F. and Malkoff, D. E. (1964). *Expl. molec. Pathol.* **3**, 583.

Braun, R. and Evans, T. E. (1969). *Biochim. biophys. Acta* **182**, 511.

Bretthauer, R. K., Marcus, L., Chaloupka, J., Halvorson, H. O. and Bock, R. M. (1963). *Biochemistry, N.Y.* **2**, 1079.

Brewer, E. N., DeVries, A. and Rusch, H. P. (1967). *Biochim. biophys. Acta* **145**, 686.

Briquet, M. and Goffeau, A. (1973). *Archs int. Physiol. Biochim.* **81**, 360.

Brown, C. M. and Johnson, B. (1970). *J. gen. Microbiol.* **64**, 279.

Brown, C. M. and Johnson, B. (1971). *Antonie van Leeuwenhoek* **37**, 477.

Brown, D. D. and Dawid, I. B. (1968). *Science, N.Y.* **160**, 272.

Brown, D. H. and Novelli, G. D. (1968). *Biochem. biophys. Res. Commun.* **31**, 262.

Brown, R. C., Evans, D. A. and Vickerman, K. (1973). *Int. J. Parasitol.* **3**, 691.

Brunk, C. F. and Hanawalt, P. C. (1969). *Expl. Cell Res.* **54**, 143.

Brunner, A., Mas, J., Celis, E. and Mattoon, J. R. (1973). *Biochem. biophys. Res. Commun.* **53**, 638.

Brunt, R. V., Eisenthal, R. and Symons, S. A. (1971). *FEBS Lett.* **13**, 89.

Busiello, T., Girolamo, A., Girolamo, M., Fischer-Fantuzzi, L. and Vesco, C. (1973). *Eur. J. Biochem.* **35**, 251.

Buetow, D. E. and Buchanan, P. J. (1964). *Expl. Cell Res.* **36**, 204.

Buetow, D. E. and Buchanan, P. J. (1965). *Biochim. biophys. Acta* **96**, 9.

Buetow, D. E. and Buchanan, P. J. (1969). *Life Sci.* **8**, 1099.

Bulder, C. J. (1963). Ph.D. Thesis, Institute of Technology, Delft.

Bulder, C. J. (1964a). *Antonie van Leeuwenhoek* **30**, 1.

Bulder, C. J. (1964b). *Antonie van Leeuwenhoek* **30**, 442.

Bullock, C. R., Christian, R. A., Peters, F. F. and White, A. M. (1971). *Biochem. Pharmac.* **20**, 943.

Bulos, B. and Racker, E. (1968). *J. biol. Chem.* **243**, 3901.

Burdett, I. D. J. (1972). *Sci. Prog., Oxford* **60**, 527.

Burgos, M. H., Aoki, A. and Sacerdote, F. L. (1964). *J. Cell Biol.* **23**, 207.

Bunn, C. L., Mitchell, C. H., Lukins, H. B. and Linnane, A. W. (1970). *Proc. natn. Acad. Sci. U.S.A.* **67**, 1233.

Burton, P. R. (1970). *J. Protozool.* **17**, 295.

Burton, P. R. and Dusanic, D. G. (1968). *J. Cell Biol.* **39**, 318.

Butow, R. A. and Zeydel, M. (1968). *J. biol. Chem.* **243**, 2545.

Butow, R. A., Ferguson, M. J. and Cederbaum, A. (1973). *Biochemistry, N.Y.* **12**, 158.

Bygrave, F. L. (1969). *J. biol. Chem.* **244**, 4768.

Cain, K., Lancashire, W. E. and Griffiths, D. E. (1974). *Biochem. Soc. Trans.* **2**, 215.

Calleja, G. B. (1973). *Archs Biochem. Biophys* **154**, 382.

Calvayrac, R. (1970). *Arch. Mikrobiol.* **73**, 308.

Calvayrac, R. and Butow, R. A. (1971). *Arch. Mikrobiol.* **80**, 62.

Calvayrac, R. and Claisse, M. L. (1973). *Planta* **112**, 17.

Calvayrac, R., van Lente, F. and Butow, R. A. (1971). *Science, N.Y.* **173**, 252.

Calvayrac, R., Butow, R. A. and Lefort-Tran, M. (1972). *Expl. Cell Res.* **71**, 422.

Cameron, I. L. (1966). *Nature, Lond.* **209**, 630.

Cannon, M., Davies, J. E. and Siminez, A. (1973). *FEBS Lett.* **32**, 277.

Campbell, A. (1957). *Bact. Rev.* **21**, 265.

Cantino, E. C. (1965). *Arch. Mikrobiol.* **51**, 42.

Cantino, E. C. (1966). *In* "The Fungi" (G. C. Ainsworth and A. S. Sussman, eds.), Vol. II, pp. 283–337. Academic Press, New York.

Cantor, M. H. and Klotz, J. (1971). *Experientia* **27**, 801.

Capaldi, R. A. (1973). *Biochem. biophys. Res. Commun.* **53**, 1331.

Capaldi, R. A. and Green, D. E. (1972). *FEBS Lett.* **25**, 205.

Carafoli, E. and Lehninger, A. L. (1971). *Biochem. J.* **122**, 681.

Carafoli, E., Balcavage, W. X., Lehninger, A. L. and Mattoon, J. R. (1970). *Biochim. biophys. Acta* **205**, 18.

Carnevali, F. and Leoni, L. (1972). *Biochem. biophys. Res. Commun.* **47** 1322.

Carnevali, F., Piperno, G. and Tecce, G. (1966). *Atti Acad. naz. Lincei Rc.* **41**, (Ser 8), 194.

Carnevali, F., Morpurgo, C. and Tecce, G. (1969). *Science, N.Y.* **163**, 1331.

Carnevali, F., Falcone, C., Frontali, L., Leon, L., Macino, G. and Palleschi, C. (1973). *Biochem. biophys. Res. Commun.* **51**, 651.

Caroline, D. F., Harding, R. W., Kuwana, H., Satayanarayana, T. and Wagner, R. P. (1969). *Genetics* **62**, 487.

Cartledge, T. G. and Lloyd, D. (1972a). *Biochem. J.* **126**, 381.

Cartledge, T. C. and Lloyd, D. (1972b). *Biochem. J.* **127**, 693.

Cartledge, T. G. and Lloyd, D. (1972c). *Biochem. J.* **126**, 755.

Cartledge, T. G. and Lloyd, D. (1973). *Biochem. J.* **132**, 609.

Cartledge, T. G., Howells, L. and Lloyd, D. (1970a). *Biochem. J.* **116**, 26*p*.

Cartledge, T. G., Howells, L. and Lloyd, D. (1970b). *Biochem. J.* **116**, 40*p*.

Cartledge, T. G., Cooper, R. A. and Lloyd, D. (1971). *In* "Separations in Zonal Rotors", (E. Reid, ed.), pp. V-4.1–V-4.16. Univ. of Surrey, Guildford.

Cartledge, T. G., Lloyd, D., Erecińska, M. and Chance, B. (1972). *Biochem. J.* **130**, 739.

Casey, J., Cohen, M., Rabinowitz, M., Fukuhara, H. and Getz., G. S. (1972). *J. molec. Biol.* **63**, 431.

Cassady, W. E. and Wagner, R. P. (1968). *Genetics* **60**, 168.

Cassady, W. E. and Wagner, R. P. (1971). *J. Cell Biol.* **49**, 536.

S

Cassady, W. E., Leiter, E. H., Bergquist, A. and Wagner, R. P. (1972). *J. Cell Biol.* **53**, 66.

Casselton, L. A. and Condit, A. (1972). *J. gen. Microbiol.* **72**, 521.

Castor, L. N. and Chance, B. (1955). *J. biol. Chem.* **217**, 453.

Castor, L. N. and Chance, B. (1959). *J. biol. Chem.* **234**, 1587.

Cattell, K. J., Knight, I. G., Lindop, C. R. and Beechey, R. B. (1970). *Biochem. J.* **117**, 1011.

Chaix, P. (1961). *In* "Haematin Enzymes", (J. E. Falk, R. Lemberg, and R. K. Morton, eds.), p. 225. Pergamon Press, London.

Chakrabarti, S., Dube, D. K. and Roy, S. C. (1972). *Biochem. J.* **128**, 461.

Chance, B. (1952). Proceedings of the 2nd International Congress of Biochemistry, Paris, p. 32.

Chance, B. (1953). *J. biol. Chem.* **202**, 397.

Chance, B. (1959a) *J. biol. Chem.* **234**, 3036.

Chance, B. (1959b). *J. biol. Chem.* **234**, 3041.

Chance, B. (1964). *Meth. Enzym.* **4**, 273.

Chance, B. (1965). *J. gen. Physiol.* **49**, 163.

Chance, B. (1966). *In* "Flavins and Flavoproteins", (E. C. Slater, ed.), p. 496. Elsevier, Amsterdam.

Chance, B. (1967). *Biochem. J.* **103**, 1.

Chance, B. (1972). *FEBS Lett.* **23**, 3.

Chance, B. and Sager, R. (1957). *Plant Physiol., Lancaster* **32**, 548.

Chance, B. and Schoener, B. (1966). *In* "Flavins and Flavoproteins", (E. C. Slater, ed.), p. 510. Elsevier, Amsterdam.

Chance, B. and Spencer, E. L. (1959). *Discuss. Faraday Soc.* **27**, 200.

Chance, B. and Williams, G. R. (1956). *Adv. Enzymol.* **17**, 65.

Chang, L. T. and Tuveson, R. W. (1967). *Genetics* **56**, 801.

Changeux, J. P. and Thiery, J. (1968). *In* "Regulatory Functions of Biological Membranes", (J. Jarnefelt, ed.), p. 116. Elsevier, Amsterdam.

Chapman, C. and Bartley, W. (1968). *Biochem. J.* **107**, 455.

Chappell, J. B. (1964). *Biochem. J.* **90**, 225.

Chappell, J. B. and Greville, G. D. (1963). *Biochem. Soc. Symp.* **23**, 39.

Chappell, J. B. and Haarhoff, K. N. (1967). *In* "Biochemistry of Mitochondria", (E. C. Slater, Z. Kaniuga and L. Wojtczak, eds.), pp. 75–91. Academic Press, London and New York.

Chappell, J. B. and Hansford, R. G. (1969). *In* "Subcellular Components Preparation and Fractionation", (G. D. Birnie and S. M. Fox, eds.), pp. 43–56. Butterworths, London.

Chappell, J. B. and Robinson, B. H. (1968). *Biochem. Soc. Symp.* **27**, 123.

Chantrenne, H. (1955). *Biochim. biophys. Acta* **18**, 38.

Charret, R. (1972). *J. Microscopie, Paris*, **14**, 279.

Charret, R. and André, J. (1968). *J. Cell Biol.* **39**, 369.

Charret, R. and Charlier, M. (1973). *J. Microscopie, Paris*, **17**, 19.

Chassang-Douillet, A., Ladet, J., Boze, H. and Galzy, P. (1973). *Z. allg. Mikrobiol.* **13**, 193.

Chen, S. Y., Ephrussi, B. and Hottinguer, H. (1950). *Heredity* **4**, 337.

Chen, W. L. and Charalampous, F. C. (1969). *J. biol. Chem.* **244**, 2767.

Chen, W. L. and Charalampous, F. C. (1973). *Biochim. biophys. Acta* **294**, 329.

Chi, J. C. H. and Suyama, Y. (1968). *Bact. Proc.* **129**, 129.

Chi, J. C. H. and Suyama, Y. (1970). *J. molec. Biol.* **53**, 531.

Childs, G. E. (1973). *J. Histochem. Cytochem.* **21**, 26.

Chin, C. H. (1950). *Nature, Lond.* **165**, 926.

Chiu, A. O. S. and Suyama, Y. (1973). *Biochim. biophys. Acta* **299**, 557.

Chiu, N., Chiu, A. O. S. and Suyama, Y. (1974). *In* "Biogenesis of Mitochondria", (A. M. Kroon and C. Saccone, eds.), pp. 383–394. Academic Press, London and New York.

Christiansen, C., Bak, A. L., Stenderup, A. and Christiansen, G. (1971). *Nature, Lond.* **231**, 176.

Cifferi, O., Pansi, B., Perani, A. and Grandi, M. (1968). *J. molec. Biol.* **37**, 529.

Ciferri, O., Taboni, O., Lazar, G. and Van Etten, J. (1974). *In* "Biogenesis of Mitochondria", (A. M. Kroon and C. Saccone, eds.), pp. 107–116. Academic Press, London and New York.

Cirillo, V. P. (1956). *J. Protozool.* **3**, 69.

Cirillo, V. P. (1957). *J. Protozool.* **4**, 60.

Claisse, M. L. (1969). *Antonie van Leeuenhoek* **35** (Suppl.), 21.

Clark, A. M. (1945). *Aust. J. exp. Biol. med. Sci.* **23**, 317.

Clark, T. B. and Wallace, F. G. (1960). *J. Protozool.* **7**, 115.

Clark-Walker, G. D. (1972). *J. Bact.* **109**, 399.

Clark-Walker, G. D. (1973). *Eur. J. Biochem.* **32**, 263.

Clark-Walker, G. D. and Gleason, F. H. (1973). *Arch. Mikrobiol.* **92**, 209.

Clark-Walker, G. D. and Linnane, A. W. (1966). *Biochem. biophys. Res. Commun.* **25**, 8.

Clark-Walker, G. D. and Linnane, A. W. (1967). *J. Cell Biol.* **34**, 1.

Claude, A. (1946). *J. exp. Med.* **84**, 51.

Claude, A. (1965). *J. Cell Biol.* **27**, 146A.

Clavilier, L., Péré, G., Slonimski, P. P. and Somlo, M. (1964). Proceedings of the 6th International Congress of Biochemistry, New York, pp. 673–674.

Clavilier, L., Péré, G. and Slonimski, P. (1969). *Mol. Gen. Genet.* **104**, 195.

Clayton, D. A. and Brambl, R. M. (1972). *Biochem. biophys. Res. Commun.* **46**, 1477.

Clayton, D. A. and Vinograd, J. (1967). *Nature, Lond.* **216**, 652.

Clegg, R. A. and Garland, P. B. (1971). *Biochem. J.* **124**, 135.

Clegg, R. A. and Skyrme, J. (1973). *Biochem. J.* **136**, 1029.

Clegg, R. A., Ragan, C. I., Haddock, B. A., Light, P. A., Garland, P. B., Swann, J. C., and Bray, R. C. (1969). *FEBS Lett.* **5**, 207.

Cloud, Jr., P. E. (1965). *Science, N.Y.* **148**, 27.

Cloud, Jr., P. E. (1968). *In* "Evolution and Environment", (E. T. Drake, ed.). Yale University Press, New Haven, Conn.

Cloud, Jr., P. E., Licari, G. R., Wright, L. A., and Troxel, B. W. (1969). *Proc. natn. Acad. Sci. U.S.A.* **62**, 623.

Cobley, J. C., Grossman, S., Beinert, H. and Singer, T. P. (1973). *Biochem. biophys. Res. Commun.* **53**, 1273.

Cobon, G. S. and Haslam, J. M. (1973). *Biochem. biophys. Res. Commun.* **52**, 320.

Cocucci, M. C. and Rossi, G. (1972). *Arch. Mikrobiol.* **85**, 265.

Coen, D., Deutsch, J., Netter, P., Petrochilo, E. and Slonimski, P. P. (1970). *Symp. Soc. exp. Biol.* **24**, 449.

Cohen, M. and Rabinowitz, M. (1972). *Biochim. biophys. Acta* **281**, 192.

Cohen, S. S. (1970). *Am. Scient.* **58**, 281.

Cohen, S. S. (1973). *Am. Scient.* **61**, 437.

Cohen, L. H., Hollenberg, C. P. and Borst, P. (1970). *Biochem. biophys. Acta* **224**, 610.

Cohen, M., Casey, J., Rabinowitz, M. and Getz, G. S. (1972). *J. molec. Biol.* **63**, 440.

Colleran, E. M. and Jones, O. T. G. (1973). *Biochem. J.* **134**, 89.

Conner, R. L., Mallory, F. B., Landrey, J. R., Ferguson, K. A., Kaneshiro, E. S. and Ray, E. (1971). *Biochem. biophys. Res. Commun.* **44**, 995.

Coolsma, J. W. Th. (1971). Abstract of Communications presented at the 7th Meeting of the Federation of European Biochemical Societies at Varna, Abstract, 194.

Cooper, C. S. and Avers, C. J. (1974). *In* "Biogenesis of Mitochondria", (A. M. Kroon and C. Saccone, eds.), pp. 289–304. Academic Press. London and New York.

Cooper, R. A. and Lloyd, D. (1972). *J. gen. Microbiol.* **72**, 59.

Cooper, S. and Wuethoff, G. (1971). *J. Bact.* **106**, 709.

Coote, J. L. and Work, T. S. (1971). *Eur. J. Biochem.* **23**, 564.

Corneo, G., Moore, C., Sanadi, D. R., Grossman, L. J. and Marmur, J. (1966). *Science, N.Y.* **151**, 687.

Corneo, G., Zardi, L. and Polli, E. (1968). *J. molec. Biol.* **36**, 419.

Correns, C. (1909). *Z. Vererblehre* **1**, 291.

Cosgrove, W. and Skeen, M. (1970). *J. Protozool.* **17**, 172.

Cottrell, S. F. and Avers, C. J. (1970). *Biochem. biophys. Res. Commun.* **38**, 973.

Cottrell, S. F. and Avers, C. J. (1971). *In* "Autonomy and Biogenesis of Mitochondria and Chloroplasts", (N. K. Boardman, A. W. Linnane and R. M. Smillier, eds.), pp. 481–491. North Holland, Amsterdam.

Crabtree, H. G. (1929). *Biochem. J.* **23**, 536.

Crandall, M. (1973a). *J. gen. Microbiol.* **75**, 363.

Crandall, M. (1973b). *J. gen. Microbiol.* **75**, 377.

Criddle, R. S. and Schatz, G. (1969). *Biochemistry, N.Y.* **8**, 322.

Criddle, R. S., Bock, R. M., Green, D. E. and Tisdale, H. D. (1962). *Biochemistry, N.Y.* **1**, 821.

Cruz, F. S. and Travassos, L. R. (1970). *Arch. Mikrobiol.* **73**, 111.

Cummins, J. E. and Rusch, H. P. (1966). *J. Cell Biol.* **31**, 577.

Cummins, J. E., Rusch, H. P. and Evans, T. E. (1967). *J. molec. Biol.* **23**, 281.

Damsky, C. H., Nelson, W. M. and Claude, A. (1969). *J. Cell Biol.* **43**, 174.

Danforth, W. F. (1967). *In* "Research in Protozoology", (T. T. Chen, ed.), Vol. 1, pp. 205–306. Pergamon Press, Oxford.

Daniels, E. W. and Breyer, E. (1965). *J. Protozool.* **12**, 417.

Daniels, E. W. and Breyer, E. P. (1968). *Z. Zellforsch. mikrosk. Anat.* **91**, 159.

Darby, R. T. and Goddard, D. R. (1950). *Physiologia Pl.* **3**, 435.

Darnell, J. E. Jr. (1968). *Bact. Rev.* **32**, 262.

Davey, P. J., Yu, R. and Linnane, A. W. (1969). *Biochem. biophys. Res. Commun.* **36**, 30.

Davey, P. J., Haslam, J. M. and Linnane, A. W. (1970). *Archs Biochem. Biophys* **136**, 54.

Davidian, N., Pennial, R. and Elliot, W. B. (1969). *Archs Biochem. Biophys* **133**, 345.

Davidson, J. B. and Stancev, N. Z. (1971). *Can. J. Biochem.* **49**, 1117.

Davson, H. and Danielli, J. F. (1943). "The Permeability of Natural Membranes". Cambridge University Press, Cambridge.

Dawes, I. W. and Carter, B. L. A. (1973). *Biochem. Soc. Trans.* **2**, 224.

Dawid, I. B. (1966). *Proc. natn. Acad. Sci. U.S.A.* **56**, 269.

Dawid, I. B. (1970). *Symp. Soc. exp. Biol.* **24**, 227.

Dawid, I. B. (1972). *J. molec. Biol.* **63**, 201.

Dawid, I. B. and Chase, J. W. (1972). *J. molec. Biol.* **63**, 217.

Dawid, I. B. and Wolstenholme, D. R. (1968). *In* "Biochemical Aspect of the

Biogenesis of Mitochondria", (E. C. Slater, J. M. Tager, S. Papa and E. Quagliariello, eds.), p. 283. Bari Adriatica Editrice.

Dawid, I. B., Horak, I. and Coon, H. G. (1974). *In* "Biogenesis of Mitochondria", (A. M. Kroon and C. Saccone, eds.), pp. 255–262. Academic Press, London and New York.

Dawidowicz, K. and Mahler, H. R. (1972). *In* "Gene Expression and its Regulation", (F. T. Kenney, B. A. Hamkalo, G. Favelukes, and J. J. August, eds.), pp. 503–522. Plenum, New York.

Dawson, R. M. C. (1966). *In* "Essays in Biochemistry", (P. N. Campbell and G. D. Grenville, eds.), Vol. 2, p. 62. Academic Press, London and New York.

Dawson, R. M. C. (1973). *Sub. Cell. Biochem.* **2**, 69.

de Barreiro, C. (1967). *Biochim. biophys. Acta*, **139**, 479.

DeDeken, R. H. (1961). *Expl. Cell Res.* **24**, 145.

DeDeken, R. H. (1966a). *J. gen. Microbiol.* **44**, 149.

DeDeken, R. H. (1966b). *J. gen. Microbiol.* **44**, 157.

deDuve, C. (1967). *In* "Enzyme Cytology", (D. B. Roodyn, ed.), p. 1. Academic Press, London and New York.

deDuve, C. (1969). *Ann. N.Y. Acad. Sci.* **168**, 369.

deDuve, C. (1971). *J. Cell Biol.* **50**, 200.

deDuve, C. (1973). *Science, N.Y.* **82**, 85.

deDuve, C. and Baudhuin, P. (1966). *Physiol. Rev.* **46**, 323.

deDuve, C. and Wattiaux, R. (1966). *A. Rev. Physiol.* **28**, 435.

DeKloet, S. R., Andrean, A. G. and Mayo, V. S. (1971). *Archs Biochem. Biophys* **143**, 175.

DeKok, J. (1973). Abstract of communication presented at the 9th International Congress of Biochemistry at Stockholm. Abstract 4d10.

Delain, E., Brack, Ch., Riou, G. and Festy, B. (1971). *J. Ultrastruct. Res.* **37**, 200.

Delain, E., Brack, Ch., Lacome, A. and Riou, G. (1972). *In* "Comparative Biochemistry of Parasites", (H. Van Den Bossche, ed.), pp. 127–138. Academic Press, New York and London.

Delange, J. R., Glazer, A. N. and Smith, E. L. (1969). *J. biol. Chem.* **244**, 1385.

Demaille, J., Vignais, P. M. and Vignais, P. V. (1970). *Eur. J. Biochem.* **13**, 416.

Dennis, E. A. and Kennedy, E. P. (1972). *J. Lipid Res.* **13**, 263.

Deutsch, J., Dujon, B., Netter, P., Petrochilo, E., Slonimski, P. P., Bolotin-Fukuhara, M. and Coen, D. (1974). *Genetics, Princeton* **76**, 195.

DeVries, H., Agsteribbe, E. and Kroon, A. M. (1971). *Biochem. biophys. Acta*, **246**, 111.

Dharmalingam, K. and Jayaraman, J. (1971). *Biochem. biophys. Res. Commun.* **45**, 1115.

Dharmalingam, K. and Jayaraman, J. (1973). *Archs Biochem. Biophys* **157**, 197.

Diacumakos, E. G., Garnjobst, L. and Tatum, E. L. (1965). *J. Cell Biol.* **26**, 427.

Dickerson, R. E., Takano, T., Eisenberg, D., Kallai, O. B., Samson, L., Cooper, A. and Margoliash, E. (1971). *J. biol. Chem.* **246**, 1511.

Dixon, B. and Rose, A. H. (1964). *J. gen. Microbiol.* **35**, 411.

Dixon, H., Kellerman, G. M., Mitchell, C. H., Towers, N. H. and Linnane, A. W. (1971). *Biochem. biophys. Res. Commun.* **43**, 780.

Dohi, M., Tamura, G. and Arima, K. (1973). *Agric. Biol. Chem.*, **37**, 703.

Doran, D. J. (1957). *J. Protozool.* **4**, 182.

Doran, D. J. (1958). *J. Protozool.* **5**, 89.

Douce, R. and Bonner, W. D. Jnr. (1972). *Biochem. Biophys. Res. Commun.* **47**, 619.

Dougherty, E. C. (1957). *J. Protozool.* **4** (Suppl.), 14.

Douglas, H. C. and Hawthorne, D. C. (1964). *Genetics, Princeton* **49**, 837.

Douglas, H. C. and Pelroy, G. (1963). *Biochim. biophys. Acta* **68**, 155.

Downie, J. A. and Garland, P. B. (1973a). *Biochem. J.* **134**, 1045.

Downie, J. A. and Garland, P. B. (1973b). *Biochem. J.* **134**, 1051.

Drabikowska, A. K. and Kruszewska, A. (1972). *J. Bact.* **112**, 1112.

DuBuy, H. G., Mattern, C. F. and Riley, F. L. (1965). *Science, N.Y.* **147**, 754.

DuBuy, H. G., Mattern, C. F. and Riley, F. L. (1966). *Biochim. biophys. Acta* **123**, 298.

Duell, E. A., Inoue, S. and Utter, M. F. (1964). *J. Bact.* **88**, 1762.

Duncan, D., Eades, J., Jukina, S. R. and Micks, D. (1960). *J. Protozool.* **7**, 18.

Duncan, H. M. and Mackler, B. (1966). *J. biol. Chem.* **241**, 1694.

Dunkle, L. D. and VanEtten, J. L. (1972a). *Arch. Microbiol.* **85**, 225.

Dunkle, L. D. and Van Etten, J. L. (1972b). *In* "Spores", (H. O. Halvorson, R. Hanson, and L. L. Campbell, eds.), pp. 283–289. American Society of Microbiology, Washington, D.C.

Duntze, W., Neumann, D., Gancedo, J. M., Atzpodien, W. and Holtzer, H. (1969). *Eur. J. Biochem.* **10**, 83.

Dure, L. S., Epler, J. L. and Barrett, W. E. (1967). *Proc. natn. Acad. Sci. U.S.A.* **58**, 1883.

Dus, K., Sletten, K. and Kamen, M. D. (1968). *J. biol. Chem.* **243**, 5507.

Eakin, R. T. and Mitchell, H. K. (1970). *J. Bact.* **104**, 74.

Ebner, E. and Schatz, G. (1973). *J. biol. Chem.* **248**, 5379.

Ebner, E., Mennucci, L. and Schatz, G. (1973a). *J. biol. Chem.* **248**, 5360.

Ebner, E., Mason, T. L. and Schatz, G. (1973b). *J. biol. Chem.* **248**, 5369.

Eccleshall, T. R. and Criddle, R. S. (1972). *Fedn Proc. Fedn Am. Socs exp. Biol.* **31**, 472.

Eccleshall, T. R. and Criddle, R. S. (1974). *In* "Biogenesis of Mitochondria", (A. M. Kroon and C. Saccone, eds.), pp. 31–46. Academic Press, London and New York.

Edelman, M., Epstein, H. T. and Schiff, J. A. (1966). *J. molec. Biol.* **17**, 463.

Edelman, M., Verma, I. M. and Littauer (1970). *J. molec. Biol.* **49**, 67.

Edelman, M., Verma, I. M., Herzog, R., Galun, E. and Littauer, U. Z. (1971). *Eur. J. Biochem.* **19**, 372.

Edgar, R. S. and Lielausis, I. (1968). *J. molec. Biol.* **32**, 263.

Edwards, C. and Lloyd, D. (1973). *J. gen. Microbiol.* **79**, 275.

Edwards, D. L. and Kwiecinski, F. (1973). *J. Bact.* **116**, 610.

Edwards, D. L. and Woodward, D. O. (1969). *FEBS Lett.* **4**, 193.

Edwards, D. L., Kwiecinski, F. and Horstmann, J. (1973). *J. Bact.* **114**, 164.

Ehrenstein, G. Von and Lipmann, F. (1961). *Proc. natn. Acad. Sci. U.S.A.* **47**, 941.

Ehrlich, S. D., Thiery, J. P. and Bernardi, G. (1972). *J. molec. Biol.* **65**, 207.

Eichel, H. J. (1956). *J. biol. Chem.* **222**, 137.

Eichel, H. J. (1960). *Biochim. biophys. Acta* **43**, 364.

Eichel, H. J. and Rem, L. T. (1963). *In* "Progress in Protozoology", (J. Ludvik, J. Lom and J. Vavya, eds.), p. 148. Czechoslovakian Academy of Sciences, Prague.

Elliott, A. M. and Bak, I. J. (1964). *J. Cell Biol.* **20**, 113.

Elvehjem, C. A. (1931). *J. biol. Chem.* **90**, 111.

Emanuel, C. F. and Chaikoff, I. L. (1957). *Biochim. biophys. Acta* **24**, 254.

Emerson, R. (1927). *J. gen. Physiol.* **51**, 105.

Emerson, R. and Held, A. A. (1969). *Am. J. Bot.* **56**, 1103.

Epel, B. L. and Butler, W. L. (1969). *Science, N.Y.* **166**, 621.

Epel, B. L. and Butler, W. L. (1970a). *Pl. Physiol., Lancaster* **45**, 723.

Epel, B. L. and Butler, W. L. (1970b). *Pl. Physiol., Lancaster* **45**, 728.

Epel, B. L. and Krauss, R. W. (1966). *Biochim. biophys. Acta* **120**, 73.

Ephrussi, B. (1949). *In* "Unités biologiques doues de continuité génétique", pp. 165–180. Edition du Centre Nat. Rech. Sci., Paris.

Ephrussi, B. (1951). *Cold Spring Harb. Symp. quant. Biol.* **16**, 75.

Ephrussi, B. (1953). "Nucleo–cytoplasmic Relations in Microorganisms". Clarendon Press, Oxford.

Ephrussi, B. (1956). *Naturwissenschaften* **43**, 505.

Ephrussi, B. and Chimenes, A. M. (1949). *Annls Inst. Pasteur, Paris* **76**, 351.

Ephrussi, B. and Grandchamp, S. (1965). *Heredity* **20**, 1.

Ephrussi, B. and Slonimski, P. P. (1950a). *Biochim. biophys. Acta* **6**, 256.

Ephrussi, B. and Slonimski, P. P. (1950b). *C.r. Acad. Sci., hebd. Séanc. Paris,* Ser. D **230**, 685.

Ephrussi, B. and Tavlitzki, J. (1949). *Annls Inst. Pasteur, Paris* **76**, 419.

Ephrussi, B., Hottingner, H. and Chimenes, A. M. (1949a). *Annls Inst. Pasteur, Paris,* **76**, 351.

Ephrussi, B., L'Heritier, Ph. and Hottinguer, H. (1949b). *Annls Inst. Pasteur, Paris,* **77**, 64.

Ephrussi, B., Hottinguer, H. and Roman, H. (1955). *Proc. natn. Acad. Sci. U.S.A.* **41**, 1065.

Ephrussi, B., Slonimski, P. P., Yotsuyanagi, Y. and Tavlitzki, J. (1956). *C.r. Trav. Lab. Carlsberg Ser. Physiol.* **26**, 87.

Ephrussi, B., Jakob, H. and Grandchamp, S. (1966). *Genetics* **54**, 1.

Epler, J. L. (1969). *Biochemistry, N.Y.* **8**, 2285.

Epler, J. L. and Barnett, W. E. (1967). *Biochem. biophys. Res. Commun.* **28**, 328.

Epler, J. L., Shugart, L. R. and Barnett, W. E. (1970). *Biochemistry, N.Y.* **9**, 3575.

Erecińska, M. and Storey, B. T. (1970). *Pl. Physiol., Lancaster* **46**, 618.

Erecińska, M., Oshino, N., Loh, P. and Brocklehurst, E. (1973). *Biochim. biophys. Acta* **292**, 1.

Erickson, S. K. and Ashworth, J. M. (1969). *Biochem. J.* **113**, 567.

Erickson, R. E., Brown, K. S. Jr., Wolf, D. E. and Folkers, K. (1960). *Archs Biochem. Biophys* **90**, 314.

Estabrook, R. W. (1956). *J. biol. Chem.* **223**, 781.

Euler, H. and Fink, H. (1927). *Hoppe-Seyler's Z. physiol. Chem.* **164**, 69.

Euler, H., Fink, H. and Hellström, H. (1927). *Hoppe-Seyler's Z. physiol. Chem.* **169**, 10.

Evans, D. A. (1973). *J. Protozool.* **20**, 336.

Evans, D. A. and Brown, R. C. (1971). *Nature, Lond.* **230**, 251.

Evans, D. A. and Brown, R. C. (1972). *J. Protozool.* **19**, 365.

Evans, D. A. and Brown, R. C. (1973). *J. Protozool.* **20**, 157.

Evans, H. H. and Evans, T. E. (1970). *J. biol. Chem.* **245**, 6436.

Evans, I., Linstead, D., Rhodes, P. M. and Wilkie, D. (1973). *Biochim. biophys. Acta* **312**, 323.

Evans, T. E. (1966). *Biochem. biophys. Res. Commun.* **22**, 678.

Evans, T. E. and Suskind, D. (1971). *Biochim. biophys. Acta* **228**, 350.

Fairfield, S. A. and Barnett, W. E. (1971). *Proc. natn. Acad. Sci. U.S.A.* **68**, 2972.

Famintzin, A. (1907). *Biol. Zbl.* **27**, 353.

Fang, M. and Butow, R. A. (1970). *Biochem. biophys. Res. Commun.* **41**, 1579.

Fassaden-Raden, J. M. and Hack, A. M. (1972). *Biochemistry, N.Y.* **11**, 4609.

Fauman, M. and Rabinowitz, M. (1972). *FEBS Lett.* **28**, 317.

Fauman, M., Rabinowitz, M. and Getz, G. S. (1969). *Biochim. biophys. Acta* **182**, 355.

Fauman, M. A., Rabinowitz, M. and Swift, H. H. (1973). *Biochemistry, N.Y.* **12**, 124.

Fauré-Frémiet, E. (1910a). *Archs Anat. micros. Morph.* **11**, 457.

Fauré-Frémiet, E. (1910b). *Anat. Anz.* **36**, 186.

Faures, M. and Fukuhara, H. (1965). *Archs int. Physiol. Biochim.* **73**, 308.

Faures, M. and Genin, C. (1969). *Mol. Gen. Genet.* **104**, 264.

Faye, G., Fukuhara, H., Grandchamp, C., Lazowska, J., Michel, F., Casey, J., Getz, G. S., Locker, J., Rabinowitz, M., Bolotin-Fukuhara, M., Coen, D., Deutsch, J., Dujon, B., Netter, P. and Slonimski, P. P. (1973). *Biochimie* **55**, 779.

Fernandez-Moran, H. (1962). *Circulation* **26**, 1039.

Federman, M. and Avers, C. J. (1967), *J. Bact.* **94**, 1236.

Ferdouse, M., Rickard, P. A. D., Moss, F. J. and Blanch, H. W. (1972). *Biotech. Bioengn.* **14**, 1004.

Fincham, R. J. S. and Day, P. R. (1963). "Fugal Genetics". Blackwell, Oxford.

Fink, H. (1932). *Hoppe-Seyler's Z. physiol. Chem.* **210**, 197.

Fink, H. and Berwald, E. (1933), *Biochem. Z.* **258**, 141.

Firkin, F. C. and Linnane, A. W. (1968). *Biochem. biophys. Res. Commun.* **32**, 398.

Firkin, F. and Linnane, A. W. (1969). *FEBS Lett.* **2**, 330.

Fitzgerald, P. H. (1963). *Heredity* **18**, 47.

Flavell, R. A. and Jones, I. G. (1970). *Biochem. J.* **116**, 811.

Flavell, R. A. and Jones, I. G. (1971a). *FEBS Lett.* **14**, 354.

Flavell, R. A. and Jones, I. G. (1971b). *Biochim. biophys. Acta* **323**, 255.

Flavell, R. A. and Trampé, P. O. (1973). *Biochim. biophys. Acta* **308**, 101.

Flavell, R. B. (1971). *Nature, Lond.* **230**, 504.

Flavell, R. B. (1972). *Biochem. Genet.* **6**, 275.

Flavell, R. B. and Fincham, J. R. S. (1968a). *J. Bact.* **95**, 1056.

Flavell, R. B. and Fincham, J. R. S. (1968b). *J. Bact.* **95**, 1063.

Flavell, R. B. and Woodward, D. O. (1970a). *Eur. J. Biochem.* **17**, 284.

Flavell, R. B. and Woodward, D. O. (1970b). *Eur. J. Biochem.* **13**, 548.

Flavell, R. B. and Woodward, D. O. (1971). *J. Bact.* **107**, 853.

Fletcher, M. J. and Sanadi, D. R. (1961). *Biochim. biophys. Acta* **51**, 356.

Flickinger, C. J. (1968). *Protoplasma* **66**, 139.

Flickinger, C. J. (1973). *Expl. Cell Res.* **80**, 31.

Forde, B. G. and John, P. C. L. (1973). *Expl. Cell Res.* **79**, 127.

Forrester, I. T., Nagley, P. and Linnane, A. W. (1970). *FEBS Lett.* **11**, 59.

Forrester, I. T., Watson, K. and Linnane, A. W. (1971). *Biochem. biophys. Res. Commun.* **43**, 409.

Foucher, M., Verdière, J., Lederer, F. and Slonimski, P. P. (1972). *Eur. J. Biochem.* **31**, 139.

Fouquet, H. (1973a). *J. Protozool.* **20**, 328.

Fouquet, H. (1973b). *J. Protozool.* **20**, 331.

Foury, F. and Goffeau, A. (1972). *Biochem. biophys. Res. Commun.* **48**, 153.

Fowler, L. R. and Hatefi, Y. (1961). *Biochem. biophys. Res. Commun.* **5**, 203.

Franke, W. W. and Kartenbeck, J. (1971). *Protoplasma* **73**, 35.

Fraser, R. S. S., Creanor, J. and Mitchison, J. M. (1973). *Nature, Lond.* **244**, 222.

Freeman, R. B. (1970). *Can. J. Biochem.* **48**, 469.

Freeman, K. B. and Haldar, D. (1967). *Biochem. biophys. Res. Commun.* **28**, 8.

Freeman, K. B. and Haldar, D. (1968). *Can. J. Biochem.* **46**, 1003.

Freeman, K. B., Haldar, D. and Work, T. S. (1967). *Biochem. J.* **105**, 947.

Freeman, K. B., Mitra, R. S. and Bartoov, B. (1973). *Sub-Cell Biochem.* **2**, 183.

Fujita, A. and Kodama, T. (1934). *Biochem. Z.* **272**, 186.

Fukami, H. M., Light, P. A. and Garland, P. B. (1970). *FEBS Lett.* **7**, 132.

Fukuhara, H. (1965). *Biochem. biophys. Res. Commun.* **18**, 297.

Fukuhara, H. (1966). *J. molec. Biol.* **17**, 334.

Fukuhara, H. (1967). *Proc. natn Acad. Sci. U.S.A.* **58**, 1065.

Fukuhara, H. (1968). *In* "Biochemical Aspects of the Biogenesis of Mitochondria", (E. C. Slater, J. M. Tager, S. Papa and E. Quagliariello, eds.), p. 303, Bari, Adriatica Editrice.

Fukuhara, H. (1969). *Eur. J. Biochem.* **11**, 135.

Fukuhara, H. (1970). *Mol. Gen. Genet.* **107**, 58.

Fukuhara, H. and Sels, A. A. (1966). *J. molec. Biol.* **17**, 319.

Fukuhara, H. and Kujawa, C. (1970). *Biochem. biophys. Res. Commun.* **41**, 1002.

Fukuhara, H., Faures, M. and Genin, C. (1969). *Mol. Gen. Genet.* **104**, 264.

Fukuhara, H., Lazowska, J., Michel, F., Faye, G., Michaelis, E., Petrochilo, E. and Slonimski, P. (1974). *In* "Biogenesis of Mitochondria", (A. M. Kroon and C. Saccone, eds.), pp. 177–178. Academic Press, London and New York.

Fuller, R. C., Smillie, R. M., Rigopoulos, N. and Yount, V. (1961). *Archs Biochem. Biophys* **91**, 197.

Fulton, J. D. and Spooner, D. F. (1959). *Expl. Parasit.* **8**, 137.

Fulton, J. D. and Spooner, D. F. (1960). *Expl. Parasit.* **2**, 293.

Gaertner, F. H. and Leef, J. L. (1970). *Biochem. biophys. Res. Commun.* **41**, 1192.

Gaitskhoki, V. S., Kisselev, O. I. and Neifakh, S. A. (1973). *FEBS Lett.* **31**, 93.

Gale, P. H., Erickson, R. E., Page, A. C. Jr. and Folkers, K. (1964). *Archs Biochem. Biophys* **104**, 169.

Galleotti, T., Kováč, L. and Hess, B. (1968). *Nature, Lond.* **218**, 194.

Gallo, M., Roche, B., Aubert, L. and Azoulay, E. (1973). *Biochimie* **55**, 195.

Galzy, P. and Bizeau, C. (1965). *Arch. Mikrobiol.* **52**, 353.

Garland, P. B. (1970). *Biochem. J.* **118**, 329.

Garbus, J., DeLuca, H. F., Loomans, M. E. and Strong, F. M. (1963). *J. biol. Chem.* **238**, 59.

Garnjobst, L., Wilson, J. F. and Tatum, E. L. (1965). *J. Cell Biol.* **26**, 413.

Garrod, D. and Ashworth, J. M. (1973). *Symp. Soc. gen. Microbiol.* **23**, 407.

Gause, G. F. and Kusakova, L. I. (1970). *Experientia* **26**, 209.

Gear, A. R. L. (1965a). *Biochem. J.* **95**, 118.

Gear, A. R. L. (1965b). *Biochem. J.* **97**, 532.

Gear, A. R. L. and Bednarek, J. M. (1972). *J. Cell Biol.* **54**, 325.

Gel'man, N. S., Lukoyanova, M. A. and Ostrovskii, D. N. (1967). "Respiration and Phosphorylation of Bacteria", Planum, New York.

Genevois, L. (1927). *Biochem. Z.* **186**, 461.

Georgopoulous, S. G. and Sisler, H. D. (1970). *J. Bact.* **2103**, 745.

Gezelius, K. (1959). *Expl. Cell Res.* **18**, 425.

Ghosh, A. K. and Bhattacharyya, S. N. (1967). *Biochim. biophys. Acta* **136**, 19.

Ghosh, A. K. and Bhattacharyya, S. N. (1971). *Biochim. biophys. Acta* **245**, 335.

Ghosh, A., Charalampous, F., Sison, Y. and Borer, R. (1960). *J. biol. Chem.* **235**, 2522.

Gibbs, S. P. (1970). *J. Cell Biol.* **46**, 599.

Gibson, I. (1970). *Symp. Soc. exp. Biol.* **24**, 379.

Gibson, I., Chance, M. and Williams, J. (1971). *Nature New Biology* **224**, 75.

Gilardi, A., Djavadi-Ohaniance, L., Labbe, P. and Chaix, P. (1971). *Biochim. biophys. Acta* **234**, 446.

Giles, G. L. and Sarafis, U. (1971). *Cytobios* **4**, 61.

Gillberg, B. O., Zetterberg, G. and Swanbeck, G. (1967). *Nature, Lond.* **214**, 415.

Gillham, N. W. (1965). *Proc. natn. Acad. Sci. U.S.A.* **54**, 1560.

Gillham, N. W. and Boynton, J. E. (1972). *J. Cell Biol.* **55**, 88a.

Gillham, N. W. and Fifer, W. (1968). *Science, N.Y.* **162**, 683.

Gillie, O. J. (1970). *J. gen. Microbiol.* **61**, 379.

Gingold, E. B., Saunders, G. W., Lukins, H. B. and Linnane, A. W. (1969). *Genetics* **62**, 735.

Gittleson, S. M., Alper, R. E. and Conti, S. F. (1969). *Life Sci.* **8**, 591.

Giorgio, A. J., Cartwright, G. E. and Wintrobe, M. M. (1963). *J. Bact.* **86**, 1037.

Glaumann, H. and Dallner, G. (1970). *J. Cell Biol.* **47**, 34.

Gleason, F. H. (1968). *Pl. Physiol., Lancaster* **43**, 597.

Gleason, F. H. and Unestam, T. (1968a). *J. Bact.* **95**, 1599.

Gleason, F. H. and Unestam, T. (1968b). *Physiologia Pl.* **21**, 556.

Gloor, U., Isler, O., Morton, R. A., Rüegg, R. and Wiss, O. (1958). *Helv. chim. Acta* **41**, 2357.

Goffeau, A. (1969). *In* "Electron Transport and Energy Conservation", (J. M. Tager S. Papa, E. Quagliariello and E. C. Slater, eds.), p. 343. Adriatice Editrice, Bari.

Goffeau, A., Colson, A. M., Landry, Y. and Foury, F. (1972a). *Biochem. biophys. Res. Commun.* **48**, 1448.

Goffeau, A., Landry, Y., Colson, A. M. and Foury, F. (1972b). Abstracts of Communications presented at the 9th Meeting of the Federation of European Biochemical Societies at Amsterdam, Abstract 610.

Goffeau, A., Landry, Y., Foury, F., Briquet, M. and Colson, A. M. (1973a). Abstracts of Communications presented at the 9th International Congress of Biochemistry, Stockholm, Abstract 5m9.

Goffeau, A., Landry, Y., Foury, F., Briquet, M. and Colson, A. M. (1973b). *J. biol. Chem.* **248**, 7097.

Goldring, E. S., Grossman, L. I., Krupnick, D., Cryer, D. R. and Marmur, J. (1970a). *Fedn Proc. Fedn ZAm. Socs exp. Biol.* **29**, 725.

Goldring, E. S., Grossman, L. I., Krupnick, D., Cryer, D. R. and Marmur, J. (1970b). *J. molec. Biol.* **52**, 323.

Goldring, E. S., Grossman, L. I. and Marmur, J. (1971). *J. Bact.* **107**, 377.

Goldstein, S., Moriber, L. and Hershenov, B. (1964). *Am. J. Bot.* **51**, 679.

Gómez-Puyou, A., Gómez-Puyou, M. T., Alvarez, P. and Mattoon, J. R. (1973). Abstracts of Communications presented at the 9th International Congress at Biochemistry, Stockholm, Abstract 4j5.

González-Cadavid, N. F. and Campbell, N. F. (1967). *Biochem. J.* **105**, 443.

González-Cadavid, N. F., Bravo, M. and Campbell, P. N. (1968). *Biochem. J.* **107**, 523.

Gordon, P. and Rabinowitz, M. (1973). *Biochemistry, N.Y.* **12**, 116.

Gordon, P. A. and Stewart, P. R. (1969). *Biochim. biophys. Acta* **177**, 358.

Gordon, P. A. and Stewart, P. R. (1971). *Microbios* **4**, 115.

Gordon, P. A. and Stewart, P. R. (1972). *J. gen. Microbiol.* **72**, 231.

Gordon, P. A., Lowdon, M. J. and Stewart, P. R. (1972a). *J. Bact.* **110**, 504.

Gordon, P. A., Lowdon, M. J. and Stewart, P. R. (1972b). *J. Bact.* **110**, 511.

Görts, C. P. M. (1971). *Antonie van Leeuwenhoek* **37**, 161.

Görts, C. P. M. and Hasilík, A. (1972a). *Antonie van Leeuwenhoek* **38**, 454.

Görts, C. P. M. and Hasilík, A. (1972b). *Eur. J. Biochem.* **29**, 282.

Gosling, J. P. and Duggan, P. F. (1971). *J. Bact.* **106**, 908.

Goss, W. A., Deitz, W. H. and Cook, T. M. (1965). *J. Bact.* **89**, 1068.

Gouhier, M. and Monuolou, J. C. (1973). *Mol. gen. Genet,* **122**, 149.

Gowdridge, B. (1956). *Genetics* **41**, 780.

Granboulan, N. and Scherrer, K. (1969). *Eur. J. Biochem.* **9**, 1.

Grandi, M. and Küntzel, H. (1970). *FEBS Lett.* **10**, 25.

Grandi, M., Helms, A. and Küntzel, H. (1971). *Biochem. biophys. Res. Commun.* **44**, 864.

Granick, S. (1966). *J. biol. Chem.* **241**, 1359.

Grant, P. T. and Sargent, J. R. (1960). *Biochem. J.* **76**, 229.

Grant, P. T., Sargent, J. R. and Ryley, J. F. (1961). *Biochem. J.* **81**, 200.

Grant, W. D. and Poulter, R. T. M. (1973). *J. molec. Biol.* **73**, 439.

Graves, L. B. Jr., Trelease, R. N., Grill, A. and Becker, W. M. (1972). *J. Protozool.* **19**, 527.

Gray, P. P. and Rogers, P. L. (1971a), *Biochim. biophys. Acta* **230**, 393.

Gray, P. P. and Rogers, P. L. (1971b), *Biochim. biophys. Acta* **230**, 401.

Green, D. E. and Fleischer, S. (1962). *In* "Horizons in Biochemistry", (M. Kasha and B. Pullman, eds.), p. 381. Academic Press, New York.

Green, D. E. and Tzagoloff, A. (1966). *Archs Biochem. Biophys* **116**, 293.

Green, D. E., Asai, J., Harris, R. A. and Penniston, J. T. (1968a). *Proc. natn. Acad. Sci. U.S.A.* **59**, 830.

Green, D. E., Asai, J., Harris, R. A. and Penniston, J. T. (1968b). *Archs Biochem. Biophys* **125**, 684.

Greenawalt, J. W., Hall, D. O. and Wallis, O. C. (1967). *Meth. Enzym.* **10**, 142.

Greenawalt, J. W., Rossi, C. S. and Lehninger, A. L. (1964). *J. Cell Biol.* **23**, 21.

Gregolin, A. and Magaldi, G. (1961). *Biochim. biophys. Acta* **54**, 62.

Greksák, M., Haricová, M. and Weissová, K. (1971). Abstracts of Communications presented at the 7th Meeting of the Federation of European Biochemical Societies at Varna, Abstract 637.

Griffiths, A. J. (1970). *In* "Advances in Microbial Physiology", (A. H. Rose and J. F. Wilkinson, eds.), Vol. 4, pp. 105–129. Academic Press, London and New York.

Griffiths, A. J., Lloyd, D., Roach, G. I. and Hughes, D. E. (1967). *Biochem. J.* **103**, 21P.

Griffiths, D. E. (1974). *In* "Mitochondria Biogenesis and Bioenergetics", (S. G. Vanden Burgh, P. Borst, L. L. M. van Deenen, J. C. Riemersma, E. C. Slater and J. M. Tager, eds.). FEBS Symp. No. 28, pp. 95–104. North Holland, American Elsevier, Amsterdam.

Griffiths, D. E., Avner, P. R., Lancashire, W. E. and Turner, J. R. (1972). *In* "The Biochemistry and Biophysics of Mitochondrial Membranes", (E. Carafoli, A. L. Lehninger and N. Siliprandi, eds.), pp. 505–521. Academic Press, New York and London.

Griffiths, D. E., Houghton, R. L. and Lancashire, W. E. (1974). *In* "Biogenesis of Mitochondria", (A. M. Kroon and C. Saccone, eds.), pp. 214–224. Academic Press, London and New York.

Grimm, P. W. and Allen, P. J. (1954). *Pl. Physiol., Lancaster* **29**, 369.

Grimmelikhuijzen, C. J. P. and Slater, E. C. (1973). *Biochim. biophys. Acta* **305**, 67.

Grimstone, A. V. (1959). *J. biophys. biochem. Cytol.* **6**, 369.

Grimstone, A. V. and Cleveland, L. R. (1964). *Proc. R. Soc.* Ser. B, **159**, 668.

Grindle, M. (1973). *Mol. Gen. Genet.* **120**, 283.

Grivell, L. A. (1967). *Biochem. J.* **105**, 44c.

Grivell, L. A. (1974). *In* "Biogenesis of Mitochondria", (A. M. Kroon and C. Saccone, eds.), pp. 275–288. Academic Press, London and New York.

Grivell, L. A. and Reijnders, L. (1973). Proceedings of the Conference on Yeast Genetics, Karpacz, 1973.

Grivell, L. A. and Walg, H. L. (1972). *Biochem. biophys. Res. Commun.* **49**, 1452.

Grivell, L. A., Reijnders, L. and deVries, H. (1971a). *FEBS Lett.* **16**, 1459.

Grivell, L. A., Reijnders, L. and Borst, P. (1971b). *Biochim. biophys. Acta* **247**, 91.

Grivell, L. A., Reijnders, L. and Borst, P. (1971c). *Eur. J. Biochem.* **19**, 64.

Grivell, L. A., Netter, P., Borst, P. and Slonimski, P. P. (1973). *Biochim. biophys. Acta* **312**, 358.

Groot, G. S. P. (1974). *In* "Biogenesis of Mitochondria", (A. M. Kroon and C. Saccone, eds.), pp. 443–452. Academic Press, London and New York.

Groot, G. S. P., Kováč, L. and Schatz, G. (1971). *Proc. natn. Acad. Sci. U.S.A.* **68**, 308.

Groot, G. S. P., Rouslin, W. and Schatz, G. (1972). *J. biol. Chem.* **247**, 1735.

Gross, V. J. and Smith, D. G. (1972). *Microbios* **6**, 139.

Gross, J. R., McCoy, M. T. and Gilmore, E. B. (1968). *Proc. natn. Acad. Sci. U.S.A.* **61**, 253.

Gross, N. J., Getz, G. S. and Rabinowitz, M. (1968). *In* "Biochemical Aspects on the Biogenesis of Mitochondria", (E. C. Slater, J. M. Tager, S. Papa and E. Quagliariello, eds.), p. 161. Adriatica Editrice, Bari.

Grossman, L. I., Goldring, E. S. and Marmur, J. (1969). *J. molec. Biol.* **46**, 367.

Grossman, L. I. Cryer, D. R. Goldring, E. S. and Marmur, J. (1971). *J. molec. Biol.* **62**, 565.

Grossman, S., Cobley, J., Hogue, P. K., Kearney, E. B. and Singer, T. P. (1973). *Archs Biochem. Biophys* **158**, 744.

Gualerzi, C. (1969). *Ital. J. Biochem.* **18**, 418.

Guarnieri, M., Mattoon, J. R., Balcavage, W. X. and Payne, C. (1970). *Analyt. Biochem.* **34**, 39.

Guerin, B. and Sulkowski, E. (1966). *Biochim. biophys. Acta* **129**, 193.

Guerineau, M., Grandchamp, C., Yotsuyanagi, Y. and Slonimski, P. P. (1968). *C.r. hebd. Séanc. Acad. Sci., Paris*, Ser. D, **226**, 1884.

Guerineau, M., Grandchamp, C., Paoletti, C. and Slonimski, P. (1971). *Biochem. biophys. Res. Commun.* **42**, 550.

Guillory, R. J. (1964). *Biochim. biophys. Acta* **89**, 197.

Gunge, N., Sugimura, T. and Iwasaki, M. (1967). *Genetics* **57**, 213.

Guttes, E. and Guttes, S. (1964). *Science, N.Y.* **145**, 1057.

Guttes, E., Guttes, S. and Devi, R. V. (1969). *Experientia* **24**, 66.

Guttes, E. W., Hanawalt, P. C. and Guttes, S. (1967). *Biochim. biophys. Acta* **142**, 181.

Guttes, S., Guttes, E. and Hadek, R. (1966). *Experientia* **22**, 452.

Guttman, H. N. and Eisenman, R. N. (1965). *Nature, Lond.* **207**, 1280.

Hackenbrock, C. R. (1966). *J. Cell Biol.* **30**, 269.

Hackenbrock, C. R. (1968a). *Proc. natn. Acad. Sci. U.S.A.* **61**, 598.

Hackenbrock, C. R. (1968b). *J. Cell Biol.* **37**, 345.

Haddock, B. A. and Garland, P. B. (1971). *Biochem. J.* **124**, 155.

Hagihara, B., Horio, T., Yamashita, J., Nozaki, M. and Okunuki, K. (1956a). *Nature, Lond.* **178**, 629.

Hagihara, B., Horio, T., Nozaki, M., Sekuzu, I., Yamashita, J. and Okunuki, K. (1956b). *Nature, Lond.* **178**, 631.

Hajra, A. K. and Agranoff, B. W. (1967). *J. biol. Chem.* **242**, 1074.

Halbreich, A. and Rabinowitz, M. (1971). *Proc. natn. Acad. Sci. U.S.A.* **68**, 294.

Hall, D. O. and Greenawalt, J. W. (1964). *Biochem. biophys. Res. Commun.* **17**, 565.

Hall, D. O. and Greenawalt, J. W. (1967). *J. gen. Microbiol.* **48**, 419.

Hall, J. D. and Crane, F. L. (1971). *J. Cell Biol.* **48**, 420.

Hall, R. H. (1941). *Physiol. Zool.* **14**, 193.

Hamburger, K. and Zeuthen, E. (1957). *Expl. Cell Res.* **13**, 443.

Hammond, R. C., Wright, M. and Whittaker, P. A. (1974). *Biochem. Soc. Trans.* **2**, 218.

Handley, L. and Caten, C. E. (1973). *Heredity* **31**, 136.

Handwerker, A., Schweyen, R. J., Wolf, K. and Kaudewitz, F. (1973). *J. Bact.* **113**, 1307.

Harris, C. C. and Leone, C. A. (1966). *J. Cell Biol* **28**, 405.

Hartwell, L. H. (1970). *A. Rev. Genet.* **4**, 373.

Haskins, F. A., Tissières, A., Mitchell, H. K. and Mitchell, M. B. (1953). *J. biol. Chem.* **200**, 819.

Haslam, J., Davey, P. J. and Linnane, A. W. (1968). *Biochem. biophys. Res. Commun.* **33**, 368.

Haslam, J. M., Proudlock, J. W. and Linnane, A. W. (1971). *Bioenergetics* **2**, 351.

Haslam, J. M., Perkins, M. and Linnane, A. W. (1973a). *Biochem. J.* **134**, 935.

Haslam, J. M., Spithill, T. W., Linnane, A. W. and Chappell, J. B. (1973b). *Biochem. J.* **134**, 949.

Hassal, K. A. (1967). *Nature, Lond.* **215**, 521.

Hassal, K. A. (1969). *Physiologia Pl.* **22**, 304.

Hatefi, Y., Haavik, A. G., Fowler, L. R. and Griffiths, D. E. (1962). *J. biol. Chem.* **237**, 2661.

Hauber, J. and Singer, T. P. (1967). *Eur. J. Biochem.* **3**, 107.

Hawker, L. E. and Abbott, P. M. (1963). *J. gen. Microbiol.* **32**, 295.

Hawker, L. E. and Hendy, R. J. (1963). *J. gen. Microbiol.* **33**, 43.

Hawkes, R. B. and Holberton, D. V. (1973). *Expl. Cell Res.* **78**, 481.

Hawkins, S. E. and Willis, L. R. (1969). *Expl Cell Res.* **54**, 275.

Hawley, E. S. and Greenawalt, J. W. (1970). *J. biol. Chem.* **245**, 3574.

Hawley, E. S. and Wagner, R. P. (1967). *J. Cell Biol.* **35**, 489.

Hawthorne, D. C. and Mortimer, R. K. (1960). *Genetics* **45**, 1085.

Hawthorne, D. C. and Mortimer, R. K. (1968). *Genetics* **60**, 735.

Hebb, C. R. and Slebodnik, J. (1958). *Expl Cell Res.* **14**, 286.

Hebb, C. R., Slebodnik, J., Singer, T. P. and Bernath, P. (1959). *Archs Biochem. Biophys* **83**, 10.

Heilporn, V. and Limbosch, S. (1971). *Biochim. biophys. Acta* **240**, 94.

Heitz, E. (1959). *Z. Naturf.* **14b**, 179.

Heller, J. and Smith, E. L. (1966). *J. biol. Chem.* **241**, 3158.

Henry, S. A. and Keith, A. D. (1972). *Chem. Phys. Lipids* **7**, 245.

Henson, C. P., Weber, C. N. and Mahler, H. R. (1968a). *Biochemistry, N.Y.* **7**, 4431.

Henson, C. P., Perlman, P., Weber, C. N. and Mahler, H. R. (1968b). *Biochemistry, N.Y.* **7**, 4445.

Herman, A. I. and Griffin, P. S. (1968). *J. Bact.* **96**, 457.

Herzfeld, F. (1970). *Z. Physiol. Chem.* **351**, 658.

Heslot, H., Goffeau, A. and Louis, C. (1970a). *J. Bact.* **104**, 473.

Heslot, H., Louis, C. and Goffeau, A. (1970b). *J. Bact.* **104**, 482.

Hewley, F. P. and Myers, J. (1971). *Pl. Physiol., Lancaster* **47**, 373.

Higgins, E. S. and Friend, W. H. (1968). *Can. J. Biochem.* **46**, 1515.

Hilker, D. M. and White, A. G. C. (1959). *Expl. Parasit.* **8**, 534.

Hill, G. C. (1972). *In* "Comparative Biochemistry of Parasites", (H. Van den Bossche, ed.), pp. 395–416. Academic Press, New York and London.

Hill, G. C. and Anderson, W. A. (1969). *J. Cell Biol.* **41**, 547.

Hill, G. C. and Anderson, W. A. (1970). *Expl Parasit.* **28**, 356.

Hill, G. C. and Cross, G. A. (1973). *Biochim. biophys. Acta* **305**, 590.

Hill, G. C. and White, D. C. (1968). *J. Bact.* **95**, 2151.

Hill, G. C., Chan, S. K. and Smith, L. (1971a). *Biochim. biophys. Acta* **253**, 78.

Hill, G. C., Gutteridge, W. E. and Mathewson, N. W. (1971b). *Biochim. biophys. Acta* **243**, 225.

Hill, R. D., Ford, S., Byington, K. H., Tzagloff, A. and Boyer, P. D. (1968). *Archs Biochem. Biophys* **127**, 756.

Hinkelman, W. and Kraepelin, G. (1970). *Arch. Mikrobiol.* **74**, 258.

Hirano, T. and Lindegren, C. C. (1961). *J. Ultrastruct. Res.* **5**, 321.

Hirano, T. and Lindegren, C. C. (1963). *J. Ultrastruct. Res.* **8**, 322.

Hiyama, J., Nishimura, M. and Chance, B. (1959). *Pl. Physiol., Lancaster* **44**, 527.

Hoare, C. A. (1954). *J. Protozool.* **1**, 28.

Hofer, H. W., Pette, D., Schwab-Stey, H. and Schwab, D. (1972). *J. Protozool.* **19**, 532.

Hoffman, H.-P. and Avers, C. J. (1973). *Science, N.Y.* **181**, 749.

Hoffman, R. M. and Raper, J. R. (1971). *Science, N.Y.* **171**, 418.

Hoffman, R. M. and Raper, J. R. (1972). *J. Bact.* **110**, 780.

Hogeboom, G. H., Schneider, W. C. and Palade, G. H. (1948). *J. biol. Chem.* **172**, 619.

Hohl, H. R., Miura-Santo, L. Y. and Cotter, D. A. (1970). *J. Cell Sci.* **7**, 285.

Hollenberg, C. P. and Borst, P. (1971). *Biochem. Res. Commun.* **45**, 1250.

Hollenberg, C. P., Borst, P., Thuring, R. W. J. and Van Bruggen, E. F. J. (1969). *Biochim. biophys. Acta* **186**, 417.

Hollenberg, C. P., Borst, P. and Van Bruggen, E. F. J. (1970a). *Biochim. biophys. Acta* **209**, 1.

Hollenberg, C. P., Rikes, W. F. and Borst, P. (1970b). *Biochim. biophys. Acta* **201**, 13.

Hollenberg, C. P., Borst, P. and Van Bruggen, E. F. J. (1972a). *Biochim. biophys. Acta* **277**, 35.

Hollenberg, C. P., Borst, P., Flavell, R., Van Kreijl, C. F., Van Bruggen, E. F. J. and Arnberg, A. C. (1972b). *Biochim. biophys. Acta* **277**, 44.

Holt, C. E. and Gurney, E. G. (1969). *J. Cell Biol.* **40**, 484.

Holtzer, H. (1968). *In* "Aspects of Yeast Metabolism", (A. K. Mills, ed.), p. 155. Blackwell, Oxford.

Hommersand, H. and Thimann, K. V. (1965). *Pl. Physiol., Lancaster* **40**, 1220.

Horgen, P. A. and Griffin, D. H. (1971a). *Nature New Biology* **234**, 17.

Horgen, P. A. and Griffin, D. H. (1971b). *Proc. natn. Acad. Sci. U.S.A.* **68**, 338.

Horn, P. and Wilkie, D. (1966). *Heredity, Lond.* **21**, 625.

Horne, R. W. and Whittaker, V. P. (1962). *Z. Zellforsch. mikosk. Anat.* **58**, 1.

Horwitz, H. B. and Holt, C. E. (1971). *J. Cell Biol.* **49**, 546.

Hosono, K., Aida, K. and Uemura, T. (1972). *J. gen. appl. Microbiol., Tokoy* **18**, 189.

Houghton, J. A. (1970). *Genet. Res.* **16**, 285.

Houghton, J. A. (1971). *Genet. Res.* **17**, 237.

Houghton, R. L., Skipton, M. D., Watson, K. and Griffiths, D. E. (1973). *Biochem. Soc. Trans.* **1**, 1110.

Houghton, R. L., Lancashire, W. E. and Griffiths, D. E. (1974). *Biochem. Soc. Trans.* **2**, 210.

Howell, N., Zuiches, C. A. and Munkres, K. D. (1971). *J. Cell Biol.* **50**, 721.

Howell, N., Trembath, M. K., Linnane, A. W. and Lukins, H. B. (1973). *Mol. Gen. Genet.* **122**, 37.

Huang, M., Biggs, D. R., Clark-Walker, G. D., and Linnane, A. W. (1966). *Biochim. biophys. Acta* **114**, 434.

Hughes, A. E. and Wilkie, D. (1972). *Heredity, Lond.* **28**, 117.

Hughes, D. E., Lloyd, D. and Brightwell, R. (1970). *Symp. Soc. gen. Microbiol.* **20**, 295.

Hughes, D. E., Wimpenny, J. W. T. and Lloyd, D. (1971). *Methods Microbiol.* **5B**, 1.

Hülsmann, W. C. (1962). *Biochim. biophys. Acta* **58**, 417.

Humphrey, B. A. and Humphrey, S. F. (1948). *J. exp. Biol.* **25**, 123.

Hungate, R. E. (1967). "The Rumen and its Microbes". Academic Press, London and New York.

Hunter, M. J. and Commerford, S. L. (1961). *Biochim. biophys. Acta* **47**, 580.

Hurley, L. S., Theriault, L. L. and Dreosti, I. E. (1970). *Science, N.Y.* **170**, 1316.

Hutchens, J. O. (1940). *J. Cell comp. Physiol.* **16**, 265.

Hyams, J. and Davies, D. R. (1972). *Mutation Research* **14**, 381.

Ibrahim, N. G., Stuchell, R. N. and Beattie, D. S. (1973). *Eur. J. Biochem.* **36**, 519.

Illingworth, R. F., Rose, A. H. and Beckett, A. (1973). *J. Bact.* **113**, 373.

Ishidate, K., Kawaguchi, K. and Tagawa, K. (1969). *J. Biochem., Tokyo* **65**, 385.

Iwasa, K. (1960a). *J. Biochem., Tokyo* **47**, 445.

Iwasa, K. (1960b). *J. Biochem., Tokyo* **47**, 584.

Iwashiwa, A. and Rabinowitz, M. (1969). *Biochim. biophys. Acta* **178**, 283.

Jacobs, N. J., MacLosky, E. R. and Jacobs, J. M. (1967). *Biochim. biophys. Acta* **148**, 645.

Jakob, H. (1965). *Genetics* **52**, 75.

Jakovcic, S., Geta, G. S., Rabinowitz, M., Jacob, H. and Swift, H. (1971). *J. Cell Biol.* **48**, 490.

Janki, R. M., Aithal, H. N., McMurray, W. C. and Tustanoff, E. R. (1974). *Biochem. biophys. Res. Commun.* **56**, 1078.

Jayaraman, J. (1969). *J. Scient. Res.* **28**, 441.

Jayaraman, J. and Sastry, P. S. (1971). *Indian J. Biochem. Biophys.* **8**, 278.

Jayaraman, J., Cotman, C., Mahler, H. R. and Sharp, C. W. (1966). *Archs Biochem. Biophys* **116**, 224.

Jinks, J. L. (1963). *In* "Methodology in Basic Genetics", (W. J. Burdette, ed.), pp. 325–354. Holden Day, San Francisco.

Jinks, J. L. (1964). "Extrachromosomal Inheritance", (Foundation of Modern Genetics Series). Prentice Hall, Englewood Cliffs.

John, P. C. L., McCullough, W., Atkinson, A. W. Jr., Forde, B. G. and Gunning, B. E. S. (1973). *In* "The Cell Cycle in Development and Differentiation", (M. Balls and F. Billett, eds.), p. 61. Cambridge University Press, Cambridge.

Johnson, B., Nelson, S. J. and Brown, C. M. (1972). *Antonie van Leeuwenhoek* **38**, 129.

Johnson, J. M. and Paltauf, F. (1970). *Biochim. biophys. Acta* **218** 431.

Jollow, D., Kellerman, G. M. and Linnane, A. W. (1968). *J. Cell Biol.* **37**, 221.

Jonah, M. and Erwin, J. A. (1971). *Biochim. biophys. Acta* **231**, 80.

Jones, D., Bacon, J. S. D., Farmer, V. C. and Webley, D. M. (1969). *Soil Biol. Biochem.* **1**, 145.

Jones, M. S. and Jones, O. T. G. (1969). *Biochem. J.* **113**, 507.

Jones, M. S. and Jones, O. T. G. (1970). *Biochem. biophys. Res. Commun.* **41**, 1072.

Jungalwala, F. B. and Dawson, R. M. C. (1970a). *Eur. J. Biochem.* **12**, 399.

Jungalwala, F. B. and Dawson, R. M. C. (1970b). *Biochem. J.* **117**, 481.

Kadenbach, B. (1967a). *Biochim. biophys. Acta* **134**, 430.

Kadenbach, B. (1967b). *Biochim. biophys. Acta* **138**, 651.

Kadenbach, B. (1968). *In* "Biochemical Aspects of the Biogenesis of Mitochondria", (E. C. Slater, J. M. Tager, S. Papa and E. Quagliariello, eds.), p. 415. Adriatica Editrice, Bari.

Kadenbach, B. (1969). *Eur. J. Biochem.* **10**, 312.

Kadenbach, B. (1970). *Eur. J. Biochem.* **12**, 392.

Kadenbach, B. (1971a). *In* "Autonomy and Biogenesis of Mitochondria and Chloroplasts", (N. K. Boardman, A. W. Linnane, and R. M. Smillie, eds.), pp. 360–371. North Holland, Amsterdam.

Kadenbach, B. (1971b). *Biochem. biophys. Res. Commun.* **44**, 724.

Kaempfer, R. (1969). *Nature, Lond.* **222**, 950.

Kagawa, Y. and Racker, E. (1966). *J. biol. Chem.* **241**, 2475.

Kahn, V. and Blum, J. J. (1967). *Biochemistry, N.Y.* **6**, 817.

Kamen, M. D. and Horio, T. (1970). *A. Rev. Biochem.* **39**, 673.

Kaplan, D. M. and Criddle, R. S. (1970). *Biochim. biophys. Acta* **222**, 611.

Kaplan, D. M. and Woodward, D. O. (1973). *Analyt. Biochem.* **52**, 102.

Karakashian, S. J., Karakashian, M. and Rudzinska, M. (1968). *J. Protozool.* **15**, 113.

Karol, M. H. and Simpson, M. V. (1968). *Science, N.Y.* **162**, 470.

Karst, F. and Lacroute, F. (1973). *Biochem. biophys. Res. Commun.* **52**, 741.

Kasamatsu, H., Robberson, D. L. and Vinograd, J. (1971). *Fedn Proc. Fedn Am. Socs exp. Biol.*, **30**, 1177.

Katoh, T. and Sanukida, S. (1965). *Biochem. biophys. Res. Commun.* **21**, 373.

Kattermann, R. and Slonimski, P. P. (1960). *C.r. hebd. Séanc. Acad. Sci., Paris*, Ser. D, **250**, 220.

Katz, R. (1971). *FEBS Lett.* **12**, 153.

Katz, R., Kilpatrick, L. and Chance, B. (1971). *Eur. J. Biochem.* **21**, 301.

Kawaguchi, K., Ishidate, K. and Tagawa, K. (1969). *J. Biochem., Tokyo* **66**, 21.

Kawai, K. and Mizushima, H. (1973). *J. Biochem., Tokyo* **74**, 183.

Kawakami, N. (1961). *Expl. Cell Res.* **25**, 179.

Kawakita, M. (1970a). *Pl. Cell Physiol., Tokyo* **11**, 377.

Kawakita, M. (1970b). *J. Biochem., Tokyo* **68**, 625.

Kawakita, M. (1971). *J. Biochem., Tokyo* **69**, 35.

Keiding, J. and Westergaard, O. (1971). *Expl. Cell Res.* **64**, 317.

Keilin, D. (1925). *Proc. R. Soc.* Ser. B. **98**, 312.

Keilin, D. (1930). *Proc. R. Soc.* Ser. B. **106**, 418.

Keilin, D. (1953). *Nature, Lond.* **172**, 390.

Keilin, D. (1966). "The History of Cell Respiration and Cytochrome". Cambridge University Press, Cambridge.

Keilin, D. and Hartree, E. F. (1939). *Proc. R. Soc.* Ser. B. **127**, 167.

Keilin, D. and Hartree, E. F. (1949). *Nature, Lond.* **164**, 254.

Keilin, D. and Hartree, E. F. (1953). *Nature, Lond.* **171**, 413.

Keilin, D. and Ryley, J. F. (1953). *Nature, Lond.* **172**, 451.

Keilin, D. and Tissières, A. (1953). *Nature, Lond.* **172**, 393.

Keilin, D. and Tissières, A. (1954). *Biochem. J.* **57**, XXIX.

Keith, A. D., Resnick, M. R. and Haley, A. B. (1969). *J. Bact.* **98**, 415.

Keith, A. D., Bulfield, G. and Snipes, W. (1970). *Biophys. J.* **10**, 618.

Keith, A. D., Wisnieski, B. J., Henry, S. and Williams, J. C. (1973). *In* "Lipids and Biomembranes of Eukaryotic Microorganisms", (J. A. Erwin, ed.), pp. 259–304. Academic Press, New York and London.

Kellems, R. E. and Butow, R. A. (1972). *J. biol. Chem.* **247**, 8043.

Kellems, R., Allison, V. F. and Butow, R. A. (1974). *In* "Biogenesis of Mitochondria", (A. M. Kroon and C. Saccone, eds.), pp. 511–524. Academic Press, London and New York.

Kellenberger, E. (1966). "Principles of Biomolecular Organization", p. 192. Ciba Foundation Symposium.

Kellerman, G. M., Biggs, D. R. and Linnane, A. W. (1969). *J. Cell Biol.* **42**, 378

Kellerman, G. M., Griffiths, D. E., Hansby, J. E., Lamb, A. J. and Linnane, A. W. (1971). *In* "Autonomy and Biogenesis of Mitochrondria and Chloroplasts", (N. K. Boardman, A. W. Linnane and R. M. Smillie, eds.), pp. 346–360. North Holland, Amsterdam.

Kessler, D. (1969). *J. Cell Biol.* **43**, 68a.

Keyhani, E. (1972). *J. Microscopie, Paris* **15**, 343.

Keyhani, E. (1973). *Expl. Cell Res.* **81**, 73.

Keyhani, E. and Chance, B. (1971). *FEBS Lett.* **17**, 127.

Keyhani, J., Keyhani, E. and Goodgal, S. (1972a). Abstracts of Communications presented at the 8th Meeting of the Federation of European Biochemical Societies at Amsterdam. Abstract A661.

Keyhani, J., Keyhani, E. and Goodgal, S. H. (1972b). *Eur. J. Biochem.* **27**, 527.

Kidder, G. W. and Goddard, D. R. (1965). *Pl. Physiol., Lancaster*, **40**, 552.

Kikuchi, G. and Barron, E. S. G. (1959). *Archs Biochem. Biophys* **84**, 96.

Kim, I.-C. and Beattie, D. S. (1973). *Eur. J. Biochem.* **36**, 509.

King, T. E. and Takemori, S. (1964). *J. biol. Chem.* **239**, 3559.

Kirby, K. S. (1965). *Biochem. J.* **96**, 266.

Kirk, J. (1970). *Nature, Lond.* **226**, 182.

Kirschner, R. H., Wolstenholme, D. R. and Gross, N. J. (1968). *Proc. natn. Acad. Sci. U.S.A.* **60**, 1466.

Kislev, N. and Eisenstadt, J. M. (1972). *Eur. J. Biochem.* **31**, 226.

Kislev, N., Selsky, M. I., Norton, C. and Eisenstadt, J. M. (1972). *Biochim. biophys. Acta* **287**, 256.

Kitsutani, S., Sawada, K., Osumi, M. and Nagahisa, M. (1970). *Pl. Cell Physiol., Tokyo* **11**, 107.

Kiyasu, U. Y., Pieringer, R. A., Paulus, H. and Kennedy, E. P. (1963). *J. biol. Chem.* **238**, 2293.

Klagsbrun, M. (1973). *J. biol. Chem.* **248**, 2612.

Kleese, R. A., Grotbeck, R. C. and Snyder, J. R. (1972a) *J. Bact.* **112**, 1023.

Kleese, R. A., Grotbeck, R. C. and Snyder, J. R. (1972b). *Can. J. Genet. Cytol.* **14**, 713.

Klein, H. P., Eaton, N. R. and Murphy, J. C. (1954). *Biochim. biophys. Acta* **13**, 591.

Knight, E. Jr. (1969). *Biochemistry, N.Y.*, **8**, 5089.

Knowles, J. K. C. (1971). *Expl. Cell Res.* **70**, 223.

Knowles, J. K. C. and Tait, A. (1972). *Molec. Gen. Genet.* **117**, 53.

Kobayashi, S. (1965). *J. Biochem., Tokyo* **58**, 444.

Kolarov, J., Šubík, J. and Kováč, L. (1972a). *Biochim. biophys. Acta* **267**, 457.

Kolarov, J. Šubík, J. and Kováč, L. (1972b). *Biochim. biophys. Acta* **267**, 465.

Komai, H. and Capaldi, R. A. (1973). *FEBS Lett.* **30**, 273.

Kormančíková, V., Kováč, L. and Vidová, M. (1969). *Biochim. biophys. Acta* **180**, 9.

Kotelnikova, A. V. and Zvjagilskaja (1966). *In* "Symposium on Mitochondrial Structure and Function", Moscow, 1965, p. 66. Acad. Sci. USSR, Moscow.

Kováč, L. (1972). *FEBS Lett.* **22**, 270.

Kováč, L. and Hrusovská, E. (1968). *Biochim. biophys. Acta* **153**, 43.
Kováč, L. and Weissová, K. (1968). *Biochim. biophys. Acta* **153**, 55.
Kováč, L., Lachowitz, T. M. and Slonimski, P. P. (1967a). *Science, N.Y.* **158**, 1564.
Kováč, L., Subík, J., Russ, G. and Kollár, K. (1967b). *Biochim. biophys. Acta* **144**, 94.
Kováč, L., Bendárová, H. and Greksák, M. (1968). *Biochim. biophys. Acta* **153**, 32.
Kováč, L., Hrusovska, E. and Smigán, P. (1970a). *Biochim. biophys. Acta* **205**, 520.
Kováč, L., Šmigán, P., Hrusovská, E. and Hess, B. (1970b). *Archs Biochem. Biophys* **139**, 370.
Kováč, L., Groot, C. S. P. and Racker, E. (1972). *Biochim. biophys. Acta* **256**, 55.
Kováčová, V., Irmlerová, J. and Kováč, L. (1968). *Biochim. biophys. Acta*, **162**, 157.
Kováčová, V., Vlček, D. and Miadoková, E. (1969). *Folia Microbiol.* **14**, 554.
Kowallik, W. (1967). *Pl. Physiol., Lancaster* **42**, 672.
Kowallik, W. (1968). *Planta* **79**, 122.
Kraepelin, G. (1967). *Z. allg. Mikrobiol.* **7**, 287.
Kraepelin, G. (1972). *Z. allg. Mikrobiol.* **12**, 235.
Kraml, J. and Mahler, H. R. (1967). *Immunochem.* **4**. 213.
Krawiec, S. and Eisenstadt, J. M. (1970a). *Biochim. biophys. Acta* **217**, 120.
Krawiec, S. and Eisenstadt, J. R. (1970b). *Biochim. biophys. Acta* **217**, 132.
Krebs, W., Schwab, D. and Schwab-Stey, H. (1972). *J. Ultrastruct. Res.* **38**, 605.
Krikso, I., Gordon, J. and Lipmann, F. (1969). *J. biol. Chem.* **244**, 6117.
Kröger, A. and Klingenberg, M. (1973). *Eur. J. Biochem.* **34**, 358.
Kroon, A. M. (1963). *Biochim. biophys. Acta* **72**, 391.
Kroon, A. M. (1969). *In* "Inhibitors, Tools in Cell Research", (Th. Bücher, and H. Sies, eds.), p. 159. Springer Verlag, Heidelberg.
Kroon, A. M., Arendzen, A. J. and de Vries, H. (1974). *In* "Biogenesis of Mitochondria", (A. M. Kroon and C. Saccone, eds.), pp. 395–404. Academic Press, London and New York.
Kruis, K. and Satava, J. (1918). "Ovývoji a klíčeni spör jakož i sexualitě kvasinek". Nakl. C. Acad., Praha.
Kubowitz, F. and Hass, E. (1932). *Biochem. Z.* **255**, 247.
Küenzi, M. T. and Fiechter, A. (1969). *Arch. Mikrobiol.* **64**, 396.
Küenzi, M. T., Tingle, M. A. and Halvorson, H. O. (1974). *J. Bact.* **117**, 80.
Kuff, E. L. and Schneider, W. C. (1954). *J. biol. Chem.* **206**, 677.
Kung, C. (1970). *J. gen. Microbiol.* **61**, 371.
Küntzel, H. (1969a). *Nature, Lond.* **222**, 142.
Küntzel, H. (1969b). *FEBS Lett.* **4**, 140.
Küntzel, H. (1969c). *J. molec. Biol.* **40**, 315.
Küntzel, H. and Noll, H. (1967). *Nature, Lond.* **215**, 1340.
Küntzel, H. and Schäfer, J. P. (1971). *Nature New Biology* **231**, 265.
Küntzel, H., Barath, Z., Ali, I., Kind, J. and Althus, H. H. (1973). *Proc. natn. Acad. Sci. U.S.A.* **70**, 1574.
Küntzel, H., Ali, I. and Blassey, C. (1974). *In* "Biogenesis of Mitochondria", (A. M. Kroon and C. Saccone, eds.), pp. 71–78. Academic Press, London and New York.
Kuriyama, Y. and Luck, D. J. L. (1973). *J. molec. Biol.* **73**, 425.
Kuriyama, Y. and Luck, D. (1974). *In* "Biogenesis of Mitochondria", (A. M. Kroon and C. Saccone, eds.). Academic Press, London and New York.
Kuroiwa, T. (1973a). *Expl. Cell Res.* **78**, 351.
Kuroiwa, T. (1973b). *J. Electron Miscosc.* **22**, 45.
Kusel, J. P. and Storey, B. T. (1972). *Biochem. biophys. Res. Commun.* **46**, 501.

Kusel, J. P. and Storey, B. T. (1973a). *Biochim. biophys. Acta* **305**, 570.

Kusel, J. P. and Storey, B. T. (1973b). *Biochim. biophys. Acta* **314**, 164.

Kusel, J. P. and Weber, M. M. (1965). *Biochim. biophys. Acta* **98**, 632.

Kusel, J. P. and Weber, M. M. (1968). *Bact. Proc.* **68**, 139.

Kusel, J. P., Moore, K. E. and Weber, M. M. (1967). *J. Protozool.* **14**, 283.

Kusel, J. P., Suriano, J. R. and Weber, M. M. (1969). *Archs Biochem. Biophys* **133**, 293.

Kusel, J. P., Boveris, A. and Storey, B. T. (1973). *Archs Biochem. Biophys.* **158**, 799.

Kutzleb, R., Schweyen, R. and Kaudewitz, F. (1973). *Mol. Gen. Genet.* **125**, 91.

Kuzela, S. and Fečíková, H. (1970). *Experientia* **26**, 940.

Kuzela, S. and Grečna, E. (1969). *Experientia* **25**, 776.

Kuzela, S., Kolarov, J. and Krempaský, V. (1973). *Biochem. biophys. Res. Commun.* **54**, 9.

Labbe, P. (1971). *Biochimie* **53**, 1001.

Labbe, P., Dechateaubodeau, G. and Labbe-Bois, R. (1972). *Biochimie* **54**, 513.

Lacroute, F. (1963). *C.r. hebd. Séanc. Acad. Sci., Paris*, Ser. D. **257**, 4213.

Lai, C. J. and Weisblum, B. (1971). *Proc. natn. Acad. Sci. U.S.A.* **68**, 856.

Laird, A. K., Nygaard, O. and Ris, H. (1952). *Cancer Res.* **12**, 276.

Lamb, A., Clarke Walker, G. and Linnane, A. W. (1968). *Biochem. biophys. Acta* **161**, 415.

Lambowitz, A. M. and Slayman, C. W. (1971). *J. Bact.* **108**, 1087.

Lambowitz, A. M. and Slayman, C. W. (1972). *J. Bact.* **112**, 1020.

Lambowitz, A. M., Slayman, C. W., Slayman, C. L. and Bonner, W. D. Jr. (1972a). *J. biol. Chem.* **247**, 1536.

Lambowitz, A. M., Smith, E. W. and Slayman, C. W. (1972b). *J. biol. Chem.* **247**, 4850.

Lambowitz, A. M., Smith, E. W. and Slayman, C. W. (1972c). *J. biol. Chem.* **247**, 4859.

Lampen, J. O., Arnow, P. M., Borowska, Z. and Laskin, A. I. (1972). *J. Bact.* **84**, 1152.

Lancashire, W. E. and Griffiths, D. E. (1971). *FEBS Lett.* **17**, 209.

Lancashire, W. E., Houghton, R. L. and Griffiths, D. E. (1974). *Biochem. Soc. Trans.* **2**, 213.

Landry, Y. and Goffeau, A. (1972). *Archs int. Physiol. Biochim.* **80**, 604.

Lane, B. G. and Tamaoki, T. (1967). *J. molec. Biol.* **27**, 335.

Lang, N. J. (1963a). *Am. J. Bot.* **50**, 280.

Lang, N. J. (1963b). *J. Protozool.* **10**, 333.

Langcake, P., Beechey, R. B., Lindop, C. R., Wickins, S. G. A., Leworthy, D. P. Wiggins, D. E. and Broughall, J. M. (1974). *Biochem. Soc. Trans.* **2**, 202.

Lansman, R. A., Rowe, M. J. and Woodward, D. O. (1974). *Eur. J. Biochem.* **41**, 15.

Larsen, W. J. (1970). *J. Cell Biol.* **47**, 373.

Lascelles, J. (1964). *In* "Tetrapyrrole Biosynthesis and its Regulation", p. 89. W. A. Benjamin Inc., New York and Amsterdam.

Laskowski, W. (1954). *Heredity, Lond.* **8**, 79.

Laub, R. and Thirion, J. (1972). *Archs. int. Physiol. Biochim.* **80**, 197.

Lauquin, G. and Vignais, P. V. (1973). *Biochim. biophys. Acta* **305**, 534.

Lauquin, G., Vignais, P. V. and Mattoon, J. R. (1973). *FEBS Lett.* **35**, 198.

Laurent, M. and Steinert, M. (1970). *Proc. natn. Acad. Sci. U.S.A.* **66**, 419.

Lebeault, J. M., Roche, B., Duvnjak, Z. and Azoulay, E. (1969). *J. Bact.* **100**, 1218.

Lederer, F. (1972). *Eur. J. Biochem.* **31**, 144.

Lederer, F., Simon, A.-M. and Verdière, J. (1972). *Biochem. biophys. Res. Commun.* **47**, 55.

Lee, C.-P., Sottocasa, G. L. and Ernster, L. (1967). *Meth. Enzym.* **10**, 33.

Leedale, G. F., Meeuse, B. J. D. and Pringsheim, E. G. (1965). *Arch. Mikrobiol.* **50**, 68.

Leedale, G. F. and Buetow, D. E. (1970). *Cytobiologie* **1**, 195.

Lehninger, A. L. (1951). *J. biol. Chem.* **190**, 345.

Leigh, J. S. Jr. and Wilson, D. F. (1972). *Biochem. biophys. Res. Commun.* **48**, 1266.

Leiter, E. H., Labrie, D. A., Bergquist, A. and Wagner, R. P. (1971). *Biochem. Genet.* **5**, 549.

Lemberg, M. (1969). *Physiol. Rev.* **49**, 48.

Lemberg, R. and Legge, J. W. (1949). "Haematin Compounds and Bile Pigments". Wiley–Interscience, New York.

Lemberg, R. and Barrett, J. (1973). "Cytochromes". Academic Press, London and York.

Lemoigne, M., Aubert, J. P. and Millet, J. (1954). *Annls Inst. Pasteur, Paris* **87**, 427.

Lenaz, G., Haard, N. F., Silman, H. I. and Green, D. E. (1968a). *Archs Biochem. Biophys* **128**, 293.

Lenaz, G., Haard, N. H., Lauwers, A., Allman, D. W. and Green, D. E. (1968b). *Archs Biochem. Biophys* **126**, 746.

Lenaz, G., Littarrn, G. R. and Castelli, A. (1969). *FEBS Lett.* **2**, 198.

Leon, S. A. and Mahler, H. R. (1968). *Archs Biochem. Biophys* **126**, 305.

Lerman, L. S. (1963). *Proc. natn. Acad. Sci. U.S.A.* **49**, 94.

Lester, R. L. and Crane, F. L. (1959). *J. biol. Chem.* **234**, 2169.

Lester, R. L., Crane, F. L. and Hatefi, Y. (1958). *J. Am. chem. Soc.* **80**, 4751.

Lester, R. L., Hatefi, Y., Widmer, C. and Crane, F. L. (1959). *Biochim. biophys. Acta* **33**, 169.

Levate, W. W., Dyer, J. R., Springer, C. M. and Bentley, R. (1962). *J. biol. Chem.* **237**, PC 2715.

Levy, M. R. and Elliott, A. M. (1968). *J. Protozool.* **15**, 208.

Lietman, P. S. (1970). *Molec. Pharmacol.* **7**, 122.

Light, P. A. (1972a). *FEBS Lett* **19**, 319.

Light, P. A. (1972b). *J. appl. Chem. Biotechnol.* **22**, 509.

Light, P. A. and Garland, P. B. (1971). *Biochem. J.* **124**, 123.

Light, P. A., Ragan, C. I., Clegg, R. A. and Garland, P. B. (1968). *FEBS Lett.* **1**, 4.

Lindblom, G. P. (1961). *J. Protozool.* **8**, 139.

Lindegren, C. C. (1956). *C.r. Trav. Lab. Carlsberg.* **26**, 253.

Lindegren, C. C. (1971). *J. Biol. Phychol.* **13**, 3.

Lindegren, C. C. (1972). *J. Biol. Psychol.* **14**, 38.

Lindegren, C. C. and Lindegren, G. (1943). *Proc. natn. Acad. Sci. U.S.A.* **29**, 306.

Lindegren, C. C. and Lindegren, G. (1971). *Nature, Lond.* **234**, 297.

Lindegren, C. C. and Lindegren, G. (1973). *Antonie van Leeuwenhoek* **39**, 351.

Lindegren, C. C., Nagai, S. and Nagai, H. (1958). *Nature, Lond.* **182**, 446.

Lindenmayer, A. and Estabrook, R. W. (1958). *Archs Biochem. Biophys* **78**, 66.

Lindenmayer, A. and Smith, L. (1964). *Biochim. biophys. Acta* **93**, 445.

Lingens, F., Oltmanns, O. and Bacher, A. (1967). *Z. Naturf. B.* **22**, 755.

Linn, S. and Lehman, I. R. (1965a). *J. biol. Chem.* **240**, 1287.

Linn, S. and Lehman, I. R. (1965b). *J. biol. Chem.* **240**, 1294.

Linn, S. and Lehman, I. R. (1966). *J. biol. Chem.* **241**, 2694.

Linnane, A. W. (1965). *In* "Oxidases and Related Redox Systems", (T. E. King, H. S. Mason and M. Morrison, eds.), p. 1102. Wiley, New York.

Linnane, A. W. and Haslam, J. M. (1971). *In* "Current Topics in Cellular Regulation", (B. L. Horecker and E. R. Stadtman, eds.), Vol. 2, p. 101. Academic Press, New York and London.

Linnane, A. W., Vitols, E. and Nowland, P. G. (1962). *J. Cell. Biol.* **13**, 345.

Linnane, A. W., Lamb, A. J., Christodoulon, C. and Lukins, H. B. (1968a). *Proc. natn. Acad. Sci. U.S.A.* **59**, 1288.

Linnane, A. W., Biggs, D. R., Huang, M. and Clark-Walker, D. R. (1968b). *In* "Aspects of Yeast Metabolism", (R. K. Mills, ed.), pp. 217–243. Blackwell, Oxford.

Linnane, A. W., Saunders, G. W., Gingold, E. B. and Lukins, H. M. (1968c). *Proc. natn. Acad. Sci. U.S.A.* **59**, 903.

Linnane, A. W., Haslam, J. M. and Forrester, I. T. (1972)a. *In* "Biochemistry and Biophysics of Mitochondria Membranes", (G. F. Azzone, E. Carafoli, A. L. Lehninger, E. Quagliariello and N. Silliprandi, eds.), pp. 523–539. Academic Press, New York.

Linnane, A. W., Haslam, J. M., Lukins, H. B. and Nagley, P. (1972b). *A. Rev. Microbiol.* **26**, 163.

Linnane, A. W., Bunn, C. L., Howell, N., Molloy, P. L. and Lukins, H. B. (1973). *In* "Biochemistry of Gene Expression in Higher Organisms", (J. K. Pollak and W. Lee, eds.), pp. 425–442. Australian and New Zealand Book Co. Ltd., Artarman, N.S.W. 69.

Linnane, A. W., Howell, N. and Lukins, H. B. (1974). *In* "Biogenesis of Mitochondria", (A. M. Kroon and C. Saccone, eds.), pp. 193–214. Academic Press, London and New York.

Linstead, D., Evans, I. and Wilkie, D. (1974). *In* "Biogenesis of Mitochondria", (A. M. Kroon and C. Saccone, eds.), pp. 179–192. Academic Press, London and New York.

Lizardi, P. M. and Luck, D. J. L. (1971). *Nature New Biology* **229**, 140.

Lizardi, P. M. and Luck, D. J. L. (1972). *J. Cell. Biol.* **54**, 56.

Lloyd, D. (1965). *Biochim. biophys. Acta* **110**, 425.

Lloyd, D. (1966a). *Expl. Cell Res.* **45**, 120.

Lloyd, D. (1966b). *Phytochem.* **5**, 527.

Lloyd, D. (1969). *Symp. Soc. gen. Microbiol.* **19**, 299.

Lloyd, D. (1974). *In* "Biochemistry and Physiology of the Algae", (W. D. P. Stewart, ed.), pp. 505–529. Blackwell, Oxford.

Lloyd, D. and Cartledge, T. G. (1974). *In* "Methodological Developments in Biochemistry", (Reid, E., ed.), Vol. 4. Longman, London.

Lloyd, D. and Chance, B. (1968). *Biochem. J.* **107**, 829.

Lloyd, D. and Chance, B. (1972). *Biochem. J.* **128**, 1171.

Lloyd, D. and Griffiths, A. J. (1968). *Expl. Cell Res.* **51**, 291.

Lloyd, D. and Turner, G. (1968). *J. gen. Microbiol.* **50**, 421.

Lloyd, D. and Venables, S. E. (1967). *Biochem. J.* **104**, 639.

Lloyd, D., Evans, D. A. and Venables, S. E. (1968). *Biochem. J.* **109**, 897.

Lloyd, D., Howells, L. and Cartledge, T. G. (1970a). *Biochem. J.* **116**, 25P.

Lloyd, D., Evans, D. A. and Venables, S. E. (1970b). *J. gen. Microbiol.* **61**, 33.

Lloyd, D., Brightwell, R., Venables, S. E., Roach, G. I. and Turner, G. (1971a). *J. gen. Microbiol.* **65**, 209.

Lloyd, D., Turner, G., Poole, R. K., Nicholl, W. G. and Roach, G. I. (1971b). *Sub-Cell. Biochem.* **1**, 91.

Lorenz, B., Kleinow, W. and Weiss, H. (1974). *Z. Physiol. Chem.* **355**, 300.

Loschen, G., Azzi, A. and Flotte, L. (1973). *FEBS Lett.* **33**, 84.

Lövlie, A. (1963). *C.r. Trav. Lab. Carlsberg.* **33**, 377.

Lowdon, M. J., Gordon, P. A. and Stewart, P. R. (1972). *Arch. Mikrobiol.* **85**, 355.

Luck, D. J. L. (1963a). *Proc. natn. Acad. Sci. U.S.A.* **49**, 233.

Luck, D. J. L. (1963b). *J. Cell Biol.* **16**, 483.

Luck, D. J. L. (1965a). *J. Cell Biol.* **24**, 445.

Luck, D. J. L. (1965b). *J. Cell Biol.* **24**, 461.

Luck, D. J. L. (1967). *Meth. Enzym.* **12**, 465.

Luck, D. J. L. and Reich, E. (1964). *Proc. natn. Acad. Sci. U.S.A.* **52**, 931.

Luha, A. A. and Whittaker, P. A. (1972). *Biochem. J.* **127**, 43P.

Luha, A. A., Sarcoe, L. E. and Whittaker, R. A. (1971). *Biochem. biophys. Res. Commun.* **44**, 396.

Lui, N. S. T., Roels, O. A., Trout, M. E. and Anderson, O. R. (1968). *J. Protozool.* **15**, 536.

Lukins, H. B., Tham. S. H., Wallace, P. G. and Linnane, A. W. (1966). *Biochem. biophys. Res. Commun.* **23**, 363.

Lukins, H. B., Jollow, D., Wallace, P. G. and Linnane, A. W. (1968). *Aust. J. exp. Biol. med. Sci.* **46**, 651.

Lukins, H. B., Tate, J. R., Saunders, G. W. and Linnane, A. W. (1973). *Mol. Gen. Genet.* **120**, 17.

Lusena, C. W. and Depocas, F. (1966). *Can. J. Biochem.* **44**, 497.

Luzikov, V. N. (1973). *Sub-Cell. Biochem.* **2**, 1.

Luzikov, V. N., Zubatov, A. S. and Rainina, E. I. (1970). *FEBS Lett.* **11**, 233.

Luzikov, V. N., Zubatov, A. S., Rainina, E. I. and Bakeyeva, L. E. (1971). *Biochim. biophys Acta* **245**, 321.

Lwoff, A. (1934). *Zentbl. Bakt. Parasit Kde.* Abt. 1 Orig. **130**, 498.

Mackler, B. and Haynes, B. (1973). *Biochim. biophys. Acta* **292**, 88.

Mackler, B., Collipp, P. J., Duncan, H. M., Rao, N. A. and Huennekens, F. M. (1962). *J. biol. Chem.* **237**, 2968.

Mackler, B., Douglas, H. C., Will, S., Hawthorne, D. C. and Mahler, H. R. (1965). *Biochemistry, N.Y.* **4**, 2016.

MacLennan, D. H. (1970). *In* "Current Topics in Membranes and Transport", (F. Bronner and A. Kleinzeller, eds.), p. 177. Academic Press, New York and London.

MacQuillen, A. M. and Halvorson, H. O. (1962). *J. Bact.* **84**, 31.

Mager, J. (1960). *Biochim. biophys. Acta* **38**, 150.

Maguigan, W. H. and Walker, E. (1940). *Biochem. J.* **34**, 804.

Mahler, H. R. (1973). *J. Supramolec. Struct.* **1**, 449.

Mahler, H. R. (1974). *In* "Molecular Cytogenetics", (B. Hamkalo, ed.) Plenum Press, New York.

Mahler, H. R. and Dawidowicz, K. (1973). *Proc. natn. Acad. Sci. U.S.A.* **70**, 111.

Mahler, H. R. and Perlman, P. S. (1971). *Biochemistry, N.Y.* **10**, 2979.

Mahler, H. R. and Perlman, P. S. (1972a). *J. Supramolec. Struct.* **1**, 105.

Mahler, H. R. and Perlman, P. S. (1972b). *Archs Biochem. Biophys* **148**, 115.

Mahler, H. R. and Perlman, P. S. (1973a). *Mol. Gen. Genet.* **121**, 285.

Mahler, H. R. and Perlman, P. S. (1973b). *Mol. Gen. Genet.* **121**, 295.

Mahler, H. R., Mackler, B., Grandchamp, S. and Slonimski, P. P. (1964a). *Biochemistry, N.Y.* **3**, 668.

Mahler, H. R., Mackler, B., Slonimski, P. P. and Grandchamp, S. (1964b). *Biochemistry, N.Y.* **3**, 677.

Mahler, H. R., Neiss, G., Slonimski, P. P. and Mackler, B. (1964c). *Biochemistry, N.Y.* **3**, 893.

Mahler, H. R., Perlman, P., Henson, C. and Weber, C. (1968). *Biochem. biophys· Res. Commun.* **31**, 474.

Mahler, H. R., Perlman, P. S. and Mehrotra, B. D. (1971a). *In* "Autonomy and Biogenesis of Mitochondria and Chloroplasts", (A. K. Boardman, A. W. Linnane, and R. M. Smillie, eds.), pp. 492–511. North Holland, Amsterdam.

Mahler, H. R., Perlman, P. S., Slonimski, P., Deutsch, M. J., Fukuhara, A. and Faye, C. (1971b). *Fedn Proc. Fedn Am. Socs exp. Biol.* **30**, 1149.

Mahler, H. R., Dawidowicz, K. and Feldman, F. (1972). *J. biol. Chem.* **247**, 7439.

Mahler, H. R., Feldman, F., Phan, S. H., Hamill, P. and Dawidowicz, K. (1974). *In* "Biogenesis of Mitochondria", (A. M. Kroon and C. Saccone, eds.), pp. 423–442. Academic Press, London and New York.

Mahoney, M. and Wilkie, D. (1962). *Proc. R. Soc.* Ser. B. **156**, 524.

Makman, R. S. and Sutherland, E. W. (1965). *J. biol. Chem.* **240**, 1309.

Malhotra, S. K. (1968). *Nature, Lond.* **219**, 1267.

Malhotra, S. K. and Eakin, R. T. (1967). *J. Cell Sci.* **2**, 205.

Mandelstam, J. and McQuillen, K. (1973). "Biochemistry of Bacterial Growth", Blackwell, Oxford.

Manning, J. E., Wolstenholme, D. R., Ryan, R. S., Hunter, J. A. and Richards, O. C. (1971). *Proc. natn. Acad. Sci. U.S.A.* **68**, 1169.

Manton, I. (1959). *J. mar. biol. Ass. U.K.* **38**, 319.

Manton, I. and Parke, M. (1960). *J. mar. biol. Ass. U.K.* **39**, 275.

Manton, I., Rayns, D. G., Ettl, H. and Parke, M. (1965). *J. mar. biol. Ass. U.K.* **45**, 241.

Marchant, R. and Smith, D. G. (1968a). *J. gen. Microbiol.* **50**, 391.

Marchant, R. and Smith, D. G. (1968b). *Biol. Rev.* **43**, 459.

Marcovitch, H. (1951). *Annls Inst. Pasteur, Paris* **81**, 452.

Margoliash, E. (1962). *J. biol. Chem.* **237**, 2161.

Margulis, L. (1970). "The Origin of Eukaryotic Cells". Yale University Press, New Haven.

Marquardt, H. (1962a). *Z. Naturf*, **17b**, 42.

Marquardt, H. (1962b). *Z. Naturf*, **17b**, 689.

Marquardt, H. (1963). *Arch. Mikrobiol.* **46**, 308.

Marmur, J. and Doty, P. (1962). *J. molec. Biol.* **5**, 109.

Marmur, J., Cahoon, M. E., Shimura, Y. and Vogel, H. J. (1963). *Nature, Lond.* **197**, 1228.

Maroudas, N. G. and Wilkie, D. (1966). *Biochim. biophys. Acta* **166**, 681.

Mason, T. L. and Schatz, G. (1973a). *J. biol. Chem.* **248**. 1346.

Mason, T. L. and Schatz, C. (1973b). *J. biol. Chem.* **248**, 1355.

Mason, T. L., Hooper, A. and Cunningham, W. (1970). *J. Cell. Biol.* **47**, 130a.

Mathre, D. E. (1971). *Pesticide Biochem. Physiol.* **1**, 216.

Matile, Ph. (1971). *Cytobiol* **3**, 324.

Matile, Ph. and Bahr, G. F. (1968). *Expl. Cell Res.* **52**, 301.

Matile, Ph., Moor, H. and Robinow, C. F. (1969). *In* "The Yeasts", (A. H. Rose and J. S. Harrison, eds.), Vol. 1, pp. 219–302. Academic Press, London and New York.

Matsunaka, S., Morita, S. and Conti, S. F. (1966a). *Pl. Physiol., Lancaster* **41**, 1364.

Matsunaka, S., Morita, S. and Conti, S. F. (1966b). *Pl. Physiol., Lancaster* **41**, 1370.

Mattoon, J. R. and Balcavage, W. X. (1967). *Meth. Enzym.* **10**, 135.

Mattoon, J. R. and Sherman, F. (1966). *J. biol. Chem.* **241**, 4330.

Mazia, D. (1965). *Symp. Soc. gen. Microbiol.* **15**, 379.

McCashland, B. W. (1956). *J. Protozool.* **3**, 131.
McCashland, B. W., March, W. R. and Kronschnabel, J. M. (1957). *Growth* **21**, 21.
McClary, D. O. and Bowers, W. D. Jr. (1968). *J. Ultrastruct. Res.* **25**, 37.
McCully, E. K. and Robinow, C. F. (1971). *J. Cell Sci.* **9**, 475.
McDougall, K. J. and Pittenger, T. H. (1966). *Genetics* **54**, 551.
McHale, D., Green, J. and Diplock, A. T. (1962). *Nature, Lond.* **196**, 1293.
McLaughlin, P. J. and Dayhoff, M. O. (1970). *Science, N.Y.* **168**, 1469.
McLaughlin, P. J. and Dayhoff, M. O. (1973). *J. molec. Evol.* **2**, 99.
McLean, J. R., Cohn, G. L., Brandt, I. K. and Simpson, M. V. (1958). *J. biol. Chem.* **233**, 657.
McMurray, W. C. and Dawson, R. M. C. (1969). *Biochem. J.* **112**, 91.
Mehrotra, D. B. and Mahler, H. (1968). *Archs Biochem. Biophys* **128**, 685.
Melnick, J. L. (1942). *J. biol. Chem.* **141**, 269.
Mereschowsky, C. (1910). *Biol. Zbl.* **30**, 278.
Meselson, F. and Stahl, F. W. (1964). *Proc. natn. Acad. Sci. U.S.A.* **52**, 931.
Meyer, R. R. (1973). *J. theor. Biol.* **38**, 647.
Meyer, R. R. and Simpson, M. V. (1969). *Biochem. biophys. Res. Commun.* **34**, 238.
Meyer, R. R., Boyd, C. R., Rein, D. C. and Keller, S. J. (1972). *Expl. Cell Res.* **70**, 233.
Meyer, T. E. and Cusanovich, M. A. (1972). *Biochem. biophys. Acta* **267**, 383
Michaelis, G., Douglass, S., Tsai, M. and Criddle, R. S. (1971). *Biochem. Genet.* **5**, 487.
Michaelis, G., Douglass, S., Tsai, M. J., Burchel, K. and Criddle, R. S. (1972). *Biochemistry, N.Y.* **11**, 2026.
Michaelis, G., Petrochilo, E. and Slonimski, P. P. (1973). *Mol. Gen. Genet.* **123**, 51.
Michel, R. and Neupert, W. (1973). *Eur. J. Biochem.* **36**, 53.
Michel, R. and Neupert, W. (1974). *In* "Biogenesis of Mitochondria", (A. M. Kroon and C. Saccone, eds.). Academic Press, London and New York.
Michel, R., Schweyen, R. J. and Kaudewitz, F. (1971). *Mol. Gen. Genet.* **111**, 235.
Millis, A. J. T. and Suyama, Y. (1972). *J. biol. Chem.* **247**, 4063.
Mitchell, C. H., Bunn, C. L., Davey, P. J. and Linnane, A. W. (1970). *Proc. Aust. biochem. Soc.* **3**, 9.
Mitchell, C. H., Bunn, C. L., Lukins, H. B. and Linnane, A. W. (1971). *Proc. Aust. biochem. Soc.* **4**, 67.
Mitchell, C. H., Bunn, C. L., Lukins, H. B. and Linnane, A. W. (1973). *Bioenergetics* **4**, 161.
Mitchell, H. K. and Mitchell, M. B. (1956). *J. gen. Microbiol.* **14**, 84.
Mitchell, M. B. and Mitchell, H. K. (1952). *Proc. natn. Acad. Sic. U.S.A.* **38**, 442.
Mitchell, M. B., Mitchell, H. K. and Tissières, A. (1953). *Proc. natn. Acad. Sci. U.S.A.* **39**, 606.
Mitchison, J. M. (1971). "The Biology of the Cell Cycle". Cambridge University Press, Cambridge.
Miura, T. and Yanagita, T. (1972). *J. Biochem., Tokyo* **72**, 141.
Miyake, S. and Sugimura, T. (1968). *J. Bact.* **96**, 1997.
Miyake, S., Iwamoto, Y., Nagao, M., Sugimura, T. and Ohsumi, M. (1972). *J. Bact.* **109**, 409.
Mohar-Betancour, O. and Garcia-Hernandez, M. (1973). Abstracts of Communications presented at the 9th International Congress of Biochemistry at Stockholm, Abstract 5j18.
Mok, T. C. K., Rickard, P. A. D. and Moss, F. J. (1969). *Biochim. biophys. Acta* **172**, 438.

Molloy, P. L., Howell, N., Plummer, D. T., Linnane, A. W. and Lukins, H. B. (1973). *Biochem. biophys. Res. Commun.* **52**, 9.

Molzahn, S. W. and Woods, R. A. (1972). *J. gen. Microbiol.* **72**, 339.

Montenecourt, B. S., Langsam, M. and Dubin, D. B. (1970). *J. Cell Biol.* **46**, 245.

Moor, H. (1964). *Z. Zellforch. mikrosk. Anat.* **62**, 546.

Moor, H. and Mühlethaler, K. (1963). *J. Cell Biol.* **17**, 609.

Moore, J., Cantor, M. H., Sheeler, P. and Kahn, W. (1970). *J. Protozool.* **17**, 671.

Moore, R. T. and McAlear, J. H. (1962). Abstract of Communications presented at the 8th International Congress of Microbiology at Montreal, Abstract 31.

Morales, N. M., Schaefer, F. W., Keller, S. J. and Meyer, R. R. (1972). *J. Protozool.* **19**, 667.

Morgan, N. A. and Griffiths, A. J. (1972). *J. Protozool.* **19**, (Suppl.), A 137.

Morgan, N. A., Howells, L., Cartledge, T. G. and Lloyd, D. (1973). *In* "Methodological Developments in Biochemistry", (E. Reid, ed.), Vol. 3, pp. 219–232. Longman, London.

Morimoto, H. and Halvorson, H. O. (1971). *Proc. natn. Acad. Sci. U.S.A.* **68**, 324.

Morimoto, H., Scragg, A. H., Nekhorocheff, J., Villa, V. and Halvorson, H. O. (1971). *In* "Autonomy and Biogenesis of Mitochondria and Chloroplasts", (N. K. Boardman, A. W. Linnane and R. M. Smillie, eds.), pp. 282–292. North Holland, Amsterdam.

Morita, T. and Mifuchi, I. (1970). *Biochem. biophys. Res. Commun.* **38**, 191.

Morpurgo, G., Serlupi-Crescenzi, G., Tecce, G., Valente, F. and Venettacci, D. (1964). *Nature, Lond.* **201**, 897.

Morré, D. J., Merritt, W. D. and Lembi, C. A. (1971). *Protoplasma*, **73**, 43.

Mortimer, R. K. and Hawthorne, D. C. (1966). *A. Rev. Microbiol.* **20**, 151.

Moss, F. J., Rickard, P. A. D., Beech, G. A. and Bush, F. E. (1969). *Biotech. Bioengng.* **11**, 561.

Moss, F. J., Rickard, P. A. D., Bush, F. E. and Craiger, P. (1971). *Biotech. Bioengng.* **13**, 63.

Mounolou, J.-C. (1967). Ph.D. Thesis, Faculté des Sciences d l'Universite de Paris.

Mounolou, J.-C. (1973). *FEBS Lett.* **29**, 275.

Mounolou, J.-C. and Perrodin, G. (1968). *C.r. hebd. Séanc Acad. Sci., Paris,* Ser. D, **267**, 1286.

Mounolou, J.-C., Jakob, H. and Slonimski, P. P. (1966). *Biochem. biophys. Res. Commun.* **24**, 218.

Mounolou, J.-C., Perrodin, G. and Slonimski, P. P. (1968). *In* "Biochemical Aspects of the Biogenesis of Mitochondria", (E. C. Slater, J. M. Tager, S. Papa and E. Quagliariello, eds.), pp. 133–154. Adriatica Editrice, Bari.

Moustacchi, E. (1969). *Mutation Res.* **7**, 171.

Moustacchi, E. (1971). *Mol. Gen. Genet.* **114**, 50.

Moustacchi, E. (1972). *Biochim. biophys. Acta* **277**, 61.

Moustacchi, E. (1973). *J. Bact.* **115**, 805.

Moustacchi, E. and Enteric, S. (1970). *Mol. Gen. Genet.* **109**, 69.

Moustacchi, E. and Williamson, D. H. (1966). *Biochem. biophys. Res. Commun.* **23**, 56.

Mühlpfordt, H. (1963a). *Z. Tropenmed. Parasit.* **14**, 357.

Mühlpfordt, H. (1963b). *Z. Tropenmed. Parasit.* **14**, 475.

Muldoon, J. J., Evans, T. E., Nygaard, P. F. and Evans, H. H. (1971). *Biochim. biophys. Acta* **247**, 310.

Müller, M. (1969). *J. Protozool*, **16**, 428.

Müller, M. (1972). *J. Cell Biol.* **52**, 478.

Müller, M. (1973). *J. Cell Biol.* **57**, 453.

Müller, M. and Møller, K. M. (1969). *Eur. J. Biochem.* **9**, 424.

Müller, M., Baudhuin, P. and deDuve, C. (1966). *J. Cell Physiol.* **68**, 165.

Müller, M., Hogg, J. F. and deDuve, C. (1968). *J. biol. Chem.* **234**, 5385.

Munkres, K. D. and Woodward, D. O. (1966). *Proc. natn. Acad. Sci. U.S.A.* **55**, 1217.

Munkres, K. D., Benveniste, K., Gorski, J. and Zuiches, C. A. (1970). *Proc. natn. Acad. Sci. U.S.A.* **67**, 263.

Munkres, K. D., Swank, R. T. and Sheir, G. S. (1971). *In* "Autonomy and Biogenesis of Mitochondria and Chloroplasts", (N. K. Boardman, A. W. Linnane and R. M. Smillie, eds.), pp. 152–161. North Holland, Amsterdam.

Murray, D. R. and Linnane, A. W. (1972). *Biochem. biophys. Res. Commun.* **49**, 855.

Murray, R. G. E. (1962). *In* "Microbial Classification", (G. C. Ainsworth and P. H. A. Sneath, eds.), p. 119. Cambridge University Press, Cambridge.

Muthukrishnan, S., Padmanaban, G. and Sarma, P. S. (1969). *J. biol. Chem.* **244**, 4241.

Muthukrishnan, S., Malathi, K. and Padmanaban, G. (1972). *Biochem. J.* **129**, 31.

Myer, H. and Olivera-Musacchio, M. De. (1960). *J. Protozool.* **7**, 124.

Nagai, S. (1969). *Mutat. Res.* **8**, 557.

Nagai, S., Yanagishima, Y. and Nagai, H. (1961). *Bact. Rev.* **25**, 404.

Nagao, M. and Sugimura, T. (1965). *Biochim. biophys. Acta* **103**, 353.

Nagata, T. (1972). *Histochemie* **32**, 163.

Nagley, P. and Linnane, A. W. (1970). *Biochem. biophys. Res. Commun.* **39**, 989.

Nagley, P. and Linnane, A. W. (1972a). *J. molec. Biol.* **66**, 181.

Nagley, P. and Linnane, A. W. (1972b). *Cell Differentiation* **1**, 143.

Nagley, P., Gingold, E. B., Lukins, H. B. and Linnane, A. W. (1973). *J. molec. Biol.* **78**, 335.

Nagley, P., Gingold, E. B. and Linnane, A. W. (1974). *In* "Biogenesis of Mitochondria", (A. M. Kroon and C. Saccone, eds.), pp. 157–168. Academic Press, London and New York.

Nakamura, M. and Ohnishi, S. (1972). *Biochem. biophys. Res. Commun.* **46**, 926.

Nass, M. M. K. (1967). *In* "Organizational Biosynthesis", (H. J. Vogel, J. O. Lampen and V. Bryson, eds.), p. 503. Academic Press, New York.

Nass, M. M. K. (1969a). *J. molec. Biol.* **42**, 529.

Nass, M. M. K. (1969b). *Science, N.Y.* **165**, 25.

Nass, M. M. K. (1970). *Proc. natn. Acad. Sci. U.S.A.* **67**, 1926.

Nass, M. M. K. and Buck, C. A. (1970). *J. molec. Biol.* **54**, 187.

Nass, M. M. K. and Nass, S. (1963a). *J. Cell Biol.* **19**, 593.

Nass, M. M. K., Nass, S. and Afzelius, B. A. (1965). *Expl. Cell Res.* **37**, 516.

Nass, S. (1969). *Int. Rev. Cytol.* **25**, 55.

Nass, S. and Nass, M. M. K. (1963b). *J. Cell Biol.* **19**, 613.

Nath, V. and Dutta, G. P. (1962). *Int. Rev. Cytol.* **13**, 323.

Nathans, D. and Lipmann, F. (1961). *Proc. natn. Acad. Sci. U.S.A.* **47**, 497.

Neal, W. K., Hoffmann, H.-P., Avers, C. J. and Price, C. A. (1970). *Biochem. biophys. Res. Commun.* **38**, 414.

Neal, W. K., Hoffman, H.-P. and Price, C. A. (1971). *Pl. Cell Physiol., Tokyo* **12**, 181.

Neff, R. J., Neff, R. H. and Taylor, R. E. (1958). *Physiol. Zool.* **31**, 73.

Negrotti, T. and Wilkie, D. (1968). *Biochim. biophys. Acta* **153**, 341.

Neiderpruem, D. J. and Hackett, D. P. (1961). *Pl. Physiol., Lancaster* **36**, 79.

Neilands, J. B. (1952). *J. biol. Chem.* **197**, 701.

Neubert, D. and Helge, H. (1965). *Biochem. biophys. Res. Commun.* **18**, 600.

Neubert, D., Oberdisse, E. and Bass, R. (1968). *In* "Biochemical Aspects of the Biogenesis of Mitochondria", (E. C. Slater, J. M. Tager, S. Papa and E. Quagliariello, eds.), p. 103. Adriatica Editrice, Bari.

Neupert, W. and Ludwig, S. D. (1971). *Eur. J. Biochem.* **19**, 523.

Neupert, W., Brdiczka, D. and Bücher, Th. (1967). *Biochem. biophys. Res. Commun.* **27**, 488.

Neupert, W., Sebald, W., Schwab, A. J., Pfaller, A. and Bücher, Th. (1969a). *Eur. J. Biochem.* **10**, 585.

Neupert, W., Sebald, W., Schwab, A. J. Pfaller, A., Massinger, P. and Bücher, Th. (1969b). *Eur. J. Biochem.* **10**, 589.

Neupert, W., Massinger, P. and Pfaller, A. (1971). *In* "Autonomy and Biogenesis of Mitochondria and Chloroplasts", (N. K. Boardman, A. W. Linnane and R. M. Smillie, eds.), pp. 328–338. North Holland, Amsterdam.

Newton, B. A. (1967). *J. gen. Microbiol.* **48**, iv.

Newton, B. A. (1968). *A. Rev. Microbiol.* **22**, 109.

Newton, B. A. and Burnett, J. K. (1972). *In* "Comparative Physiology of Parasites", (H. Van Den Bossche, ed.), pp. 127–138. Academic Press, New York and London.

Newton, B. A., Steinert, M. and Borst, P. (1973a). *Trans. R. Soc. trop. Med. Hyg.* **87**, 259.

Newton, B. A., Cross, G. A. M. and Baker, J. R. (1973b). *Symp. Soc. gen. Microbiol.* **23**, 339.

Nicklowitz, W. (1957). *Expl. Cell Res.* **13**, 591.

Nilsson, J. (1970). *C.r. Trav. Lab. Carlsberg* **38**, 87.

Ninnemann, H. (1972). *FEBS Lett.* **24**, 181.

Ninomiya, H. and Suzuoki-Z. (1952). *J. Biochem.*, *Tokyo* **39**, 321.

Nishi, A. and Scherbaum, O. H. (1962). *Biochim. biophys. Acta* **65**, 419.

Noll, H. (1970). *Symp. Soc. exp. Biol.* **24**, 419.

Nordstrom, K. (1967). *J. gen. Microbiol.* **48**, 277.

Nosoh, Y. and Takamiya, A. (1962). *Pl. Cell Physiol.*, *Tokyo* **3**, 53.

Novick, R. P. (1969). *Bact. Rev.* **33**, 210.

Novikoff, A. B. (1961). *In* "The Cell", (J. Brachet, and A. E. Mirsky, eds.), Vol. 2, pp. 299–421, Academic Press, New York.

Novikoff, A. B., Podber, E., Ryan, J. and Noe, E. (1953). *J. Histochem. Cytochem.* **1**, 27.

Nyns, E. J. and Hamaide, M. C. (1972). *Archs int. Physiol. Biochim.* **80**, 978.

O'Brien, T. W., Denslow, N. D. and Martin, G. R. (1974). *In* "Biogenesis of Mitochondria", (A. M. Kroon and C. Saccone, eds.), pp. 347–356. Academic Press, London and New York.

Ogur, M., Coker, L. and Ogur, S. (1964). *Biochem. biophys. Res. Commun.* **14**, 193.

Ogur, M., Roshaumanesh, A. and Ogur, S. (1965). *Science, N.Y.* **147**, 1590.

Ohaniance, L. and Chaix, P. (1966). *Biochim. biophys. Acta* **128**, 288.

Ohnishi, T. (1970). *Biochem. biophys. Res. Commun.* **41**, 344.

Ohnishi, T. (1972). *FEBS Lett.* **24**, 305.

Ohnishi, T., Kawaguchi, K. and Hagihara, B. (1966a). *J. biol. Chem.* **241**, 1797.

Ohnishi, T., Sottocasa, G. and Ernster, L. (1966b). *Bull. Soc. Chim. biol.* **48**, 1189.

Ohnishi, T., Kröger, A., Heldt, H. W., Pfaff, E. and Klingenberg, M. (1967). *Eur. J. Biochem.* **1**, 301.

Ohnishi, T., Schleyer, H. and Chance, B. (1969). *Biochem. biophys. Res. Commun.* **36**, 487.

Ohnishi, T., Asakura, T., Wohlrab, H., Yonetani, T. and Chance, B. (1970). *J. biol. Chem.* **245**, 901.

Ohnishi, T., Asakura, T., Yonetani, T. and Chance, B. (1971). *J. biol. Chem.* **246**, 5960.

Ohnishi, T., Asakura, T., Wilson, D. F. and Chance, B. (1972a). *FEBS Lett.* **21**, 59.

Ohnishi, T., Panebianco, P. and Chance, B. (1972b). *Biochem. biophys. Res. Commun.* **49**, 99.

Ohta, J. (1954). *J. Biochem., Tokyo* **41**, 489.

Okunuki, K. (1961). *Adv. Enzymol.* **23**, 29.

Oltmanns, O. (1971). *Mol. Gen. Genet.* **111**. 300.

Oltmanns, O. and Lingens, F. (1967). *Z. Naturf.* B **22**, 751.

Omura, T. and Sato, R. (1964). *J. biol. Chem.* **239**, 2370.

Orme-Johnson, N. R., Orme-Johnson, W. H., Hansen, R. E., Beinert, H. and Hatefi, Y. (1971). *Biochem. biophys. Res. Commun.* **44**, 446.

Osafune, T. (1973). *J. Electron Microsc.* **22**, 51.

Osafune, T., Mihara, S., Hase, E. and Ohkuro, I. (1972a). *Pl. Cell Physiol., Tokyo* **13**, 211.

Osafune, T., Mihara, S., Hase, E. and Ohkuro, I. (1972b). *Pl. Cell Physiol., Tokyo* **13**, 981.

Oshino, R., Oshino, N. and Chance, B. (1971). *FEBS Lett.* **19**, 96.

Oshino, R., Asakura, T., Tamura, M., Oshino, N. and Chance, B. (1972a). *Biochem. biophys. Res. Commun.* **46**, 1055.

Oshino, R., Oshino, N., Tamura, M., Kobilinsky, L. and Chance, B. (1972b). *Biochim. biophys. Acta* **273**, 5.

Oshino, R., Oshino, N., Chance, B. and Hagihara, B. (1973). *Eur. J. Biochem.* **35**, 23.

O'Sullivan, J. and Casselton, P. J. (1973). *J. gen. Microbiol.* **75**, 333.

Osumi, M. (1965). *Bot. Mag., Tokyo* **78**, 231.

Osumi, M. (1969). *Jap. Wom. Univ. J.* **16**, 69.

Osumi, M. and Katoh, T. (1967). *Jap. Wom. Univ. J.* **14**, 67.

Osumi, M. and Sando, N. (1969). *J. Electron Microsc.* **18**, 47.

Osumi, M. and Kitsutani, S. (1971). *J. Electron Miscrosc.* **20**, 23.

Osumi, M. and Ubukata, K. (1970). *Jap. Wom. Univ. J.* **17**, 44.

Osumi, M., Sando, N. and Miyake, S. (1966). *Jap.Wom.Univ.J.* **13**, 70.

Osumi, M., Ichinokawa, K. and Hirosawa, T. (1971). *Jap. Wom. Univ. J.* **18**, 65.

Pace, D. M. (1945). *Biol. Bull. mar. biol. Lab., Woods Hole* **89**, 76.

Packer, L. (1970). *Fedn. Proc. Fedn Am. Socs exp. Biol.* **29**, 1533.

Packer, L. (1972). *Bioenergetics* **3**, 115.

Packer, L., Williams, M. A. and Criddle, R. S. (1973). *Biochim. biophys. Acta,* **292**, 92.

Packter, N. M. and Glover, J. (1960). *Nature, Lond.* **187**, 413.

Padilla, G. M., Cameron, I. L. and Elrod, L. H. (1966). *In* "Cell Synchrony" (I. L. Cameron and G. M. Padilla, eds.), p. 269. Academic Press, New York and London.

Pajot, P. and Claisse, M. (1973). Abstracts of Communications presented at the 9th International Congress of Biochemistry at Stockholm. Abstract 416.

Palade, G. E. (1953). *J. Histochem. Cytochem.* **1**, 188.

Paléus, S. and Tuppy, H. (1961). *In* "Haematin Enzymes", (J. E. Falk, R. Lemberg and R. K. Morton eds.), pp. 362–369. Pergamon Press, New York.

Paltauf, F. and Johnston, J. M. (1970). *Biochim. biophys. Acta* **218**, 424.

Paltauf, F. and Schatz, G. (1969). *Biochemistry, N.Y.* **8**, 335.

Paoletti, C., Couder, H. and Guerineau, M. (1972). *Biochem. biophys. Res. Commun.* **48**, 950.

Pappas, G. D. (1959). *Ann. N.Y. Acad. Sci.* **78**, 448.

Pappas, G. D. and Brandt, P. W. (1959). *J. biophys. biochem. Cytol.* **6**, 85.

Parisi, B. and Cella, R. (1971). *FEBS Lett.* **14**, 209.

Parker, J. H. and Mattoon, J. R. (1968). *Genetics* **60**, 210.

Parker, J. H. and Mattoon, J. R. (1969). *J. Bact.* **100**, 647.

Parker, J. H. and Sherman, F. (1969). *Genetics* **62**, 9.

Parker, J. H., Trimble, I. R. Jr. and Mattoon, J. R. (1968). *Biochem. biophys. Res. Commun.* **33**, 590.

Parks, L. W. and Starr, P. R. (1963). *J. cell. comp. Physiol.* **61**, 61.

Parson, W. W. and Rudney, H. (1966). *Biochemistry, N.Y.* **5**, 1013.

Parsons, D. F. (1967). *Meth. Enzym.* **10**, 655.

Parsons, D. F. and Williams, G. R. (1967). *Meth. Enzym.* **10**, 443.

Parsons, D. F., Williams, G. R. and Chance, B. (1966). *Ann. N.Y. Acad. Sci.* **137**, 643.

Parsons, J. A. (1965). *J. Cell Biol.* **25**, 641.

Parsons, J. A. and Rustad, R. C. (1968). *J. Cell Biol.* **37**, 683.

Penman, S. (1970). *Cold Spring Harb. Symp. quant. Biol.* **35**, 561.

Perini, F., Schiff, J. A. and Kamen, M. D. (1964). *Biochim. biophys. Acta* **88**, 91

Perasso, R. (1973). *Expl. Cell. Res.* **81**, 15

Perasso, R. and Adoutte, A. (1974). *J. Cell Sci.* **14**, 475.

Perkins, M., Haslam, J. M. and Linnane, A. W. (1972). *FEBS Lett.* **25**, 271

Perkins, M., Haslam, J. M. and Linnane, A. W. (1973). *Biochem. J.* **134**, 923.

Perlish, J. S. and Eichel, H. J. (1968). *J. Protozool.* **15** (Suppl.), 14.

Perlish, J. S. and Eichel, H. J. (1969). *J. Protozool.* **16** (Suppl.), 12.

Perlish, J. S. and Eichel, H. J. (1971). *Biochem. biophys. Res. Commun.* **44**, 973.

Perlman, P. S. and Mahler, H. R. (1970a). *Archs Biochem. Biophys* **136**, 245.

Perlman, P. S. and Mahler, H. R. (1970b). *Bioenergetics* **1**, 113.

Perlman, P. S. and Mahler, H. R. (1971a). *Nature New Biology* **231**, 12.

Perlman, P. S. and Mahler, H. R. (1971b). *Biochem. biophys. Res. Commun.* **44**, 261.

Perlman, R. and Pastan, I. (1968a). *Biochem. biophys. Res. Commun.* **30**, 656.

Perlman, R. and Pastan, I. (1968b). *J. biol. Chem.* **243**, 5420.

Perlman, S. and Penman, S. (1970). *Biochem. biophys. Res. Commun.* **40**, 941.

Pette, D. (1966). *In* "Regulation of Metabolic Processes in Mitochondria", (J. M. Tager, S. Papa, E. Quagliariello and E. C. Slater, eds.), p. 28. Elsevier, Amsterdam.

Pettigrew, G. W. (1972). *FEBS Lett.* **22**, 64.

Pettigrew, G. and Meyer, T. (1971). *Biochem. J.* **125**, 46P.

Pickett, J. M. (1967). Carnegi Institute Year Book, p. 197, Washington.

Pickett-Heaps, J. D. (1971). *Planta* **100**, 357.

Pihl, E. and Bahr, G. F. (1970). *Expl. Cell Res.* **63**, 391.

Pinto da Costa, S. O. and Bacila, M. (1973). *J. Bact.* **115**, 461.

Piperno, G. Fonty, G. and Bernardi, G. (1972). *J. molec. Biol.* **65**, 191.

Pirani, A., Tiboni, O. and Cifferi, O. (1971). *J. molec. Biol.* **55**, 107.

Pitelka, D. R. (1963). "Electron Microscopic Structure of Protozoa", MacMillan, New York.

Pittenger, T. H. (1956). *Proc. Natn. Acad. Sci. U.S.A.* **42**, 747.

Plattner, H. and Schatz, G. (1969). *Biochemistry, N.Y.* **8**, 339.

Plattner, H., Salpeter, M. M., Saltzgaber, J. and Schatz, G. (1970). *Proc. Natn. Acad. Sci. U.S.A.* **66**, 1252.

Plattner, H., Salpeter, M., Saltzgaber, J., Rouslin, W. and Schatz, G. (1971). *In* "Autonomy and Biogenesis of Mitochondria and Chloroplasts", (N. K. Boardman, A. W. Linnane, and R. M. Smillie, eds.) pp. 175–184. North Holland, Amsterdam.

Plesnicar, M., Bonner, W. D. and Storey, B. T. (1967). *Pl. Physiol.* **42**, 366.

Plummer, D. T., Murray, D. R. and Linnane, A. W. (1974). *Biochem. Soc. Trans.* **2**, 221.

Pollack, J. K. and Munn, E. A. (1970). *Biochem. J.* **117**, 913.

Polakis, E. S. and Bartley, W. (1965). *Biochem. J.* **97**, 284.

Polakis, E. S., Bartley, W. and Meek, G. A. (1964). *Biochem. J.* **90**, 369.

Polakis, E. S., Bartley, W. and Meek, G. A. (1965). *Biochem. J.* **97**, 298.

Ponta, H., Ponta, U. and Wintersberger, E. (1971). *FEBS Lett.* **18**, 204.

Poole, R. K. and Lloyd, D. (1972). *Arch. Microbiol.* **88**, 257.

Poole, R. K. and Lloyd, D. (1973). *Biochem. J.* **136**, 195.

Poole, R. K. and Lloyd, D. (1974). *Biochem. J.* **144**, 141.

Poole, R. K., Lloyd, D. and Kemp, R. B. (1973). *J. gen. Microbiol.* **77**, 209.

Poole, R. K., Lloyd, D. and Chance, B. (1974). *Biochem. J.* **138**, 201.

Poole, R. K., Nicholl, W. G., Howells, L. and Lloyd, D. (1971a). *J. gen. Microbiol.* **68**, 283.

Poole, R. K., Nicholl, W. G., Turner, G., Roach, G. I. and Lloyd, D. (1971b). *J. gen. Microbiol.* **67**, 161.

Porra, R. J., Irving, E. A. and Tennick, A. M. (1972a). *Archs Biochem. Biophys* **149**, 563.

Porra, R. J., Barnes, R. and Jones, O. T. G. (1972b). *Hoppe-Seyler's Z. physiol. Chem.* **353**, 1365.

Portier, P. (1918). "Les Symbiotes". Masson, Paris.

Prescott, D. M., Bollum, F. J. and Kluss, B. C. (1962). *J. Cell Biol.* **13**, 172.

Pretlow, T. P. and Sherman, F. (1967). *Biochim. biophys. Acta* **148**, 629.

Proudlock, J. W., Haslam, J. M. and Linnane, A. W. (1969). *Biochem. biophys. Res. Commun.* **37**, 847.

Proudlock, J. W., Haslam, J. M. and Linnane, A. W. (1971). *Bioenergetics* **2**, 327.

Puglisi, P. P. and Algeri, A. A. (1971). *Mol. Gen. Genet.* **110**, 110–117.

Puglisi, P. P. and Algeri, A. A. (1974). *In* "Biogenesis of Mitochondria", (A. M. Kroon and C. Saccone, eds.), pp. 169–176. Academic Press, London and New York.

Puglisi, P. P. and Cremona, T. (1970). *Biochem. biophys. Res. Commun.* **39**, 461.

Puglisi, P. P. and Zennaro, E. (1971). *Experientia* **27**, 963.

Pullman, M. E. and Monroy, G. C. (1963). *J. biol. Chem.* **238**, 3762.

Rabinowitz, M. and Plaut, W. (1962). *J. Cell Biol.* **15**, 535.

Rabinowitz, M. and Swift, H. (1970). *Physiol. Rev.* **50**, 376.

Rabinowitz, M., Getz, G. S., Casey, J. and Swift, H. (1969). *J. molec. Biol.* **41**, 381.

Rabinowitz, M., Casey, J., Gordon, P., Locker, J., Hsu, H. and Getz, S. (1974). *In* "Biogenesis of Mitochondria". (A. M. Kroon and C. Saccone, eds.), pp. 89–106. Academic Press, London and New York.

Racker, E. and Kandrach, A. (1971). *J. biol. Chem.* **246**, 7069.

Raff, R. A. and Mahler, H. R. (1972). *Science, N.Y.* **177**, 575.

Raff, R. A. and Mahler, H. R. (1973). *Science, N.Y.* **180**, 516.

Ragan, C. I. and Garland, P. B. (1969). *Eur. J. Biochem.* **10**, 399.

Ragan, C. I. and Garland, P. B. (1971). *Biochem. J.* **124**, 171.

Ragan, C. I. and Racker, E. (1973). *J. biol. Chem.* **248**, 2563.

Raison, J. K. and Smillie, R. M. (1969). *Biochim. biophys. Acta* **180**, 500.

Rank, G. H. (1970a). *Can. J. Genet. Cytol.* **12**, 129.

Rank G. H. (1970b). *Can. J. Genet. Cytol.* **12**, 340.

Rank, G. H. (1973). *Heredity, Lond.* **30**, 265.

Rank, G. H. and Bech-Hansen, N. T. (1972a). *Can. J. Microbiol.* **18**, 1.

Rank, G. H. and Bech-Hansen, N. T. (1972b). *Genetics* **72**, 1.
Rank, G. H. and Pearson, C. (1969). *Can. J. Genet. Cytol.* **11**, 716.
Rao, S. S. and Grollman, A. P. (1967). *Biochem. biophys. Res. Commun.* **29**, 696.
Raut, C. (1953). *Expl. Cell Res.* **4**, 295.
Raven, P. H. (1970). *Science, N.Y.* **169**, 641.
Ray, D. S. and Hanawalt, P. C. (1964). *J. molec. Biol.* **9**, 812.
Ray, D. S. and Hanawalt, P. C. (1965). *J. molec. Biol.* **11**, 760.
Ray, H. N. and Malhotra, M. N. (1960). *Nature, Lond.* **188**, 870.
Ray, S. K. and Cross, G. A. M. (1972). *Nature New Biology*, **237**, 174.
Read, C. P. and Rothman, A. H. (1955). *Am. J. Hyg.* **61**, 249.
Reich, E. and Luck, D. J. L. (1966). *Proc. natn. Acad. Sci. U.S.A.* **55**, 1600.
Reich, K. (1955). *Physiol. Zool.* **28**, 145.
Reijnders, L. and Borst, P. (1972). *Biochem. biophys. Res. Commun.* **47**, 126.
Reijnders, L., Kleisen, C. M., Grivell, L. A. and Borst, P. (1972). *Biochim. biophys. Acta* **272**, 396.
Reijnders, L., Sloof, P. and Borst, P. (1973a). *Eur. J. Biochem.* **35**, 266.
Reijnders, L., Sloof, P., Sival, J. and Borst, P. (1973b). *Biochim. biophys. Acta* **324**, 320.
Reilly, C. and Sherman, F. (1965). *Biochim. biophys. Acta* **95**, 640.
Rendi, R. and Ochoa, S. (1962). *J. biol. Chem.* **237**, 3707.
Renger, H. C. and Wolstenholme, D. R. (1970). *J. Cell Biol.* **47**, 689.
Renger, H. C. and Wolstenholme, D. R. (1971). *J. Cell Biol.* **50**, 533.
Renger, H. C. and Wolstenholme, D. R. (1972). *J. Cell Biol.* **54**, 346.
Resnick, M. A. and Mortimer, R. K. (1966). *J. Bact.* **92**, 597.
Reusch, Jr. V. M. and Burger, M. M. (1973). *Biochim. biophys. Acta* **300**, 79.
Rhodes, P. M. and Wilkie, D. (1973). *Biochem. Pharmac.* **22**, 1047.
Richards, O. C., Ryan, R. S. and Manning, J. E. (1971). *Biochim. biophys. Acta* **238**, 190.
Richmond, M. H. (1970). *Symp. Soc. gen. Microbiol.* **20**, 249.
Richter, D. (1971). *Biochemistry, N.Y.* **10**, 4422.
Richter, D. (1973). *J. Bact.* **115**, 52.
Richter, D. and Lipmann, F. (1970). *Biochemistry, N.Y.* **9**, 5065.
Richter, D., Hameister, H., Petersen, K. G. and Klink, F. (1968). *Biochemistry, N.Y.* **7**, 3753.
Richter, D., Herrlich, P. and Schweiger, M. (1972). *Nature New Biology* **238**, 74.
Rickard, P. A. D., Moss, F. J., Phillips, D. and Mok, T. C. K. (1971). *Biotech. Bioengng.* **13**, 169.
Ridley, S. M. and Leach, R. M. (1970). *Nature, Lond.* **227**, 463.
Rifkin, M. R. and Luck, D. J. L. (1971). *Proc. natn. Acad. Sci. U.S.A.* **68**, 287.
Rifkin, M. R., Wood, D. L. and Luck, D. J. L. (1967). *Proc. natn. Acad. Sci. U.S.A.* **58**, 1025.
Ringrose, P. S. and Lambert, R. W. (1973). *Biochim. biophys. Acta* **299**, 374.
Riou, G. C. R. (1967). *C.r. hebd. Séanc Acad. Sci., Paris Ser. D.* **265**, 2004.
Riou, G. and Delain, E. (1969a). *Proc. natn. Acad. Sci. U.S.A.* **62**, 210.
Riou, G. and Delain, E. (1969b). *Proc. natn. Acad. Sci. U.S.A.* **64**, 618.
Riou, G. and Paoletti, C. (1967). *J. molec. Biol.* **27**, 377.
Riou, G. and Pautrizel, R. (1967). *C.r. hebd. Séanc. Acad. Sci., Paris*, Ser. D. **265**, 61.
Riou, G. and Pautrizel, R. (1969). *J. Protozool.* **16**, 509.
Riou, G., Pautrizel, R. and Paoletti, C. (1966). *C.r. hebd. Séanc. Acad. Sci., Paris*, Ser. D. **262**, 2376.
Riou, G., Brack, C., Festy, B. and Delain, E. (1970). *J. Parasitol.* **56**, 464.

Rippa, M. (1961). *Archs Biochem. Biophys* **94**, 333.

Ris, H. (1961). *Can. J. Cytol.* **3**, 95.

Ritchie, D and Hazeltine, P. (1953). *Expl. Cell Res.* **17**, 58.

Robberson, D., Aloni, Y., Attardi, G. and Davidson, N. (1971). *J. molec. Biol.* **60**, 473.

Roberts, C. T. Jr. and Orias, E. (1973). *Genetics* **73**, 259.

Robertson, J. D. (1960). *Prog. Biophys. biophys. Chem.* **10**, 343.

Robertson, J. D. (1964). *In* "Cellular Membranes in Development", (Lock, M. ed.), pp. 1–82. Academic Press, New York.

Rogers, P. J. and Stewart, P. R. (1973a). *J. Bact.* **115**, 88.

Rogers, P. J. and Stewart, P. R. (1973b). *J. Gen. Microbiol.* **79**, 205.

Rogers, P. J. Preston, B. N., Titchener, E. B. and Linnane, A. W. (1967). *Biochem. biophys. Res. Commun.* **27**, 405.

Rohatgi, K. and Krawiec, S. (1973). *J. Protozool.* **20**, 425.

Roman, H. (1956). *C.r. Trav. Lab. Carlsberg, Ser. Physiol.* **26**, 299.

Roodyn, D. B. (1962). *Biochem. J.* **85**, 177.

Roodyn, D. B. (1966). *In* "Regulation of Metabolic Processes in Mitochondria", (J. M. Tager, J. M. Papa, E. Quadliariello, and E. C. Slater, eds.), p. 562. Elsevier, Amsterdam.

Roodyn, D. B. and Wilkie, D. (1967). *Biochem. J.* **103**, 3c.

Roodyn, D. B. and Wilkie, D. (1968). "Biogenesis of Mitochondria". Methuen, London.

Roodyn, D. B., Suttie, J. W. and Work, T. S. (1961). *Biochem. J.* **83**, 29.

Ross, E., Mason, T. L., Poyton, R. O. and Schatz, S. (1974). *In* "Biogenesis of Mitochondria", (A. M. Kroon and C. Saccone, eds.), Academic Press, London and New York.

Rossi, M. and Woodward, D. O. (1973). (In Press.)

Rouiller, C. H. (1960). *Int. Rev. Cytol.* **9**, 227.

Rouslin, W. and Schatz, G. (1969). *Biochem. biophys. Res. Commun.* **37**, 1002.

Rousseau, P. and Halvorson, H. O. (1973). *Can. J. Microbiol.* **19**, 547.

Rowe, M. J., Lansman, R. A. and Woodward, D. O. (1974). *Eur. J. Biochem.* **41**, 25.

Rowlands, R. T. and Turner, G. (1973). *Mol. Gen. Genet.* **126**, 201.

Rowlands, R. T. and Turner, G. (1974a). *Biochem. Soc. Trans.* **2**, 230.

Rowlands, R. T. and Turner, G. (1974b). *Mol. Gen. Genet.* In Press.

Rubin, M. S. and Tzagoloff, A. (1973a). *J. biol. Chem.* **248**, 4269.

Rubin, M. S. and Tzagoloff, A. (1973b). *J. biol. Chem.* **248**, 4279.

Rudzinska, M. A. and Trager, W. (1959). *J. biophys. biochem. Cytol.* **6**, 103.

Ryley, J. F. (1951). *Biochem. J.* **49**, 577.

Ryley, J. F. (1952). *Biochem. J.* **52**, 483.

Ryley, J. F. (1955). *Biochem. J.* **59**, 353.

Ryley, J. F. (1956). *Biochem. J.* **62**. 215.

Ryley, J. F. (1962). *Biochem. J.* **85**, 211.

Sagan, L. (1967). *J. theor. Biol.* **14**, 225.

Sager, R. (1972). "Cytoplasmic Genes and Organelles". Academic Press, New York.

Sager, R. and Palade, G. E. (1957). *J. biophys. biochem. Cytol.* **3**, 463.

Sager, R. and Ramanis, Z. (1964). *Genetics* **50**, 282.

Sager, R. and Ramanis, Z. (1965). *Proc. natn. Acad. Sci. U.S.A.* **53**, 1053.

Sager, R. and Ramanis, Z. (1970). *Proc. natn. Acad. Sci. U.S.A.* **65**, 593.

Sala, F. and Küntzel, H. (1970). *Eur. J. Biochem.* **15**, 280.

Sanders, H. K., Mied, P. A., Briquet, M., Hernandez-Rodriguez, J., Gottal, R. F. and Mattoon, J. R. (1973). *J. molec. Biol.* **80**, 17.

Sano, S. and Granick, S. (1969). *J. biol. Chem.* **236**, 1173.

Sargent, D. F. and Taylor, C. P. S. (1972). *Pl. Physiol., Lancaster* **49**, 775.

Sato, H. (1960). *Anat. Rec.* **138**, 381.

Sato, N., Ohnishi, T. and Chance, B. (1972). *Biochim. biophys. Acta* **275**, 288.

Sato, T. and Tamiya, H. (1937), *Cytologia, Tokyo, Fujii jub.* 1133.

Sauer, H. W. (1973). *Symp. Soc. gen. Microbiol.* **23**, 375.

Saunders, G. W., Gingold, E. B., Trembath, M. K., Lukins, H. B. and Linnane, A. W. (1971). *In* "Autonomy and Biogenesis of Mitochondria and Chloroplasts", (N. K. Boardman and A. W. Linnane and R. M. Smillie eds.), pp. 185–193. Amsterdam, North Holland.

Saunders, J. P. M., Flavell, R. A., Borst, P. and Mol, J. N. M. (1973). *Biochim. biophys. Acta* **312**, 441.

Schäfer, K. P. and Küntzel, H. (1972). *Biochem. biophys. Res. Commun.* **46**, 1312.

Schäfer, K. P., Bugge, G., Grandi, M. and Küntzel, H. (1971). *Eur. J. Biochem.* **212**, 478.

Schatz, G. (1965). *Biochem. biophys. Res. Commun.* **12**, 448.

Schatz, G. (1967). *Meth. Enzym.* **10**, 197.

Schatz, G. (1968). *J. biol. Chem.* **243**, 2192.

Schatz, G. and Racker, E. (1966). *Biochem. biophys. Res. Commun.* **22**, 579.

Schatz, G. and Saltzgaber, J. (1969a). *Biochem. biophys. Res. Commun.* **37**, 996.

Schatz, G. and Saltzgaber, J. (1969b). *Biochim. biophys. Acta* **180**, 186.

Schatz, G., Haslbrunner, E. and Tuppy, H. (1964). *Biochem. biophys. Res. Commun.* **15**, 127.

Schatz, G., Racker, E., Tyler, D. D., Gonze, J. and Estabrook, R. W. (1966). *Biochem. biophys. Res. Commun.* **22**, 585.

Schatz, G., Penefsky, H. S. and Racker, E. (1967). *J. biol. Chem.* **242**, 2552.

Schatz, G., Groot, G. S. P., Mason, T., Rouslin, W., Wharton, D. C. and Saltzgaber, J. (1972). *Fedn Proc. Fedn Am. Socs exp. Biol.* **31**, 21.

Schildkraut, C. L., Marmur, J. and Doty, P. (1962a). *J. molec. Biol.* **4**, 430.

Schildkraut, C. L., Mandel, M., Levisohn, S., Smith-Sonneborn, J. E. and Marmur, J. (1962b). *Nature, Lond.* **196**, 795.

Schimmer, O. and Arnold, C. G. (1970a). *Mol. Gen. Genet.* **107**, 281.

Schimmer, O. and Arnold, C. G. (1970b). *Mol. Gen. Genet.* **107**, 366.

Schimmer, O. and Arnold, C. G. (1970c). *Mol. Gen. Genet.* **108**, 33.

Schmitt, H. (1969). *FEBS Lett.* **4**, 234.

Schmitt, H. (1970). *Eur. J. Biochem.* **17**, 278.

Schmitt, H. (1971). *FEBS Lett.* **15**, 186.

Schmitt-Verhulst, A. M., Bex, F. and Sels, A. A. (1973). *Eur. J. Biochem.* **36**, 185.

Schnaitman, C. and Greenawalt, J. W. (1968). *J. Cell Biol.* **38**, 158.

Schneider, W. C. (1963). *J. biol. Chem.* **238**, 3572.

Schneider, W. C. and Hogeboom, G. H. (1951). *Cancer Res.* **11**, 1.

Schnepf, E. and Brown, Jr. R. M. (1972). *In* "Results and Problems in Cell Differentiation", (J. Reinert, and H. Urspring eds.), Vol. 2, pp. 299–322. Springer-Verlag, Berlin.

Schonbaum, G. R., Bonner, W. D., Storey, B. T. and Bahr, J. T. (1971). *Pl. Physiol., Lancaster* **47**, 124.

Schopf, J. W. (1967). *In* "McGraw-Hill Year Book of Science and Technology". p. 47.

Schori, L., Ben-Shaul, Y. and Edelman, M. (1970). *Israel J. Chem.* **8**, p. 117

Schötz, F., Bathelt, H., Arnold, C-G and Schimmer, O. (1972). *Protoplasma* **75**, 229.

T

Schuel, H., Berger, E. R., Wilson, J. R. and Schuel, R. (1969). *J. Cell Biol.* **43**, 125a.

Schuit, K. E. and Balzer, R. H. Jr. (1971). *Expl. Cell Res.* **65**, 408.

Schuster, F. L. (1963). *J. Protozool.* **10**, 313.

Schuster, F. L. (1965). *Expl. Cell Res.* **39**, 329.

Schutgens, R. B. H. (1971). Abstracts of Communications presented at the 7th Meeting of the Federation of European Biochemical Societies at Varna. Abstract 442.

Schutgens, R. B. H., Reijnders, L., Hoekstra, S. P. and Borst, P. (1973). *Biochim. biophys. Acta* **308**, 372.

Schuurmans-Steckhoven F. M. A. H. (1966a). *Archs Biochem. Biophys* **115**, 555.

Schuurmans-Steckhoven F. M. A. H. (1966b). *Archs Biochem. Biophys* **115**, 569.

Schwab, A. J. (1973). *FEBS Lett.* **35**, 63.

Schwab, A. J. (1974). *In* "Biogenesis of Mitochondria", (A. M. Kroon and C. Saccone, eds.), pp. 591–594. Academic Press, London and New York.

Schwab, A. J., Sebald, W. and Weiss, H. (1972). *Eur. J. Biochem* **30**, 511.

Schwab, D. and Schwab-Stey, H. (1972). *Z. Naturf.* **27b**, 1571.

Schwab, R., Sebald, W. and Kaudewitz, F. (1971). *Mol. Gen. Genet.* **110**, 361.

Schwab-Stey, H, Schwab, D. and Krebs, W. (1971). *J. Ultrastruct. Res.* **37**, 82.

Schwaier, R., Nashed, N. and Zimmerman, F. K. (1968). *Mol. Gen. Genet.* **102**, 290.

Schweitzer, E. and Bolling, H. (1970). *Proc. natn. Acad. Sci. U.S.A.* **67**, 660.

Schweitzer, and Castorph, H. (1971). *Hoppe-Seyler's Z. physiol. Chem.* **382**, 377.

Schweyen, R. and Kaudewitz, F. (1970). *Biochem. biophys. Res. Commun.* **38**, 728.

Schweyen, R. J. and Kaudewitz, F. (1971). *Biochem. biophys. Res. Commun.* **44**, 1351.

Scopes, A. W. and Williamson, D. H. (1964). *Expl. Cell. Res.* **35**, 361.

Scott, W. A. and Mitchell, H. K. (1969). *Biochemistry, N.Y.* **8**, 4282.

Scragg, A. H. (1971). *Biochem. biophys. Res. Commun.* **45**, 701.

Scragg, A. H. (1974). *In* "Biogenesis of Mitochondria", (A. M. Kroon and C. Saccone, eds.), pp. 47–58. Academic Press, London and New York.

Scragg, A. H., Morimoto, H., Villa, V. Nekhorocheff, J. and Halvorson, H. O. (1971), *Science, N.Y.* **171**, 908.

Sebald, W. (1974). *In* "Biogenesis of Mitochondria", (A. M. Kroon and C. Saccone, eds.), Academic Press, London and New York.

Sebald, W., Bücher, Th., Olbrich, B. and Kaudewitz, F. (1968). *FEBS Lett.* **1**, 235.

Sebald, W., Schwab, A. J. and Bücher, Th. (1969). *FEBS Lett.* **4**, 243.

Sebald, W., Birkmayer, G. D., Schwab, A. J. and Weiss, H. (1971). *In* "Autonomy and Biogenesis of Mitochondria and Chloroplasts," (N. K. Boardman, A. W. Linnane and R. M. Smillie, eds,), pp. 339–345. North Holland, Amsterdam.

Sebald, W., Weiss, H. and Jackl, G. (1972). *Bur. J. Biochem.* **30**, 413.

Sebald, W., Machleidt, W. and Otto, J. (1973). *Eur. J. Biochem.* **38**, 311.

Sedar, A. W. and Porter, K. R. (1955). *J. biophys. biochem. Cytol.* **1**, 583.

Sedar, A. W. and Rudzinska, M. A. (1956). *J. biophys. biochem. Cytol.* **2** (Suppl.). 331.

Sekuzu, I., Mizushima, H. and Okunuki, K. (1964). *Biochim. biophys. Acta* **85**, 516.

Sekuzu, I., Mizushima, H., Hirota, S., Yubisui, T., Matsumara, M. and Okunuki, K. (1967). *J. Biochem., Tokyo* **62**, 110.

Sels, A. A. (1969). *Biochem. biophys. Res. Commun.* **34**, 740.

Sels, A. A. and Cocriamount (1968). *Biochem. biophys. Res. Commun.* **32**, 192.

Sels, A. and Jakob, H. (1967). *Biochem. biophys. Res. Commun.* **28**, 453.

Sels, A. A. and Verhulst, A-M. (1971). *Eur. J. Biochem.* **19**, 115–123.

Sels, A. A. and Verhulst, A-M. (1972). *Archs. int. Physiol. Biochim.* **80**, 205.

Senior, A. E. and McLennan, D. H. (1970). *J. biol. Chem.* **245**, 5086.

Shakespeare, P. G. and Mahler, H. R. (1971). *J. biol. Chem.* **246**, 7649.

Shakespeare, P. G. and Mahler, H. R. (1972). *Archs. Biochem. Biophys.* **151**, 496.

Shannon, C., Rao, A., Douglass, S. and Criddle, R. S. (1972). *Supramolec. Structure* **1**, 145.

Shannon, C., Enns, R., Wheelis, L., Burchiel, K. and Criddle, R. S. (1973). *J. biol. Chem.* **248**, 3004.

Sharp, C. W., Mackler, B., Douglas, H. C., Palmer, G. and Felton, S. P. (1967). *Archs Biochem. Biophys* **122**, 810.

Sharpless, T. K. and Butow, R. A. (1970a). *J. biol. Chem.* **245**, 50.

Sharpless, T. K. and Butow, R. A. (1970b). *J. biol. Chem.* **245**, 58.

Sheldon, R., Jurale, C. and Kates, J. (1972), *Proc. natn. Acad. Sci., U.S.A.* **69**, 417.

Shephard, E. H. and Hübscher, G. (1969). *Biochem. J.* **113**, 429.

Sherald, J. L. and Sisler, H. D. (1970). *Pl. Physiol., Lancaster* **46**, 180.

Sherald, J. L. and Sisler, H. D. (1972). *Pl. Cell. Physiol.* **13**, 1039.

Sherman, F. (1959). *J. cell. comp. Physiol.* **54**, 37.

Sherman, F. (1963). *Genetics* **48**, 375.

Sherman, F. (1964). *Genetics* **49**, 39.

Sherman, F. (1965). *In* "Mécanisms de Regulation des Activitiés Cellularies chez les Microorganisms", p. 465. C.N.R.S. Paris.

Sherman, F. (1967). *Meth. Enzym.* **10**, 610.

Sherman, F. and Ephrussi, B. (1962). *Genetics* **47**, 695.

Sherman, F. and Slonimski, P. O. (1964). *Biochim. biophys. Acta* **90**, 1.

Sherman, F. and Stewart, J. W. (1971). *A. Rev. Genet.* **5**, 257.

Sherman, F. and Stewart, J. W. (1973). *In* "Biochemistry of Gene Expression in Higher Organisms", (J. K. Pollak and J. W. Lee, eds.), pp. 56–86. Australian and New Zealand Book Co. Ltd., Artarmon, N.S.W.

Sherman, F., Tabor, H. and Campbell, W. (1965). *J. molec. Biol.* **13**, 21.

Sherman, F., Stewart, J., Margoliash, E., Parker, J. and Campbell, W. (1966). *Proc. Natn. Acad. Sci. U.S.A.* **55**, 1498.

Sherman, F., Stewart, J. W., Parker, J. H., Inhaber, E., Shipman, N. A., Putterman, G. J., Gardisky, R. L. and Margoliash, E. (1968). *J. biol. Chem.* **243**, 5446.

Sherman, F., Stewart, J. W., Parker, J. H., Putterman, G. J., Agrawel, B. B. L. and Margoliash, E. (1970). *Symp. Soc. exp. Biol.* **24**, 85.

Sherman, F., Liebman, S. W., Stewart, J. W. and Jackson, M. (1973). *J. molec Biol.* **78**, 157.

Shimizu, I., Nagai, J., Hatanaka, H., Saito, E. and Katsuki, H. (1971). *J. Biochem., Tokyo* **70**, 175.

Shimizu, I., Nagai, J., Hatanaka, H. and Katsuki, H. (1973). *Biochim. biophys. Acta* **296**, 310.

Shinagawa, Y., Inouye, A., Ohnishi, T. and Hagihara, B. (1966). *Expl. Cell. Res.* **43**, 301.

Shrago, E., Shug, A. L. and Ferguson, S. M. (1971). *Int. J. Biochem.* **2**, 312.

Siegel, M. R. and Sisler, H. D. (1965). *Biochim. biophys. Acta* **103**, 558.

Sierra, M. F. and Tzagoloff, A. (1973). *Proc. natn. Acad. Sci. U.S.A.* **70**, 3155.

Simonson, D. H. and van Wagendonk, W. J. (1952). *Biochem. biophys. Acta* **9**, 515.

Simpson, L. (1972). *Int. Rev. Cytol.* **32**, 139.

Simpson, L. (1973). *J. Protozool.* **20**, 2.

Simpson, L. and Braly, P. (1970). *J. Protozool.* **17**, 511.

Simpson, L. and daSilva, A. (1971). *J. molec. Biol.* **56**, 443.

Sinohara, H. (1973). *Biochim. biophys. Acta* **299**, 662.

Sjostrand, F. S. and Barajas, L. (1968). *J. Ultrastruct. Res.* **25**, 121.

Sjostrand, F. S. and Barajas, L. (1970). *J. Ultrastruct. Res.* **32**, 293.

Skipton, M. D., Watson, K., Houghton, R. L. and Griffiths, D. E. (1973). *Biochem. Soc. Trans.* **1**, 1107.

Slonimski, P. (1949). *Annls. Inst. Pasteur, Paris* **76**, 510.

Slonimski, P. (1952). Resherches sur la formation des enzymes respiratoires chez la levure. Thesis, Faculté des Sciences Paris.

Slonimski, P. P. (1953a). *Actual. biochim.* No. 17.

Slonimski, P. P. (1953b). *Symp. Soc. gen. Microbiol.* **3**, 76.

Slonimski, P. P. (1956). Abstracts of Communications presented at the 3rd International Congress of Biochemistry at Paris, Abstract 242.

Slonimski, P. P. and Ephrussi, B. (1949). *Annals. Inst. Pasteur, Paris* **77**, 47.

Slonimski, P. P. and Hirsch, H. M. (1952). *C.r. hebd. Séanc. Acad. Sci., Paris*, Ser. D. **235**, 741.

Slonimski, P. P., Archer, R., Péré, G., Sels, A. and Somlo, M. (1965). *In* "Mécanismes de Regulation des Activites Cellulaires chez les Microorganisms", p. 435. Centre National de la Recherche Scientifique, Paris.

Slonimski, P. P., Perrodin, G. and Croft, J. H. (1968). *Biochem. biophys. Res. Commun.* **30**, 232.

Smith, D., Tauro, P., Schweizer, E. and Halvorson, H. O. (1968). *Proc. natn. Acad. Sci. U.S.A.* **60**, 936.

Smith, D. G. and Marchant, R. (1968). *Arch. Mikrobiol* **60**, 262.

Smith, D. G., Marchant, R., Maroudas, N. G. and Wilkie, D. (1969). *J. gen. Microbiol.* **56**, 47.

Smith, D. G., Wilkie, D. and Srivastava, K. C. (1972). *Microbios* **6**, 231.

Smith, J. D., Abelson, J. N., Clark, B. F. C., Goodman, H. M. and Brenner, S. (1966). *Cold Spring Harb. Symp. quant. Biol.* **31**, 479.

Smith, M. H. (1963). *Biochim. biophys. Acta* **71**, 370.

Smith, M. H., George, P. and Preer, J. R. (1962). *Archs. Biochem. Biophys.* **99**, 313.

Smith, R. W. and Koffler, H. (1971). *Adv. Microbiol. Physiol.* **6**, 219.

Smith-Johannsen, H. and Gibbs, S. P. (1972). *J. Cell. Biol.* **52**, 598.

So, A. G. and Davie, E. W. (1963). *Biochemistry, N.Y.* **2**, 132.

Sober, H. A. (1970). "Handbook of Biochemistry including selected Data for Molecular Biology," Chemical Rubber Co. Cleveland, Ohio.

Somlo, M. (1965). *Biochim. biophys. Acta* **97**, 183.

Somlo, M. (1966). *Bull. Soc. Chim. biol.* **48**, 247.

Somlo, M. (1968). *Eur. J. Biochem.* **5**, 276.

Somlo, M. (1970). *Archs. Biochem. Biophys.* **136**, 122.

Somlo, M. (1971). *Biochimie* **53**, 819.

Somlo, M. and Fukuhara, H. (1965). *Biochem. biophys. Res. Commun.* **19**, 587.

Sonneborn, T. M. (1950). *J. exp. Zool.* **113**, 87.

Sonneborn, T. M. (1970). *In* "Methods in Cell Physiology", (D. Prescott, ed.), Vol. 4, p. 241. Academic Press, London and New York.

Sonnenshein, G. E. and Holt, C. E. (1968). *Biochem. biophys. Res. Commun.* **33**, 361.

Soto, E. F., Pasquini, J. M., Placido, R. and La Torre, J. L. (1969), *J. Chromat.* **41**, 400.

Sottocasa, G. L., Kuylenstierna, B., Ernster, L. and Bergstrand, A. (1967). *J. Cell. Biol.* **32**, 415.

South, D. J. and Mahler, H. R. (1968). *Nature, Lond.* **218**, 1226.

Spenser, C., Symons, S. A. and Brunt, R. V. (1971). *Arch. Mikrobiol.* **75**, 246.

Speziali, G. A. G. and van Wijk, R. (1971). *Biochim. biophys. Acta* **235**, 466.

Srb, A. M. (1963). *Symp. Soc. exp. Biol.* **17**, 175.

Sribney, M. (1968). *Archs Biochem. Biophys* **126**, 954.

Srivastava, H. K. (1971). *FEBS Lett.* **16**, 189.

Srivastava, H. K. and Bowman, I. B. R. (1972). *Comp. Biochem. Physiol.* **40**, 973.

Stahn, R., Maier, K-P. and Hannig, K. (1970). *J. Cell. Biol.* **46**, 576.

Stanier, R. Y. (1961). *Anals. Inst. Pasteur, Paris* **101**, 297.

Stanier, R. Y. (1970). *Symp. Soc. gen. Microbiol.* **20**, 1.

Stanier, R. Y. and van Neil, C. B. (1962). *Arch. Mikrobiol.* **42**, 17.

Stanier, R. Y., Adelberg, E. and Douderoff, M. (1963). "The Microbiol World". Prentice-Hall, Englewood Cliffs, N.J.

Starr, P. R. and Parks, L. W. (1962). *J. Bact.* **83**, 1042.

Stegeman, W. J., Cooper, C. S. and Avers, C. J. (1970). *Biochem. biophys. Res. Commun.* **39**, 69.

Steinert, M. (1969). *FEBS Lett.* **5**, 291.

Steinert, M. and Steinert, G. (1962). *J. Protozool.* **9**, 203.

Steinert, M. and Van Assel, S. (1967a). *J. Cell. Biol.* **34**, 489.

Steinert, M. and Van Assel, S. (1967b). *Archs int. Physiol. Biochim.* **75**, 370.

Steinert, M., Van Assel, S., Borst, P., Mol, J. N. M., Kleisen, C. M. and Newton, B. A. (1973). *Expl. Cell. Res.* **76**, 175.

Steinschneider, A. (1969). *Biochim. biophys. Acta* **186**, 405.

Stekhoven, F., Waitkus, R. F. and van Moerker, Th. B. (1972). *Biochemistry, N.Y.* **11**, 1144.

Stevens, B. J., Curgy, J. J., Ledoigt, G. and André, J. (1974). *In* "Biogenesis of Mitochondria", (A. M. Kroon and C. Saccone, eds.), pp. 327–336. Academic Press, London and New York.

Stevenson, J., Hayward, P. J., Hemming, F. W. and Morton, R. A. (1962). *Nature, Lond.* **196**, 1291.

Stewart, J. W. and Sherman, F. (1973). *J. molec. Biol.* **78**, 169.

Stewart, J. W., Margoliash, E. and Sherman, F. (1966). *Fedn. Proc. Fedn. Am. Socs. exp. Biol.* **25**, 2587.

Stewart, J. W., Sherman, F., Shipman, N. H. and Jackson, M. (1971). *J. biol. Chem.* **246**, 7429.

Stich, W. and Eisgruber, H. (1951). *Hoppe-Seyler's Z. physiol. Chem.* **287**, 19.

Stockem, W. (1968). *Histochemie* **15**, 160.

Stoekenius, W. (1963). *J. Cell. Biol.* **17**, 443.

Stone, G. E., Miller, O. L. Jr. and Prescott, D. M. (1964). *J. Protozool.* **11**, (Suppl.) 24.

Storck, R. and Morrill, R. C. (1971). *Biochem. Genet.* **5**, 467.

Storrie, B. and Attardi, G. (1973). *J. Cell Biol.* **56**, 833.

Strauss, P. R. (1972). *J. Cell. Biol.* **53**, 312.

Stuart, K. D. (1970). *Biochem. biophys. Res. Commun.* **39**, 1045.

Stuchell, R. N., Weinstein, B. I. and Beattie, D. S. (1973). *FEBS Lett.* **37**, 23.

Subík, J. and Kolarov, J. (1970). *Folia microbiol., Praha* **15**, 448.

Subík, J., Kužela, S., Kolarov, J., Kováč, L. and Lachowicz, T. M. (1970). *Biochim. biophys. Acta* **205**, 513.

Subík, J., Kolarov, J. and Kováč, L. (1972a). *Biochem. biophys. Res. Commun.* **49**, 192.

Subík, J., Kováč, L. and Kolarov, J. (1972b). *Biochim. biophys. Acta* **283**, 146.

Sugimura, T. and Rudney, H. (1960). *Biochim. biophys. Acta* **37**, 560.

Sugimura, T., Okabe, K. and Rudney, H. (1964). *Biochim. biophys. Acta* **82**, 350.

Sugimura, T., Okabe, K., Nagao, M. and Gunge, N. (1966a). *Biochim. biophys. Acta* **115**, 267.

Sugimura, T., Okabe, K. and Imamura, A. (1966b). *Nature, Lond.* **212**, 304.

Sussman, A. S. (1965). *In* "The Fungi", (G. C. Ainsworth and A. S. Sussman, eds.), p. 733. Academic Press, New York.

Sussman, R. and Rayner, E. P. (1971). *Archs Biochem. Biophys.* **144** 127.

Suyama, Y. (1966). *Biochemistry, N.Y.* **5**, 2214.

Suyama, Y. (1967). *Biochemistry, N.Y.* **6**, 2829.

Suyama, Y. (1969). *In* "Atti del Seminario di Studi Biologici", (Quagliariello, E. eds.), Vol. 4, p. 83. Adriatica Edtrice, Bari, Italy.

Suyama, Y. and Eyer, J. (1967). *Biochem. biophys. Res. Commun.* **28**, 746.

Suyama, Y. and Eyer, J. (1968). *J. biol. Chem.* **243**, 320.

Suyama, Y. and Miura, K. (1968). *Proc. natn. Acad. Sci. U.S.A.* **60**, 235.

Suyama, Y. and Preer, J. R. Jr. (1965). *Genetics* **52**, 1051.

Suzuki, K. (1969). *Science, N.Y.* **163**, 81.

Suzuoki-Z. and Suzuoki-T. (1951a). *J. Biochem., Tokyo* **38**, 237.

Suzuoki-Z. and Suzuoki-T. (1951b). *Nature, Lond.* **168**, 610.

Swank, R. T., Sheir, G. I. and Munkres, K. D. (1971). *Biochemistry, N.Y.* **10**, 3924.

Swank, R. T., Sheir, G. I. and Munkres, K. D. *Biochemistry, N.Y.* **10**, 3931.

Swanljung, P., Swanljung, H. and Partis, M. (1972). *Biochem. J.* **128**, 479.

Swanson, R. F. (1971). *Nature, Lond.* **231**, 31.

Swick, R. W., Stange, J. L., Nance, S. L. and Thomson, J. F. (1967). *Biochemistry, N.Y.* **6**, 737.

Swift, H., Rabinowitz, M. and Getz, G. (1967). *J. Cell Biol.* **35**, 131A.

Swift, H., Rabinowitz, M. and Getz, G. (1968). *In* "Biochemical Aspects of the Biogenesis of Mitochondria", (E. G. Slater, J. M. Tager, S. Papa and E. Quagliariello, eds.), p. 3. Adriatic Editrice, Bari.

Syrett, P. J. (1951). *Ann. Bot.* **15**, 473.

Tabak, H. H. and Cooke, W. B. (1968). *Bot. Rev.* **34**, 124.

Taeter, H. (1972). Ph.D. Thesis, University of Louvain.

Tait, A. (1968). *Nature, Lond.* **219**, 941.

Tait, A. (1970a). *Biochem. Genet.* **4**, 461.

Tait, A. (1970b). *Nature, Lond.* **225**, 181.

Tait, A. (1972). *FEBS Lett.* **24**, 117.

Taketomi, T. (1961). *Z. allg. Mikrobiol.* **1**, 331.

Takeuchi, I. (1960). *Devl. Biol.* **2**, 343.

Talbert, D. M. and Sorokin, C. (1971). *Arch. Mikrobiol.* **78**, 281.

Tamiya, H. (1928). *Acta phytochim, Tokyo* **4**, 77.

Tandler, B., Erlandson, R. A., Smith, A. L. and Wynder, E. L. (1969). *J. Cell Biol.* **41**, 477.

Tang, I. S., French, C. S. (1933). *Chin. J. Physiol.* **1**, 353.

Tavlitzki, J. (1949). *Annls. Inst. Pasteur, Paris* **76**, 497.

Tempest, D. W. (1970). *In* "Advances in Microbial Physiology", (A. H. Rose and J. F. Wilkinson eds.), Vol. 4 pp. 223–249. Academic Press, London and New York.

Terenzi, H. F. and Storck, R. (1969a). *J. Bact.* **97**, 1248.

Terenzi, H. F. and Storck, R. (1969b). *Mycologia*, **61**, 894.

Ter Schegget, J. and Borst, P. (1971a). *Biochim. biophys. Acta* **246**, 239.

Ter Schegget, J. and Borst, P. (1971b). *Biochim. biophys. Acta* **246**, 249.

Tewari, J. R., Tu, J. C. and Malhotra, S. K. (1972). *Cytobios* **5**, 261.

Tewari, K. K., Jayaraman, J. and Mahler, H. R. (1965). *Biochem. biophys. Res, Commun.* **23**, 56.

Tewari, K. K., Votsch, W. and Mahler, H. R. (1966a). *J. molec. Biol.* **20**, 453.

Tewari, K. K., Votsch, W., Mahler, H. R. and Mackler, B. (1966b). *J. molec. Biol* **20**, 453.

Thirion, J. and Laub, R. (1972a). *Archs. int. Physiol. Biochim.* **80**, 199.

Thirion, J. and Laub, R. (1972b). *Archs. int. Physiol Biochim.* **80**, 280

Thomas, D. Y. and Scragg, A. H. (1973). *Eur. J. Biochem.* **37**, 585.

Thomas, D. Y. and Wilkie, D. (1968a). *Biochem. biophys. Res. Commun.* **30**, 368.

Thomas, D. Y. and Wilkie, D. (1968b). *Genet. Res.* (Cambridge). **11**, 33.

Thomas, D. Y. and Williamson, D. H. (1971). *Nature New Biology* **233**, 196.

Thompson, E. D., Starr, P. R. and Parks, L. W. (1971). *Biochem. biophys. Res. Commun.* **43**, 1304.

Threlfall, D. R. and Goodwin, T. W. (1964). *Biochem. J.* **90**, 40p.

Threlfall, D. R., Williams, B. L. and Goodwin, T. W. (1965). *In* "Progress in Protozoology", p. 141. Exerpta Medica Foundation, Amsterdam.

Thyagarajan, T. R., Conti, S. F. and Naylor, H. B. (1961). *Expl. Cell Res.* **25**, 216.

Tibbs, J. and Marshall, B. J. (1969). *Biochim. biophys. Acta* **172**, 382.

Tiboni, O., Parisi, B., Perani, A. and Cifferi, O. (1970). *J. molec. Biol.* **47**, 467.

Tissières, A. and Mitchell, H. K. (1954). *J. biol. Chem.* **208**, 241.

Tissières, A., Mitchell, H. K. and Haskins, F. A. (1953). *J. biol. Chem.* **205**, 423.

Toner, J. J. and Weber, M. M. (1967). *Biochem. biophys. Res. Commun.* **28**, 821.

Toner, J. J. and Weber, M. M. (1972). *Biochem. biophys. Res. Commun.* **46**, 652.

Torch, R. (1955). *J. Protozool.* **2**, 167.

Tottmar, S. O. C. and Ragan, C. I. (1971). *Biochem. J.* **124**, 853.

Towers, N. R., Dixon, H., Kellerman, G. M. and Linnane, A. W. (1972). *Archs Biochem. Biophys* **151**, 361.

Towers, N. R., Kellerman, S. M. and Linnane, A. W. (1973a). *Archs Biochem. Biophys* **155**, 159.

Towers, N. R., Kellerman, G. M., Raison, J. K. and Linnane, A. W. (1973b). *Biochim. biophys. Acta* **299**, 153.

Trager, N. and Rudzinska, M. A. (1964). *J. Protozool.* **11**, 133.

Trembath, M. K., Bunn, C. L., Lukins, H. B. and Linnane, A. W. (1973). *Mol. Gen. Genet.* **121**, 35.

Tripodi, G., Pizzolongo, P. and Giannattasio, M. (1972). *J. Cell Biol.* **55**, 530.

Tsai, M., Michaelis, G. and Criddle, R. S. (1971). *Proc. natn. Acad. Sci. U.S.A.* **68**, 473.

Tsekos, I. (1972). *Arch. Mikrobiol.* **85**, 138.

Tuppy, H. and Birkmayer, G. D. (1969). *Eur. J. Biochem.* **8**, 237.

Tuppy, H. and Swetly, P. (1968). *Biochim. biophys. Acta* **153**, 293.

Tuppy, H. and Wildner, G. (1965). *Biochem. biophys. Res. Commun.* **20**, 733.

Tuppy, H., Swetly, P. and Wolff, I. (1968). *Eur. J. Biochem.* **5**, 339.

Turian, G. (1962). *Pathologia Microbiol.* **23**, 687.

Turner, G. (1969). Ph.D. Thesis, University of Wales.

Turner, G. (1973). *Eur. J. Biochem.* **40**, 201.

Turner, G. and Lloyd, D. (1971). *J. gen. Microbiol.* **67**, 175.

Turner, G., Lloyd, D. and Chance, B. (1969). *Biochem. J.* **114**, 91p.

Turner, G., Lloyd, D. and Chance, B. (1971). *J. gen. Microbiol.* **65**, 359.

Tustanoff, E. R. and Ainsworth, P. J. (1973). Abstracts of Communications presented at the 9th International Congress of Biochemistry at Stockholm. Abstract 4i7.

Tustanoff, E. R. and Bartley, W. (1964a). *Biochem. J.* **91**, 595.

Tustanoff, E. R. and Bartley, W. (1964b). *Can. J. Biochem.* **42**, 651.

Tyler, D. D. (1973). Abstracts of Communications presented at the 9th International Congress of Biochemistry at Stockholm. Abstract 4e28.

Tzagoloff, A. (1969a). *J. biol. Chem.* **244**, 5020.

Tzagoloff, A. (1969b). *J. biol. Chem.* **244**, 5027.

Tzagoloff, A. (1970). *J. biol. Chem.* **245**, 1545.

Tzagoloff, A. (1971). *J. biol. Chem.* **246**, 3050.

Tzagoloff, A. (1972). *Bioenergetics* **3**, 39.

Tzagoloff, A. and Akai, A. (1972). *J. biol. Chem.* **247**, 6517.

Tzagoloff, A. and Meagher, P. (1971). *J. biol. Chem.* **246**, 7328.

Tzagoloff, A. and Meagher, P. (1972). *J. biol. Chem.* **247**, 594.

Tzagoloff, A., Akai, A. and Sierra, M. F. (1972). *J. biol. Chem.* **247**, 6511.

Tzagoloff, A., Rubin, M. S. and Sierra, M. F. (1973). *Biochim. biophys. Acta* **401**, 71.

Tzagoloff, A., Akai, A. and Rubin, M. S. (1974). *In* "Biogenesis of Mitochondria", (A. M. Kroon and C. Saccone, eds.), pp. 405–422. Academic Press, London and New York.

Uchida, A. and Suda, K. (1973). *Mutation Res.* **19**, 57.

Ulrich, J. T. and Mathre, D. E. (1972). *J. Bact.* **110**, 628.

Unestam, T. and Gleason, F. H. (1968). *Physiologia Pl.* **21**, 573.

Utter, M. F., Keech, D. B. and Nossal, P. M. (1958). *Biochem. J.* **68**, 431.

Utter, M. F., Duell, E. A. and Bernofsky, C. (1968). *In* "Aspects of Yeast Metabolism", (A. K. Mills, ed.), p. 197. Blackwell, Oxford.

Uzzell, T. and Spolsky, C. (1973). *Science, N.Y.* **180**, 516.

Vakirtzi-Lemonias, C., Kidder, G. W. and Dewey, V. C. (1963). *Comp. Biochem. Physiol.* **8**, 331.

Van Assel, S. and Steinert, M. (1971). *Expl. Cell Res.* **65**, 353.

Van Bruggen, E. F. J., Borst, P., Ruttenberg, G. J. C. M., Gruber, M. and Kroon, A. M. (1968). *Biochem. biophys. Acta* **161**, 402.

Van De Vijver, G. (1966). *Enzymologia* **31**, 382.

Van Etten, J., Parisi, B. and Cifferi, O. (1966). *Nature, Lond.* **212**, 932.

Van Kreije, C. F., Borst, P., Flavell, R. and Hollenberg, C. P. (1972). *Biochim. biophys. Acta* **277**, 61.

Van Schijndel, B. C., Reitsema, A. and Scherphof, G. L. (1974). *In* "Biogenesis of Mitochondria", (A. M. Kroon and C. Saccone, eds.), pp. 541–544. Academic Press, London and New York.

Van Wijk, R. and Konijn, J. A. (1971). *FEBS Lett.* **13**, 184.

Vary, M. J., Edwards, C. L. and Stewart, P. R. (1969). *Archs Biochem. Biophys* **130**, 235.

Vary, M. J., Stewart, P. R. and Linnane, A. W. (1970). *Archs Biochem. Biophys* **141**, 430.

Verdière, J. and Lederer, F. (1971). *FEBS Lett.* **19**, 72.

Verma, I. M., Edelman, M., Hertzberg, M. and Littauer, U. Z. (1970). *J. molec. Biol.* **52**, 137.

Verma, I. M., Edelman, M. and Littauer, U. Z. (1971a). *Eur. J. Biochem.* **19**, 124

Verma, I. M., Kay, C. M. and Littauer, U. Z. (1971b). *FEBS Lett.* **12**, 317.

Vesco, C. and Penman, S. (1969). *Proc. natn. Acad. Sci., U.S.A.* **62**, 218.

Vickerman, K. (1960). *Nature, Lond.* **188**, 249.

Vickerman, K. (1965). *Nature, Lond.* **208**, 762.

Vickerman, K. (1971). *In* "Ecology and Physiology of Parasites", (A. M. Fallis, ed.) pp. 58–91. University of Toronto Press, Toronto and Buffalo.

Vidová, M. and Kováč, L. (1972). *FEBS Lett.* **22**, 347.

Vignais, P. M., Vignais, P. V. and Lehninger, A. L. (1963). *J. biol. Chem.* **239**, 2011.

Vignais, P. V. and Huet, J. (1970). *Biochem. J.* **116**, 26P.

Vignais, P. V., Huet, J. and André, J. (1969). *FEBS Lett.* **3**, 177.

Vignais, P. V., Huet, J. and André, J. (1970. *Biochem. J.* **116**, 26p–27p.

Vignais, P. V., Stevens, B. J., Huet, J. and André, J. (1972). *J. Cell Biol.* **54**, 468.

Vignais, P. V., Huet, J. and de Jerphanion, M. B. (1973). Abstracts of Communications presented at the 9th International Congress of Biochemistry at Stockholm. Abstract 5m8.

Villa, V. D. and Stork, R. (1968). *J. Bact.* **96**, 184.

Villanueva, J. R. and Garcia Acha, I. (1971). *Meth. Microbiol.* **4**, 665.

Vitols, E. V. and Linnane, A. W. (1961). *J. biophys. biochem. Cytol.* **9**, 701.

Vitols, E. V., North, R. J. and Linnane, A. W. (1961). *J. biophys. Biochem. Cytol.* **9**, 689.

Vogel, F. S. and Kemper, L. (1965). *Lab. Invest.* **14**, 1868.

Volkova, T. M. and Meissel, M. N. (1967). *Mikrobiologiya* **36**, 656.

Volland, C. and Chaix, P. (1970). *Bull. Soc. Chim. biol.* **52**, 581.

von Brandt, T. (1951). *In* "Biochemistry and Physiology of Protozoa", (A. L. Woff, ed.), Vol. I, pp. 178–234. Academic Press, New York.

von Dach, H. (1942). *Biol. Bull. mar. biol. Lab., Woods Hole* **82**, 356.

von Jagow, G. and Klingenberg, M. (1970). *Eur. J. Biochem.* **12**, 583.

von Jagow, G. and Klingenberg, M. (1972). *FEBS Lett.* **24**, 278.

von Jagow, G., Weiss, H. and Klingenberg, M. (1973). *Eur. J. Biochem.* **33**, 140.

von Meyenburg, H. K. (1969). *Arch. Mikrobiol.* **66**, 289.

Wagner, R. P. and Bergquist, A. (1963). *Proc. natn. Acad. Sci. U.S.A.* **49**, 892.

Wagner, R. P., Somers, C. E. and Bergquist, A. (1960). *Proc. natn. Acad. Sci. U.S.A.* **46**, 708.

Wagner, R. P., Bergquist, A. Brotzman, B., Eakin, E. A., Clark, C. H. and LePage, R. N. (1967). *In* "Organizational Biosynthesis", (H. J. Vogel, J. O. Lampen and V. Bryson, eds.). p. 267. Academic Press, New York.

Wainio, W. W. (1970). "The Mammalian Mitochondrial Respiratory Chain", p. 313. Academic Press, New York and London.

Wakabayashi, K. (1972). *Antibiot., Tokyo* **25**, 475.

Wakabayashi, K. and Gunge, N. (1970). *FEBS Lett.* **6**, 302.

Wakabayashi, K. and Kamei, S. (1973a). Abstracts of Communications presented at the 9th International Congress of Biochemistry at Stockholm. Abstract 5m 11.

Wakabayashi, K. and Kamei, S. (1973b). *FEBS Lett.,* **33**, 263.

Wakiyama, S. and Ogura, Y. (1972). *J. Biochem., Tokyo* **71**, 295.

Waldron, C. (1973). *Heredity, Lond.* **31**, 135.

Waldron, C. and Roberts, C. F. (1973). *J. gen. Microbiol.* **78**, 379.

Wallin, J. E. (1927). "Symbioticism and the Origin of Species."

Wallace, P. G. and Linnane, A. W. (1964). *Nature, Lond.* **201**, 1191.

Wallace, P. G., Huang, M. and Linnane, A. W. (1968). *J. Cell Biol.* **37**, 207.

Wallis, O. C., Ottolenghi, P. and Whittaker, P. A. (1972). *Biochem. J.* **127**, 46p.

Warburg, O. (1919). *Biochem. Z.* **100**, 230.

Warburg, O. and Hass, E. (1934). *Naturwissenschaften* **22**, 207.

Warburg, O. and Negelein (1929). *Biochem. Z.* **214**, 64.

Ward, J. M. and Nickerson, W. J. (1958). *J. gen. Physiol.* **41**, 703.

Ward, K. A., Marzuki, S. and Haslam, J. M. (1973). *In* "Biochemistry of Gene Expression in Higher Organisms", (J. K. Pollak and J. W. Lee, eds.), pp. 105–116. Australin and New Zealand Book Co. Ltd., Artarmon, N.S.W.

Waring, M. J. (1965). *J. molec. Biol.* **13**, 269.

Waring, M. J. (1968). *Nature, Lond.* **219**, 1320.

Watrud, L. S. and Ellingboe, A. H. (1973a). *J. Bact.* **115**, 1151.

Watrud, L. S. and Ellingboe, A. H. (1973b). *J. Cell Biol.* **59**, 127.

Watson, K. (1972). *J. Cell Biol.* **55**, 721.

Watson, K. and Linnane, A. W. (1972). *J. Bioenergetics* **3**, 235.
Watson, K. and Smith, J. E. (1967a). *J. Biochem., Tokyo* **61**, 527.
Watson, K. and Smith, J. E. (1967b). *Biochem. J.* **104**, 332.
Watson, K. and Smith, J. E. (1968). *J. Bact.* **96**, 1546.
Watson, K., Paton, W. and Smith, J. E. (1969). *Can. J. Microbiol.* **15**, 975.
Watson, K., Haslam, J. M. and Linnane, A. W. (1970). *J. Cell Biol.* **46**, 88.
Watson, K., Haslam, J. M., Veitch, B. and Linnane, A. W. (1971). *In* "Autonomy and Biogenesis of Mitochondria and Chloroplasts", (N. K. Boardman, N. R. A. W. Linnane and R. M. Smillie, eds.), pp. 162–174. North Holland, Amsterdam.
Watson, K., Bertoli, E. and Griffiths, D. E. (1973a). *FEBS Lett.* **30**, 120.
Watson, K., Bertoli, E. and Griffiths, D. E. (1973b). *Biochem. Soc. Trans.* **1**, 1129.
Wattiaux, R. (1974). *In* "Methodological Developments in Biochemistry", Vol. 4, (E. Reid, ed.). Longman, London.
Wattiaux, R. and Wattiaux-DeConinck, G. (1970). *Biochem. biophys. Res. Commun.* **40**, 1185.
Wattiaux, R., Wattiaux-DeConinck, G. and Ronveaux-Dupal, M-F. (1971). *Eur. J. Biochem.* **22**, 31.
Webster, D. A. and Hackett, D. P. (1965). *Pl. Physiol., Lancaster* **40**, 1091.
Webster, D. A. and Hackett, D. P. (1966). *Pl. Physiol., Lancaster* **41**, 599.
Weeks, C. O. and Gross, S. R. (1971). *Biochem. Genet.* **5**, 505.
Weinbach, E. C. and Garbus, J. (1966). *J. biol. Chem.* **241**, 169.
Weinbach, E. C. and Garbus, J. (1969). *Nature, Lond.* **221**, 1016.
Weiner, E. R. and Ashworth, J. M. (1970). *Biochem. J.* **118**, 505.
Weislogel, P. O. and Butow, R. A. (1970). *Proc. natn. Acad. Sci. U.S.A.* **67**, 52.
Weislogel, P. O. and Butow, R. A. (1971). *J. biol. Chem.* **246**, 5113.
Weiss, B. (1965). *J. gen. Microbiol.* **39**, 85.
Weiss, H. (1972). *Eur. J. Biochem.* **30**, 469.
Weiss, H. (1974). *Biochem. Soc. Trans.* **2**, 185.
Weiss, H. and Ziganke, B. (1974a). *Eur. J. Biochem.* **41**, 63.
Weiss, H. and Ziganke, B. (1974b). *In* "Biogenesis of Mitochondria", (A. M. Kroon and C. Saccone, eds.), pp. 491–500. Academic Press, London and New York.
Weiss, H., von Jagow, G., Klingenberg, M. and Bücher, Th. (1970). *Eur. J. Biochem.* **14** 75.
Weiss, H., Sebald, W. and Bücher, Th. (1971). *Eur. J. Biochem.* **22**, 19.
Weiss, H., Sebald, W., Schwab, A. J., Kleinow, W. and Lorenz, B. (1973). *Biochimie* **55**, 815.
Welsiger, R. A. and Fridowich, I. (1973a). *J. biol. Chem.* **248**, 3582.
Welsiger, R. A. and Fridowich, I. (1973b). *J. biol. Chem.* **248**, 4793.
Werry, P. A. Th. J. and Wanka, F. (1972). *Biochim. biophys. Acta* **287**, 232.
Werzbitski, P. W. (1910). *Centralbl. Bakt. I. Abt. Orig.* **53**, 303.
Wesley, R. D. and Simpson, L. (1973a). *Biochim. biophys. Acta* **319**, 237.
Wesley, R. D. and Simpson, L. (1973b). *Biochim. biophys. Acta* **219**, 254.
Wesley, R. D. and Simpson, L. (1973c). *Biochim. biophys. Acta* **319**, 267.
West, D. J. and Woodward, D. O. (1973). *J. Bact.* **113**, 637.
Westergaard, O. (1970). *Biochim. biophys. Acta* **213**, 36.
Westergaard, O. and Pearlman, R. E. (1969). *Expl. Cell Res.* **54**, 308.
Westergaard, O., Marcker, K. A. and Keiding, J. (1970). *Nature, Lond.* **227**, 708.
Wheeldon, L. W. and Lehninger, A. L. (1966). *Biochemistry, N.Y.* **5**, 3533.
Whittaker, P. A. (1969). *Microbios* **2**, 195.
Whittaker, P. A. and Wallis, O. C. (1971). *Biochem. J.* **125**, 82p.

Whittaker, P. A. and Wright, M. (1972). *Biochem. biophys. Res. Commun.* **48**, 1455.

Whittaker, P. A., Hammond, R. C. and Luha, A. A. (1972). *Nature New Biology*, **238**, 266.

Wiemken, A., Von Meyenburg, H. K. and Matile, Ph. (1970). *Acta Fac. med. Univ. brun.* **37**, 47.

Wilgram, G. F. and Kennedy, E. P. (1963). *J. biol. Chem.* **238**, 2615.

Wilkie, D. (1963). *J. molec. Biol.* **7**, 527.

Wilkie, D. (1964). "The Cytoplasm in Heredity", Methuen, London.

Wilkie, D. (1970a). *Symp. Soc. exp. Biol.* **24**, 71.

Wilkie, D. (1970b). *Symp. Soc. gen. Microbiol.* **20**, 381.

Wilkie, D. (1970c). *J. molec. Biol.* **47**, 107.

Wilkie, D. (1972). *In* "Mitochondria Biogenesis and Bioenergetics", (S. G. Van den Burgh, P. Borst, L. L. M. van Deenan, J. C. Riemersma, E. C. Slater and J. M. Slater, eds.), *FEBS Symp.* No. 28. pp. 85–94. North Holland, Amsterdam.

Wilkie, D. and Lewis, D. (1963). *Genetics* **48**, 1701.

Wilkie, D. and Maroudas, N. G. (1969). *Genet. Res, Cambridge* **13**, 107.

Wilkie, D. and Thomas, D. Y. (1973). *Genetics* **73**, 367.

Wilkie, D., Saunders, G. and Linnane, A. W. (1967). *Genet. Res., Cambridge* **10**, 199.

Williams, S. G. (1971). *Aust. J. biol. Sci.* **24**, 1181.

Williamson, D. H. (1969). *Antonie van Leeuwenhoek* **35** Suppl. C3.

Williamson, D. H. (1970). *Symp. Soc. exp. Biol.* **24**, 247.

Williamson, D. H. and Moustacchi, E. (1971). *Biochem. biophys. Res. Commun.* **42**, 195.

Williamson, D. H., Maroudas, N. G. and Wilkie, D. (1971a). *Mol. Gen. Genet.* **111**, 209.

Williamson, D. H., Moustacchi, E. and Fennell, D. (1971b). *Biochim. biophys. Acta* **238**, 369.

Wilson, B. W. and James, T. W. (1966). *In* "Cell Synchrony", (I. L. Cameron and G. M. Padilla, eds.), p. 236. Academic Press, New York and London.

Wilson, D. F. and Leigh, J. S. Jr. (1972). *Archs Biochem. Biophys* **150**, 154.

Wilson, D. F., Linsay, J. G. and Brocklehurst, E. S. (1972). *Biochim. biophys. Acta* **256**, 277.

Wilson, D. F., Erecińska, M., Dutton, P. L. and Tzudsuki, T. (1970). *Biochem. biophys. Res. Commun.* **41**, 1273.

Wilson, E. B. (1928). "The Cell in Development and Heredity." MacMillan, New York.

Wilson, J. F. (1961). *Am. J. Bot.* **48**, 46.

Wilson, J. F. (1963). *Am. J. Bot.* **50**, 780.

Wilson, M. A. and Cascarno, J. (1972). *Biochem. J.* **129**, 209.

Wilson, S. B. (1970). *Biochim. biophys. Acta* **223**, 383.

Winge, O, (1935). *C.r. Trav. Lab. Carlsberg* **21**, 77.

Wintersberger, E. (1964a). *Biochem. Z.* **341**, 409.

Wintersberger, E. (1964b). *Z. physiol. Chem.* **336**, 285.

Wintersberger, E. (1965). *Biochem. Z.* **341**, 409.

Wintersberger, E. (1966a). *Biochem. biophys. Res. Commun.* **25**, 1.

Wintersberger, E. (1966b). *In* "Regulation of Metabolic Processes in Mitochrondria", (J. M. Tager, S. Papa, E. Quagliariello, E. and E. C. Slater, eds.), p. 439. Elsevier, Amsterdam.

Wintersberger, E. (1967). *Z. physiol. Chem.* **348**, 1701.

Wintersberger, E. (1970). *Biochem. biophys. Res. Commun.* **40**, 1179.

Wintersberger, E. (1972). *Biochem. biophys. Res. Commun.* **48**, 1287.
Wintersberger, E. and Vienhauser, G. (1968). *Nature, Lond.* **220**, 699.
Wintersberger, E. and Wintersberger, U. (1970a). *FEBS Lett.* **6**, 58.
Wintersberger, E. and Wintersberger, U. (1970b). *Eur. J. Biochem.* **13**, 11.
Wirtz, K. W. A. and Zilversmitt, D. B. (1968). *J. biol. Chem.* **243**, 3596.
Wisnieski, B. J. and Kiyomoto, R. K. (1972). *J. Bact.* **109**, 186.
Wisnieski, B. J., Keith, A. D. and Resnick, M. A. (1970). *J. Bact.* **101**, 160.
Witt, I., Kronau, R. and Holtzer, H. (1966). *Biochim. biophys. Acta* **128**, 33.
Wohlrab, H. and Jacobs, E. E. (1967). *Biochem. biophys. Res. Commun.* **28**, 1003.
Wolf, K., Sebald-Althus, M., Schweyen, R. J. and Kaudewitz, (1971). *Mol. Gen. Genet.* **110**, 101.
Wolf, K., Dujon, B. and Slonimski, P. P. (1973). *Mol. Gen. Genet.* **125**, 53.
Wolstenholme, D. R. and Gross, N. J. (1968). *Proc. natn. Acad. Sci. U.S.A.* **61**, 245.
Wong, D. T., Horng J,-S. and Gordee, R. S. (1971). *J. Bact.* **106**, 168.
Wood, D. D. and Luck, D. J. L. (1969). *J. molec. Biol.* **41**, 211.
Wood, D. D. and Luck, D. J. L. (1971). *J. Cell Biol.* **51**, 249.
Woods, R. A. (1971). *J. Bact.* **108**, 69.
Woods, R. A. and Hogg, J. (1969). *Heredity, Lond.* **24**, 516.
Woodward, D. O. and Munkres, K. D. (1966). *Proc. natn. Acad. Sci. U.S.A.* **55**, 872.
Woodward, D. O. and Munkres, K. D. (1967). *In* "Organizational Biosynthesis", (H. J. Voegel, J. O. Lampen and V. Bryson, eds.), p. 489. Academic Press, New York.
Woodward, D. O., Edwards, D. L. and Flavell, R. B. (1970) *Symp. Soc. exp. Biol.* **24**, 55.
Work, T. S. (1968). *In* "Biochemical Aspects of the Biogenesis of Mitochondria" (E. C. Slater, J. M. Tager, S. Papa and E. Quagliariello, eds.), p. 272. Adriatica Editrice, Bari.
Wrigglesworth, J. M., Packer, L. and Branton, D. (1970). *Biochim. biophys. Acta* **205**, 125.
Wright, R. E. and Lederberg, J. (1957). *Proc. natn. Acad. Sci. U.S.A.* **413**, 919.
Yamanaka, T., Nakajima, H. and Okunuki, K. (1962). *Biochim. biophys. Acta* **63**, 510.
Yamanaka, T., Nagata, Y. and Okunuki, K. (1968). *J. Biochem., Tokyo* **63**, 753.
Yamashita, S. and Racker, E. (1969). *J. biol. Chem.* **244**, 1220.
Yamazaki, I., Nakajima, R., Honma, H. and Tamura, M. (1968). *In* "Structure and Function of Cytochromes", (K. Okunuki, M. D. Kamen and I. Sekuzo, eds.), pp. 552–559. University of Tokyo Press, Tokyo.
Yang, S. and Criddle, R. S. (1969). *Biochim. biophys. Res. Commun.* **35**, 429.
Yang, S. and Criddle, R. S. (1970). *Biochemistry, N,Y.* **9**, 3063.
Yaoi, Y. (1967). *J. Biochem., Tokyo* **61**, 54.
Yčas, M. (1954). *Expl. Cell Res.* **10**, 746.
Yčas, M. (1956). *Expl. Cell Res.* **11**, 1.
Yčas, M. and Drabkin, D. L. (1957). *J. biol. Chem.* **224**, 921.
Yčas, M. and Starr, T. J. (1953). *J. Bact.* **65**, 83.
Yonetani, T. (1960). *J. biol. Chem.* **235**, 845.
Yonetani, T. and Ohnishi, T. (1966). *J. biol. Chem.* **241**, 2983.
Yonetani, T. and Ray, G. S. (1965). *J. biol. Chem.* **240**, 4503.
Yoo, B. Y., Calleja, G. B. and Johnson, B. F. (1973). *Arch. Mikrobiol.* **91**, 1.
Yoshida, Y. and Kumaoka, H. (1969). *Biochim. biophys. Acta* **189**, 461.
Yost, Jr., F. J. and Fridowitch, I. (1973). *J. biol. Chem.* **248**, 4905.

Yotsuyanagi, Y. (1962a). *J. Ultrastruct. Res.* **7**, 121.

Yotsuyanagi, Y. (1962b). *J, Ultrastruct. Res.* **7**, 141.

Yotsuyanagi, Y. (1966). *C.r. hebd. Séanc. Acad. Sci., Paris*, Ser. D. **262**, 1348.

Yu, R., Lukins, H. B. and Linnane, A. W. (1968). *In* "Biochemical Aspects of the Biogenesis of Mitochondria", (E. C. Slater, J. M. Tager, S. Papa and E. Quagliariello, eds.), p. 359. Adriatica Editrice, Bari.

Yu, R., Poulson, R. and Stewart, P. R. (1972a). *Mol. Gen. Genet.* **114**, 325.

Yu, R., Poulson, R. and Stewart, P. R. (1972b). *Mol. Gen. Genet.* **114**, 339.

Yumsui, T., Mizushima, H., Sekuzu, I. and Okunuki, K. (1966). *Biochim. biophys. Acta* **118**, 442.

Zagon, I. S. (1970). *J. Protozool*, **17**, 664.

Zalokar, M. (1959). *Am. J. Bot.* **46**, 555.

Zeuthen, E. (1953). *J. Embryol. exp. Morph.* **1**, 239.

Zollinger, W. D. and Woodward, D. O. (1972). *J. Bact.* **109**, 1001.

Zylber, E. and Penman, S. (1969). *J. molec. Biol.* **46**, 201.

Zylber, E. A., Perlman, S. and Penman, S. (1971). *Biochim. biophys. Acta* **240**, 588.

Zylber, E., Vesco, C. and Penman, S. (1969). *J. molec. Biol.* **44**, 195.

Index